Lecture Notes in Energy

Volume 32

Lecture Notes in Energy (LNE) is a series that reports on new developments in the study of energy: from science and engineering to the analysis of energy policy. The series' scope includes but is not limited to, renewable and green energy, nuclear, fossil fuels and carbon capture, energy systems, energy storage and harvesting, batteries and fuel cells, power systems, energy efficiency, energy in buildings, energy policy, as well as energy-related topics in economics, management and transportation. Books published in LNE are original and timely and bridge between advanced textbooks and the forefront of research. Readers of LNE include postgraduate students and non-specialist researchers wishing to gain an accessible introduction to a field of research as well as professionals and researchers with a need for an up-to-date reference book on a well-defined topic. The series publishes single and multi-authored volumes as well as advanced textbooks.

More information about this series at http://www.springer.com/series/8874

Masakazu Sugiyama · Katsushi Fujii
Shinichiro Nakamura
Editors

Solar to Chemical Energy Conversion

Theory and Application

Springer

Editors
Masakazu Sugiyama
Department of Electrical Engineering and Information Systems, School of Engineering
The University of Tokyo
Tokyo
Japan

Shinichiro Nakamura
Nakamura Laboratory
RIKEN Research Cluster for Innovation, RIKEN
Wako
Japan

Katsushi Fujii
Global Solar Plus Initiative
The University of Tokyo
Tokyo
Japan

ISSN 2195-1284 ISSN 2195-1292 (electronic)
Lecture Notes in Energy
ISBN 978-3-319-25398-5 ISBN 978-3-319-25400-5 (eBook)
DOI 10.1007/978-3-319-25400-5

Library of Congress Control Number: 2015954610

Springer Cham Heidelberg New York Dordrecht London

Printed on acid-free paper

Springer International Publishing AG Switzerland is part of Springer Science+Business Media (www.springer.com)

Preface

This book provides both the theoretical background and state-of-the-art review of solar-to-chemical energy conversion: the most important technology for energy storage, which is vital for sustainable human life in the future. The theoretical background starts with the concept of chemical potential and equilibrium in a molecular system and solid-state system, especially semiconductor. All chemical reactions, including the reactions at the interface between an electrolyte and a semiconductor (and/or metal) surface, are driven by the extent of nonequilibrium, or the difference in chemical potential, as described in the text. On such theoretical basis, a variety of technologies for solar-to-chemical energy conversion are discussed. Chemical, electrochemical and photoelectrochemical approaches are described for converting solar energy into hydrogen or other hydrocarbon species as energy storage media. Photosynthesis is the most sophisticated system of solar-to-chemical energy conversion developed by nature. Its up-to-date understanding and the way to implement its mechanism in an energy-efficient manner are discussed, including the use of algae for engineered photosynthesis. The broad-spectrum description in this book will provide a basis for the research and development of chemical energy storage in the coming decades.

Contents

Introduction—Solar to Chemical Energy Conversion

Masamichi Fujihira

First, let me start this introduction by briefly describing the history of the invention of the piston steam engine during the Industrial Revolution in Great Britain.

The Industrial Revolution [1] was a transition to new manufacturing processes from about 1760 to sometime between 1820 and 1840. This transition included going from hand production methods to machines and new chemical manufacturing and iron production processes. It improved the efficiency of water power, and promoted the increasing use of steam power and the development of machine tools. For the readers of this book—***Solar to Chemical Energy Conversion***, the most important aspect of the Industrial Revolution is without doubt the transition **from "*wood and other bio-fuels*" to "*coal*"**.

The first successful piston steam engine was introduced by **Thomas Newcomen** before 1712. His steam engines were extremely inefficient by modern standards, but helped expand coal mining by allowing mines to go deeper. Despite their disadvantages, Newcomen engines were reliable and easy to maintain, and continued to be used in coalfields until the early decades of the 19th century.

A fundamental change in working principles was brought about by Scotsman **James Watt.** In close collaboration with Englishman **Matthew Boulton**, by 1778, he had succeeded in perfecting his steam engine, which incorporated a series of radical improvements (for further details, please refer to Refs. [1, 2]).

In relation to the long-term and difficult themes which this book deals with, i.e. ***Solar to Chemical Energy Conversion***, I would like to emphasize that although the first invention by Newcomen triggered the start of the transition, it was the Watt and Boulton engines that truly contributed to the Industrial Revolution. Through this example, I would like to stress that younger generations have a better chance than their predecessors to make breakthroughs to reach such a long-term achievement as solar-to-chemical energy conversion. The fact that we use Watt as the physical unit of ***power*** today surely supports this view.

M. Fujihira (✉)
Department of Biomolecular Engineering,
Tokyo Institute of Technology, Tokyo, Japan
e-mail: mfujihira@jcom.home.ne.jp

M. Sugiyama et al. (eds.), *Solar to Chemical Energy Conversion*,
Lecture Notes in Energy 32, DOI 10.1007/978-3-319-25400-5_1

Secondly, I would like to introduce the world view, ***Spaceship Earth***. This term [3] usually expresses concern over the use of limited resources available on earth and encourages humankind to act as a harmonious crew working toward the greater good. The earliest known use is mentioned in a passage in Henry George's best known work, *Progress and Poverty* (1879). After it was rephrased several times by others, the phrase was popularized by Buckminster Fuller in his book in 1968 called *Operating Manual for Spaceship Earth* [4]. This quote, referring to fossil fuels, reflects his approach:

> …we can make all of humanity successful through science's world-engulfing industrial evolution provided that we are not so foolish as to continue to exhaust in a split second of astronomical history the orderly energy savings of billions of years' energy conservation aboard our Spaceship Earth. These energy savings have been put into our Spaceship's life-regeneration-guaranteeing bank account for use only in self-starter functions.

I hope, within the next half a century or the next century at the longest, we will succeed in producing artificial fuels at reasonably low costs, by the development of the research fields discussed in this book. Examples of artificial fuels include H_2 produced from water and CH_4 and CH_3OH from CO_2 and water by utilizing solar energy at their extremely high power conversion efficiencies without the expense of fossil fuels. During the Industrial Revolution described in the beginning of this introduction, Newcomen as well as Watt and Boulton used coal mainly as the energy source of their piston steam engine. In the coming century, we will probably not be allowed to use fossil fuels anymore, regardless of whether we agree with the view proposed by Buckminster Fuller in 1968.

Thirdly, I would like to point out that a century is not long enough to make breakthroughs for such long-term research themes. To clarify the grounds for speculating the time necessary for the achievement of such long-term research goals, I would like to introduce an international conference deeply related to these long-term research themes. We called the conference "***the International Conference on Photochemical Conversion and Storage of Solar Energy (IPS)***". The conference has been held every 2 years since the first one (IPS-0) in Boston in September 1974. Fortunately, I was involved in this international conference as a member of the international organization committee from IPS-10 in 1994 to IPS-12 in 1998.

When the conference started in 1974, I was 30 years old and had just come back to the University of Tokyo from Ohio State University after a 2 year post-doctoral experience in the USA [5]. I thus had little background knowledge of why the conference was organized. Probably or hopefully the conference was motivated by: (1) oil crisis which started in 1973, (2) Nature's publication of water photolysis on TiO_2 reported by Honda and Fujishima in 1972 [6], and (3) the Spaceship Earth view written in 1968 described above.

On the occasion of the 10th International Conference (IPS-10) in 1994, the Chairman of IPS-1, Bolton, was asked to write a historical perspective [7]. In 2014, IPS-20 was held in Berlin. All the past host cities of the conference are listed on the conference website [8]. From the comparison between Bolton's historical perspective and the IPS-20 program, we can clearly see the progress made from IPS-10 in 1994 to IPS-20 in 2014. I hope someone in the present international organization committee will one day describe another historical perspective of these two decades (IPS-11–IPS-20). Only by witnessing the long-term progress made in a particular field can we envision the future prospects of that field correctly.

Finally, I would like to mention that Japan is still suffering from the 9.0 magnitude Great East Japan Earthquake which hit the Tohoku area at 14:46 on Friday March 11, 2011. A year after the disaster, a new conference "International Workshop on Solar-Chemical Energy Storage" was held in Sendai, the capital city of the Tohoku area, gathering world leading researchers of solar to chemical energy conversion. After the conference, some of the participants of the Sendai Conference saw the devastation of the earthquake during their visit to the coastal area by bus. The conference motivated the organizers to compile this book in the hope to succeed in some of the long-term difficult themes faced, through concentrated efforts to overcome the hardships caused by this serious disaster, especially energy related problems. To end, I would like to express my gratitude to the organizers of the Sendai Conference, Masakazu Sugiyama, Katsushi Fujii, and Shinichiro Nakamura for their efforts in realizing such a wonderful meeting and completing this significant book, Solar to Chemical Energy Conversion.

References

1. https://en.wikipedia.org/wiki/Industrial_Revolution
2. Carvill J (1981) Famous names in engineering, 1st edn. Butterworth-Heinemann Ltd, London. ISBN-13: 978-0408005401
3. https://en.wikipedia.org/wiki/Spaceship_Earth
4. Fuller B (1963) Operating manual for spaceship earth. E.P. Dutton & Co., New York. ISBN: 0-525-47433-1. The quotation is from Section 8: The regenerative landscape
5. Bard AJ, Stratmann M (eds) (2007) Encyclopedia of electrochemistry. In: Fujihira M, Rubinstein I, Rusling JF (eds) Modified electrodes, vol 10. Wiley-VCH Verlag GmbH & Co. KGaA, Weinheim. ISBN: 987-3-527-30402-8
6. Fujishima A, Honda K (1972) Electrochemical photolysis of water at a semiconductor electrode. Nature 238:37–38
7. Bolton JR (1995) Sol Energy Mater Sol Cells 38:543–554
8. http://www.helmholtz-berlin.de/events/ips20/

Part I
Fundamental Background

Thermodynamics for Electrochemistry and Photoelectrochemistry

Katsushi Fujii

Abstract Electrochemistry and photoelectrochemistry are based on thermodynamics, especially for its equilibrium. In this chapter, the thermodynamics for the electrochemistry and photoelectrochemistry are discussed.

1 Gibbs Free Energy and Chemical Reaction

Thermodynamics is an important tool for understanding the theoretical background of electrochemical reactions. It is an established rule of thumb and defines macroscopic variables, such as internal energy, entropy, and pressure, and other textbooks should be referenced for its details [1]. This chapter focuses on thermodynamics used for electrochemical and photoelectrochemical reactions.

According to the first law of thermodynamics for a closed system, any net increase in the internal energy U is described using heat δQ entering the system and the work δW done by the system;

$$\delta Q = \mathrm{d}U + \delta W. \tag{1}$$

This shows that heat entering the system is used for the net increase in the internal energy and the work done by the system. This is the analogous to the law of energy conservation, i.e., the energy entering the system is used for the net increase in the potential energy and the work done by the system (The energy for thermodynamic "work" is accounted for by changes in the external generalized mechanical constraints on the system. The work is a generalization of the concept of mechanical work in physics.). Usually, this is expressed as follows:

K. Fujii (✉)
Global Solar Plus Initiative, The University of Tokyo, 4-6-1 Komaba, Meguro-Ku, Tokyo 153-8904, Japan
e-mail: k.fujii@rcast.u-tokyo.ac.jp

M. Sugiyama et al. (eds.), *Solar to Chemical Energy Conversion*, Lecture Notes in Energy 32, DOI 10.1007/978-3-319-25400-5_2

$$\mathrm{d}U = \delta Q - \delta W. \tag{2}$$

The letter d indicates an exact differential, expressing that internal energy U is a property of the state of the thermodynamic system. The internal energy depends only on the original state and the final state, and not upon the path taken. In contrast, the Greek delta (δ) in this equation reflects the fact that the heat transfer and work transfer are not properties of the initial and final states of the system. Thus, heat and work depend on the path taken and are not state functions. (The work is changed by the path taken in the real world, that is, the existence of friction etc. This means the process in reality can be close to the reversible process but is a non-reversible process.) For a reversible process in a closed system, δW has the following relationship of differential equation using pressure (P) and volume (V) under constant pressure condition (This pressure can be changed to variable in some cases. In this chapter, the equations are treated as the chemical reaction occurring under standard conditions (25 °C, 101.3 kPa for chemical reaction).);

$$\delta W = P\mathrm{d}V. \tag{3}$$

Enthalpy is defined as the entering heat used for the net increase of internal energy and the work done by the system, which does not depend on the path. Enthalpy (H) consists of the internal energy of the system (U) plus the work done by the system, that is the product of pressure (P) and volume (V) of the system, which are all functions of the state of the system;

$$\mathrm{d}H = \mathrm{d}U + P\mathrm{d}V. \tag{4}$$

The unit of enthalpy is the energy unit like "joule." Enthalpy changes with its energy transfer, i.e., increases with receiving heat from the environment. It should be noted that enthalpy includes internal energy change, thus this change is observed as system phase changes, reactions, and heat capacitances.

The other important thermodynamic concept is entropy. It should be noted that entropy is included in internal energy U and cannot be used for the work done by the system. Entropy (S) is a measure of disorder of the system. According to statistical mechanics, entropy is defined by Boltzmann's entropy formula with the quantity Ω, which is the number of microstates corresponding to a given macrostate;

$$S = k_B \ln \Omega, \tag{5}$$

where, k_B is the Boltzmann's constant. The other definition from thermodynamics according to the Clausius theorem is;

$$\mathrm{d}S \leq \frac{\delta Q}{T}, \tag{6}$$

where, Q is heat entering into a closed system and T is absolute temperature. This clearly shows that entropy changes upon heat transfer. The equality establishes only in a closed system at the same temperature under the reversible process. Entropy is also the state function of a system and the unit is the energy/temperature like "joule/kelvin."

Thermodynamic free energy is a state function like internal energy U and is defined as the energy converted to do work. (The free energy is the internal energy of a system minus the amount of energy that cannot be used to perform work. This unusable energy is given by the entropy of a system multiplied by the temperature of the system.) The unit is energy/amount of substance, that is, "joule/mole." Free energy can be voluntarily reduced, that is, a chemical reaction occurs spontaneously if the change in free energy is negative. Furthermore, when the free energy takes a constant value of minimum, the system is in equilibrium. Free energy is defined in particular, as Helmholtz energy (F) under constant temperature and volume;

$$\mathrm{d}F = \mathrm{d}U - T\mathrm{d}S = \mathrm{d}H - T\mathrm{d}S - p\mathrm{d}V. \tag{7}$$

Gibbs energy (G) is defined under constant temperature and pressure;

$$\mathrm{d}G = \mathrm{d}U - T\mathrm{d}S + p\mathrm{d}V = \mathrm{d}F + p\mathrm{d}V = \mathrm{d}H - T\mathrm{d}S. \tag{8}$$

Since chemical reaction is usually treated under the condition of constant temperature and pressure, Gibbs energy is used. When the chemical reaction contains its pressure change, Helmholtz energy has to be used. It should be noted that internal energy cannot be determined but free energy, enthalpy, and entropy can be measured. Thus, chemical reaction can be discussed by using free energy.

Standard formation Gibbs energy ($\Delta_f G^0$, where the ^{0} expresses the system under the standard condition; 25 °C, 101.3 kPa) is the change of Gibbs free energy that accompanies the formation of 1 mol of that substance from its component elements at their standard states (the most stable form of the element at 25 °C, 101.3 kPa). Thus, the standard molar Gibbs energy of formation of element is defined as zero (0). Standard formation Gibbs energy is defined from Eq. (8) as

$$\Delta_f G^0 = \Delta_f H^0 - T\Delta_f S^0 \tag{9}$$

where, $\Delta_f H^0$ is standard formation enthalpy and $\Delta_f S^0$ is standard formation entropy.

Standard formation Gibbs energy also tells us about the spontaneity of a reaction;

$\Delta_f G^0 < 0$, favored reaction, stabilizing direction, spontaneous, exergonic,
$\Delta_f G^0 = 0$, Neither the forward nor the reverse reaction proceeds, at equilibrium,
$\Delta_f G^0 < 0$, disfavored reaction, instabilizing direction, nonspontaneous, endergonic.

One of the simplest substance formation, water generation from hydrogen and oxygen reaction, is used as an example here. The reaction under the standard condition is described as follows:

$$H_2(g) + \frac{1}{2}O_2(g) \rightarrow H_2O(\ell). \tag{10}$$

The standard formation enthalpy ($\Delta_f H^0$) is obtained due to the phase change and the reaction;

$$\Delta_f H^0 = -285.83\,[\text{kJ/mol}]. \tag{11}$$

The standard formation entropy ($\Delta_f S^0$) also exists;

$$\begin{gathered} -T\Delta_f S^0 = 48.70\,[\text{kJ/mol}] \\ (\text{in this condition}, \Delta_f S^0 = -163.34\,[\text{J/mol/K}]). \end{gathered} \tag{12}$$

As a result, the total energy change is expressed by the change of the Gibbs energy (standard formation Gibbs energy, $\Delta_f G^0$);

$$\Delta_f G^0 = \Delta_f H^0 - T\Delta_f S^0 = -237.13\,[\text{kJ/mol}]. \tag{13}$$

Since the $\Delta_f G^0$ for the stable standard substance with a singular element is defined as zero, the $\Delta_f G^0$s of H_2 and O_2 in Eq. (10) are zero. The lower Gibbs energy indicates the more stable condition. Thus, H_2O is more stable than H_2 and O_2 because the Gibbs energy of H_2O is lower than those of H_2 and O_2. The free-energy diagram associated with this reaction is shown in Fig. 1.

It should be noted here that the Gibbs energy change does not contain the activation energy (E_a). As known well, the reaction of H_2 and O_2 does not occur under the standard condition due to the existence of a large activation energy. The discussion of electrochemical reactions is based on this Gibbs energy change, which means it does not contain the activation energy. Unknown energies including activation energy for electrochemical reactions are expressed as "overpotential."

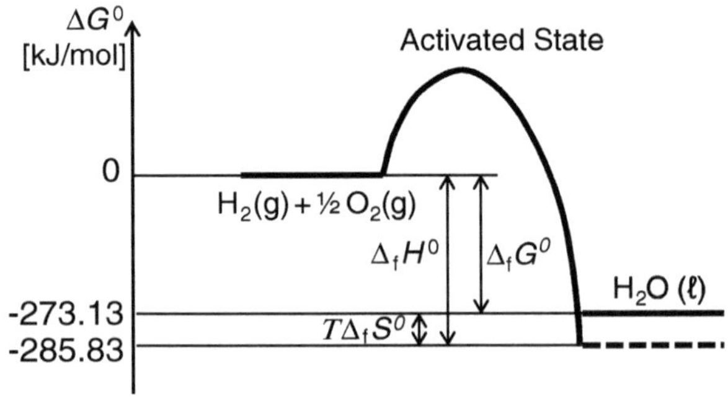

Fig. 1 Schematic diagram for H_2 and O_2 forms H_2O. The relationship of Gibbs energy ($\Delta_f G^0$), enthalpy ($\Delta_f H^0$), and entropy term ($T\Delta_f S^0$) are shown here

2 Equilibrium Criteria for Chemical and Electrochemical Systems

Ideal chemical equilibrium condition can be expressed by using standard formation Gibbs energy ($\Delta_f G^0$). An equilibrium chemical reaction is;

$$pP + qQ + \cdots \text{(reactants)} = xX + yY + \cdots \text{(products)}. \tag{14}$$

The ideal equilibrium condition of this reaction is expressed as;

$$\begin{gathered} p\Delta_f G^0(p) + q\Delta_f G^0(q) + \cdots = x\Delta_f G^0(x) + y\Delta_f G^0(y) + \cdots, \\ \sum_{\text{reactant}} p\Delta_f G^0(p) = \sum_{\text{product}} x\Delta_f G^0(x). \end{gathered} \tag{15}$$

This expresses that the sum of Gibbs energy equals between the products and the reactants when the reaction is under equilibrium.

Considering the free energy for nonideal mixed substance, driving force of reaction is weakened in mixed substances compared with pure substances due to the concentration decrease. In order to express this nonideality, chemical potential (μ_i^0) of *i*th component is introduced. When pressure is constant, chemical potential in a mixture is expressed as the partial molar Gibbs energy. The pure component is estimated as 1 mol using this expression. The definition for ideal condition of n mole *i*th component in a mixture is;

$$\mu_i^{0\phi} = \frac{\partial \Delta_f G^0}{\partial n_i} = \frac{\partial}{\partial n_i} \sum_j n_j \Delta_f G^0(n_j). \tag{16}$$

where, n_i is the molar ratio of *i*th component in a mixture. This $\mu_i^{0\phi}$ is standard chemical potential. Describing this deviation for nonideal mixed substance from the ideal one, activity (*a*) is introduced. Nonideal chemical potential (μ_i^0) is expressed with activity as;

$$\mu_i^0 = \mu_i^{0\phi} + RT \ln a, \tag{17}$$

where, R is gas constant. The definition of this activity (*a*) without any interaction between molecules is the molar ratio of a substance in the mixture. That is,

gas	partial pressure (p_i [kPa]) divided by standard pressure ($p_0 = 101.3$ [kPa]),
solute	molar concentration (m_i [mol/kg or mol/L]) divided by standard condition ($m_0 = 1$ [mol/kg or mol/L]),
solvent	solvent for diluted solution is defined as $a = 1$,
solid	pure solid is defined as $a = 1$,
electron	electrons in a material is defined as $a = 1$.

Using the chemical potential, the equilibrium reaction of Eq. (14) for nonideal mixed substance is expressed instead of Eq. (15) as;

$$\begin{aligned} p\mu_p^{0\phi} + pRT\ \ln a_p + q\mu_q^{0\phi} + qRT\ \ln a_q + \cdots \\ = x\mu_x^{0\phi} + xRT\ \ln a_x + y\mu_y^{0\phi} + yRT\ \ln a_y + \cdots. \end{aligned} \tag{18}$$

From the definition of a general chemical reaction, the temperature and pressure are constant, thus;

$$\begin{aligned} -\left((xRT\ \ln a_x + yRT\ \ln a_y + \cdots) - (pRT\ \ln a_p + qRT\ \ln a_q + \cdots)\right) \\ = \left(x\mu_x^{0\phi} + y\mu_y^{0\phi} + \cdots\right) - \left(p\mu_p^{0\phi} + q\mu_q^{0\phi} + \cdots\right) = \Delta_r G^0, \end{aligned}$$

$$-RT\ \ln\left(\frac{a_x^x a_y^y \ldots (\text{product})}{a_p^p a_q^q \ldots (\text{reactant})}\right) = \Delta_r G^0. \tag{19}$$

where, $\Delta_r G^0$ is standard reaction Gibbs energy. Equilibrium constant (K) is defined by using Eq. (19) as;

$$\Delta_r G^0 = -RT\ \ln\left(\frac{a_x^x a_y^y \ldots (\text{product})}{a_p^p a_q^q \ldots (\text{reactant})}\right) = -RT\ \ln K. \tag{20}$$

This Eq. (20) is called "low of mass action."

Electrode potential also plays an important role in electrochemical reaction. The basis for an electrochemical reaction is always a redox reaction (reduction and oxidation reaction), which can be broken down into two half-reactions: oxidation at anode (loss of electron) and reduction at cathode (gain of electron). The driving force of the redox reaction is the electricity due to electric potential difference between two electrodes. This potential difference is created by the applied potentials of the two metal electrodes in an electrolyte. The electron energy at an electrode under equilibrium condition can be defined for a certain half-reaction including ions and electrons as follows. The ideal half-reaction can be described as;

$$pP + qQ + \cdots (\text{reactants/oxides}) + ne^- = xX + yY + \cdots (\text{products/redactants}) \tag{21}$$

Important point of this Eq. (21) is that P, Q, X, Y, … represent not only substances but also ions. The equilibrium condition of Eq. (21) using the Faraday constant (elementary charge for 1 mol electron; $F = 96{,}485$ [°C/mol]) and standard electrode potential (oxidation reduction potential, redox potential; ϕ^0(Ox/Red) under standard condition) is;

$$p\Delta_f G^0(p) + q\Delta_f G^0(q) + \cdots + n\Delta_f G^0(e) = x\Delta_f G^0(x) + y\Delta_f G^0(y) + \cdots,$$

$$\phi^0(Ox/Red) = \frac{-n\Delta_f G^0(e)}{nF} = \frac{-\left(\sum_{\text{product/Red}} x\Delta_f G^0(x) - \sum_{\text{reactant/Ox}} p\Delta_f G^0(p)\right)}{nF} = \frac{-n\Delta_r G^0}{nF}. \quad (22)$$

It should be noted that the sign of ϕ and that of $\Delta_f G$ are opposite from the definition.

The electrode potential may not be obtained empirically. The galvanic cell (composed of only two electrodes, anode, and cathode) potential results from a pair of electrodes. Thus, only one empirical value is available in a pair of electrodes and it is not possible to determine the value for each electrode in the pair. A reference electrode, for which the potential is defined or agreed upon by convention, needed to be established. The most important reference electrode is the standard hydrogen electrode (SHE) (or normal hydrogen electrode (NHE)), where the half-reaction can be described as;

$$2H^+(\text{in } \ell) + 2e^- = H_2(g). \quad (23)$$

The Gibbs energy of H^+ is defined conventionally as;

$$\Delta_f G^0(H^+) = 0.00. \quad (24)$$

The equilibrium of hydrogen gas and ion in aqueous solution is;

$$\Delta_f G^0(H^+) - 2F\phi^0(H^+/H_2) = \Delta_f G^0(H_2). \quad (25)$$

From definitions;

$$\Delta_f G^0(H^+) = 0.00, \Delta_f G^0(H_2) = 0.00,$$

thus,

$$\phi^0(H^+/H_2) = 0.00\,[V]. \quad (26)$$

This standard electrode potential of Eq. (26) is used as a reference for electrochemical reactions. Any electrode, for which the electrode potential is not yet known, can be paired with SHE to form a galvanic cell. The galvanic cell potential gives the unknown electrode's potential. Using this process, any electrode with an unknown potential can be paired with either the SHE or another electrode for which the potential has already been derived and that unknown value can be established. From electrochemical notation, the potential needs to clarify the value with "+" or "−", and the unit with "V versus (reference electrode)" like "V versus SHE" due to the relative measurement method.

One simple reaction is the chemical reaction with electron transfer. The example of half-reactions are;

$$
\begin{aligned}
&5\mathrm{Fe(CN)}_6^{3-} + 5e^- = 5\mathrm{Fe(CN)}_6^{4-},\\
&\phi^0\left(\mathrm{Fe(CN)}_6^{3-}/\mathrm{Fe(CN)}_6^{4-}\right) = +0.36\,[\mathrm{V\ vs.\ SHE}],\\
&\mathrm{MnO}_4^- + 8\mathrm{H}^+ + 5e^- = \mathrm{Mn}^{2+} + 4\mathrm{H_2O},\\
&\phi^0\left(\mathrm{MnO}_4^-/\mathrm{Mn}^{2+}\right) = +1.51\,[\mathrm{V\ vs.\ SHE}].
\end{aligned}
\tag{27}
$$

Electron moves from the lower to the higher electrode potential, thus;

$$5\mathrm{Fe(CN)}_6^{4-} + \mathrm{MnO}_4^- + 8\mathrm{H}^+ = 5\mathrm{Fe(CN)}_6^{3-} + \mathrm{Mn}^{2+} + 4\mathrm{H_2O}. \tag{28}$$

Similar to this Eq. (28), an electron in electrochemical reaction moves from/to electrode. That is, an electron moves from an electrode when the electrode potential is more negative (electrode energy is more positive) than a half-reaction under equilibrium condition (reduction), and an electron moves to electrode when the electrode potential is more positive (electrode energy is more negative) than a half-reaction under equilibrium condition (oxidation). The details is discussed in Sects. 4 and 5.

For nonideal conditions, electrochemical potential has to be introduced in order to express the nonideality similar to chemical reaction. The electrochemical potential ($\tilde{\mu}_i$) is defined using the valence of ion (z_i) as;

$$\tilde{\mu}_i^0 = \mu_i^{0\phi} + RT\ \ln a_i + z_i F\phi_i. \tag{29}$$

Electrochemical potential of electron is;

$$\tilde{\mu}_e^0 = \mu_e^{0\phi} + RT\ \ln a_e - F\phi_e. \tag{30}$$

Nonideal equilibrium condition of Eq. (21) is expressed using electrochemical potential instead of Eq. (22) as;

$$
\begin{aligned}
&p\mu_p^{0\phi} + pRT\ \ln a_p + pz_pF\phi^0 + q\mu_q^{0\phi} + qRT\ \ln a_q + qz_qF\phi^0 + \cdots + n\mu_e^{0\phi} + RT\ \ln a_e - nF\phi^0\\
&\quad = x\mu_x^{0\phi} + xRT\ \ln a_x + xz_xF\phi^0 + y\mu_y^{0\phi_1} + yRT\ \ln a_y + yz_yF\phi^0 + \cdots.
\end{aligned}
\tag{31}
$$

That is;

$$
\begin{aligned}
&-\big((xRT\ \ln a_x + yRT\ \ln a_y + \cdots(\text{redactant})) - (pRT\ \ln a_p + qRT\ \ln a_q + \cdots(\text{oxide}))\big)\\
&\quad - \big((xz_x + yz_y + \cdots) - (pz_p + qz_q + \cdots)\big)F\phi^0 + \left(n\mu_e^{0\varphi} + nRT\ \ln a_e\right) - nF\phi^0\\
&\quad = \left(x\mu_x^{0\phi} + y\mu_y^{0\phi} + \cdots\right) - \left(p\mu_p^{0\phi} + q\mu_q^{0\phi} + \cdots\right).
\end{aligned}
\tag{32}
$$

The charge has to be balanced in the chemical equation in Eq. (32), thus,

$$-\left(\left(xz_x + yz_y + (\text{reductants})\right) - \left(pz_p + qz_q + (\text{oxides})\right)\right)F\phi^0 - nF\phi^0 = 0. \quad (33)$$

From the relationship between chemical potentials and standard reaction of Gibbs energy in Eq. (19) and the definition of standard electrode potential of Eq. (22);

$$\left(x\mu_x^{0\phi} + y\mu_y^{0\phi} + \cdots\right) - \left(p\mu_p^{0\phi} + q\mu_q^{0\phi} + \cdots\right) = \Delta_r G^0 = -nF\phi^0(Ox/Red). \quad (34)$$

Equation (32) can be organized with the relationship of Eqs. (33) and (34),

$$RT \ln \frac{a_p^p a_q^q \ldots (\text{oxide})}{a_x^x a_y^y \ldots (\text{redactant})} + n\left(\mu_e^{0\phi} + RT \ln a_e\right) = -nF\phi^0(Ox/Red). \quad (35)$$

The definition of electrode potential and Eq. (35) derive;

$$\phi^0 = -\frac{n\left(\mu_e^{0\phi} + RT \ln a_e\right)}{nF} = \phi^0(Ox/Red) + \frac{RT}{nF} \ln \frac{a_p^p a_q^q \cdots (\text{oxide})}{a_x^x a_y^y \cdots (\text{redactant})}. \quad (36)$$

This Eq. (36) indicates that the electrode potential changes when the activity of the elements in the mixture changes even under standard condition. This relationship is called Nernst equation.

3 Marcus Theory and Gerischer Model

The thermodynamic treatment of chemical reaction shows the equilibrium condition but it cannot treat the rate of chemical reaction. Activation energy is required for the discussion of the rate. Marcus theory explains the activation energy of chemical reaction and the rates of electron transfer reactions [2]. The rate at which an electron can move or jump from one chemical species (called the electron donor) to another (called the electron acceptor). It was originally explained as outer sphere electron transfer reactions, in which the two chemical species only change in their charge with an electron jumping (e.g., the oxidation of an ion like Fe^{2+}/Fe^{3+}), but do not undergo large structural changes. It was extended to include inner sphere electron transfer contributions, in which a change of distances or geometry in the solvation or coordination shells of the two chemical species is taken into account; e.g., the Fe–O distances in $Fe(H_2O)^{2+}$ and $Fe(H_2O)^{3+}$ are different. Adiabatic (continuous free energy change of reacting substance during reaction) and nonadiabatic (discontinuous free energy change existing of reacting substance during reaction) reactions are usually discussed in chemical reactions. The Marcus theory discusses adiabatic reactions, and the Gerischer model applied for electrochemical reaction can take into account nonadiabatic reactions.

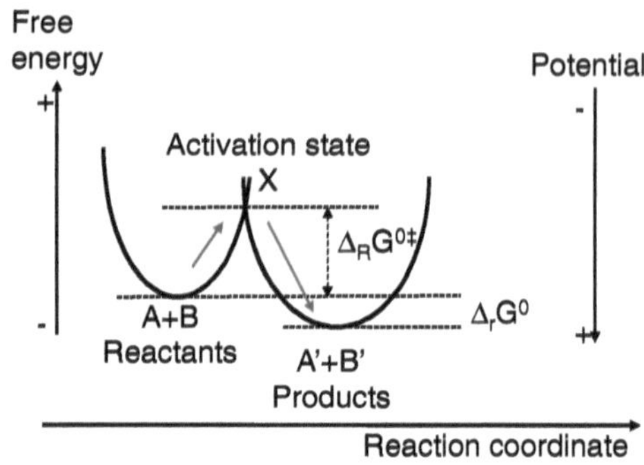

Fig. 2 Schematic diagram for reaction coordinate. The direction of energy and potential are opposite

Reaction coordinate (an abstract one-dimensional coordinate to represent progress along a reaction pathway) has to be introduced for the discussion of the chemical reaction rate as shown in Fig. 2. The chemical reaction process starts from the configuration of reactants (A + B), passes through via activation state (X), and reaches to the configuration of products (A′ + B′). The reaction coordinate typically choses to follow the path along the gradient (path of shallowest ascent/deepest descent) of potential energy from reactants to products. For this configuration, energy needs to reach the activation state (X) from the configuration of reactants (A + B), and it goes naturally from activation state to the configuration of reactants (A′ + B′) with certain probability. For common chemical reaction, the energy from A + B to X is given by thermal motion. The chemical reaction probability P_{cr} from A + B to X is described with Boltzmann distribution because the fraction of A + B with the activation energy $\Delta_{\mathrm{R}}G^{0\ddagger}$ out of entire ensemble of A + B follows Boltzmann distribution;

$$P_{\mathrm{cr}} = \exp\left(\frac{-\Delta_R G^{0\ddagger}}{RT}\right), \tag{37}$$

where R is gas constant, and T is absolute temperature. Reaction speed (v_{ct}) from A to B is described with the concentration of reactants A + B ($c_{\mathrm{A+B}}$) and P_{cr};

$$v_{\mathrm{cr}} = kc_{A+B}, \quad k = k_0 P_{\mathrm{cr}} = k_0 \exp\left(\frac{-\Delta_R G^{0\ddagger}}{RT}\right), \tag{38}$$

where k_0 is a constant related to the probability of the process from X to the configuration of products (A′ + B′).

For the Marcus theory, the reaction coordinate path is selected as a parabolic shape of the Gibbs energy of activation. Since the reaction coordinate describes the minimum energy path from the reactants to the products, the real path may not be simplified as parabolic shape. The activation energy of the Marcus theory is

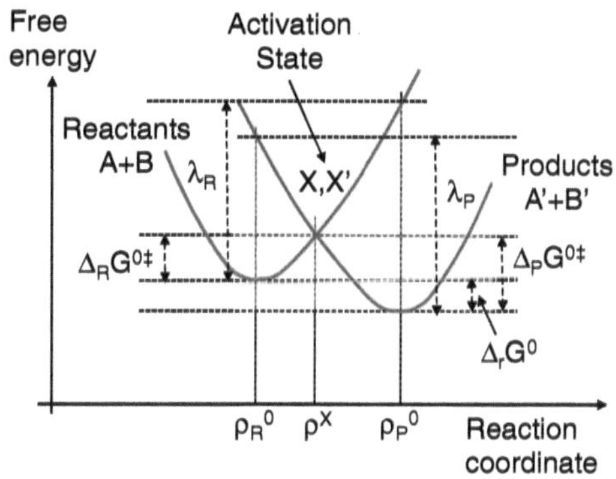

Fig. 3 Schematic diagram of chemical reaction based on Marcus Theory using reaction coordinate

calculated when the parabolic potential curves of Gibbs energy shapes have the same curvature for both reactants and products;

$$\Delta_R G^{0\ddagger} = \frac{(\Delta_r G^0 + \lambda)^2}{4\lambda} \tag{39}$$

where λ is reorganization energy and $\lambda = \lambda_R = \lambda_P$ in Fig. 3 for this case. This reorganization energy is considered the energy required for the reorganization process of the molecule after electron transfer is finished. The electron transfer occurs very fast compared with the molecular reorganization, that is, the electron transfer occurs without any nuclear motion in the molecule (Franck–Condon principle). It should be noted that the energy of activation state calculated from the Marcus theory is that for the adiabatic reaction ($\Delta_R G^{0\ddagger}$) and not for the nonadiabatic reaction (λ).

Quantum mechanical treatments considering the molecular wave function overlap were also proposed in order to develop more detailed treatment of electron transport for chemical reactions. Famous models include Dogonadze and Levich treatment, which takes into account the Landu–Zener formula of transition theory, and the theory proposed by Hopfield, which introduces the Fermi's golden rule for the transition. Detailed descriptions can be found in text books [2].

For electrochemical reaction, equilibrium condition for oxidant (*Ox*) and reductant (*Red*) is described as;

$$Ox + ne^- \leftrightarrow Red. \tag{40}$$

This equilibrium condition can be established with a certain electrode potential of electron. For this condition, the reaction speed of reduction and oxidation are the same, i.e., the minimum points of reaction coordinates for both reactants and products are the same. For the situation shown in Fig. 4, the activation energies from $Ox + ne^-$ to *Red* and *Red* to $Ox + ne^-$ are the same as;

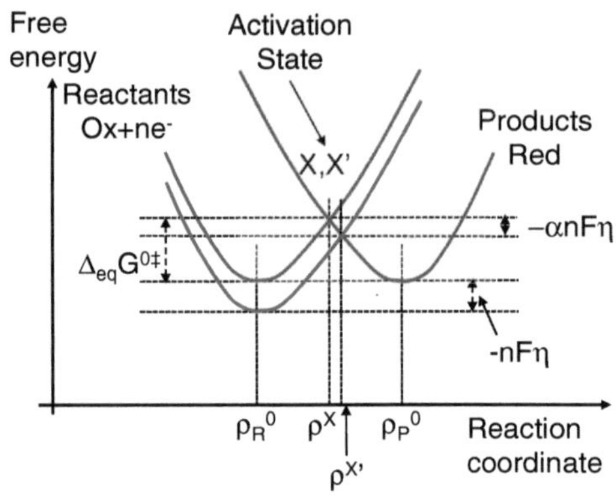

Fig. 4 Schematic diagram of electrochemical equilibrium and the changes with applied bias η

$$\Delta_{\text{eq}}G^{0\ddagger} = \Delta_c G^{0\ddagger} = \Delta_a G^{0\ddagger}, \tag{41}$$

where, $\Delta_{\text{eq}}G^{0\ddagger}$, $\Delta_c G^{0\ddagger}$, and $\Delta_a G^{0\ddagger}$ are equilibrium, cathodic, and anodic activation energy, respectively. As a result, the chemical reaction rate of $Ox + ne^-$ to Red (v_c) and Red to $Ox + ne^-$ (v_a) are the same. Thus;

$$v_c = k_c c_{Ox} = v_a = k_a c_{Red}, \tag{42}$$

where, k_c and k_a are cathodic (reduced) and anodic (oxidized) rate constant, C_{Ox} and c_{Red} are the concentration of oxidant and reductant. Using Eq. (38);

$$k_{\text{eq}} = k_c = k_a = k_0 \exp\left(\frac{-\Delta_{\text{eq}}G^{0\ddagger}}{RT}\right) \tag{43}$$

Oxidation and reduction current can be estimated from Eq. (43);

$$i_0 = nFk_{\text{eq}}c_{Ox} \exp\left(\frac{-\Delta_{\text{eq}}G^{0\ddagger}}{RT}\right) = nFk_{\text{eq}}c_{Red} \exp\left(\frac{-\Delta_{\text{eq}}G^{0\ddagger}}{RT}\right). \tag{44}$$

where, i_0 is exchange current density.

The equilibrium condition of Eq. (40) can be changed by bias. When positive bias (η) is applied to the equilibrium condition, the electron energy changes by $-nF\eta$. The α of this change is applied to anodic reaction (oxidation) and $1 - \alpha$ is applied to cathodic reaction (reduction). This can be understood from Fig. 4. Thus, the anodic current (i_a) and cathodic current (i_c) change with bias;

$$i_c = nFk_{\text{eq}}c_{Ox}\exp\left(\frac{-\Delta_{\text{eq}}G^{0\ddagger} - (1-\alpha)nF\eta}{RT}\right),$$
$$i_a = nFk_{\text{eq}}c_{Red}\exp\left(\frac{-\Delta_{\text{eq}}G^{0\ddagger} + \alpha nF\eta}{RT}\right). \tag{45}$$

This is called Butler–Volmar equation. The measured current (i) is the sum of cathodic and anodic currents. Considering the current direction;

$$\begin{aligned} i &= i_a - i_c \\ &= nFk_{\text{eq}}\left(\exp\left(\frac{-\Delta_{\text{eq}}G^{0\ddagger}}{RT}\right)\right)\left(c_{Ox}\exp\left(\frac{-(1-\alpha)nF\eta}{RT}\right) - c_{Red}\exp\left(\frac{\alpha nF\eta}{RT}\right)\right). \end{aligned} \tag{46}$$

This can be observed by Tafel plot in experiments when the applied bias is large enough. It can be described using the measured current (i);

$\eta = a - b\,\log|i|$, in the equation,

$$b = \frac{RT}{(1-\alpha)nF} \quad \text{or} \quad b = \frac{RT}{\alpha nF}. \tag{47}$$

Under equilibrium condition of Eq. (46), this can be expressed as;

$$\exp\left(\frac{nF\eta}{RT}\right) = \frac{c_{Ox}}{c_{Red}}, \tag{48}$$

here, $\eta = \phi^0 - \phi^0(\text{Ox/Red})$, thus, Eq. (48) is;

$$\phi^0 = \phi^0(Ox/Red) + \frac{RT}{nF}\ln\left(\frac{c_{Ox}}{c_{Red}}\right). \tag{49}$$

This Eq. (49) is exactly the same as the Nernst equation of Eq. (36).

The Gerischer model uses a different approach from the Marcus theory. In this model, the charge transfer is considered in terms of electric energies in solid and of energy levels in solution. The energy level in solution is much more complex than in solid state physics because of the effect of the polar solvent surrounding the ions. As a result, the Gerischer model is only applicable for weak interactions between the redox system and the electrode, that is, mainly for nonadiabatic conditions.

The energy diagram for reduced and oxidized species in polar solvent is shown in Fig. 5. The energy levels involved in an electron transfer differ from the thermodynamic equilibrium value of $E_{F,\text{redox}}$ due to the interaction with the solvent. (Electron energy is used for the Gerischer model instead of chemical potential because the model is based on an electron transfer like in solid. Thus, this electron

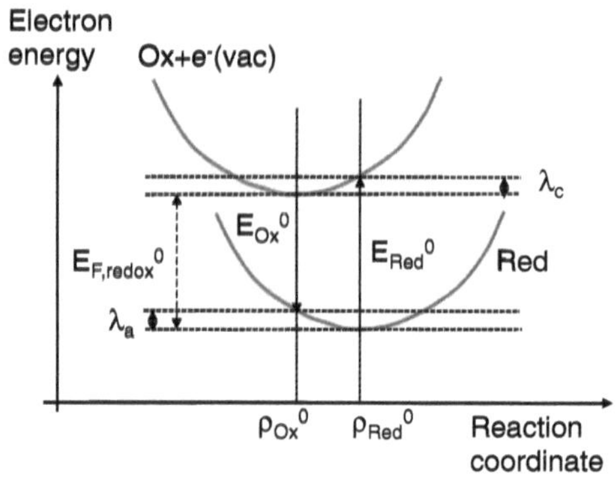

Fig. 5 Schematic diagram of the Gerischer model using reaction coordinate (*left*) and based on the electron energy of vacuum (*right*)

energy of $E_{F,\mathrm{redox}}$ is the same as the equilibrium state of electrochemical potential ($\tilde{\mu}_i$). It should be noted here that the signs of electron energy and electrochemical energy are opposite.) The electron energy of E^0_{Red} is required to transfer an electron from the reduced state (Red) to the vacuum, and the energy is leading to the formation of the oxidation state (Ox). This electron transfer is expected to be very fast compared with the reorganization of the solvation shells and the solvent dipoles (Franck—Condon principle). Thus, the reorganization of solvation shells and solvent dipoles occur after the electron transfer step. The reorganization continues until the oxidation spices reaches its equilibrium state with changing reorganization energy λ_c. The opposite electron transfer process to capture an electron from vacuum level into the oxidation state (Ox) is similar. For this case, the electron transfer energy and reorganization energy are expressed as E^0_{Ox} and λ_a. The electron transfer energy of E^0_{Red} and E^0_{Ox} are similar to ionization energy (I^0) and electronic affinity (A^0). In addition;

$$E^0_{Red} - \lambda_c = E^0_{Ox} + \lambda_a = E^0_{F,\mathrm{redox}}. \tag{50}$$

Assuming harmonic oscillation for the fluctuation of the solvent molecules near the equilibrium position of Red and Ox, the energy curves (E_{Red} and E_{Ox}) have parabolic shapes, that is;

$$\begin{aligned} E_{Red}(\rho) &= \gamma_{Red}\left(\rho - \rho^0_{Red}\right)^2, \\ E_{Ox}(\rho) &= \gamma_{Ox}\left(\rho - \rho^0_{Ox}\right)^2, \end{aligned} \tag{51}$$

where, γ_{Red} and γ_{Ox} are force constants for Red and Ox, ρ^0_{Red} and ρ^0_{Ox} are the position at which energy takes the minimum for the Red and Ox, respectively. For simplicity, the force constants and reorganization energy are assumed to be the same ($\gamma_{Red} = \gamma_{Ox} = \gamma$, $\lambda_{Red} = \lambda_{Ox} = \lambda$). The distribution functions (W_{Red} and W_{Ox}) of solvation states are given by;

$$W_{Red}(\rho) = W^0 \exp\left(\frac{-\left(E_{Red}(\rho) - E_{Red}(\rho^0_{Red})\right)}{kT}\right),$$
$$W_{Ox}(\rho) = W^0 \exp\left(\frac{-\left(E_{Ox}(\rho) - E_{Ox}(\rho^0_{Ox})\right)}{kT}\right). \tag{52}$$

Equation (52) can be converted into forms with electronic energy dependence alone using Eq. (51), here the integration over an entire energy is unity by normalizing pre-exponential factor;

$$W_{Red}(E) = (4kT\lambda)^{-1/2}\exp\left(\frac{-\left(E - E^0_{F,\mathrm{redox}} - \lambda\right)^2}{4kT\lambda}\right),$$
$$W_{Ox}(E) = (4kT\lambda)^{-1/2}\exp\left(\frac{-\left(E - E^0_{F,\mathrm{redox}} + \lambda\right)^2}{4kT\lambda}\right). \tag{53}$$

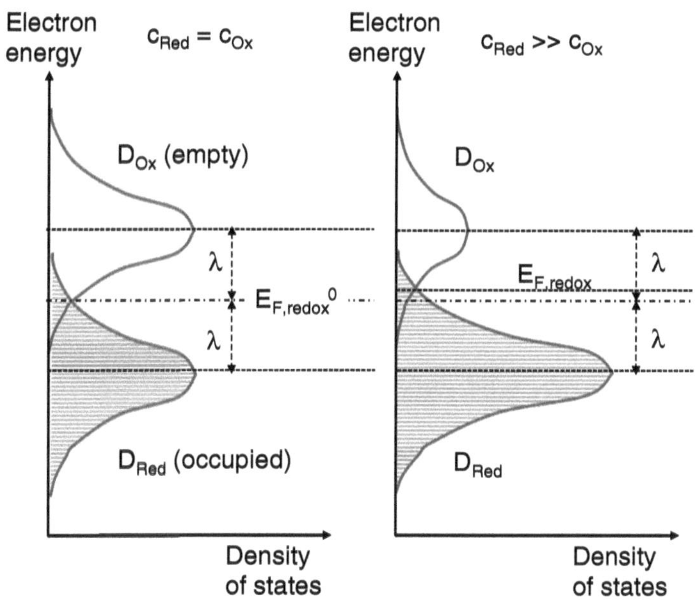

Fig. 6 Schematic electron energies of a redox system versus density of states by Gerischer model

The densities of electronic states are proportional to the concentration of the reduced (c_{Red}) and oxidized species (c_{Ox}), thus the total distribution (D_{Red}(E) and D_{Ox}(E)) is;

$$D_{Red} = c_{Red} W_{Red}(E), D_{Ox} = c_{Ox} W_{Ox}(E). \tag{54}$$

The corresponding energy diagram is shown in Fig. 6.

4 Equilibrium and Non-equilibrium in Electrochemical Reactions

Electrochemistry is one of the chemistry branches that studies chemical reactions which take place at the interface of an electron conductor and an ionic conductor. The electron conductor is an electrode and is usually made of a metal or a semi-conductor. The ionic conductor is an electrolyte and is usually made of an ionic solution or an ionic liquid. The reaction involves electron transfer between the electrode and the electrolyte. Electrochemical reaction is both a chemical reaction driven by an external voltage as in electrolysis and a voltage created by a chemical reaction as in battery. A chemical reaction where electrons are transferred between molecules are called oxidation/reduction (redox) reaction. In general, the situations of electrochemistry are that the oxidation and reduction reactions are separated in space or time with connections of external electric circuit. The oxidation and reduction reactions are paired and cannot be isolated under real situation.

Equilibrium between an electrode and an electrolyte is important to understand the electrochemical reaction. The equilibrium condition for chemical reactions is usually discussed using chemical potential but the definition of equilibrium is different in electron conductor. In order to connect the equilibrium condition between chemical reaction and electron conductor, the discussion is started from the concept of the Fermi level. The Fermi level is the hypothetic potential energy for an electron inside an electron conductor.

In solid state physics, work function is the minimum thermodynamic work needed to remove an electron from a solid surface to a point in the vacuum immediately outside the solid surface as shown in Fig. 7. That is, "work" is an energy. Here "immediately" means that the final electron position is far from the surface on the atomic scale, but still so close to the solid that the energy of an electron is influenced by ambient electric fields in the vacuum. The work function is not a characteristic of a bulk material, but rather a property of the surface of the material (depending on crystal facet and contamination). When the total energy of the solid with $N + 1$ electrons is defined as $E_{tot}(N + 1)$, the energy of the solid with one electron removed can be described as $E_{tot}(N) + E_{vac}$, where E_{vac} is the energy of the removed electron. The work function W_F is;

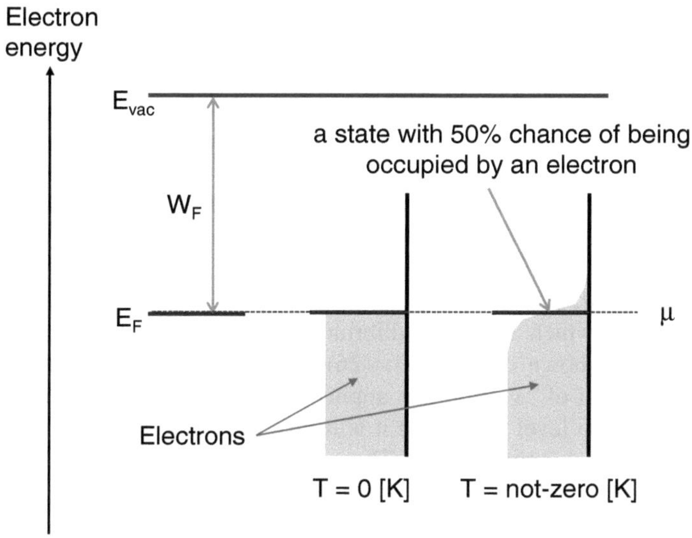

Fig. 7 Relationship among vacuum level (E_{vac}), the Fermi level (E_F), chemical potential (μ), and work function (W_F)

$$W_F = (E_{\mathrm{tot}}(N) + E_{\mathrm{vac}}) - E_{\mathrm{tot}}(N+1) = -\frac{\partial E_{\mathrm{tot}}(N)}{\partial N} + E_{\mathrm{vac}}. \tag{55}$$

The total energy of the solid ($E_{\mathrm{tot}}(N)$) can be treated as the Gibbs energy in the discussion of chemical reaction. Thus, in the case of $N \gg 1$, from the definition of chemical potential (Eq. (16));

$$\frac{\partial E_{\mathrm{tot}}(N)}{\partial N} = \mu. \tag{56}$$

The electron energy which is defined as the work function can be treated as the Fermi energy (E_F), which is the free energy of a single electron in solid under equilibrium. The energy defined as the work function at the temperature of 0 K is;

$$W_F = -\mu_{\mathrm{at\,0\,K}} = -E_{F\,\mathrm{at\,0\,K}}. \tag{57}$$

The chemical potential is almost the same as the Fermi energy even around room temperature.

For the strict definition for the finite temperature is little different. For a system of identical fermions (like electrons), the average number of fermions in a single-particle state i (n_i), is given by the Fermi–Dirac (F–D) distribution as shown in Fig. 7,

$$n_i = \frac{1}{1 + \exp\left(\frac{\varepsilon_i - \mu}{k_B T}\right)} \tag{58}$$

where k_B is the Boltzmann's constant, T is the absolute temperature, ε_i is the energy of the single-particle at state i, and μ is the chemical potential [3]. For the case of electrons in an electron conductor, μ, which is the point of symmetry, is typically called the Fermi level (E_F) or total electrochemical potential ($\tilde{\mu}$). That is, the average number of fermions (n_i) is 0.5 when the $\varepsilon_i = E_F = \tilde{\mu}$.

As we discussed, the standard point of electrode potential is standard hydrogen electrode (SHE), which is the equilibrium between proton (H^+) and hydrogen molecule (H_2) as shown in Eqs. (23)–(26). The electrochemical potential is measured using the unit of "V" with the applied bias. For work function, the standard point is the vacuum level (E_{vac}) and it is measured by the unit of "eV" due to the definition for one electron energy. The positive/negative directions of electrode potential and work function are opposite due to the definition. Although the vacuum level is of no practical reference, the question arises where the SHE can be quantitatively related to the vacuum level. Although many reports exist for the value, the value of $E_{SHE} = -4.5$ eV for the SHE is used mostly [4].

Work function (W_F) depends on the configurations of atoms at the surface of the material. For example, the work function for polycrystalline silver is 4.26 eV, but it varies for different crystal faces as (100) surface: 4.64 eV, (110) surface: 4.52 eV, (111) surface: 4.74 eV for silver crystals [5]. Examples of work function is shown in Fig. 8. Standard hydrogen electrode is also shown in Fig. 8.

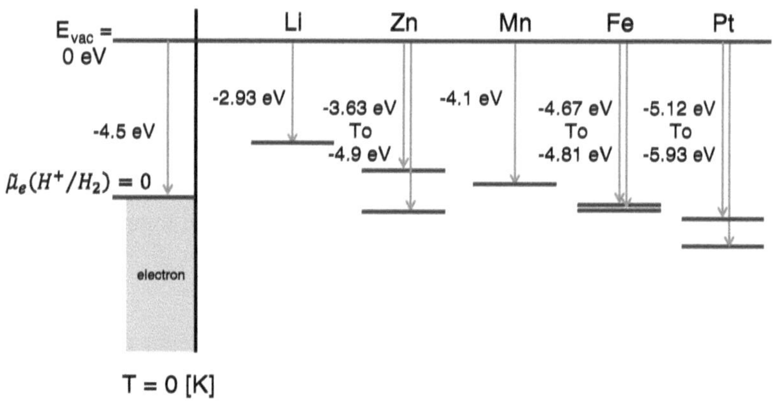

Fig. 8 Examples of work functions. Work function has positive value for the definition but the energy shown here is the value versus the electron energy of vacuum level as zero in order to compare the standard hydrogen electrode. The position for standard hydrogen electrode (SHE) is also shown

The work function (W_F) with existing electrostatic potential can be defined as;

$$W_F = -e\phi - E_F, \tag{59}$$

where, e is elementary charge, ϕ is the electrostatic potential in the vacuum nearby the surface, and E_F is the Fermi level (electrochemical potential of electrons) inside the material. The term $-e\phi$ is the energy of an electron in the vacuum nearby the surface which is almost the same as the electrostatic potential of inside the material. The work function is thus defined as the thermodynamic work required to remove an electron from the material to a state in the vacuum nearby the surface.

The chemical potentials or the Fermi levels are the same when the materials are in equilibrium. This means that the electrostatic potential in the vacuum near the surface changes at the interface when two different material are contacted. Thus, charge accumulation occurs at the interface. This charge accumulation forms to prevent charged carrier flow of the electron energies before contact, that is, electrons flow from high to low energy states of electron, for example. This is because a small amount of charged particles (electrons) move from its relatively high energy metal to relatively low energy metal. The thickness of the spatial distribution of the charges (space charge region) are defined by the Poisson equation. Two different kinds of metal contacts for example are shown in Fig. 9. The space charge region for metals are negligible because a huge number of charged particle exist in metals. This means the electrostatic potential at the interface changes abruptly.

In case of the equilibrium contact between a metal and an electrolyte, the situation of the space charge region is different form that for the two-metal contacts as shown in Fig. 10. The depletion layer thickness (the thickness of the space charge

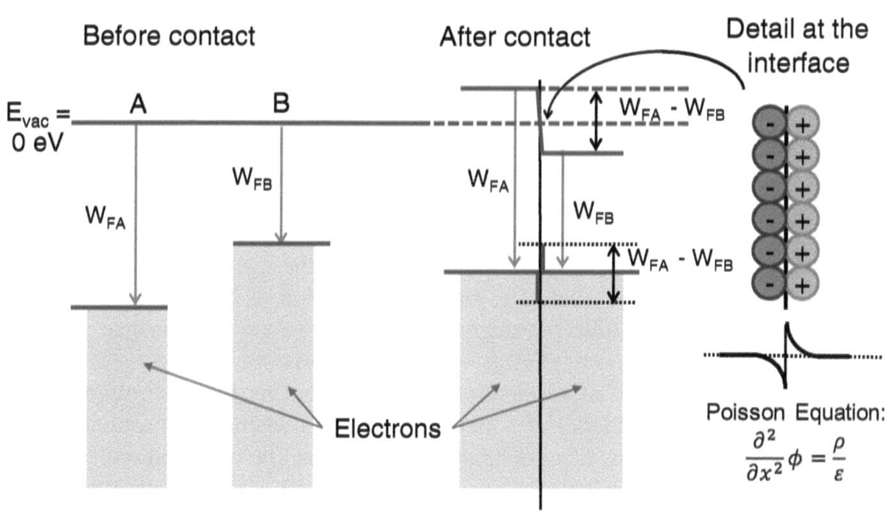

Fig. 9 Schematic diagram of contact between two different metals (*A* and *B*) contact. The Fermi levels (electrochemical potentials) become to be the same and the electrostatic potentials in the vacuum nearby the surfaces change to be different after the contact

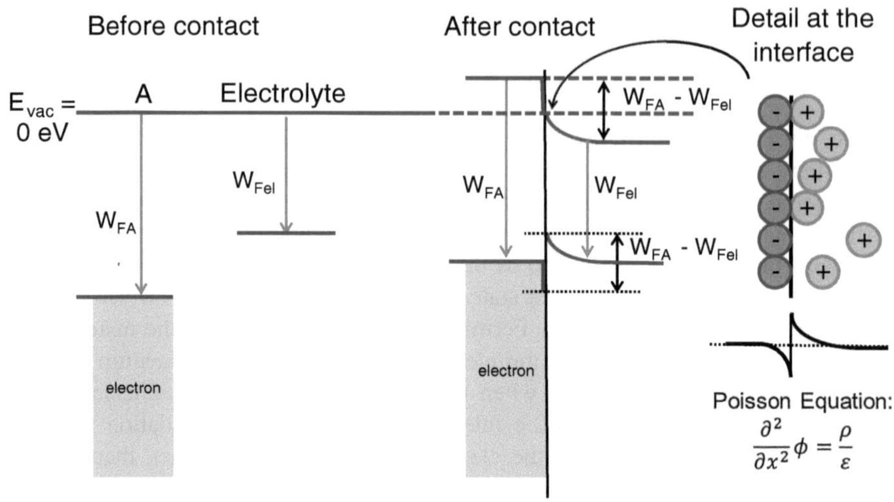

Fig. 10 Schematic diagram for metal and electrolyte (A and el) contact. The space charge region for the electrolyte is much larger than that for the metal due to the difference of the number of charged particles

region) for the electrolyte can be observed due to the density of charged particles in an electrolyte is much smaller than that in a metal. The charged particle concentration in an electrolyte is about 10^{20} cm^{-3} for 1 mol/L solvent ionic solution whereas that for the metal is over 10^{22} cm^{-3}. The Mott–Schottky plot ($1/C^2$ vs. bias where; C is the capacitance of the interface) is useful technique to know the characteristics of the depletion layer in the electrolyte.

The surface potential of a metal is generally pinned due to absorbed molecules of an electrolyte on the surface of the metal for the case of metal–electrolyte contact. (This means that the surface potential does not move as long as the surface ion condition does not change because the surface potential is defined by the coulomb interaction of the absorbed ions and the charges in the metal.) The effect is negligible, however, due to the huge number of charged particles in the metal as discussed. This means that the thickness of the depletion layer of metal is negligible thin, thus the potential of the metal is the same from the inside to very close to the surface, that is, the Fermi level. The surface orientation of a metal affects the ion absorption due to its surface structure change, however, the surface potential change of the metal is also negligible in the same way.

The chemical potential of an electrolyte was defined by the electrolyte composition, whereas the Fermi level of a metal can proceed when an appropriate bias is applied to the metal. Thus, electrochemical reaction can be occurred with the bias. Since the space charge region of a metal is too thin and can be neglected, it looks like surface potential change when bias is applied to a metal. When the anodic potential is more positive than the oxidized reaction in an electrolyte and the cathodic potential is more negative than the reduced reaction in an electrolyte,

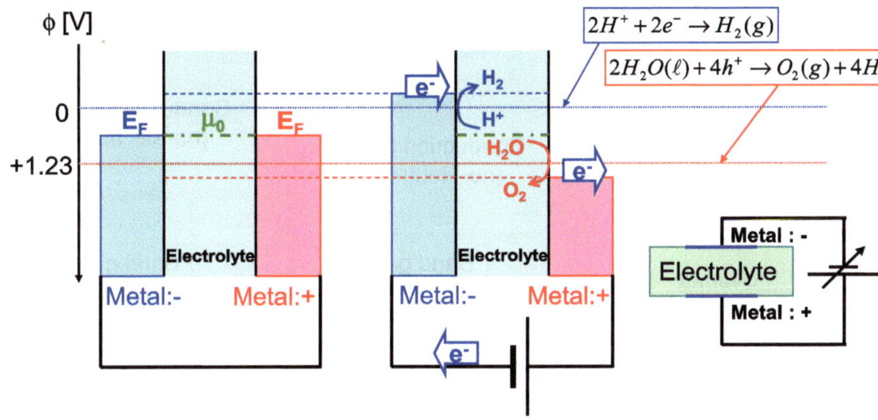

Fig. 11 Schematic diagram for electrochemical reaction. Water splitting is as an example for the reaction. The depletion layer is not shown in this graph

electrochemical reaction start as shown in Fig. 11. During the reaction, ion moves in the electrolyte and keeps the charge balance. This is electrochemical reaction (electrolysis) driven by an external bias. The characteristic of the electrochemical reaction is the high chemical reaction driving energy. The potential at 298 K for standard condition for chemical reaction is about 26 meV. A few volts for electrochemical reaction is commonly used for an electrochemical reaction. When spontaneous electrochemical reactions occur due to the difference of the redox potential, voltage is applied to the electrode, which is called battery.

5 Electrochemical and Photoelectrochemical Reactions

For energy conversion processes, not only electrochemical reaction but also photoelectrochemical reaction is important [6]. Photoelectrochemical reaction is basically the same as the electrochemical reaction but it is much more complicated due to the usage of semiconductor electrode which has a band gap between the valence band and conduction band [6]. The difference among metal, semiconductor, and insulator based on the electron energy is shown in Fig. 12. The conductivity of the metal comes from the overlap of conduction band and valence band. Semiconductor and insulator have gaps between the conduction band and valence band, thus, the conductivity is much lower than that of metal. The difference of the semiconductor and insulator is just the band gap size, that is, a large band gap material is generally called as a insulator.

The most important elements for semiconductors are silicon (Si) and germanium (Ge). The Si and Ge have four valence electrons in their outermost shell (s^2p^2 electron configuration) which gives them the ability to generate sp^3 orbital hybridisation. The

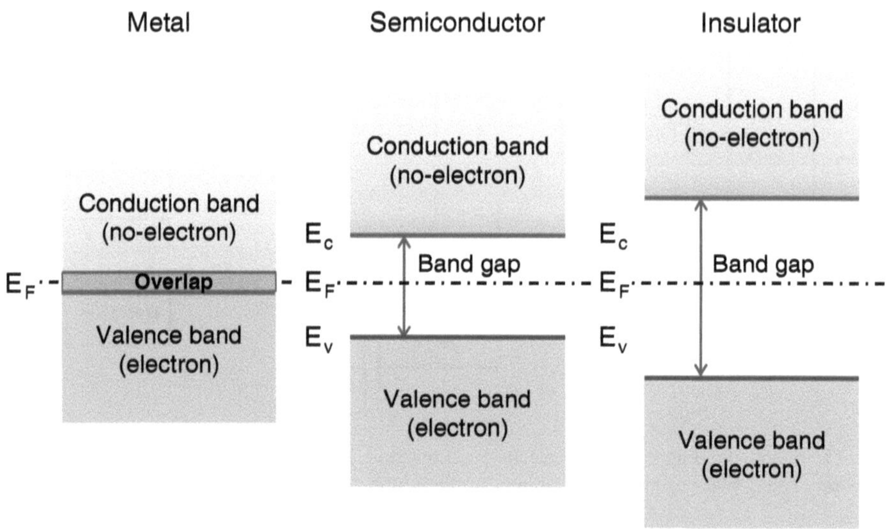

Fig. 12 Schematic electron energy diagram for metal, semiconductor, and insulator. The E_F shows the Fermi level, and that for semiconductor is an intrinsic case

sp^3 hybridisation has the lowest energy in sp^x hybridisation and the four bonds are equivalent. This allows the formation of diamond-like crystal structure with pure covalent bonds. This also results in the gaining or loss of electrons of the constituent atoms equally at the same time. Binary compounds, particularly between elements in Groups III and V such as gallium arsenide (GaAs), groups II and VI such as zinc serenade (ZnSe), and between different group IV elements such as silicon carbide (SiC) can be form covalent crystals in the same manner. Some oxide, organics, and amorphous materials also have semiconductor-like properties.

A pure semiconductor is a poor electrical conductor due to the right number of electrons filling in the valence band. Through doping technique (adding a small amount of impurity), a semiconductor can be modified to have excess electrons (becoming an n-type semiconductor) or a deficiency of electrons (becoming a p-type semiconductor). This is equivalent to adding an excess of electrons to the covalent bond in a crystal or to make a deficiency of electrons in the covalent bonds in the crystal as shown in Fig. 13. Doping an element with excess electron(s) (donor) makes n-type semiconductor and doping an element with deficient electron (s) (acceptor) makes p-type semiconductor. Group-V elements are donors and group-III elements are acceptors for the semiconductor consisting of group-VI elements for example. In both cases, the semiconductor becomes much more conductive. Electrons in conduction band are an origin of conductivity for n-type semiconductors and holes (deficient of electrons) has an origin of conductivity for p-type semiconductors. The polarity of n- and p-type and doping control of semiconductor electrode affect photoelectrochemical properties.

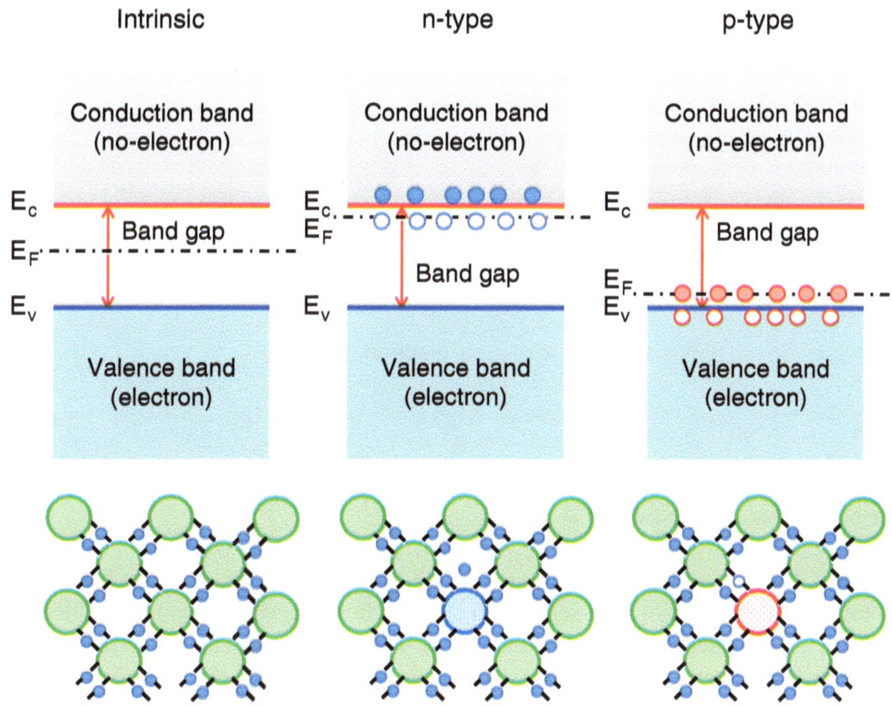

Fig. 13 Schematic diagram for intrinsic, n-type, and p-type semiconductors. The E_c, E_F, E_v show the *bottom* of conduction band, the Fermi level, and the *top* of valence band, respectively. The crystal structure model is also shown with a doped impurity to make n- and p-type are also shown

For metal, the Fermi level is located inside the conduction band, indicating that the band is partially filled. This means the Fermi level of metal can be controlled by external bias. For an insulator and a semiconductor, the Fermi level is located in the band gap, indicating that an empty conduction band is located at far above the Fermi energy as shown in Fig. 12. The definition of work function of semiconductor is the same as the other materials but the position is in the band gap, where electron cannot be allowed to exist. Therefore, the treatment of semiconductor electrode for electrochemical reaction is inevitably different from a metal electrode.

The electron affinity of a semiconductor (E_{ea}) is defined as the energy obtained by moving an electron from the vacuum just outside the semiconductor (E_{vac}) to the bottom of the conduction band (E_c) just inside the semiconductor. Here "just outside" means that the initial electron position is far from the surface on the atomic scale, but still close to the solid to be influenced by the ambient electric field in the vacuum, which is similar to the "immediately" for the work function. (It has to be noted that the definition of "electron affinity" for molecule or atom is different.) The electron affinity can be contrasted with the work function (W_F). The work function is the thermodynamic work that can be obtained by reversibly, isothermally moving an electron from the material surface to the vacuum immediately outside the solid

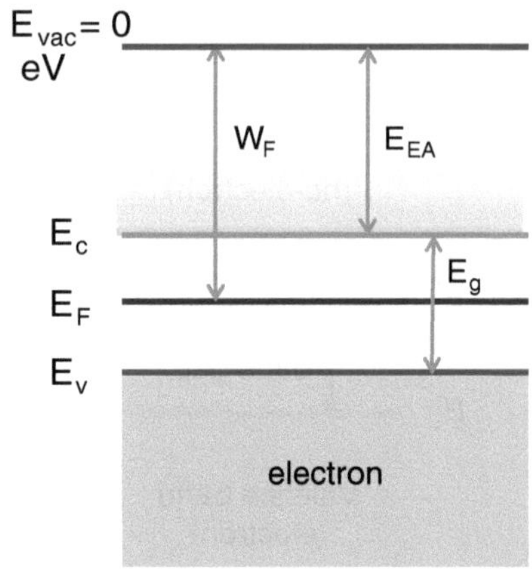

Fig. 14 Relationship among vacuum level (E_{vac}), the Fermi level (E_F), work function (W_F), electron affinity (E_{ea}), the *bottom* of the conduction band (E_c), the *top* of the valence band (E_v), and the band gap of semiconductor (E_g)

surface. The location is the Fermi level as discussed, not the conduction band edge. While the work function of a semiconductor can be changed by doping, the electron affinity ideally does not change with doping. Thus, it is almost an intrinsic value for a material. However, the electron affinity depend on the surface condition (e.g., crystal surface and surface chemical bonds).

For a semiconductor, the ideal energy for an electron from the surface to the vacuum immediately outside the semiconductor surface is "electron affinity (E_{ea}) plus band gap of semiconductor (E_g)." This is the case for intrinsic semiconductor. (It is much more complicated in reality because of the existence of the doping atoms for example.) The relationship is shown in Fig. 14.

The situation of the equilibrium at the contact between a semiconductor and an electrolyte is different from the contact between a metal and an electrolyte. It is the same that the chemical potentials or the Fermi levels are the same when the materials are in equilibrium. It is also the same that the charge accumulation forms to prevent charged carrier flow. The charged carrier in semiconductor is electron for n-type and hole for p-type, and is excited from neutral electron or hole trap (neutral donor or acceptor) thermally. The charged carrier can move in semiconductor, however, the donor or acceptor cannot move. Thus, the charge to form depletion layer is decided to be positive-charged donor for n-type and negative-charged acceptor for p-type semiconductors. As a result, the polarity of depletion layer is defined by this charged donor or acceptor. This situation is completely different from the metal case. In addition, the depletion layer thickness inside the semiconductor is significant because the number of charged particles in a semiconductor is much smaller than that in an electrolyte. The charged particle concentration in an electrolyte is about 10^{20} cm^{-3} for 1 mol/L ionic solution whereas that for the semiconductor is typically at the level of 10^{17} cm^{-3} depending on doping concentration. It has to be noted that for the contact between a metal and an electrolyte,

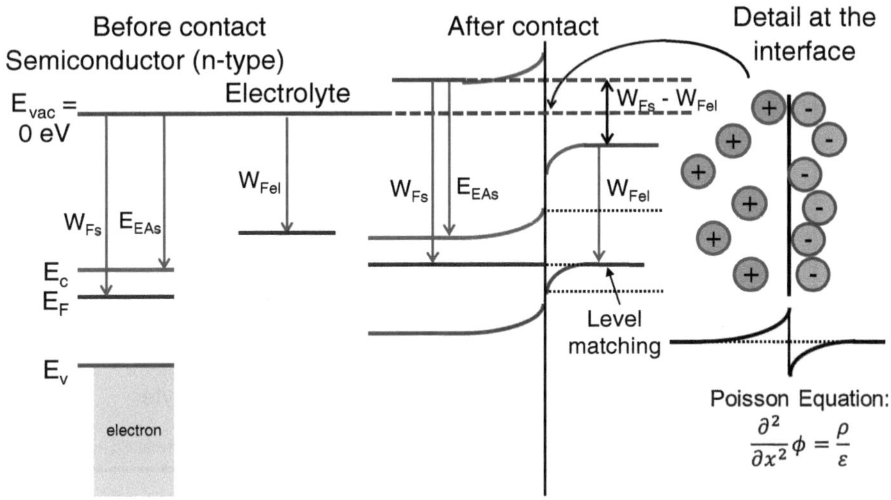

Fig. 15 Schematic diagram for n-type semiconductor and electrolyte (s and el) contact. The space charge region for the semiconductor is much larger than that for the electrolyte due to the difference of the number of charged particles

the depletion layer can be observed only in the electrolyte region. It is caused by the difference of the density of charged particles.

The position of the band edge energies of a semiconductor at the semiconductor/electrolyte interface is generally pinned due to absorbed molecules of an electrolyte on the surface of the semiconductor. This affects much of the depletion layer of the semiconductor because the density of charged particle is not high, whereas the pinned surface energy for a metal does not affect so much owning to a very thin depletion layer. The Mott-Schottky plot ($1/C^2$ vs. bias where C is the capacitance of the interface) is useful technique to know the characteristics of the depletion layer in the semiconductor [7]. This measurement indicates that not only the depleted donor/acceptor concentration but also the flatband potential, which is associated with the band bending in the depletion region inside the semiconductor.

The equilibrium contact condition of a semiconductor and an electrolyte is shown in Fig. 15. This graph displays the case for n-type semiconductor. The depletion layer curve is opposite where p-type semiconductor is used. It should be noted here that the carrier for n-type semiconductor is electrons in conduction band (not the electrons in valence band) and that for p-type semiconductor is holes in valence band (not the holes in conduction band). The current–bias characteristic for semiconductor–electrolyte electrode interface shows diode-like characteristics due to this carrier distribution.

An interesting property of the electrochemical reaction using semiconductor electrode is photoelectrochemical reaction without bias. The photoelectrochemical reaction occurs under illuminated light, where the energy is larger than the band gap energy of semiconductor, even without external bias as shown in Fig. 16.

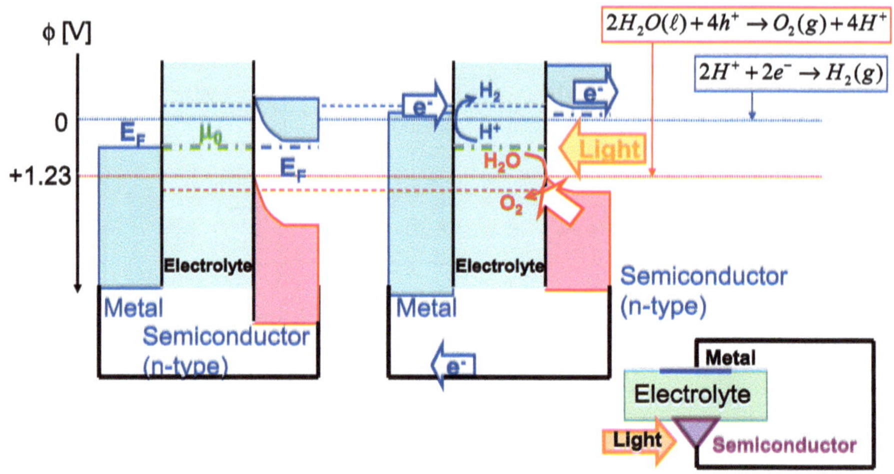

Fig. 16 Schematic diagram for photoelectrochemical reaction using n-type semiconductor

The keypoint of this reaction is the generated minority carrier by light absorption, that is, holes for n-type semiconductor and electrons for p-type semiconductor, respectively. The minority carrier in semiconductors can move to the interface between a semiconductor and electrolyte due to the electric field in the depletion layer, which is formed by the semiconductor–electrolyte contact. The minority carrier promotes photoelectrochemical reaction at the surface of the semiconductor, that the reaction is oxidation for n-type semiconductor and reduction for p-type

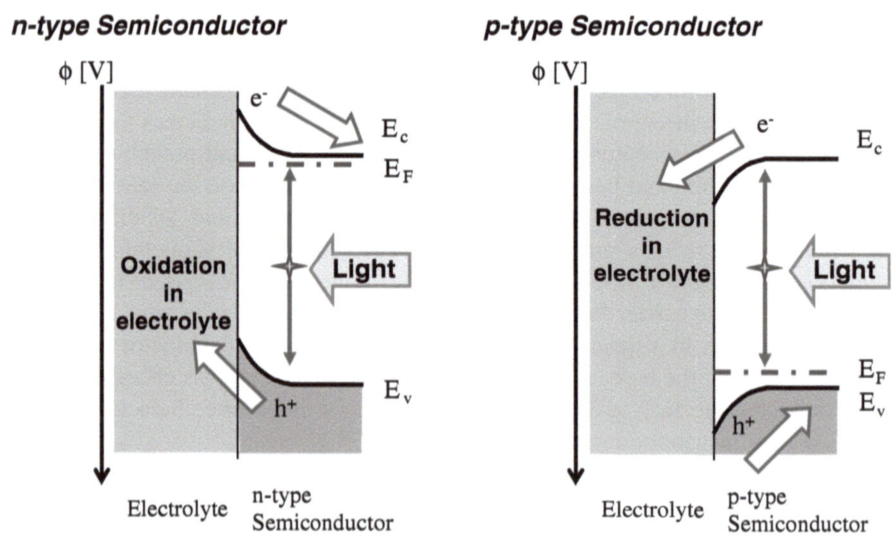

Fig. 17 Schematic diagram for photoelectrochemical reaction for n-type and p-type semiconductor electrode without bias

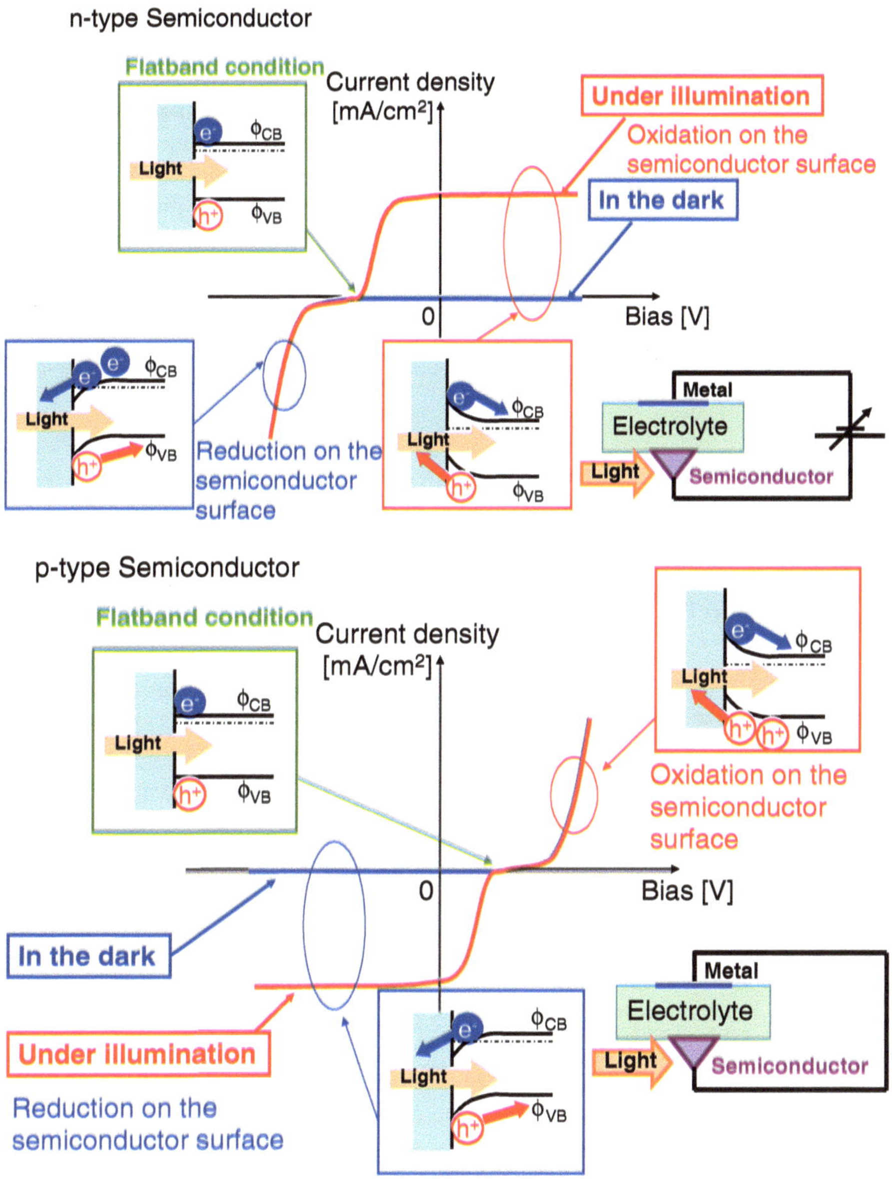

Fig. 18 Schematic current–bias diagram for semiconductor–electrolyte–metal configuration. The difference for n-type and p-type semiconductors are also shown

semiconductor, respectively. The differences in the photoelectrochemical reaction between the n-type and p-type semiconductors are shown in Fig. 17. This reaction without bias is the basic reaction for the energy conversion by photoelectrochemical reaction.

The current–bias characteristics is opposite between an n-type semiconductor and a p-type semiconductor as shown in Fig. 18. Current is not observed at the reverse bias region without light illumination. Photocurrent is observed in this reverse bias region when the light, where the energy exceeds the band gap energy, is illuminated. Band diagrams corresponding to the applied bias are also illustrated in Fig. 18. The band bending inside the semiconductor can be changed by the applied bias but the energies of conduction/valence band edges at the surface of the semiconductor cannot be changed generally because the existence of the absorbed molecule at the semiconductor surface is not changed by the applied bias. As a result, band bending in the depletion layer is changed by the applied bias. This is the cause of the dependence of current density on the applied bias. The flatband condition, where the band edges are flat, corresponds to the bias at which current density is zero.

References

1. Reference for Thermodynamics. For example Fermi E (1956) Thermodynamics. Dover Books on Physics, Dover Publications. ISBN: 0-486-60361-X; Prigogine I, Defay R (1954) Chemical thermodynamics. Prentice Hall Press, Englewood Cliffs. ISBN: 0-582-46283-5)
2. For example Marcus RA (1956) On the theory of oxidation-reduction reactions involving electron transfer I. J Chem Phys **24**:966–978; Marcus RA (1956) Electrostatic free energy and other properties of states having nonequilibrium polarization I. J Chem Phys 24:979–989; Memming R (2001) Semiconductor electrochemistry, Chap. 6. Wiley-VCH, New York, pp 112–150. ISBN: 3-527-30147-X
3. Sze SM (1981) Physics of semiconductor devices, Chap. 1, 2nd edn. Wiley, New York, pp. 7–60. ISBN: 0-471-05661-8
4. Lohmann F (1967) Fermi-Niveau und Flachbandpotential von Molekülkristallen aromatischer Kohlenwasserstoffe. Z Naturforsch a 22:843–844 (in German); Memming R (2011) Semiconductor eectrochemistry, Chap. 3. Wiley-VCH, New York, pp. 46–60. ISBN: 3-527-30147-X
5. Dweydari AW, Mee CHB (1975) Work function measurements on (100) and (110) surfaces of silver. Phys Stat Sol A 27:223–230
6. Memming R (2011) Semiconductor electrochemistry, Chap. 7. Wiley-VCH, New York, pp. 151–40. ISBN: 3-527-30147-X
7. Nozik AJ, Memming R (1996) Physical chemistry of semiconductor-liquid interfaces. J Phys Chem 100:13061–13078

Fundamentals of Semiconductors for Energy Harvesting

Masakazu Sugiyama

Abstract The most important feature of a semiconductor for the purpose of energy harvesting, such as photovoltaic cells and photocatalysts, is that the number densities of electrons and holes in a semiconductor is much smaller than in a metal and those densities can be easily altered by external excitation such as light irradiation. From a thermodynamic viewpoint, even though the internal energies of electrons and holes are fixed at the band-edge energies of the conduction and the valance bands, respectively, their *free energies*, which is equivalent to quasi Fermi levels, can be altered significantly by the logarithm of electron/hole density. The larger density of electrons (holes) reduces their entropy and their free energy gets closer to the internal energy, i.e., the conduction (valence) band edge energy. The free energy difference between electrons and holes in a semiconductor is the potential to exert work externally, and it corresponds to the terminal voltage for a photovoltaic cell. Such accumulation of electrons and holes upon the absorption of photons is the common working principle for photovoltaic cells and photocatalysts. The following properties are therefore important for a semiconductor to serve as a good material for photovoltaic cells and photocatalysts:

(1) strong light absorption per volume;
(2) slow recombination between electrons and holes, which necessitates as small density of crystal defects as possible not only inside the semiconductor but also at the surface;
(3) efficient transport of electrons and holes to different regions that accepts electrons or holes selectively.

M. Sugiyama (✉)
Department of Electrical Engineering and Information Systems, School of Engineering, The University of Tokyo, 7-3-1 Hongo, 113-8656 Bunkyo-Ku Tokyo, Japan
e-mail: sugiyama@ee.t.u-tokyo.ac.jp

M. Sugiyama et al. (eds.), *Solar to Chemical Energy Conversion*,
Lecture Notes in Energy 32, DOI 10.1007/978-3-319-25400-5_3

1 Electrons and Holes in a Semiconductor

A characteristic of a semiconductor is the existence of bands as a result of hybridization among the orbitals with valence (highest energy) electrons as shown in Fig. 1. It should be emphasized that a band is the collection of quantum states, the wave function of which spreads over an entire crystal. Fermi-Dirac distribution of electrons in the bands, which is a thermodynamic principle as will be described in Sect. 8, makes one of the bands in a semiconductor, which is called as the valence band, almost fully occupied by electrons and another band just above the valence band is made almost empty and it is called as the conduction band. At a temperature higher than zero kelvin, for the sake of minimizing the total free energy of the system, some electrons occupy the states in the conduction band and correspondingly some vacant states exist in the valence band. An electron surrounded by unoccupied states in the conduction band move among neighboring quantum states upon external force, which is the quantum-mechanical view of electrons' motion in a semiconductor. Similarly, a vacancy in the valence band, which is surrounded by the states occupied by electrons, switch from a state to a neighboring one upon external force; this is the quantum-mechanical view of holes' motion. For electrons and holes to achieve equilibrium in a certain spatial range of a semiconductor, they have to migrate in the range. For the thermodynamics of electrons and holes in a semiconductor, therefore, our interest is focused on the electrons and holes that can move spatially. Those mobile electrons and holes are often referred to as *carriers* since they carry charges. This is why we focus on a smaller number of electrons in the conduction band and neglect abundant electrons in the lower-energy bands including the valence band. In the same manner, we focus on a small number of holes in the valance band and neglect all the vacant quantum states that exist in the conduction band and the bands with higher energies.

A characteristic of semiconductors is that the concentrations of electrons and holes can be altered by doping. In this chapter, we refer to concentration as the

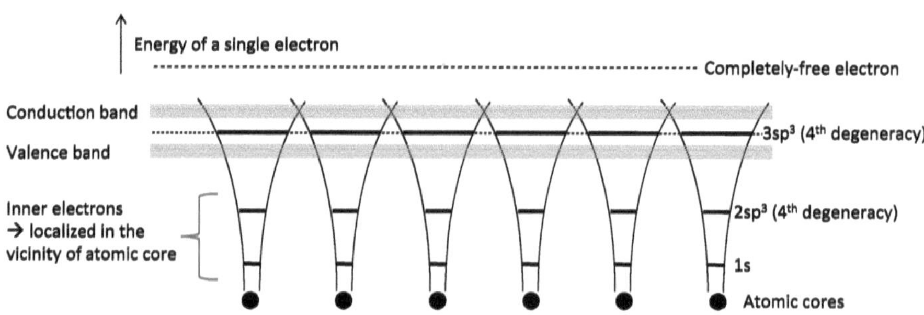

Fig. 1 A schematic energy diagram of atomic quantum states and bands in a silicon crystal. One s- and three p- orbitals form four degenerated sp^3 hybrid orbitals. The orbitals with the highest energy (sp^3) interfere with each other due to the spatial overlap of orbitals and a set of stabilized states (the valance band) and un-stabilized states (the conduction band) are formed

number density in a unit volume. Fermi-Dirac distribution of electrons imposes a law of mass action in which the product between electron concentration n and hole concentration p is determined by several material parameters as well as temperature,

$$np = N_c N_v \exp\left(-\frac{E_g}{kT}\right), \tag{1}$$

where N_c (N_v) are the effective density of states in the conduction (valance) band, E_g is the bandgap, k is the Boltzmann constant and T is absolute temperature. Approximately, the effective density of states for electrons (holes) in the conduction (valence) band is the maximum number of quantum states that electrons (holes) can occupy. It is, as well as the bandgap, a parameter unique to a semiconductor material. For example, the values for GaAs at 300 K is as follows: $N_c = 4.7 \times 10^{17}$ cm^{-3}, $N_v = 9.0 \times 10^{18}$ cm^{-3} and $E_g = 1.42$ eV, resulting in the concentrations of electrons n and holes p if no impurity exist, $n = p$=4.7×10^6 cm^{-3}. The value is called intrinsic carrier concentration n_i where "intrinsic" means a pure semiconductor without any impurity.

The relationship in Eq. (1) also holds true when dopant atoms exist in a semiconductor. Adding atoms with extra valance electrons to a semiconductor (n-type doping) increases the concentration of electrons according to the concentration of dopant atoms, and the concentration of holes decreases to keep the relationship in Eq. (1) as depicted in Fig. 2. Similarly, the dopant atoms with insufficient valance electrons increases the concentration of holes (p-type doping) and the concentration

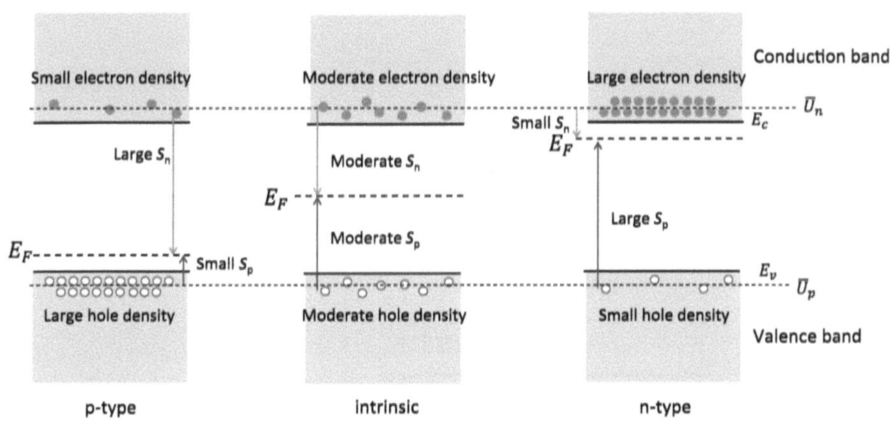

Fig. 2 Concentrations of electrons and holes in a semiconductor with different doping conditions. The relevant energies are also shown; E_c the conduction-band edge, E_v the valence-band edge, E_F the Fermi level (for both electrons and holes), $\bar{U}_n$ the averaged internal energy of elctrons, $\bar{U}_p$: the averaged internal energy of holes. S_n denotes entropy for electrons and S_p is entropy for holes

of electrons decreases accordingly. Such an interaction between the concentrations means equilibrium between electrons and holes. The situation is analogous to H^+ and OH^- ions in an electrolyte.

2 Free Energy of Electrons in a Semiconductor: An Intuitive Picture

A variety of semiconductor heterostructures can be fabricated by combining semiconductors with different bandgaps and doping concentrations. A typical example of a silicon pn junction is shown in Fig. 3. It should be emphasized that a n-type-doped semiconductor contains the same density of free electrons and positive ions of dopant atoms, resulting in the neutral charge density. The ions are bound to the lattice of silicon atoms. Similarly, a p-type semiconductor contains both free holes and negative ions of dopant atoms. At the pn junction, electrons and holes migrate and combine with each other, and they disappear. As a result,

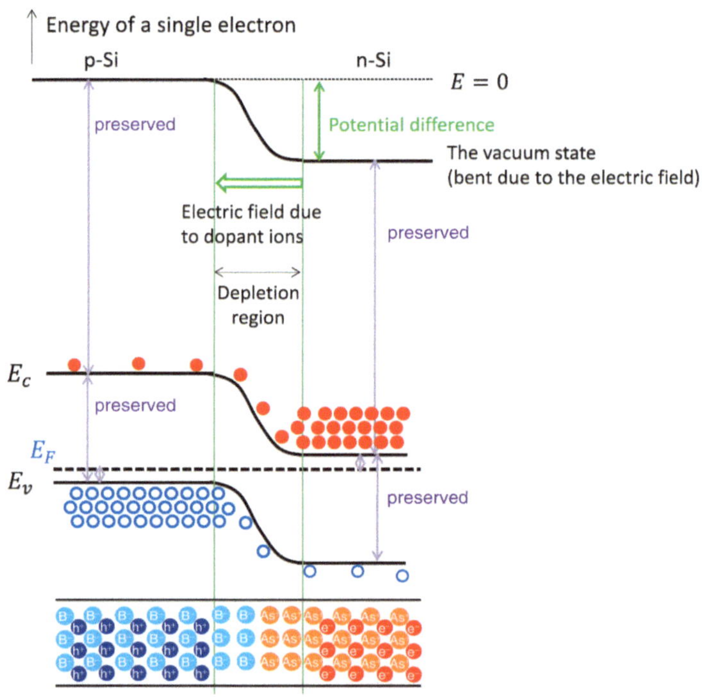

Fig. 3 The band lineup of a silicon pn junction. The *bottom cartoon* depicts the distribution of dopant ions (B^- and As^+) as well as electrons and holes. At the junction, depletion region is formed where electrons and holes are absent due to recombination and an electric field emerges as a result of negative and positive charges of dopant ions

the negative ions in the p-region and the positive ions in the n-region, which are immobile because they compose chemical bonds with the host silicon atoms, forms charge distribution and an electric field emerges. Since the energy of an electron, which is the vertical axis of the band diagram, is affected not only by the existence of neighboring atomic cores but also by such an electric field spreading over a large number of atomic cores, the energies of all the electron states (and the vacuum state as the origin of electrons' energy states) changes as a function of the position in the pn junction. This is a well-known band bending that emerges when different kinds of semiconductors (and metals) are put together. The band bending prevents electrons and holes from both migrating to the junction front and disappearing by recombination. As a result of these two competing processes at the junction front, the recombination of electrons and holes and the emergence of the electric field that prevents the additional supply of electrons and holes, establish equilibrium for the concentrations of electrons and holes.

Such equilibrium can be described in terms of thermodynamics. The criteria of thermodynamic equilibrium is the balance of free energy. It is therefore vital to obtain the free energies for electrons and holes. In general, free energy F for an ensemble of particles is described as

$$F = E - TS, \tag{2}$$

where E is internal energy, T is temperature and S is entropy. Conventional discussion on the behavior of electrons and holes employs energy per a particle, not for a mole of particles as is often employed in chemistry, and here we also discuss all the energies per a particle. Let us neglect the term associated with a work by volume expansion and use Helmholtz free energy (this is not exactly correct as we can see in the following section).

According to the Fermi-Dirac distribution of electrons, a majority of electrons and holes exist at the states with the energy close to the band edges. More precisely, the averaged energy of electrons in the conduction band is

$$E_{avr}(\text{electrons}) = E_c + \frac{3}{2}kT. \tag{3}$$

The detailed derivation is found in Sect. 9. The term $(3/2)kT$ can be regarded as the averaged kinetic energy of a particle moving in 3-dimensional space and this term is negligibly small compared with E_c since kT is approximately 1/40 eV at room temperature while E_c takes an order of eV. In this section, we will neglect the term $(3/2)kT$.

The entropy for an electron approximately takes the form

$$S = -k\ln\left(\frac{n}{N_c}\right). \tag{4}$$

This can be derived based on the definition of entropy in statistical mechanics:

$$S = \left(\frac{1}{n}\right) k \ln W, \tag{5}$$

where W is the number of possible arrangements for n electrons to occupy their places out of N_c sites, with n and N_c as numbers in a unit volume. The denominator n is for the purpose of obtaining entropy per a single electron.

To obtain Eq. (4) from Eq. (5), a well-known mathematical relationship is used

$$W = \frac{N_c!}{(N_c - n)!n!}, \tag{6}$$

with Stirling's approximation:

$$\ln x! \approx x \ln(x) - x. \tag{7}$$

In most cases, n is much smaller than N_c and an approximation $N_c - n \approx N_c$ is used.

The discussion above leads to the expression for the free energy of an electron:

$$E_{Fn} = E_c + kT \ln\left(\frac{n}{N_c}\right), \tag{8}$$

where the subscript n of E_{Fn} denotes electrons. Equation (8) actually is well known in solid state physics as Fermi level for electrons. This is not at all a coincidence but has a profound meaning; Fermi level is equivalent to free energy if we apply thermodynamics to the electrons in a semiconductor!

It is important to exaggerate the difference between internal energy and free energy. For an electron in a semiconductor, its internal energy is fixed at the edge of the conduction band E_c (if we neglect the contribution of kinetic energy $(3/2)kT$). On the other hand, the free energy of an electron can be altered by electron concentration n by adding dopant atoms for example. The larger electron concentration is, the closer its free energy approached to its internal energy by the reduction of entropy term TS. In other words, increasing electron concentration boosts the free energy through the reduction in entropy.

3 The Energy of Holes

It is important to recognize that the energy of a hole takes the opposite sign with respect to the energy of an electron that has been removed to make the hole. Let us consider two states with and without a hole as in Fig. 4. The energy of a hole is

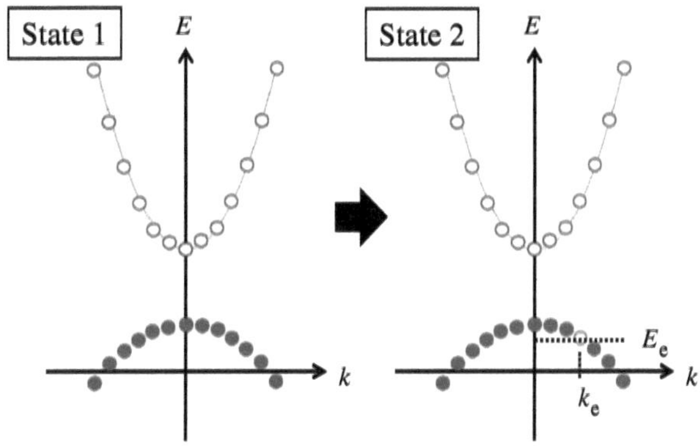

Fig. 4 Two states with and without a hole. In the state 2, an electron has been removed, leading to the appearance of a hole in the valance band

obtained by taking the difference in the total energy between the state 2 with a hole and the state 1 without a hole.

$$E_h = \sum_{state\,2} E - \sum_{state\,1} E. \tag{9}$$

The only difference between the two states is the absence of an electron in the state 2 and the total energy of the state 2 can be equalized to that of the state 1 by adding the energy of a missing electron E_e.

$$\sum_{state\,1} E = \sum_{state\,2} E + E_e. \tag{10}$$

From Eqs. (9) and (10), it is clear that the sign of energy is opposite between an electron and a hole:

$$E_h = -E_e. \tag{11}$$

This complexity stems from the concept of holes: a hole is not a real particle but a virtual one describing the vacancy of an electron. As a result, both the internal energy and the free energy of a hole should be accompanied by a negative sign, or these energies increases downwards in an energy diagram, when the energy of a hole is discussed simultaneously with that of an electron.

We can now discuss on the free energy of a hole in analogy to the one for an electron. The internal energy of a hole is approximately the energy of the valance band edge E_v. The entropy of a hole can be described using the hole concentration p and the effective density of states N_v in the valence band, which can be regarded as the maximum number of the states that holes can occupy:

$$S = -k\ln\left(\frac{p}{N_v}\right). \tag{12}$$

Then the free energy of a hole can be obtained as

$$-E_{Fp} = -E_v + kTln\left(\frac{p}{N_v}\right), \tag{13}$$

where the subscript p of E_{Fp} denotes holes. Both the free energy E_{Fp} and the internal energy E_{v} are accompanied by a negative sign because they are the energies for a hole. The entropy term, however, has the same sign as the formula for an electron (Eq. (8)) because the derivation of entropy depends solely on the combination of particles occupying the states, regardless of whether the particle is real or virtual.

Equation (13) can be converted to a famous form:

$$E_{Fp} = E_v - kTln\left(\frac{p}{N_v}\right), \tag{14}$$

which is identical to the expression of the Fermi level for holes. Fermi level is identical to free energy not only for electrons but also for holes. Interestingly, Eq. (14) can be derived from the formula of electron free energy (Eq. (8)) using the low of mass action (Eq. (1)) and the relationship

$$E_g = E_c - E_v. \tag{15}$$

This fact indicates that electron free energy and hole free energy take the same value, i.e., both particles are in equilibrium, when the law of mass action holds true.

4 Equilibrium of Electrons and Holes in a pn Junction

In a semiconductor under equilibrium, i.e., without an external work, electrons and holes take the same free energy and it changes according to doping as depicted in Fig. 2. When a junction of p- and n-type semiconductors is in equilibrium, thermodynamics imposes the equal free energy (or Fermi level) for the entire region of the structure. At the same time, apart from the junction front, i.e., outside the depletion region, the concentrations of electrons and holes never changes from the original value before forming the junction, and the free energy (or Fermi level) stays at the same position with respect to the band edges. These two requirements result in the band bending in the depletion region and the size of the band-edge offset is equal to the difference in the free energies in p- and n- regions. The exact shape of the band bending is determined by electrostatics: the distribution of charge density in the vicinity of the junction front, which emerges as a result of the depletion of electrons and holes, results in the band bending according to Poisson's equation.

As depicted in Fig. 3, most of electrons exist in the n-region because of the slope of the band-edge energy E_c accelerates the electrons whose energy is close to E_c to the direction of the n-region. However, some of the electrons in the n-region have sufficient kinetic energy to climb up the slope of E_c and migrate to the p-region. From the n-region to the p-region, electron concentration n decreases exponentially as the band-edge energy E_c increases, following

$$n = N_c \exp\left(-\frac{E_c - E_F}{kT}\right), \tag{16}$$

which is the relationship transformed from Eq. (8), with the constant free energy E_{Fn}. Hole concentration similarly follows the relationship upon a change in E_v,

$$p = N_v \exp\left(-\frac{E_F - E_v}{kT}\right), \tag{17}$$

which is converted from Eq. (14).

5 Free Energies of Electrons and Holes in a Metal and a Schottky Junction

For metals, electrons occupy the quantum states to the middle of a band and there is no bandgap on the top of the states occupied by electrons. In this situation, according to Fermi-Dirac distribution, the concentrations of electrons and holes are so large that the entropy of electrons and holes is almost zero. Accordingly, the free energy of electrons corresponds to the energy top of the quantum states that electrons occupy at 0 K. (At a temperature higher than 0 K, some electrons are excited to higher-energy states than the energy top at 0 K). Above the electron free energy, a lot of unoccupied quantum state exist and they can exchange electrons with occupied states upon external force, leading to free motion of holes as well as electrons. Therefore, we can assume that a collection of holes exist at the energy states neighboring to electrons and the free energy of holes takes the same value as that of electrons in a metal. Note that the entropy of electrons and holes in a metal is almost zero and the free energy is independent of the concentration of electrons and holes, resulting in a constant value of the free energies for electrons and holes across the spatial axis.

We can now discuss the equilibrium for a junction between a metal and a semiconductor: a Schottky junction. Similarly to the case of a semiconductor pn junction, free energy is equalized for the entire structure and the position of the free energy with respect to the band edges is preserved for a semiconductor region apart from the junction (i.e., outside the depletion region). As depicted in Fig. 5, the free energy takes the same value across the entire structure. The band bending in a semiconductor emerges in order to equalize the free energies in a metal and that in a semiconductor,

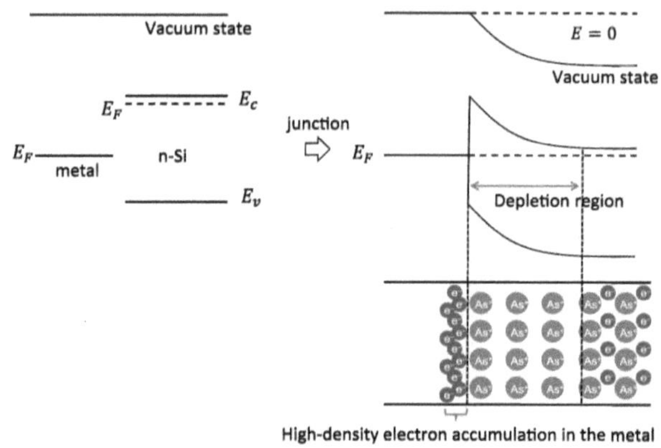

Fig. 5 The band lineup for a Schottky junction between a metal and an n-type semiconductor. The distribution of electrons, holes and ionized impurities is also shown at the *bottom schematic*

which had been different before forming the junction. The band bending is formed as a result of electron depletion in a part of semiconductor close to the junction, i.e., a depletion region. The electrons which initially existed there accumulates in the metal region close to the junction front. Nevertheless, the locally large concentration of electrons in the metal never affects the free energy in that region.

In reality, it often the case that a set of quantum states exist at the surface of a semiconductor because of the breakage of periodicity in atomic arrangement at the surface. These states often exist in the middle of the bandgap, which we call surface states or interfacial states. If electrons occupy these states, the free energy of electrons at the surface of a semiconductor is clamped to the energy of surface states because the density of such surface states is so small that the occupancy of states by electrons is large, making entropy almost zero. As a result, the electron free energy in a semiconductor aligns to the energy of surface states, which also determines the extent of band bending in the depletion region, as depicted in Fig. 6. Such a

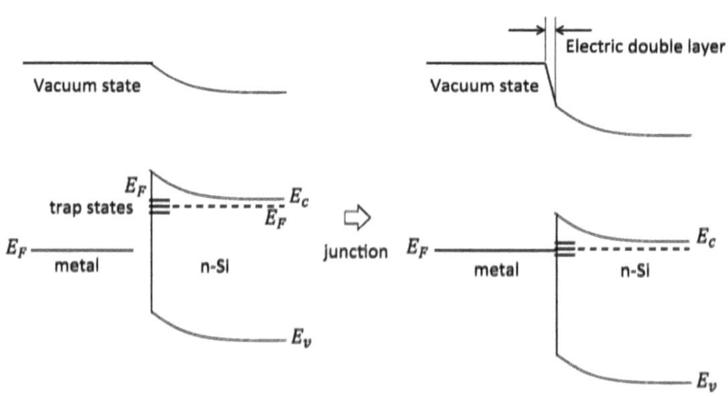

Fig. 6 The band lineup for a Schottky junction with trap states at the surface of a semiconductor

phenomena is often called as Fermi level pinning. When a semiconductor with such as surface state forms a junction with a metal, the band bending in a semiconductor is already determined by the energy of surface states and the misalignment of the electron free energy between the metal and the semiconductor is compensated by the appearance of the electric double layer at the metal/semiconductor interface. This is a very thin layer, less than a couple of nanometers, in which the opposite charges (electrons and holes) face each other between the metal and the semiconductor sandwiching the metal/semiconductor interface. The reason why we deal with such a complicated band alignment here is that the situation is quite similar to the interface between an electrolyte and a semiconductor which inevitably appears in photocatalysis and photoelectrochemistry using semiconductors.

6 Non-equilibrium in a Semiconductor

So far, we have discussed situations without external work applied to a semiconductor and the law of mass action has been always applicable. Here we will discuss situations in which some external work increases the concentrations of electrons and holes in a semiconductors from their equilibrium values.

The most readily-understandable situation may be the generation of electron-hole pairs in a semiconductor upon the absorption of photons. A photon absorbed in a semiconductor creates a pair of an electron and a hole by exciting an electron from a state in the valance band to another one in the conduction band. Such extra electrons and holes recombine and disappear as a successive event. The balance of the rates between generation and recombination determines the concentrations of electrons and holes upon light absorption by a semiconductor. Such concentrations of both electrons and holes are clearly larger than the values determined by the law of mass action (Eq. (1)). The electrons and holes in such a situation are no more in equilibrium. The increase in the concentration makes entropy smaller for both electrons and holes and the values defined by Eqs. (8) and (14) now take different values between electrons and holes. We call the value of E_{Fn} and E_{Fp} as quasi Fermi level for electrons and holes, respectively. Since electrons and holes are no longer in equilibrium, we have to define *quasi* free energies individually for electrons and holes.

The behavior of (quasi) free energies, or (quasi) Fermi levels, upon light absorption is depicted in Fig. 7. For an intrinsic semiconductor, light irradiation increases the concentrations of both electrons and holes by Δn and Δp, respectively, where $\Delta n = \Delta p$. Quasi Fermi levels approach to the band edge energies following Eqs. (8) and (14) upon an increase of electron and hole concentrations to $n + \Delta n$, $p + \Delta p$, respectively. Then a splitting $\Delta\mu$ (= $E_{Fn} - E_{Fp}$) emerges between the quasi free energies of electrons and holes. If we are able to take an electron and a hole out of the semiconductor under light irradiation while avoiding their recombination, the pair of an electron and a hole can generate external work if they can recombine at a place outside the semiconductor and release the difference of free energies $\Delta\mu$. The

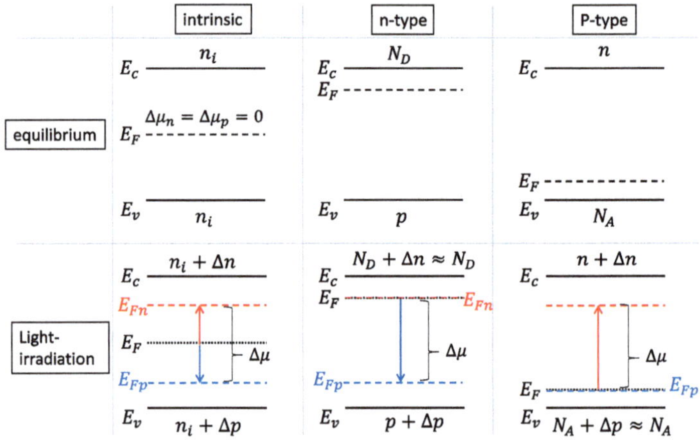

Fig. 7 The positions of Fermi levels and quasi Fermi levels for semiconductors in equilibrium and under light irradiation. For the latter (*bottom row*), the dotted lines denoted as E_F indicates the position of Fermi levels without light irradiation for reference. n and p denote electron and hole concentration, respectively, and N means the concentration of dopants with the subscripts D as donor and A as acceptor, respectively

splitting between the quasi Fermi levels for electrons and holes, $\Delta\mu$, is therefore very important as a measure of the maximum work if a pair of an electron and a hole is extracted out of a semiconductor.

For a n-type semiconductor without light irradiation, the electron concentration is determined by the dopant concentration N_D, and the hole concentration p is very small so that the product $N_D\,p$ satisfies the law of mass action. Upon light absorption, the electron and the hole concentrations increase by Δn and Δp, respectively. But Δn is much smaller than the initial electron concentration N_D and almost no change is brought about in the electron quasi Fermi level E_{Fn}. On the other hand, the hole concentration increases substantially from the initial value p, leading to a substantial shift of the hole quasi Fermi level E_{Fp} to the direction of valence-band-edge energy. For a p-type semiconductor, similarly, only the electron quasi Fermi level moves upon light absorption. In this manner, for a doped semiconductor, only the quasi Fermi level of minority carriers moves upon light absorption and the quasi Fermi level of majority (doped) carriers is almost never changed.

7 Semiconductor Device Structures for Energy Harvesting from Light

Based on the considerations on non-equilibrium of carriers in a semiconductor, let us look at a couple of structures including a semiconductor which serve for energy harvesting from light. The first example is a semiconductor photocatalyst. Figure 8 depicts its conceptual structure. Here, two metals with different Fermi levels are

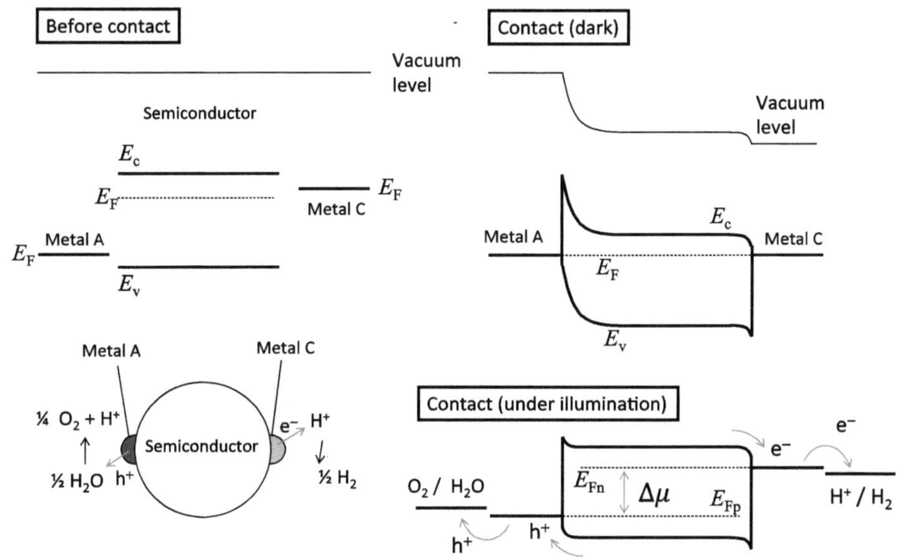

Fig. 8 The *left-bottom* schematic depicts an idealized semiconductor photocatalyst with two metals serving as selective contacts for electrons and holes. The metals also function as catalysts for the transfer of electrons and holes to the molecules in an electrolyte. Three diagrams depicts band lineup and (quasi) Fermi levels before making a contact, a contact under dark and illumination. For the band lineup under illumination, the vacuum level is omitted because its misalignment between an electrolyte and a semiconductor is not straightforward; it is affected by the electric double layer at the electrolyte/semiconductor interface and is strongly dependent on the surface. The level denotes as "H^+/H_2" is the redox potential for H_2 evolution from H^+, which is described in another chapter. Similarly, the level with the notation "O_2/H_2O" is the redox potential for O_2 evolution from H_2O

attached to a semiconductor particle which is slightly doped to be n-type. Theoretically, an intrinsic semiconductor can serve for energy harvesting from light. In reality, however, it is often difficult to obtain a genuinely intrinsic semiconductor with extremely high purity. In addition, slight doping helps deactivate crystal defects as centers for electron-hole recombination without photon emission, i.e., a process leading to energy dissipation, by filling the quantum energy states in the band gap associated with crystal defects. It is therefore realistic to apply a slight doping to a bulk of semiconductor. Figure 8 supposes such a situation in which an entire semiconductor particle is doped to be slightly n-type.

Under a dark condition, the Fermi level in the metal C is aligned to the Fermi level of the semiconductor so that the slope of E_c in the vicinity of the junction favors the transport of electron to the metal C while hole transport to the metal C is conversely blocked by the slope of E_v. The mismatch of Fermi level between the metal A and the semiconductor forms a large band bending, which blocks (favors) the transport of electrons (holes) to the metal A. Note that no net transport of carriers exist in the junction under equilibrium and the discussion on the ease of transport here is only for the carriers that exist in excess of the equilibrium concentration upon external work such as light illumination. Since the semiconductor here is n-type, almost no holes exist and we normally focus on the transport of electrons. This is why we call the junction with the metal C as "ohmic" (conductive) and the junction with the metal A as Schottky (rectifying or inslative under a zero bias condition).

Upon illumination, the generation of extra electrons and holes in the semiconductor splits the quasi Fermi levels for electrons and holes. The remaining band bending still favors the transport of electrons to the metal C and that of holes to the metal A. As a result, an electron-hole pair with the difference of free energy $\Delta\mu$ is split to the different directions, making it possible to extract work $\Delta\mu$ out of the semiconductor.

Now, the electrons in the metal C and the holes in the metal A have the free energies, i.e., quasi Fermi levels: E_{Fn} for the electrons and E_{Fp} for the holes, respectively. These values are determined by the band-edge energies of the semiconductor bulk and the concentrations of electrons and holes that exist in the semiconductor based on Eqs. (8) and (14). On the other hand, there exists a relationship between the band-edge energies of semiconductors and the standard electrode potentials for a variety of redox reactions in an electrolyte, as is described in the previous chapter. If, as is the situation in Fig. 8, the free energy of the electrons in the metal C, E_{Fn}, exists at the upper side of a redox reaction, electrons can be transferred from the metal C to the molecules in an electrolyte participating in the reaction. Note that this is a consideration in terms of thermodynamics and just the possibility of the reaction progress is predicted; nothing is known about the rate of the electron transfer. Similarly, if the free energy of the holes in the metal A, E_{Fp}, exists at the lower side of a redox reaction, holes can be transferred from the metal A to the molecules in an electrolyte participating in the reaction. It should be kept in mind that the energy of holes increases downwards in an energy diagram that depicts the energy of an electron. This is an ideal operation mode of a semiconductor photocatalyst in an electrolyte; a semiconductor particle with the surface modified by two appropriate kinds of metals can serve for water splitting upon light irradiation.

Of course, if we expand the size of the hypothetical structure in Fig. 8 and connect two electrical wires to the metals A and C, we can extract electrons and holes to an external device with the difference in the free energies $\Delta\mu$ between electron and holes. This is a photovoltaic cell using one Schottky junction (at the interface with the metal A) and the other junction that favors the transfer of the majority carriers in the semiconductor (in this case, electron) at the semiconductor/metal interface, which is often called ohmic junction.

As for a photovoltaic cell, it is more common to use a semiconductor pn junction for the purpose of separating electrons and holes to different electrodes. A typical solar cell employs a pn junction in the vicinity of a surface as depicted in Fig. 9. In this example, a thicker region is n-type but it can be p-type with an n-region at the surface. The back side is contacted to the metal C with ohmic band alignment by employing a metal with its original Fermi level a bit larger than the Fermi level of the n-type semiconductor, similarly to the case in Fig. 8. As for the metal contacting

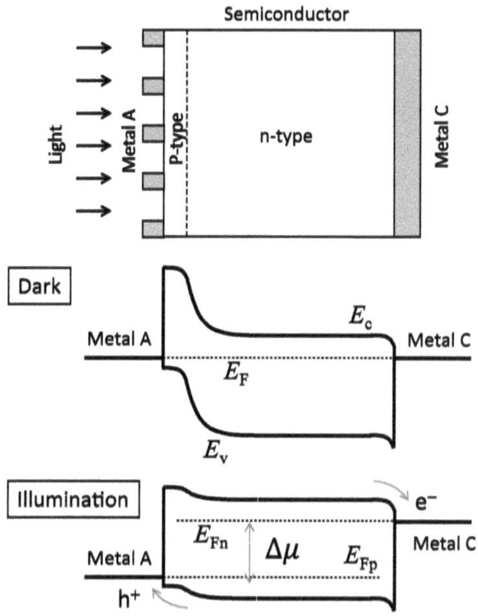

Fig. 9 *Upper schematic* depicts a typical photovoltaic cell using a semiconductor pn junction with two metals serving as contacts for electrons and holes, respectively. Bottom two diagrams depict the band lineup and (quasi) Fermi levels under dark and illumination

to the p-region of the semiconductor, the metal A is assumed to have the same Fermi level as the one in the semiconductor p-region for the simplicity of the picture. The resultant slope of the conduction-band-edge energy, E_c, favors the majority of electrons with smaller kinetic energies to migrate to the direction of the metal C. Since the electrons are in equilibrium across the device structure, a small fraction of electrons exist in the vicinity of the metal A, where E_c is much larger, but the majority exist in the n-region and the metal C. Similarly, the majority of holes prefers moving to the direction of the metal A owing to the slope of E_v.

Under illumination, similar to the case of a photocatalyst in Fig. 8, the quasi Fermi levels are split and band bending becomes smaller, but still a slope of E_c and E_v exist and it separates electrons and holes to different directions. The electrons in the metal C and the hole in the metal A have different free energies and they can make external work if they are led to an external device.

If we compare the situations in Figs. 8 and 9, it is clear that a Schottky junction and a pn junction work in a similar manner to separate electrons and holes. The maximum difference in free energy (or quasi Fermi levels) $\Delta\mu$ is obtained when intensive light absorption generates abundant electrons and holes accumulated in the conduction and the valence bands, respectively. In this situation, a large electron (hole) concentration makes it difficult for the positive donor (negative acceptor) ions in a crystal lattice to exist unaccompanied by electrons (holes), where naked ions are the source of a depletion region as depicted in Figs. 3 and 5. Such difficulty of making a depletion region makes the bands flat upon the existence of abundant electrons and holes in a semiconductor. This favors large $\Delta\mu$ as is understandable by the band-lineup charts in Figs. 8 and 9. Such a flat-band situation, however, lacks electric field inside, making it difficult to drive electrons and holes to the separated electrodes. This is called "open-circuit" situation when we can obtain the maximum $\Delta\mu$ (or voltage in the case of a photovoltaic cell) while no current is obtained.

One practical disadvantage of Schottky junction is that it is not easy to make a large band bending in the semiconductor. Even though an appropriate metal is employed which have a sufficiently large difference in Fermi level with respect to the one in a semiconductor, the existence of semiconductor surface states prevents a large band bending as expected from the difference in Fermi levels. On the other hand, it is relatively easy to have a large band bending at a pn junction and this is why a pn junction is a common structure in photovoltaic devices for the separation of electrons and holes.

8 Fermi-Dirac Distribution of Electrons in the Bands and Effective Density of States

This section can be skipped if the reader is not so familiar with solid-state physics. More detailed discussion is found in many textbooks on semiconductor physics such Refs. [1] and [2].

Here we will review a conventional theory on how electrons populate in the conduction and the valence bands and we will find a connection to the thermodynamic treatment of the concentrations of electrons and holes in a semiconductor. An ordinary discussion of electron occupancy of the quantum states in a semiconductors first assumes Fermi level, E_F. As depicted in Fig. 10, the electron density in each energy segment $n(E)$ is determined by the product between the density of states in the conduction band $D_e(E)$ and the probability of an electron occupying a state at energy E, $f(E)$.

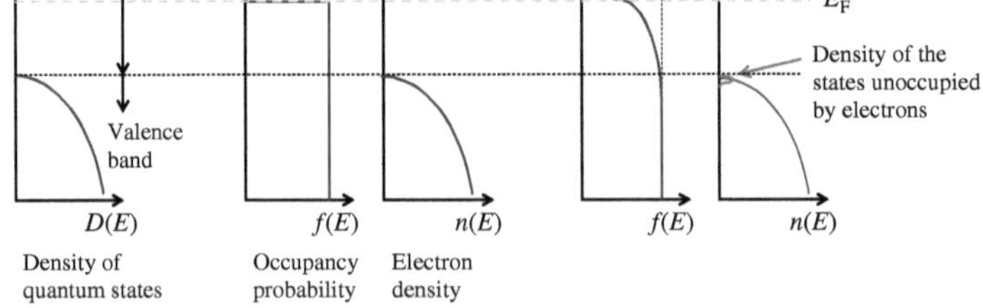

Fig. 10 Distribution of electrons along its energy in the bands of a semiconductor

$$n(E) = D_e(E)f(E), \tag{18}$$

where

$$D_e(E) = 4\pi\left(\frac{2m_e^*}{h^2}\right)^{\frac{3}{2}}(E - E_c)^{\frac{1}{2}}, \tag{19}$$

i.e., the density of states increases according to the square root of the energy distance from the edge of the conduction band, where m_e^* is the effective mass of an electron in the conduction band and h is the Planck constant. $f(E)$ takes a well-known form of Fermi-Dirac distribution with a reference energy E_F, which we call Fermi level.

$$f(E) = \frac{1}{\exp\left(\frac{E - E_F}{kT}\right) + 1}. \tag{20}$$

With E_F positioned at a certain energy in the bandgap, at 0 K, electrons occupy the valence band completely and the conduction band is completely empty. At a temperature above 0 K, the function $f(E)$ makes a small portion of the states in the conduction band occupied by electrons and some of the states in the valence band become empty, which states are regarded as holes.

The total electron concentration is obtained by integrating $n(E)$ along the energy over the conduction band.

$$n = \int_{E_c}^{\infty} D_e(E)\exp\left(-\frac{E - E_F}{kT}\right)dE = 2\left(\frac{2\pi m_e^* kT}{h^2}\right)^{\frac{3}{2}}\exp\left(-\frac{E_c - E_F}{kT}\right), \tag{21}$$

where $f(E)$ is approximated since $E - E_F \gg kT$:

$$f(E) \approx \exp\left(-\frac{E - E_F}{kT}\right), \tag{22}$$

meaning that Fermi-Dirac distribution is approximated as Boltzmann distribution if the energy of our interest is far enough from E_F. Equation (21) can be summarized as

$$n = N_c \exp\left(-\frac{E_c - E_F}{kT}\right), \tag{23}$$

with

$$N_c = 2\left(\frac{2\pi m_e^* kT}{h^2}\right)^{\frac{3}{2}}. \tag{24}$$

The intuitive interpretation of Eq. (23) is that all the electrons take the energy of E_C and the density of quantum states N_C are concentrated at the same energy. The exponential term is the probability of electrons occupying the quantum states at the energy E_C with the origin of energy taken at E_F.

Interestingly, Eq. (23) is completely the same as Eq. (16) that was converted from Eq. (8), indicating that the Fermi level for electrons distribution along the energy axis is in essence the free energy of electrons according to a thermodynamic interpretation.

In a similar manner, the concentration of holes is obtained as:

$$p = N_v \exp\left(-\frac{E_F - E_v}{kT}\right). \tag{25}$$

where N_v is the effective density of states for holes. This is equivalent to the thermodynamic expression, Eq. (14). Multiplying the concentrations of electrons n and holes p with Eqs. (23) and (25) yields the law of mass action in Eq. (1). Here the relation in Eq. (15) is used. The discussion above clarifies that the thermodynamic treatment of carrier concentrations in a semiconductor is another viewpoint of Fermi-Dirac distribution of electrons in a semiconductor.

9 Detailed Derivation of Electron/Hole Free Energies in a Semiconductor

This section can be skipped if the reader is not so familiar with solid-state physics. The detailed discussion is in Ref. [3].

Let us first obtain the averaged energy of the electrons in the conduction band.

$$\langle \epsilon_e \rangle = \frac{1}{n_e} \int_{E_c}^{\infty} E D_e(E) f(E) dE = E_c + \frac{3}{2} kT. \tag{26}$$

Similarly, the averaged energy of the holes in the valence band is:

$$-\langle \epsilon_h \rangle = \frac{1}{n_h} \int_{E_v}^{-\infty} (-E) D_h(E) f(E) dE = -E_v + \frac{3}{2} kT, \tag{27}$$

where $D_h(E)$ is the density of states in the valence band, which takes the form of Eq. (19) with a substitution of m_e with m_h, effective mass of a hole in the valence band.

Here, the tedious process of calculation is omitted but we certainly get the relationship in Eq. (3). Because the kinetic energy of electrons and holes in 3-dimensional space is $(3/2)kT$, electrons and holes in a semiconductor can be regarded as monoatomic ideal gases. Then, we can apply the precise formulation of entropy for a monoatomic ideal gas due to 3-dimensional translational motion, which is known as Sackur-Tetrode equation [4]:

$$S = k\left\{\frac{5}{2} + \ln\left(\left(\frac{2\pi m_e^* kT}{h^2}\right)^{\frac{3}{2}}\frac{1}{n}\right)\right\}, \tag{28}$$

where n is the electron concentration. Here we describe the derivation for the electrons but similar treatment is possible for holes. Using Eq. (24), we can obtain the expression

$$TS = kT\ln\left(\frac{N_c}{n}\right) + \frac{5}{2}kT. \tag{29}$$

Therefore, admitting that ϵ_e corresponds to the internal energy of an electron U, Gibbs free energy of an electron is

$$\begin{aligned} G &= U + PV - TS = \langle\epsilon_e\rangle + PV - TS = \left(E_c + \frac{3}{2}kT\right) + kT - \left(kT\ln\left(\frac{N_c}{n}\right) + \frac{5}{2}kT\right) \\ &= E_c + kT\ln\left(\frac{n}{N_c}\right) = E_{Fn}. \end{aligned} \tag{30}$$

Here, ideal gas equation $PV = kT$ is used and the last equator corresponds to Eq. (8). Now we can confirm that the Fermi level for the electrons in a semiconductor is equivalent to the Gibbs free energy for an electron in the viewpoint of thermodynamics. Similar derivation confirms that the Fermi level for holes is equivalent to the Gibbs free energy of a hole.

References

1. Ashcroft NW, Mermin ND (1976) Solid-state physics. Brooks/Cole, Belmont
2. Sze SM, Lee MK (2012) Semiconductor devices: physics and technology. Wiley, Chichester
3. Würfel P (2009) Physics of solar cells: from basic principles to advanced concepts. Wiley, Weinheim
4. Atkins P, De Paula J (2014) Atkins' Physical Chemistry. Oxford University Press, Oxford

Part II
Modeling Interface for Energy Storage: Modeling of Chemical and Electrochemical Reactions

Fundamentals of Chemical Reaction Kinetics

Shinichiro Nakamura

Abstract This chapter discusses the fundamentals of chemical reaction kinetics. Starting from conventional macroscopic aspect, the author intends to bridge it to the molecular theory, including quantum chemistry and molecular dynamics. The former is established with Arrhenius and Eyring plots, and the latter is represented by transition state theory (TST). Problems in solid surfaces and in catalytic turn over frequency (TOF) are also discussed. These will be provided as an overview of the basic concepts in accordance with the purpose of this book.

1 Arrhenius Plot and Eyring Plot

Chemical kinetics is based on experiments. The method is now established as one of the most indispensable research methods [1, 2]. Chemical kinetics enables the experimental determination of reaction rates from which rate laws and rate constants are derived. Rate laws can be derived for zero order reactions (for which reaction rates are independent of concentration), first order reactions, and second order reactions. The activation energy for a reaction is experimentally determined through the Arrhenius equation and Eyring equation. The main factors are the physical state of the reactants, the concentrations of the reactants, temperature, and whether or not any catalysts are present in the reaction. The Arrhenius equation gives the dependence of the rate constant k for a chemical reaction at the absolute temperature T (in kelvins) via the Arrhenius plot ($\ln(k)$ vs. $1/T$). Here, A is the pre-exponential factor (or simply the pre-factor), E_a is the activation energy, and K_B is the Boltzmann's constant. It is considered as an empirical relationship.

$$k = A \exp(-E_a/K_B T)$$

S. Nakamura (✉)
Nakamura Laboratory, RIKEN Research Cluster for Innovation, RIKEN,
2-1 Hirosawa, Wako, Saitama 351-0198, Japan
e-mail: snakamura@riken.jp

M. Sugiyama et al. (eds.), *Solar to Chemical Energy Conversion*,
Lecture Notes in Energy 32, DOI 10.1007/978-3-319-25400-5_4

The Eyring equation, known also as Eyring–Polanyi equation, has equivalent importance.

$$k = \frac{K_B T}{h} \exp(-\Delta G^*/RT)$$

where ΔG^* is the Gibbs energy of activation, h is Planck's constant, R is gas constant. This equation can be rewritten as:

$$k = \frac{K_B T}{h} \exp(\Delta S^*/R) \exp(-\Delta H^*/RT).$$

The linear form,

$$\ln(k/T) = \frac{-\Delta H^*}{R} \cdot \frac{1}{T} + \ln \frac{K_B}{h} + \frac{\Delta S^*}{R}$$

is used as the Eyring plot (ln(k/T) vs. $1/T$). It can provide important properties of enthalpy of activation (ΔH^*) and entropy of activation (ΔS^*).

To link the discussion with the microscopic molecular level description, in other words, to advance from thermodynamics to statistical mechanics, the partition function q plays an essential role. The relation of q to statistical mechanics is the relation of wave function to quantum mechanics, where ε_j denotes the energy of microscopic state.

$$q = \sum_j \exp(-\varepsilon_j/kT)$$

Here, it is relevant to notice the relation of the partition function with important thermodynamic properties. Energetic properties, such as internal energy ΔU, pressure ΔP and enthalpy ΔH, all depend on derivatives of the partition function q. On the other hand, entropic properties, such as Helmholtz free energy ΔF $(F = U - TS)$, entropy ΔS, and Gibbs free energy ΔG, depend directly on the value of q.

An illustrative example of a gas phase reaction is useful to show that the equilibrium constant K, in fact it is related to ΔG and can be expressed by the partition functions.

$$NOCl \longleftrightarrow NO + Cl$$

$$K = \frac{[NO][Cl]}{[NOCl]}$$

The partition function q is introduced to give a P (i), i.e., probability of the system existing in a state with energy ε_i.

$$P(i) = \exp(-\varepsilon_i/kT)/q$$

On the left and right sides of the reaction, we have

$$P(NOCL) = q_{NOCl}/q_t$$
$$P(NO + CL) = q_{NO+Cl}/q_t$$

where q_{NO+Cl} represents the state in which the two gases are coexisting. Since the two exist in the same volume of area, the property of the microscopic states is obtained by multiplication.

$$\begin{aligned} q_{NO+Cl} &= \sum \exp(-\varepsilon_{NO+Cl}/kT) \\ &= \sum \exp(-\varepsilon_{NO}/kT) \sum \exp(-\varepsilon_{Cl}/kT) \\ &= q_{NO}q_{Cl} \end{aligned}$$

By determining a reference point of energy, and introducing the distribution function per unit volume (the volume of the system is assumed to be *V*),

$$q_i^o = q_i^z \exp(-\varepsilon_i/KT),$$

where q_i^z is conventional partition function.

$$\begin{aligned} K &= \frac{N_{NO}N_{Cl}}{N_{NOCl}} = \frac{q_{NO}q_{Cl}}{q_{NOCl}} = \frac{q_{NO}^o V q_{Cl}^o V}{q_{NOCl}^o V} \\ Kc &= K/V = \frac{q_{NO}^z q_{Cl}^z}{q_{NOCl}^z} \exp(-\Delta\varepsilon/kT) \\ \Delta\varepsilon &= \varepsilon_{NO} + \varepsilon_{Cl} - \varepsilon_{NOCL} \end{aligned}$$

Thus, we can express here the equilibrium constant of macroscopic chemical reaction in terms of molecular partition functions. We are now ready to proceed to the next discussion of the transition state theory TST, from a molecular point of view.

2 Transition State Theory

Transition-state theory (TST) is an established method. There are excellent review papers which readers can refer to [3]. There are very instructive explanations in some theoretical chemistry books [4, 5]. This theory provides a basis for understanding qualitatively how chemical reactions take place. Since the Arrhenius equation is derived from empirical observations, further development was necessary to reveal the meaning of two parameters; the pre-exponential factor (*A*) and the activation energy (*Ea*). The TST successfully showed the meanings.

Transition state theory is a semi-classical theory. The quantum nature is taken into account by means of the quantization of vibrational, translational, rotational and electronic states. The microscopic rate constant consists of a function of these states. The macroscopic rate constant is an average over such microscopic rate constants weighted by the probability of finding a molecule with a given set of quantum numbers. The connection between the properties of a single molecule and the experimental conditions from a very large number of molecules is based on the statistical mechanics.

The basic ideas are: (i) rates of reaction can be determined by activated complexes (or states) existing near the saddle point (transition state) on a potential energy surface (PES). The details of the complexes are not important. (ii) The activated complexes are in equilibrium (quasi-equilibrium) with the reactant. (iii) The activated complexes turn out to be products, and one can calculate the rate of this conversion. The schematic picture is shown in Fig. 1.

The reactants undergo collisions that keep all of their degrees of freedom in thermal equilibrium. At any instant of time, some will have enough internal energy to access a transition state (TS) on the Born-Oppenheimer ground state PES. The rate is expressed in terms of the concentration of species that exist near the TS multiplied by the rate at which these species move through the TS region. The concentration of the species at the TS is written in terms of equilibrium constant expressed by partition function as shown above.

For a reaction, A + B (reactant in Fig. 1) → C (product in Fig. 1), passing through a TS denoted as AB*, as explained for NOCl example (vide supra), we can write the concentration (in volume *V*) of the species in terms of the concentration of A and B and their partition functions as

$$[AB^*] = (\frac{q_{AB^*}}{V})/\left[\left(\frac{q_A}{V}\right)\left(\frac{q_B}{V}\right)\right][A][B]$$

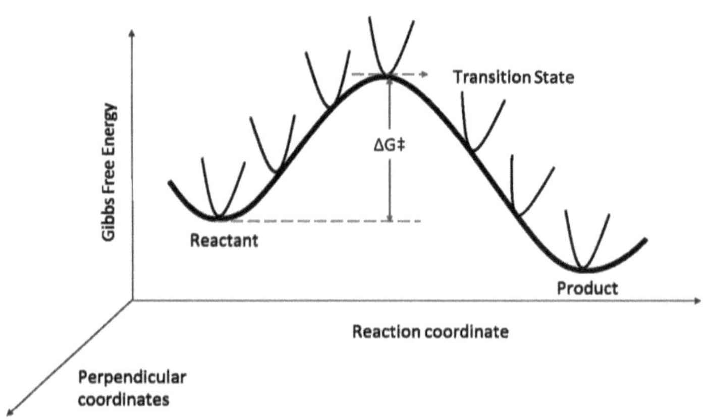

Fig. 1 Schematic picture of transition state theory, the reaction proceeds along with the reaction coordinate, the other coordinates are perpendicular to the reaction coordinate

In the partition function of the TS, the q_{AB*} contains all of the rotational, vibrational, translational, and electronic terms, *except for* one contribution for motion along the one internal coordinate corresponding to the reaction path (horizontal line of Fig. 1). In the next stage, we must consider the factor for a collision along with the reaction coordinate with a momentum p_s. It is known that the momentum factor entering into the partition function for translation along the reaction coordinate is

$$\frac{1}{h}\exp(-p_s^2/2\mu K_B T)dp_s$$

where μ is the mass factor associated with the reaction coordinate. Putting all of these elements together, we arrive at an integral of p_s and narrow TS region of ds, as following,

$$\iint \kappa\frac{1}{h}\exp(-p_s^2/2\mu K_B T)(p_s/\mu\delta s)dp_s ds$$

where κ is so-called transmission coefficient which specifies the fraction of trajectories crossing the TS that proceed to products. By integration from $p_s = (2\mu E^*)^{1/2}$ to infinitive, and the transmission coefficient being constant, we will obtain,

$$\kappa(K_B T/h)\exp(-E^*/K_B T),$$

where E^* denotes the energy of the transition state. Following the basic assumption of TST, the exponential part is now replaced by above obtained [AB*] value. Therefore, the rate of the chemical reaction is given in terms of molecular partition function as

$$k = \kappa(K_B T/h)[AB^*] = \kappa(K_B T/h)\left(\frac{q_{AB*}}{V}\right)/\left[\left(\frac{q_A}{V}\right)\left(\frac{q_B}{V}\right)\right][A][B]$$

The theory is remarkably successful for most thermal reactions. However, it has limitations [3]. For example, for photochemical reactions where the reactants do not reach thermal equilibrium before the reaction occurs, TST would not be a good model. Other example includes the case when TST is applied to each elementary step of a multi-step reaction, the theory assumes that each intermediate is enough stable to reach a Boltzmann distribution before the next step. When the intermediates are very unstable with very short lifetime, TST fails because the momentum of the reaction trajectory from the reactants to the intermediate can influence much on the product selectivity. TST assumes that unless atoms or molecules collide with enough energy to form the transition structure, the reaction does not occur. However, as quantum mechanics teaches us, for any barrier with a finite amount of energy, there is a possibility that particles can still tunnel across the barrier. There is a chance that molecules will react even if they do not collide with enough energy to pass the energy barrier. Although it is expected to be negligible in many reactions with large activation energies, it becomes of critical importance for reactions with relatively low energy barriers, such as many biological systems, since the tunneling probability increases with decreasing barrier height.

Transition state theory also fails for some reactions at high temperature because the theory assumes that the reaction systems will pass over the lowest energy saddle point on the PES. Although the description is consistent when occurring at relatively low temperatures, molecules populate higher energy vibrational modes at high temperatures. The motion becomes very complex and collisions may lead a molecule in the TS from the lowest energy saddle point to much higher excited states.

3 More Complexed Systems

The theories mentioned above have been successful in explaining the chemical kinetics of many reaction systems occurring in dilute or homogeneous environment. Such success has been based on the assumption that reaction rate constants depend only on the properties of the reactants. The effect of the surrounding environment is neglected. Nevertheless, in condensed matter especially on the solid surface of the catalyst, this is a crude approximation that is rarely valid. There are important reviews that point out some of the deviations from conventional chemical kinetic theories that result from the influence of co-adsorbates on the rate of surface chemical reactions [6].

In practice, the chemical reactions consist of complexity. There are at least two categories of the complexity. The first one is complexity due to numerous elemental steps. With combustion reactions for example, there are so many elemental reactions included that sometimes it is almost impossible to identify each reaction. Especially in many industrial researches and developments, there is no choice but to numerically solve the simultaneous differential equations. For the sake of these important but difficult problems, there are various available software codes which can be useful [7].

For the second category of the complexity in chemical kinetics, there is an example which represents its complexity with clear images. This is chemical oscillations, such as the Belousov-Zhabotinsky reaction, or more generally many reactions which appear with dissipative structure. The details of this subject are out of the scope of this chapter, but some instructive books are cited with one comment that the chemical oscillation reaction is a much more common phenomenon than is generally believed [8–10].

4 Efficiency of Catalysts

This book aims to provide a general overview on saving energy in chemical bonds from various technological points of views. In this regard, for discussing the chemical reaction kinetics, it is a catalyst that plays a central role. Therefore, concerning the catalysts, we will present here an important point to bridge experiments and molecular level approaches. A computational study of a catalytic cycle generates the energies of quantum states, whereas experiments focus on obtaining rate constants.

The most important properties for catalyst, amongst all, are activity, selectivity and efficiency. When it comes to the efficiency of a catalytic cycle, the most important property is its turnover frequency (TOF),

$$TOF = N/[C]t,$$

where N is the number of cycles per catalyst with concentration C, per time t. There are many simulations and theoretical computational studies. Based on TST, a large number of researches on catalyst design is reported. However, until recently, there has been no simple way to evaluate it from a theoretically obtained energy profile.

S.Kozuch and S.Shaik recently presented an important review paper which proposes *energetic span model* (δE in Fig. 2) that enables one to evaluate TOFs in a straightforward manner [11]. It implies a change in kinetic concepts. In a similar manner to Ohm's law, the catalytic chemical current, that is TOF, was shown by a chemical potential.

In order to design the efficient catalysts, naturally there should be neither very high barriers nor deep stable intermediates in the PES of the catalytic cycle. Even though this common sense is well accepted by catalyst researchers, there remain unanswered questions; which is more efficient, the cycle with TS of low barrier but with deep (in energy) intermediate, or the cycle with TS of high barrier and with relatively shallow (in energy) intermediate? It was not evident before the discussion of the TOF by their researches. The essence of the study by S.Kozuch et al. will be explained briefly below.

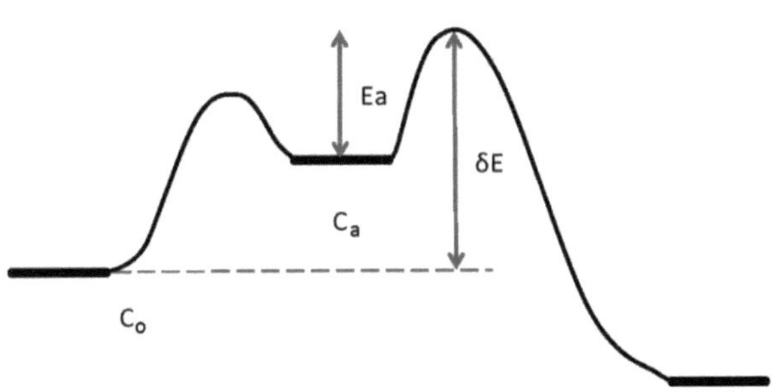

Fig. 2 Energetic Span Model, the important parameter is not the barrier but the energy span [12]

They showed that the critical property for TOF is not TS (*Ea*) but energetic span δE, following to Amatore and Jutand [12], as shown in Fig. 2. According to Arrhenius, the rate of reaction is

$$Rate = k[C_a] = (A\ \exp(-E_a/RT))[C_a].$$

Since the concentration of the active species is given by their Boltzmann distribution relative to the lowest intermediate (C_O), we have

$$[C_a] = [C_0]\exp(-\Delta E/RT)$$

By using this, we get

$$Rate = [C_0]A\ \exp(-(E_a + \Delta E)/RT))$$
$$\delta E = E_a + \Delta E$$

This equation is accurate only when the energy of the reaction (***ΔGr*** is shown in Fig. 3) approaches zero. In other words, the catalytic cycles can advance forward, only when the *ΔGr* value is positive. As a result of the turn of catalytic cycle, the energy of the system (counting the energy of a new reactant and discounting that of a leaving product) has changed. That is, the starting-point is now below the original one by the reaction energy, *ΔGr*. In this situation, do we apply the Boltzmann distribution with respect to C_O of the first or the second cycle? The answer is no. This is a typical feature of catalytic cycle that contrasts with ordinal closed thermodynamic reaction systems. According to S.Kozuch et al. the reason lies in the "Ouroboros-like" nature of a catalytic cycle: once one turnover is completed, the catalyst restarts a new cycle.

In their works, they presented an equation to evaluate the turn over frequency. The derivations were explained in the papers [12], but here we show the final form.

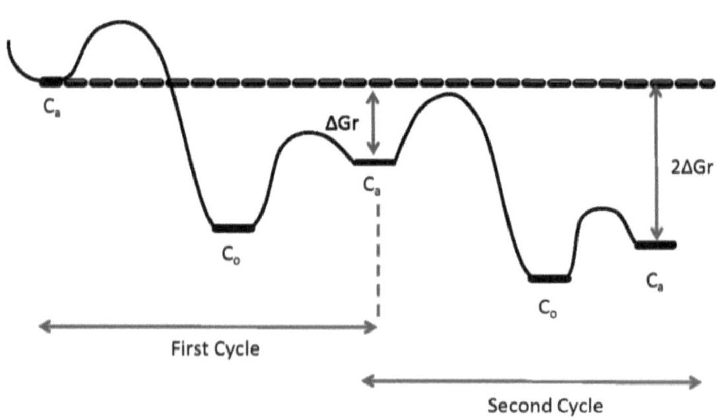

Fig. 3 Advance of catalytic cycles depends on the positive value of *ΔGr*, if zero or negative, no cycle can advances

$$TOF = \frac{K_B}{h} \frac{\exp(-\Delta G_r/RT) - 1}{\sum_{i,j=1}^{N} \exp[(T_i - I_j - \delta G'_{i,j})/RT]}$$

$$\delta G'_{i,j} = \Delta G_r, \quad if \quad i > j$$
$$= 0, \quad if \quad i \leq j$$

They introduced the criterion of $\delta G'_{i,j}$ which makes it possible to estimate TOF. This term can be the value of ΔGr, when j-th TS follows the j-th intermediate in the forward direction, or zero if the TS precedes the intermediate. The *Ti* and *Ij* denote the energy of i-th transition state and the energy of j-th intermediate, respectively (see Ref. 11 for the details).Based on their examples, they present the concept of TOF-determining transition state (TDTS) as well as TOF-determining intermediate (TDI). According to their works, there is no meaning for the rate-determining *step* term in catalytic cycle, instead, the rate-determining *state* (TDI and TDTD) define the kinetics of the catalytic cycle. Contrasting with the conventional arguments, they presented: (i) one transition state does not determine the kinetics of the cycle, and (ii) the rate determining states are not necessarily the highest TS or the lowest intermediate. This issue is still a recently proposed discussion. It will be validated with a number of experiments, and will further be improved so that the theory would become more reliable and applicable.

References

1. Knox KH (1971) Molecular thermodynamics: An introduction to statistical mechanics for chemists. Wiley, New York
2. Steinfeld JI, Francisco JS, Hase WL (1998) Chemical kinetics and dynamics. Prentice Hall, Upper Saddle River
3. Truhlar DG, Garrett BC, Klippenstein SJ (1996) Current status of transition-state theory. J Phys Chem 100:12771–12800
4. Simis J (2003) An introduction to theoretical chemistry. Cambridge University Press, Cambridge
5. Jensen F (2007) Introduction to computational chemistry. Wiley, Hoboken
6. Zaera F (2002) Kinetics of chemical reactions on solid surfaces: deviations from conventional theory. Acc Chem Rev 35:129–136
7. http://www.kintecus.com
8. Gray P, Scott SK (1994) Chemical oscillations and instabilities: non-linear chemical kinetics. Oxford University Press, Oxford
9. Kuramoto K (2003) Chemical oscillations, waves, and turbulence. Dover Publications, New York
10. Prigogine I, Kondepudi D (1999) Thermodynamique des moteurs thermique aux structures dissipatives. Edition Odule Jacob, Paris
11. Kozuch S, Shaik S (2011) How to conceptualize catalytic cycles? the energetic span model. Acc Chem Rev 44:101–110
12. Amatore C, Jutand A (1999) Mechanistic and kinetic studies of palladium catalytic systems. J Organomet Chem 576:254–278

Physical Model for Interfacial Carrier Dynamics

Mikiya Fujii, Ryota Jono and Koichi Yamashita

Abstract This chapter reviews interfacial carrier dynamics in power convergence processes of organic solar cells and dye-sensitized solar cells from the standpoint of chemical reactions. To this end, three models are reviewed for organics solar cells along with recent studies. The first model is chemical kinetics based on reaction rates estimated by Marcus theory. The second model is chemical dynamics, where quantum dynamics is introduced to understand charge carrier dynamics as chemical dynamics. The third one is a modeling of photovoltaic devices to reproduce and consider power convergence efficiencies based on drift and diffusion dynamics of carriers. For dye-sensitized solar cells, theoretical models for electron transfer from dyes to TiO_2 and charge recombination due to internal conversion are reviewed. Through reviews of these different models, we discuss current understanding and remaining problems, which should be addressed in the future, of carrier dynamics in power convergence of organic solar cells and dye-sensitized solar cells.

1 Introduction

Photovoltaics, such as organic solar cells and dye-sensitized solar cells, have attracted much interest as an alternative energy source over recent years because low manufacturing cost, flexibility, and sustainability can be expected. Developing new materials to archive high photoconversion efficiency is carried by huge amount of non-systematized knowledge and experiences of chemists because a basic principle exploring relation chemical structures and ability of photoconversion has not been revealed yet. To further improve power convergence efficiency, the fundamental understanding of the complicated process of photogeneration of free carrier is desired.

M. Fujii · R. Jono
The University of Tokyo, 7-3-1 Hongo, Bunkyo-ku, Tokyo 113-8656, Japan

K. Yamashita (✉)
Department of Chemical System Engineering, The University of Tokyo,
7-3-1 Hongo, Bunkyo-ku, Tokyo 113-8656, Japan
e-mail: yamasita@chemsys.t.u-tokyo.ac.jp

M. Sugiyama et al. (eds.), *Solar to Chemical Energy Conversion*,
Lecture Notes in Energy 32, DOI 10.1007/978-3-319-25400-5_5

2 Interfacial Carrier Dynamics in Organic Solar Cells

The mechanism of photoconversion in organic photovoltaics is basically explained with five elementary processes: (1) formation of an exciton in donor molecules by photon absorptions, (2) the exciton migrates to the interfaces between donor and acceptor molecules, (3) the exciton dissociates at the interface, which leads to the formation of free carriers (electron and hole), (4) The generated free carriers migrate as polarons involving disorder of molecular configurations to electrodes. Finally, (5) the free carries are captured to the electrodes. Besides, sometimes in the course of the photoconversion process, the electron and hole recombine with each other and then decay to the electronic ground state. Recombination in the course of the exciton dissociation process is called geminate recombination because the recombined electron and hole were generated from an identical exciton. On the other hand, recombination of the free electron and hole after the exciton dissociation is called bimolecular recombination because these free electron and hole were generated from different excitons.

Power convergence efficiencies and energy loss of organic photovoltaics depend on various chemical factors in various time and spatial scales, e.g. chemical structures in angstrom scale, exciton diffusion length in nanometer scale, miscibility and morphology of domains of the donor and acceptor in micrometer scale, etc. Especially, exciton dissociation and recombination of the electron and hole, on the course of photoconversion process, are known as key factors deciding the power conversion efficiency and energy loss. Roughly speaking, the efficiency of the charge generation and recombination of free carriers depend on chemical condition at the interface, i.e. chemical composition, molecular configuration and intermolecular packing, energy alignment of molecular orbitals, spatial delocalization of molecular orbitals, local permittivity, etc. Therefore, we, here, focus on the theoretical and computational modeling of chemical reactions at the interface such as exciton dissociation, charge transfer, and recombination reactions.

2.1 *Charge Transfer Kinetics: Marcus Theory*

Excitons are thought to be able to dissociate into free carriers with aid of energy gap of LUMOs between an electron donor and an electron acceptor after the excitons that were generated in donor domain reached the interface. In other words, the electron populating in LUMO of the donor can transfer to LUMO of the acceptors on extremely rapid time scale (femtosecond) [1]. To analyze such electron transfer reaction, the first approach is based on kinetic approaches estimating rates of charge transfer reactions from an electron donor to an acceptor. The Marcus theory, which was presented about six decade ago first and is sometimes regarded as a nonadiabatic electron transfer theory, has been successfully applied to many charge transfer reactions [2–4]. The Marcus theory has been constructed based on a similar idea with the transition state theory (TST) on the Gibbs free energy. But, the reaction coordinates used in the theory are not any 1-dimensional manifold in the configuration space and are abstract coordinates representing reorganization of distributions of molecular orientation of solvents although the reaction coordinates in the TST is 1-dimensional manifold in the configuration space. Representing charge transfer reactions along with the abstract reaction coordinates related to molecular orientation can be considered one concrete example of "slaving principle" presented by Haken [5, 6]. Namely, the fastest dynamics of charges follows the slowest rotational dynamics of solvent molecules in systems.

The Marcus theory considers the reaction (charge non-transferred) and product (charge transferred) state in charge transfer reactions as two harmonic free energy surfaces (see Fig. 1). First, systems thermally fluctuate on the reactant parabola and then stochastically reach to an intersection of the two parabolas. At this intersection point, the systems can transit to the product parabola, i.e., charge transfer reaction can occur nonadiabatically. Finally, the systems fall down to the bottom of the product parabola, i.e. the systems reach thermally equilibrated state. Then, charge transfer rates in the Marcus theory is represented as

$$k_{ET} = \frac{2\pi}{\hbar\sqrt{4\pi\lambda k_B T}} V_{\mathrm{RP}}^2 \exp\left(-\frac{\Delta G^{\ddagger}}{k_B T}\right) \text{ with } \Delta G^{\ddagger} = \frac{(\lambda + \Delta G)^2}{4\lambda} \tag{1}$$

where k_B, T, ΔG, and $\Delta G^{\ddagger}$ are Boltzmann constant, temperature, difference of free energy, and activation energy, respectively. Besides, λ is called as reorganization energy and considered as virtually released energy to equilibrate the distribution of

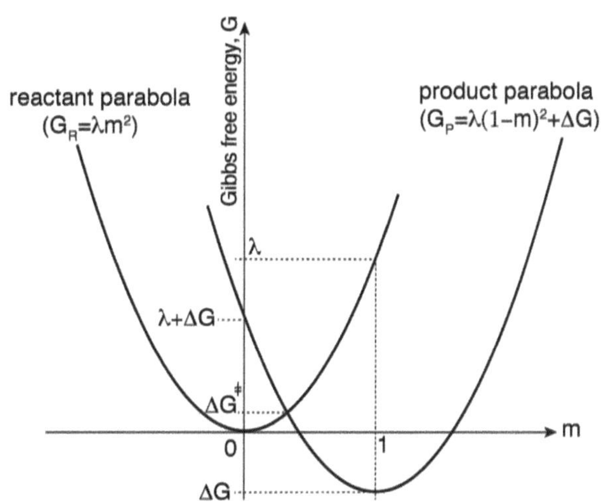

Fig. 1 Reactant and product parabolas considered in Marcus theory

molecular orientation of solvents after "sudden" transition ("vertical" transition at $m = 0$ in Fig. 1) of charge distribution,

$$\lambda = \frac{e^2}{2}\left(\frac{1}{\varepsilon_{\mathrm{op}}} - \frac{1}{\varepsilon}\right)\left(\frac{1}{r_{\mathrm{D}}} + \frac{1}{r_{\mathrm{A}}} - \frac{2}{R_{\mathrm{DA}}}\right). \tag{2}$$

Here, ε_{op} and ε are the optical dielectric constant of the medium and the static dielectric constant, respectively. r_D and r_A are the effective radii of donor and acceptor, respectively. The last term of Eq. (2) represents the Coulomb interaction between the donor and acceptor. The reorganization energy λ is, sometimes, considered as summation of reorganization energies related to molecular orientation of solvents and configurational change of solutes, which are virtually caused by "sudden" charge transfer. As shown in Eq. (1), the activation energy $\Delta G^{\ddagger}$ is determined by a balance of the reorganization energy λ and the free energy gap ΔG. V_{RP} in the expression for rate constant represents nonadiabatic electronic coupling between reactant and product. This electronic coupling can be written as a well-known generalized Mulliken- Hush (GMH) formula [7, 8]:

$$V_{\mathrm{RP}} = \frac{\mu_{\mathrm{RP}}\Delta E_{\mathrm{RP}}}{\sqrt{(\mu_{\mathrm{RP}})^2 + 4(\Delta\mu_{\mathrm{RP}})^2}}, \tag{3}$$

where ΔE_{RP} and $\Delta\mu_{\mathrm{RP}}$ correspond to the energy difference and dipole difference between the reactant and product, respectively. μ_{RP} is transition dipole moment between them. Because the GMH represents a vertical transition between adiabatic reactant and product states, the GMH can be used along with traditional quantum chemistry calculations, e.g. Configuration Interaction (CI) methods, even though diabatic representations are adopted in the Marcus theory. To apply the GMH to calculation of charge transfer rates in photovoltaics, extending the time dependent density functional theory (TDDFT) to enable calculation of transfer properties between excited states is very desirable because the transition dipole moment between donor excited and charge transfer states is required. On the other hand, diabatic representations are more straightforward than the adiabatic representation to understand chemical phenomena considered in the Marcus theory. In the diabatic representation, the electronic coupling can be written as [9]

$$V_{\mathrm{RP}} = \left(1 - S_{\mathrm{RP}}^2\right)^{-1}(H_{\mathrm{RP}} - S_{\mathrm{RP}}(H_{\mathrm{RR}} + H_{\mathrm{PP}})/2). \tag{4}$$

Besides, constrained density functional theory (CDFT) [10–12] or constrained Hartree Fock (CHF) is considered to be useful to prepare the diabatic states. The above procedure that is mixture of the diabatic representation and constrained methods can be accessible with NWCHEM package [13].

Many studies applying the Marcus theory to organic photovoltaics have been presented [14–17]. P.A. van Hal et al. have used the Marcus theory to explain experimental results that charge transfer rates of OPV4/C_{60} depend on polarity of solvents [14]. They theoretically showed that, in the more polar solvents, charge-separated state ($OPV4^{+}/C_{60}^{-}$) is stabilized with concomitant increase of the reorganization energy and the charge separation reaction is in Marcus' normal region ($0 < \lambda + \Delta G$). Their qualitative analyses lead to the reduction of the barrier for charge transfer reaction,

$$\Delta G^{\ddagger} = \frac{(\lambda + \Delta G)^2}{4\lambda}, \tag{5}$$

in the more polar solvents. They also show that charge recombination reactions are in Marcus' inverted region ($\lambda + \Delta G < 0$) and the barriers for the recombination reactions are more strongly reduced than ones for the charge separation as increase of the polarity of the solvents.

Brédas and coworkers have developed a method of calculating charge transfer and recombination rates in ideal interfacial configurations by estimating each component in the Marcus theory with quantum chemical calculations. They also investigated the effects of the intermolecular configuration between donor and acceptor molecules to the reaction rates. Their analyses of Pc/PTCDI indicated the charge transfer and recombination reactions are in Marcus' normal and inverted regions, respectively [15], as well as the study of $OPV4/C_{60}$ by P.A. van Hal et al. [14]. They showed that the both charge transfer and recombination rates strongly depend on the intermolecular geometry and local excited states. Especially, the electronic coupling is very sensitive to the intermolecular geometry because the overlap of wave functions strongly depends on the intermolecular geometry. Besides, charge transfer reaction is much faster (ca. 10^2–10^4 times) than charge recombination reactions in general. However, in some geometry, e.g. in the case of parallel configurations of pentacene and C_{60} as shown in Fig. 2, they also found that the decay of the charge transfer state is very fast and could compete with the dissociation process of the charge transfer state into mobile charge carriers [17].

Charge transfer rates at the ideal interfaces were well studied as mentioned in the previous paragraph. However, the molecular configurations and packing in realistic devices are generally disordered. Namely, the interfaces of the donor and acceptor are far from uniform, which leads to different excitation energies, charge transfer

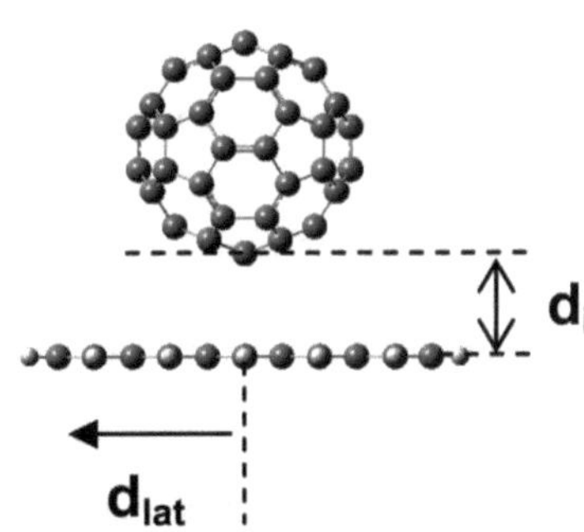

Fig. 2 A parallel configuration of pentacene and C_{60} (From Ref. [17])

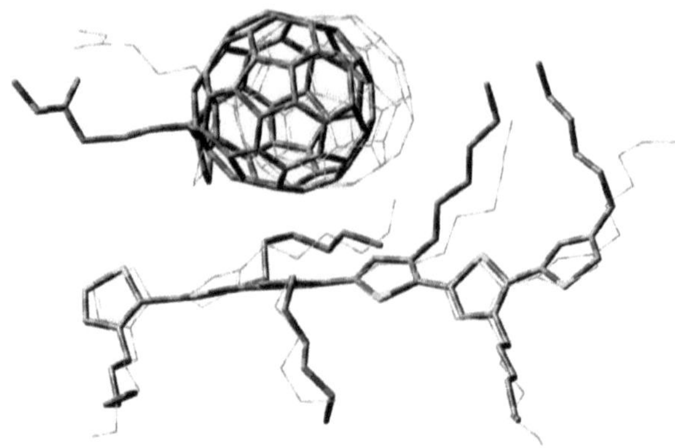

Fig. 3 Typical disordered configurations of the complex of P3HT/PCBM at the interface, which are obtained with molecular dynamics simulation. The configurations at t = 0.5 and 9.5 ns are represented in the tube frame and the wire frame, respectively. (From Ref. [18])

rates, etc. To clarify effects of different microscopic atomic arrangements on the electronic structure and charge transfer rates, Troisi et al. studied the effects in the prototypical photovoltaic interface between P3HT and PCBM by molecular dynamics simulations and quantum chemical calculations [18]. Typical disordered structures at the interface, which are obtained by the molecular dynamics simulation, are shown in Fig. 3. They found two types of interfacial configurations indicating distinct charge transfer characteristics. The first type exhibits charge transfer states with complete charge separation while the second one exhibits charge bridging states which some experimental groups had reported [19, 20]. A very broad range of charge separation ($7.7 \times 10^9 - 1.8 \times 10^{12} \mathrm{s}^{-1}$) and charge recombination ($2.5 \times 10^5 - 1.1 \times 10^{10} \mathrm{s}^{-1}$) "instantaneous" rates were also reported because various configurations exist at the interface. Especially, they found a significant correlation between the internal reorganization energy and rates of charge recombination process. Because no correlation was found in the charge separation process, they concluded that reduction of the internal reorganization energy might help reducing the (unwanted) charge recombination while the charge separation process might not decrease.

2.2 Charge Transfer Dynamics at Interfaces

Charge transfer kinetics could provide quantitative explanations of experimental results and chemical understandings of the competition of multi reaction processes as reviewed in the previous section. However, chemical mechanism of charge transfer and recombination processes has not been fully revealed yet. For example, how excitons overcome the Coulombic binding energy of a few hundred millielectronvolts, which is one order of magnitude higher than thermal fluctuation (ca.

25 meV)? How reaction path should be set as a unified notion of multiple electronic states (i.e. nonadiabatic) and nuclear configuration? Therefore, detailed analyses of chemical dynamics in each elementary process like charge transfer and recombination reactions are required to shed light on the interfacial carrier dynamics. To this goal, recent transient absorption spectroscopy and time-resolved photo luminescence are one of the most promising experimental approaches [21–23]. Theoretically, nonadiabatic quantum dynamics among multiple electronic states on the electronic devices like organic solar cells and photocatalyst is a tremendous challenge even in modern theoretical chemistry with computer resources. Although no agreement on how to analyze the nonadiabatic quantum dynamics in such devices has not been established, some studies to analyze the dynamics have been presented based on various approximations. Here, we introduce three approaches to analyze quantum dynamics on the multiple electronic states: quantum master equation, Multi Configurational Time Dependent Hartree (MCTDH), and nonadiabatic surface hopping dynamics approaches.

Fujii and Yamashita [24] investigated charge transfer dynamics as transition dynamics of quantum population between electronic eigenstates. Especially, the packing effects in organic donor–acceptor molecules were investigated. Their analyses are based on an electron and harmonic bath Hamiltonian:

$$H = \sum_k \hbar\omega_k |\Phi_k(\alpha)\rangle\langle\Phi_k(\alpha)| + \sum_m \hbar\Omega_m B_m^\dagger B_m + \sum_{k,l}\sum_m C_{klm} |\Phi_k(\alpha)\rangle\langle\Phi_k(\alpha)| \left(B_m^\dagger + B_m\right), \tag{6}$$

where $\hbar\omega_k$ is the electronic eigenenergy. $B_m^\dagger$ and B_m are creation and annihilation operators, respectively, for the mth harmonic mode of which frequency is Ω_m, C_{klm} is a coupling constant between electrons and the mth phonon mode. α is a rotating angle of the donor molecule and specify the packing of donor and acceptor molecules. Substituting Eq. (6) into the time-dependent Schrödinger equation and applying the Born–Markov approximation [25] lead to an equation of motion for the electronic density matrix that is well known as the quantum master equation (QME):

$$\dot{\rho}_{kk'}(t) = -i(\omega_k - \omega_{k'})\rho_{kk'}(t) - \sum_{ll'} \Gamma_{kk':ll'}\rho_{ll'}(t), \tag{7}$$

where $\Gamma_{kk':ll'}$ is called decay matrix. This QME was calculated to investigate charge transfer dynamics from the donor excited state in some molecular packing. Then, they revealed that the charge transfer dynamics strongly depends on the molecular packing. In Fig. 4, the charge transfer dynamics are shown for three cases of the molecular packing. While the efficient charge transfer that is suitable for photovoltaic devices is achieved in the case of $\alpha = 0$ (Fig. 4b), the charge transfer is hardly achieved in the case of $\alpha = 10$ (Fig. 4c). The charge transfer dynamics in the packing $\alpha = -10$ (Fig. 4d) is different from other two cases. A slow hole transfer

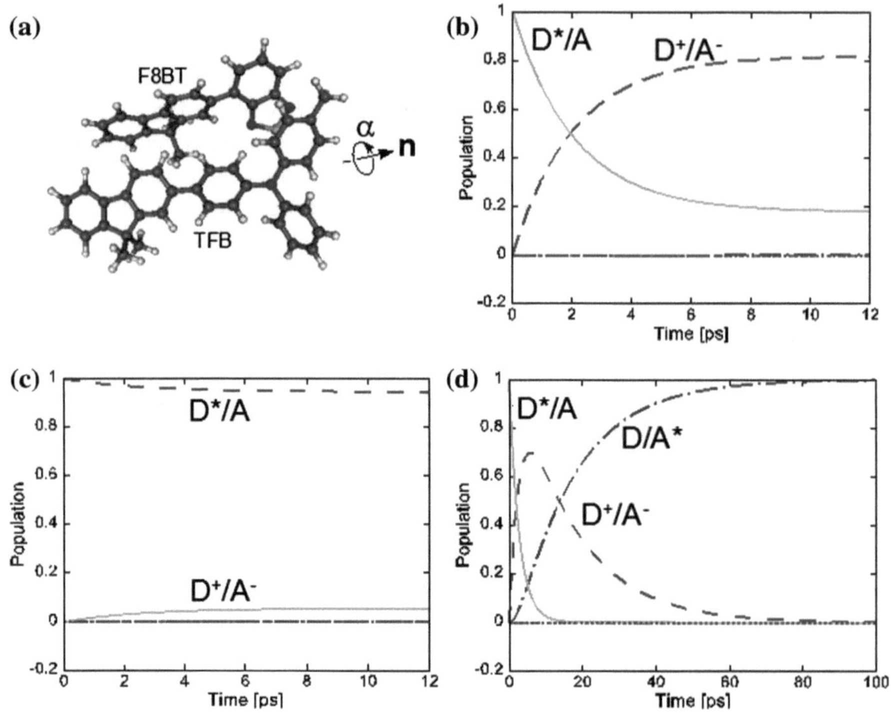

Fig. 4 Packing dependency of charge transfer dynamics. **a** A stable structure of the complex of donor (TFB) and acceptor (F8BT) molecules. **n** and α is rotating axis and angle of the donor, respectively. **b–d** Charge transfer dynamics estimated with the quantum master equation when $\alpha = 0$ (**b**),$\alpha = 10$ (**c**), and $\alpha = -10$ (**d**)

from the donor to the acceptor occurred following the charge transfer, which leads to the acceptor excited state. These differences originate in differences of energy levels of charge transfer states that depend on the molecular packing because the Coulomb interaction between hole on the donor and electron on the acceptor depends on the packing. Their results indicate a dynamical aspect of configurational dependency of the charge transfer reaction as Troisi et al. showed the disorder of configuration affects the charge transfer rates at the interface.

Tamura and et al. have also investigated charge transfer dynamics at the interface. Especially, they have attempted a molecular-level description of exciton dissociation in the terms of nonadiabatic nuclear wave packet dynamics on multiple potential energy surfaces by using MCTDH (Multi-Configurational Time-Dependent Hartree) method. MCTDH is one of the most promising methods to propagate multidimensional wave packets [26]. In the MCTDH approach, the wave packets are expanded in terms of Hartree products of single particle functions $\phi^k_{j_k}(x_k, t)$ as

$$\psi(x_1, x_2, \cdots x_f, t) = \sum_{j_1}^{n_1} \cdots \sum_{j_f}^{n_f} A_{j_1,\cdots j_f}(t) \prod_{k=1}^{f} \phi_{j_k}^{k}(x_k, t), \tag{8}$$

where $A_{j_1,\cdots j_f}(t)$ is time-dependent expansion coefficient. Equations of motion for the time evolution of the single particle functions and the expansion coefficients are given in terms of nonlinear dynamical equations derived from the Dirac-Frenkel time-dependent variational principle. Then, they applied the MCTDH to oligothiophene-fullerene complex whose potential energy surfaces consist of two diabatic surfaces for exciton (XT) and charge transfer (CT) states [27, 28] (see Fig. 5a and b). Figure 5c shows the exciton populations in the quantum dynamics calculations at the donor-acceptor distance of 3 and 3.5 Å. The differences between results for 3 and 3.5 Å arise from the distance-dependency of the nonadiabatic coupling and potential profiles shown in Fig. 5b. At 3.5 Å, an oscillatory dynamics of the exciton population due to back transfers from the CT to XT states occurs because of small ΔE (~0.007 eV). This oscillatory behavior extremely differs from chemical kinetics calculated by Marcus theory (dashed curve). In contrast to the distance of 3.5 Å, the electronic excess energy at 3 Å rapidly dissipates into the vibrational energy because of relatively large ΔE (~0.21 eV).

Jailaubekov et al. [29] have experimentally/theoretically investigated real-time dynamics of formation and relaxation of hot charge-transfer exciton at the interfaces of donor (phthalocyanine) and acceptor (fullerene). Concretely, they experimentally applied two nonlinear optical techniques (time-resolved second harmonic generation and time-resolved two-photon photoemission) to interfaces of copper-phthalocyanine/fullerene. Besides, they theoretically applied a nonadiabatic surface hopping dynamics based on QM/MM quantum chemical calculations [30, 31] to interfaces of H_2-phthalocyanine/fullerene. In the interfaces of H_2-phthalocyanine/fullerene, some charge transfer states are found as shown by density of states in Fig. 6a. Spatial distribution of electron and hole in the charge

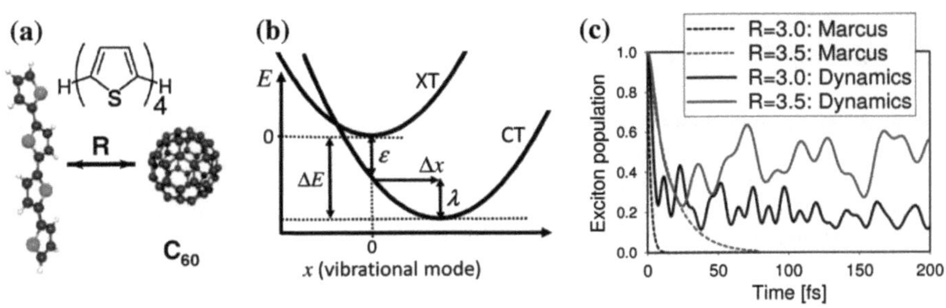

Fig. 5 **a** Molecular configuration of oligothiophene-fullerene complex. **b** Two diabatic energy surfaces. These diabatic surfaces represent exciton (XT) and charge transfer (CT) states, respectively. **c** Quantum dynamics of exciton dissociation by the MCTDH (*solid curve*). Exponential decay curves representing kinetics of exciton dissociations by Marcus theory are also shown as a reference (*dashed curves*). (From Ref. [28])

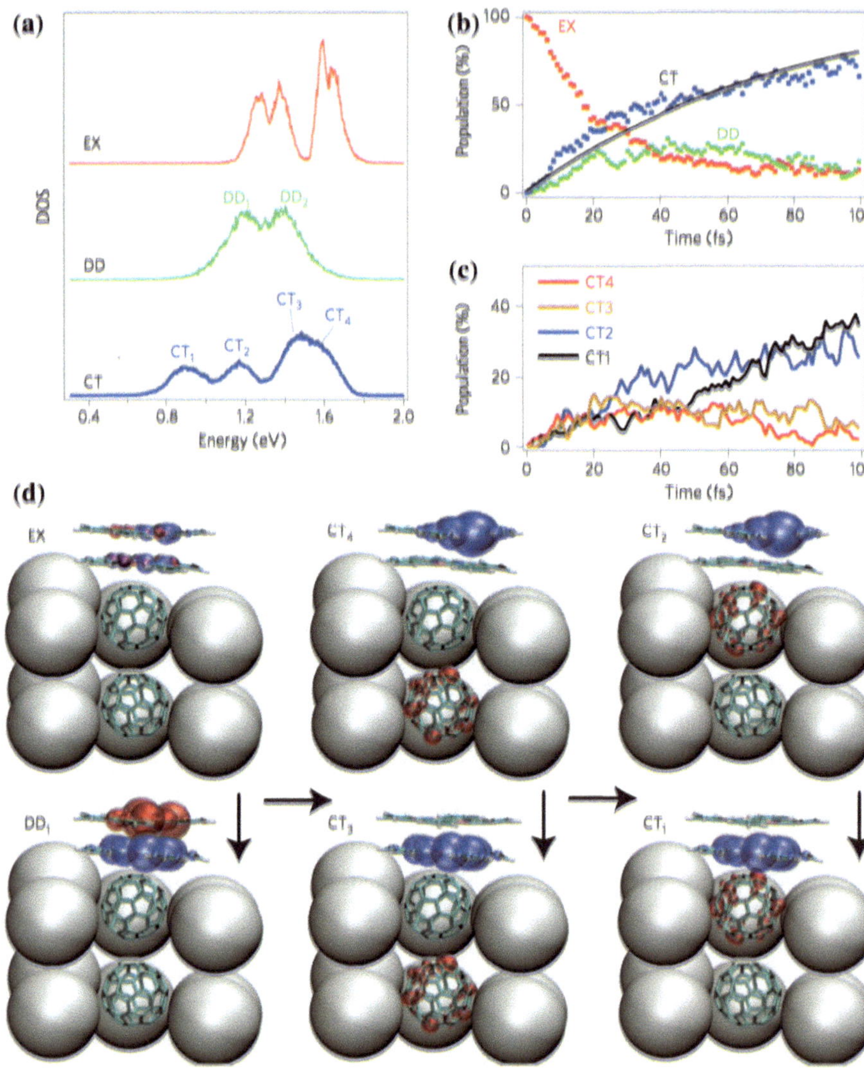

Fig. 6 **a** Time-averaged DOS distributions for the S1 exciton in donors (*red*), the interfacial CT exciton (*blue*) and the inter-donor exciton (*green*). **b** Time-dependent populations of S1 in H2Pc (*red*), the inter-donor exciton (*green*) and the total CT exciton (*blue*). **c** Time-dependent sub-populations of the four interfacial CT excitons: CT1 (*black*), CT2 (*blue*), CT3 (*orange*), CT4 (*red*). **d** Snapshots of selected excitonic state charge distributions (*red*, *electron*; *blue*, *hole*). (From Ref. [29])

transfer states are shown in Fig. 6d, where electron and hole are separated larger in the higher excited charge transfer states. Their nonadiabatic simulations showed that the charge transfer reaction from the S1 exciton in the donor to some charge transfer excitons including "hot charge transfer excitons" which are highly excited charge transfer states occur (see Fig. 6b). Surprisingly, the higher energy charge transfer states are formed at slightly faster rates initially than the lower energy charge transfer states (see Fig. 6c). As time goes on, the separated electron and hole that were in the highly excited charge-transfer states gradually approach each other at the interface. Namely, the system decays to low-lying charge transfer states from the highly excited charge-transfer states. Ultimately, the electron and hole pair may recombine from the lowest charge transfer states to the ground state. These nonadiabatic dynamics over multiple potential energy surfaces indicate that the current understanding of the charge transfer reactions at the interface, which is based on the two electronic states, is insufficient and should be reconsidered in the future.

In the present section, we have briefly reviewed charge transfer dynamics, especially quantum and/or nonadiabatic dynamics, at the donor-acceptor interfaces. These quantum dynamics could reveal interesting and unique features in elementary process of organic solar cells (e.g., strong dependency to the molecular packing, backward charge transfer and so non-exponential decay, and nonadiabatic dynamics over multiple potential energy surfaces), which cannot be addressed by Marcus theory. However, some open questions presented at the first of this section do not resolved yet. So, further detailed analyses of elementary processes like charge transfer and recombination reactions are demanded in the future.

2.3 Device Simulators of Organic Solar Cells

In the previous two sections, we focused on an elementary process, especially charge transfer reaction. However, studies of only an elementary process could not address the power convergence efficiencies because organic solar cells generate electronic current through multiple elementary processes. To consider the efficiency of power convergence efficiency, studies taking account of whole chemical processes are desired. In the remaining of this part, therefore, we briefly introduce a study of device simulator of organic solar cells, which treat whole chemical processes to generate electronic current. Koster et al. have presented the most fascinating study [32] among various device simulators [33–36]. They combined two physical aspects. The first aspect is drift and diffusion dynamics of photo-generated charge carriers represented as

$$J_n = -qn\mu_n \frac{\partial \psi}{\partial x} + qD_n \frac{\partial n}{\partial x}, \tag{9}$$

$$J_p = -qp\mu_p \frac{\partial \psi}{\partial x} - qD_p \frac{\partial p}{\partial x}, \tag{10}$$

where $J_{n(p)}$ is the electron (hole) current density, $\mu_{n(p)}$ is the electron (hole) mobility, $D_{n(p)}$ is the electron (hole) diffusion constant, and q is the elementary charge. $n(x)$ and $p(x)$ are the electron and hole density, respectively. The potential,

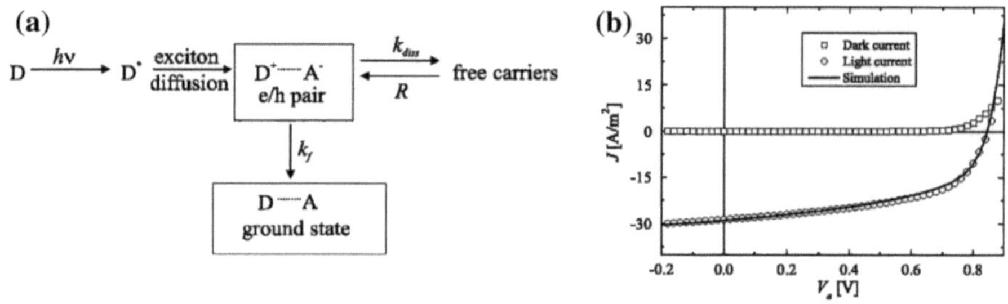

Fig. 7 **a** Schematic view of the charge carrier separation at the interface, which is considered in the Koster's device simulator. **b** Light current by the Koster's device simulator (*solid line*). Experimental dark current (*squares*) and light current (*circles*). (From Ref. [32])

$\psi(x)$, is estimated by the Poisson equation. The second aspect is the chemical kinetics considering generation and recombination of charge carriers at interfaces, which is shown in Fig. 7a. The combination of these two approaches could simulate an experimental J-V curve (Fig. 7b). So, studies based on such device simulator can be regarded as starting point to consider the power convergence efficiencies from physical and chemical properties of materials like diffusion constants, permittivity, orbital energies, band-gap, reaction rates, etc. But, in such device simulators, there are some bold approximations and parameters. Then, the parameters sometimes require unrealistic value (e.g., too small recombination rate, k_f). These approximations and parameters, therefore, should be reexamined by detailed studies of each elementary process. Namely, Complementary studies of detailed analyses of each process and overlooking analyses of whole processes are needed in the future.

3 Interfacial Carrier Dynamics in Dye-Sensitized Solar Cells

Interfacial carrier dynamics in Dye–Sensitized Solar Cells (DSSC) occur at the many interfaces such as TiO_2/electrolyte, TiO_2/dye, and dye/electrolyte. The most important and studied carrier dynamics is the electron injection from the excited state of the dye to the conduction band (CB) of the TiO_2 and is under the influence of electronic structure of the interface. The standard components of DSSC are transparent electrode, dyes adsorbed on TiO_2 nano–powder, redox couples in electrolyte, and counter electrode in the order from photo–anode [37]. Many trials improving the total conversion efficiency from light to electron have been reported in half a century. In earlier DSSC, semiconductor plates had been used to support dyes. Thereafter, Grätzel introduced TiO_2 nano–powder to enlarge surface area to support more dye molecules and reported 7.1 % for total efficiency in 1991 [38]. Now the total efficiency reaches 13 % [39]. The standard energetic model of dye–sensitized solar cells is shown in Fig. 8.

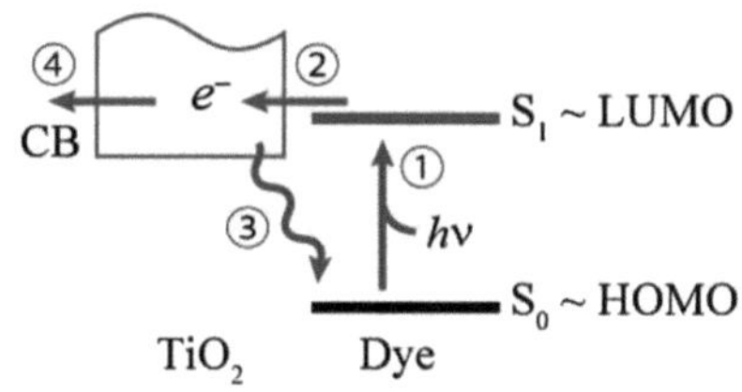

Fig. 8 The energetic model of Dye–Sensitized Solar Cells is composed of four elementally processes. *1* Photon absorption to photoexcited state. *2* Electron injection from photoexcited state of the dye to conduction band of TiO_2. *3* Charge recombination to ground state. *4* Charge separation

The dye absorbs photon to be excited, and then, an electron in the photoexcited dye is injected to the conduction band of TiO_2. If the electron in the conduction band does not recombine with a hole in dye or redox mediator in electrolyte, charge separation is achieved and electric power is generated. In this review, we introduce the electron transfer between dyes, molecules and semiconductor materials used in DSSC.

3.1 Built–in Potential at Interface Between Two Materials

Figure 9 shows the energy levels of the interface before and after forming a contact of TiO_2 and mediator in electrolyte. TiO_2 and mediator have their own Fermi levels before contacted. Here the Fermi energy of *n*–TiO_2 is supposed to be higher than the energy of lowest unoccupied molecular orbital (LUMO) of the mediator, which is approximated to be the reduction potential. To align these Fermi energies, electrons in TiO_2 migrate to the LUMO of redox mediators. As a consequence,

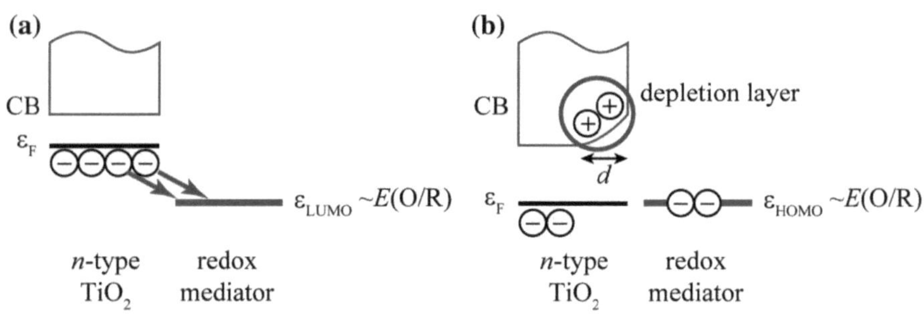

Fig. 9 Energy level picture at (**a**) before and (**b**) after forming a contact of TiO_2 and redox mediator in electrolyte. The electrons in Fermi level of *n*–type TiO_2 migrate to redox mediator to align the Fermi energies

a depletion layer is formed at the surface of TiO_2 and some amount of mediators is reduced. The depletion layer results in a potential barrier between surface and bulk of TiO_2, i.e. built–in potential, to prevent electron migrations from TiO_2 to the mediators [40]. The built–in potential can be calculated by solving Poisson–Boltzmann equation. The Poisson–Boltzmann equation and its solution are represented as follows:

$$\frac{\partial^2 \phi(x)}{\partial x^2} = -\frac{qN_\mathrm{d}}{\varepsilon\varepsilon_0}$$

$$\phi(x) = \frac{qN_d}{2\varepsilon\varepsilon_0}(x-d)^2$$

where x is the direction of the electron migration, $\phi(x)$ is the potential energy, and N_d and ε are the carrier density and the permittivity of TiO_2. The depletion width d is over 150 nm if we use $\phi(x) = 1.5\mathrm{eV}$, which is experimentally measured built–in potential [41], $N_\mathrm{d} = 10^{18}$ cm^{-3}, and $\varepsilon \sim 100$. However, TiO_2 in the active layer of DSSC is usually not bulk plate but rather nano–powder and its radius is usually smaller than 20 nm. Such small radius hinders formation of depletion layer. In nano–particles, Poisson–Boltzmann equation and its solution are rewritten in spherical coordinate as follows.

$$\frac{\partial^2 \phi(r)}{\partial r^2} - \frac{2}{r}\frac{\partial \phi(r)}{\partial r} = -\frac{qN_\mathrm{d}}{\varepsilon\varepsilon_0}$$

$$\phi(r) = \frac{qN_d}{6\varepsilon\varepsilon_0}\left[R^2 - r^2\right] + \frac{qN_d}{3\varepsilon\varepsilon_0}R^2$$

where R is radius of particle. The calculated built–in potential is only 0.04 eV if R = 20 nm is used [42]. Therefore we would not take the impact of built–in potential in the electron transfer at the surface of TiO_2 nano–powder in DSSCs into account.

3.2 Theoretical Model of Non–radiative Transition

Electron injection from excited dye to conduction band of TiO_2 is non–radiative transition. There are many theoretical models of non–radiative transition [43, 44, 45]. The most simple and primitive model deals two–level system which have the wave functions ϕ_i with eigenvalue ε_i if they are isolated system where i are Donor and Acceptor (Fig. 10a). Suppose coupling between ϕ_D and ϕ_A is represented by constant V, the time evolution of the population is oscillated between non–eigen-states ϕ_D and ϕ_A. If the initial state is started from the pure donor state ϕ_D, the population of the initial state ϕ_D and final state ϕ_A is expressed as follow.

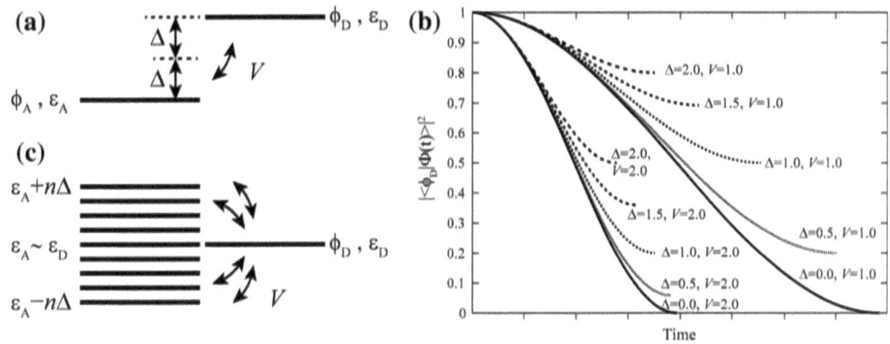

Fig. 10 **a** Energy diagram of two–level system. **b** Time evolution of the population of donor state in early short time range. **c** Energy diagram of multi–level system

$$|\langle\phi_A|\Psi(t)\rangle|^2 = \frac{V^2}{\Delta^2+V^2}\sin^2\left(\frac{\sqrt{\Delta^2+V^2}}{\hbar}t\right)$$

$$|\langle\phi_D|\Psi(t)\rangle|^2 = 1-\frac{V^2}{\Delta^2+V^2}\sin^2\left(\frac{\sqrt{\Delta^2+V^2}}{\hbar}t\right)$$

$$\Delta \equiv \frac{\varepsilon_D-\varepsilon_A}{2}$$

Figure 10b shows the time evolution of the population of donor state for each parameter Δ and V. The strong coupling between donor and acceptor states leads rapidly electron injection. The electron completely injects from donor to acceptor, if the energy difference between them is zero (Δ = 0). The amplitude $\frac{V^2}{\Delta^2+V^2}$ shows that some population of initial state will remain if $\Delta \neq 0$. If coupling parameter V is in the order of energy difference parameter, Δ, about half of population are remained in initial state ϕ_D. In very short time range for the time evolution of population, larger energy difference leads to less electron injection. The time derivative of population of acceptor state is the transition rate from donor to acceptor. At the limit of infinite time, the transition rate from donor to acceptor is rewritten with Dirac delta function $\delta(x)$, as follows.

$$k_{D\to A} = \frac{d}{dt}\lim_{t\to\infty}|\langle\phi_A|\Psi(t)\rangle|^2 = \frac{2\pi}{\hbar}V^2\delta(\varepsilon_D-\varepsilon_A)$$

This equation is called Fermi's Golden rule.

The acceptor state in DSSC is continuous electronic band as the conduction band of TiO_2, therefore the two–level model must be expanded to multi–level model (Fig. 10c). Bixon–Jortner Model gives the population of the initial state ϕ_D.

$$|\langle \phi_D | \Psi(t) \rangle|^2 = \exp\left[\frac{2\pi V^2 \rho}{\hbar} t\right]$$

where ρ is the density of the acceptor states. In this expression, the rate of the electron transfer is $2\pi V^2\rho/\hbar$ and most efficient at $\varepsilon_D = \varepsilon_A$, which agrees with Fermi's Golden rule. This model tells us that the population of the donor state decays exponentially.

3.3 Driving Force for Electron Transfer

Ruthenium bipyridine complex $[Ru(bpy)_3]^{3+/2+}$ have been well studied as a model complex for photocatalysis to split water to hydrogen. Its redox potential at the first excited state E° (D^+/D^*), which is experimentally estimated by the redox potential in ground state E° (D^+/D) and the lowest excitation energy E_{0-0}, is more negative than the reduction potential (i.e. conduction band minimum) of TiO_2. However the electron injection from Ru complex in electrolyte to TiO_2 is not efficient [46]. Chemically connected interface between dye molecule and semiconductor are introduced to induce a strong electronic coupling between dye and semiconductor. Carboxyl group is widely used due to the high overlap between 3d orbital of titanium and 2p orbital of oxygen [47]. However $[Ru(dcbpy)_3]^{3+/2+}$ complex adsorbed on TiO_2 surface does not show efficient electron transfer because its redox potential E° (D^+/D^*) is −0.8 V *vs* SCE and more positive than that of original complex $[Ru(bpy)_3]^{3+/2+}$ [48]. Systematic research reveals the redox potential of $[Ru(dcbpy)_2X_2]^{3+/2+}$, whose ligand is replaced by X = CN^- or SCN^- group, is sufficiently negative than conduction band of TiO_2 and shows efficient electron injection from photoexcited state of the ruthenium dyes to conduction band of TiO_2 [38, 49]. The transient absorption experiment shows the exponential rapid decay of photoexcited ruthenium dye with time constant less than 20 fs [50]. The overlap between density of states of the unoccupied molecular orbitals of the dye and conduction band of semiconductor is needed for efficient electron transfer [51, 52]. As mentioned in 2–1–3, the Gericher model shows the most probable electronic state of the reduced excited dye is $E^{\circ *} + \lambda$ and the rate of electron transfer from dye to conduction band is maximum when $E_{CB} < E^{\circ *} + 2\lambda$ where E_{CB} is the conduction band minimum of the semiconductor, $E^{\circ *}$ and λ are the redox potential and the reorganization energy of the photoexcited dye molecule (shown in Fig. 11). Reorganization energy of the $[Ru(phen)_3]^{3+/2+}$ redox couple is reported to be 0.25 eV, thus this energy offset is considered to be the driving force for electron transfer from the Ru complex to the conduction bands of the TiO_2.

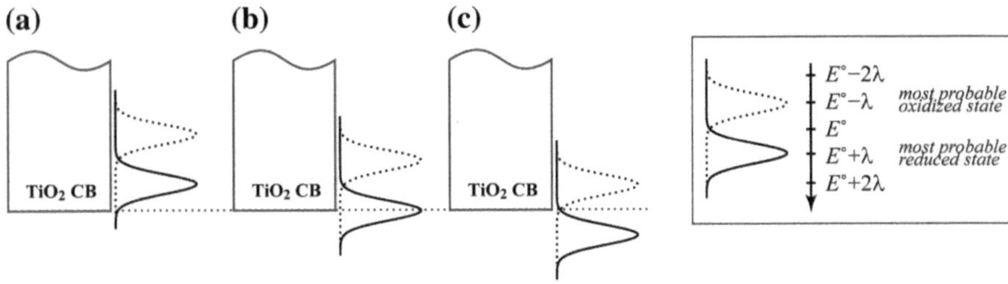

Fig. 11 Energetic levels considered in Gericher model. The built–in potential of TiO_2 conduction band (CB) is neglected. Two Gaussian distributions indicate the energy fluctuation for reduced and oxidized states and their crossing point is equal to the redox potential of adsorbed dyes. **a** Overlap between photoexcited state of adsorbed dyes and conduction band of TiO_2 is maximum ($E_{CB} \sim E^{\circ *} + 2\lambda$). **b** The energy of the most probable reduced state locates above conduction band minimum of TiO_2 is maximum ($E_{CB} \sim E^{\circ *} + \lambda$). **c** There are no overlaps between photoexcited state of adsorbed dyes and conduction band of TiO_2 ($E_{CB} \sim E^{\circ *}$)

3.4 *Electron Transfer from Dye to TiO_2*

The exponential rapid decrease of the population of photoexcited dye is also true for organic dyes. The experimentally reported injection times are 6 fs for TiO_2–alizarin system by using transient absorption technique [53]. To simulate these electron injection dynamics, we start from time–dependent Schrödinger equation (TD–SE).

$$i\hbar\frac{\partial|\Psi(\boldsymbol{r},\boldsymbol{R},t)\rangle}{\partial t} = \mathcal{H}(\boldsymbol{r},\boldsymbol{R},t)|\Psi(\boldsymbol{r},\boldsymbol{R},t)\rangle$$

where $\boldsymbol{r}$ and $\boldsymbol{R}$ are electronic and nuclear coordinates, $|\Psi(\boldsymbol{r},\boldsymbol{R},t)\rangle$ is the full wave function. The Hamiltonian of the total system $\mathcal{H}(\boldsymbol{r},\boldsymbol{R},t)$ includes nuclear kinetic energy operator and electronic Hamiltonian $\mathcal{H}^{\mathrm{el}}(\boldsymbol{r};\boldsymbol{R}(t))$.

$$\mathcal{H}(\boldsymbol{r},\boldsymbol{R},t) = \sum_{A}\frac{-\hbar^2}{2M_A}\nabla^2_{\boldsymbol{R}_A} + \mathcal{H}^{\mathrm{el}}(\boldsymbol{r};\boldsymbol{R}(t))$$

The eigenvalues and eigenstates of the $\mathcal{H}^{\mathrm{el}}(\boldsymbol{r};\boldsymbol{R}(t))$ are E_J and $|\Phi_J(\boldsymbol{r};\boldsymbol{R}(t))\rangle$. If we approximate full wave function $|\Psi(\boldsymbol{r},\boldsymbol{R},t)\rangle$ as a product of electronic wave function $|\Phi_J(\boldsymbol{r};\boldsymbol{R}(t))\rangle$ and nuclear wave function, the solution of the TD–SE can be expanded with the $|\Phi_J(\boldsymbol{r};\boldsymbol{R}(t))\rangle$ and time–dependent coefficients $C_J(t)$, which includes nuclear wave functions [54],

$$|\Psi(\boldsymbol{r},\boldsymbol{R},t)\rangle = \sum_{J} C_J(t)|\Phi_J(\boldsymbol{r};\boldsymbol{R}(t))\rangle$$

The TD–SE is rewritten in this basis.

$$i\hbar\frac{\partial}{\partial t}C_I(t) = \sum_J\left\{E_J(\boldsymbol{R}(t))\delta_{IJ} - i\hbar\frac{d\boldsymbol{R}(t)}{dt}\langle\Phi_I(\boldsymbol{r};\boldsymbol{R}(t))|\nabla_{\boldsymbol{R}}\Phi_J(\boldsymbol{r};\boldsymbol{R}(t))\rangle\right\}C_J(t)$$

The term $\langle\Phi_I(\boldsymbol{r};\boldsymbol{R}(t))|\nabla_{\boldsymbol{R}}\Phi_J(\boldsymbol{r};\boldsymbol{R}(t))\rangle$ is nonadiabatic coupling and is finite when two adiabatic states are close to each other. The Jth electronic wave function $|\Phi_J(\boldsymbol{r};\boldsymbol{R}(t))\rangle$ can be expanded by using the molecular orbitals, occupied Kohn–Sham orbitals or configuration state functions with their coefficients. The nuclear equation of motion is described by the Hamiltonian mechanics and the force acting on nuclei is expressed as follows.

$$M_A\ddot{\boldsymbol{R}}_A = -\sum_I\sum_J C_I^*(t)C_J(t)\left\langle\Phi_I(\boldsymbol{r};\boldsymbol{R}(t))|\nabla_{\boldsymbol{R}_A}\mathcal{H}^{\mathrm{el}}(\boldsymbol{r};\boldsymbol{R}(t))|\Phi_J(\boldsymbol{r};\boldsymbol{R}(t))\right\rangle$$

The interface between alizarin molecule and TiO_2 surface is investigated by using this method under some numerical approximation [55]. The nuclear motion was calculated in the ground state because the random thermal motions are regarded to be dominant in the dynamics in both ground and excited states. The electron of photoexcited state of alizarin, which lies above the conduction band minimum, injects fast. The time constant calculated by exponential fitting for the population change is 3.6 fs. On the other hand, the calculated density of states of the alizarin attached to TiO_2 shows some unoccupied orbital of the alizarin moiety is located below the conduction band minimum of TiO_2. The electron injection from these states also occurs as predicted in two–level model, but the time constant is 3–fold longer than that of electron injection from the photoexcited states whose energy is higher than the conduction band minimum. The time constant of the averaged overall electron dynamics is 7.9 fs in agreement with experiments.

3.5 *Avoiding Energy Loss: Direct Charge Transfer Transitions*

As mentioned above, the reorganization energy λ is needed as a driving force for electron transfer from a photoexcited dye to conduction band of semiconductor. It means the energy loss λ is inevitable in DSSC. Direct charge transfer transition from donor to conduction band of the semiconductor allows avoiding such energy loss, but only few materials show such transition. Catechol– and TCNQ–derivatives are organic materials to show direct charge transfer transitions to conduction band of TiO_2 (Fig. 12).

TCNQ is well known as an electron acceptor molecule. The usual electronic state of TCNQ in electrolyte is anion state and its redox potential E° (TCNQ/TCNQ$^-$) is –0.1 V *vs* NHE. The energy difference from the conduction band minimum of TiO_2 is very small, therefore TCNQ may be inadequate molecule for DSSC due to small V_{OC}. However, TCNQ forms a chemical bond between one of dicyanomethylene

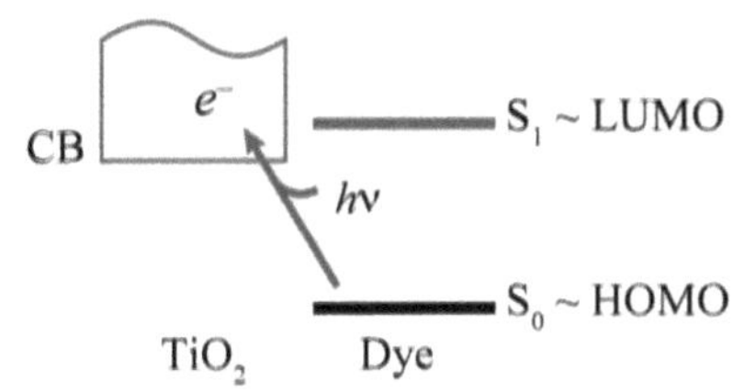

Fig. 12 The energetic model of direct electron injection from ground state to charge transfer state

carbon and oxygen at the TiO_2 surface [56]. The chemical bond changes the redox potential of the complex to more positive than free molecule in electrolyte. In fact, the reported total efficiency of the device made of TiO_2–TCNQ hybrid material is 2.2 % with 0.36 V for open–circuit voltage, and is much higher than those of TiO_2–catechol surface complex [57].

To fill a lack of the molecular level knowledge to design the materials showing interfacial charge transfer absorption and control the absorption property, the formation mechanism of TiO_2–TCNQ complex was investigated [58]. The O–C bonding and anti–bonding orbitals are a linear combination of the 2p orbital of the surface oxygen and a orbital from TCNQ. The anti–bonding orbital corresponds to the highest occupied molecular orbital (HOMO) of the TiO_2–TCNQ. The HOMO formed by the nucleophilic addition reaction is significantly delocalized on the O–C bond. The transition probability of the interfacial charge transfer transition is determined by the matrix element of the light–matter interaction ($V = \mu E$) between the wave functions of the HOMO and unoccupied orbitals of TiO_2. To induce the interfacial charge transfer transition, the spatial overlap between the HOMO and unoccupied orbitals is required. Hence, the delocalization of the HOMO contributes to the appearance of the interfacial charge transfer absorption. This electronic hybridization between TiO_2 and TCNQ results from the direct chemical bonding of TCNQ to the TiO_2 surface by the nucleophilic addition reaction with the above mentioned electronic flow.

3.6 *Charge Recombination*

Charge recombination occurs at the interface of donor and acceptor. The mechanism is classified to internal conversion. The rate of internal conversion considered in Fig. 13 is represented by Fermi's golden rule

$$k_{\mathrm{IC}} = \frac{2\pi}{\hbar}|V|^2\delta\left(\left(E_f + E_{f,vf}\right) - \left(E_i + E_{i,vi}\right)\right)$$

where E_j is the eigenvalue of the electronic wavefunction Φ_j. $E_{j,vj} = \hbar\omega\left(v_j + \frac{1}{2}\right)$ is the eigenvalue of vibrational wave function $\Theta_{j,vj}$ at the electronic state Φ_j with vibrational quantum number vj.

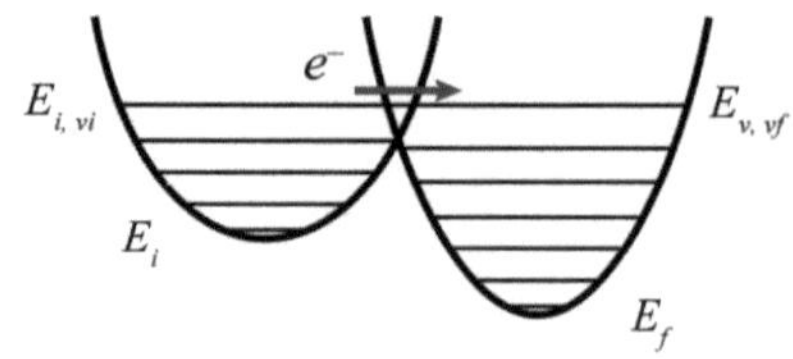

Fig. 13 Vibronic states of initial and final state. The reference energy of vibration state are each electronic state

The interaction V is expressed in the normal mode coordinates $\boldsymbol{Q} = \boldsymbol{LR}$ as follows.

$$V = \left(\frac{\hbar}{\mathrm{i}}\right)^2 \sum_k \left\langle \Phi_\mathrm{f}\Theta_{\mathrm{f,vf}} \middle| \frac{\partial \Phi_\mathrm{i}}{\partial Q_{fk}} \frac{\partial \Theta_{\mathrm{i,vi}}}{\partial Q_{fk}} \right\rangle \approx \left(\frac{\hbar}{\mathrm{i}}\right)^2 \sum_k \left\langle \Phi_\mathrm{f} \middle| \frac{\partial \Phi_\mathrm{i}}{\partial Q_{fk}} \right\rangle \left\langle \Theta_{\mathrm{f,vf}} \middle| \frac{\partial \Theta_{\mathrm{i,vi}}}{\partial Q_{fk}} \right\rangle$$

After each vibrational states at the initial electronic state is weighted with Boltzmann factor, the net rate constant of internal conversion is rewritten

$$k_{\mathrm{IC}} = \frac{2\pi}{\hbar} \sum_{kl} R_{kl} \times \left(\frac{1}{Z} \sum_{vi,vf} \exp\left[-\beta E_{i,vi}\right] P_{kl} \right) \delta\left(\left(E_f + E_{f,vf}\right) - \left(E_i + E_{i,vi}\right) \right)$$

$$R_{kl} \equiv \left(\frac{\hbar}{\mathrm{i}}\right)^2 \left\langle \Phi_f \middle| \frac{\partial \Phi_i}{\partial Q_{fk}} \right\rangle \left\langle \frac{\partial \Phi_i}{\partial Q_{fl}} \middle| \Phi_f \right\rangle$$

$$P_{kl} \equiv \left\langle \Theta_{f,vf} \middle| \frac{\partial \Theta_{i,vi}}{\partial Q_{fk}} \right\rangle \left\langle \frac{\partial \Theta_{i,vi}}{\partial Q_{fl}} \middle| \Theta_{f,vf} \right\rangle$$

where R_{kl} and P_{kl} are defined as the electronic and the vibrational parts of the internal conversion rate.

The electronic part of the internal conversion rate R_{kl} is rewritten by using Hellman–Feynmann theorem [59],

$$\left\langle \Phi_\mathrm{f} \middle| \frac{\partial \Phi_\mathrm{i}}{\partial Q_{fk}} \right\rangle = \left\langle \Phi_\mathrm{f} \middle| \frac{\partial \hat{V}}{\partial Q_{fk}} \middle| \Phi_\mathrm{i} \right\rangle / \left(E_i - E_f\right)$$

$\hat{V}$ is potential operator and depends on only the position of electron and nuclei.

$$\left\langle \Phi_\mathrm{f} \middle| \frac{\partial \hat{V}}{\partial Q_{fk}} \middle| \Phi_\mathrm{i} \right\rangle = -\sum_A \frac{Z_A e^2}{\sqrt{M_A}} \langle \Phi_\mathrm{f} | \boldsymbol{E}_{f \leftarrow i,A} | \Phi_\mathrm{i} \rangle \boldsymbol{L}_{fAk}$$

$\left\langle \Phi_\mathrm{f} \middle| \boldsymbol{E}_{f \leftarrow i,A} \middle| \Phi_\mathrm{i} \right\rangle$ is the transition electric field at atom A and is computed from wave function theory or time dependent density functional theory.

Next let's estimate the vibrational part P_{kl}. If we consider the normal modes at the initial state to be equivalent to those at final, i.e. ground state, the vibrational wave function is expressed as same form by using Hermite polynomials H_v.

$$\Theta_{i,vi} = \Theta_{f,vf} \equiv \Theta_v = \prod_k \sqrt{\frac{1}{2^v v!}} \left(\frac{1}{\pi}\right)^{\frac{1}{4}} \exp\left[-\frac{Q_k^2}{2}\right] H_v(Q_k)$$

In this expression, P_{kl} is identically $\delta_{kl}\left(v_{i_k} + \frac{1}{2}\right)$ and does not depend on individual molecules. The behavior of vibrational part of the internal conversion weighted by Boltzmann factor is temperature dependent and is effective only in the modes $\beta E_k \leq 1$, i.e. low–frequency modes [60].

The internal conversion rate of the TiO_2–TCNQ and its derivatives TiO_2–TCNE and TiO_2–TCNAQ are calculated within some numerical approximation [61]. A significant improvement in absorption with TCNQ versus TCNE is accompanied by only modest rise of the recombination rate. The recombination increases significantly with TCNAQ, likely offsetting any gain in absorption. Because a larger molecule will have a larger number of contributing lower–frequency vibrations, the increase in internal conversion/recombination with TCNAQ is not surprising. The computed trend in absorbance and recombination implies that TCNQ may provide an optimal balance between the two considered factors among the three molecules.

4 Conclusion

In this chapter, we reviewed charge carrier dynamics in power convergence processes of organic solar cells and dye-sensitized solar cells from the standpoint of chemical reactions.

For organic solar cells, charge transfer kinetics based on the Marcus theory is reviewed first as the starting point for considering interfacial carrier dynamics in organic solar cells from the standpoint of chemical reactions. Then, charge transfer dynamics as quantum and/or nonadiabatic dynamics is reviewed and revealed that interfacial chemical reactions in organic solar cells are more complex and fascinating than ones assumed in the chemical kinetics. Though the charge transfer reactions have been well studied for a long time as a one of the most important chemical reactions, the interfacial charge transfer reactions in organic solar cells are not understood sufficiently yet and need to be studied in detail. Especially, the notion of "hot charge transfer excitons and their dissociation" through multiple potential energy surfaces as well as the suppression of recombination will be a key factor to enhance the power convergence efficiency. Finally, studies of device simulators are briefly reviewed. Though current device simulators have not been sufficient to answer the demands of people who want to develop higher efficient organic solar cells yet, the deep understanding of each process and taking them into account in the device simulator will help us develop higher efficient organic solar cells in the future.

For dye-sensitized solar cells, starting from the electronic structure of the TiO_2/dye/electrolyte calculated by Poisson–Boltzmann equation, we introduced explanations for the driving force for electron transfer from the excited state of the dye to the CB of the TiO_2 which is recognized as energy loss by using the two-level system, redox reaction, time–dependent Schrödinger equation. Our studies for avoiding energy loss are also reviewed.

References

1. Bakulin AA, Martyanov DS, Paraschuk DYu, Pshenichnikov MS, van Loosdrecht PHM (2008) Ultrafast charge photogeneration dynamics in ground-state charge-transfer complexes based on conjugated polymers. J Phys Chem B 112:13730–13737
2. Marcus RA (1956) On the theory of oxidation-reduction reactions involving electron transfer. I. J Chem Phys 24:966–978
3. Marcus RA, Sutin N (1985) Electron transfers in chemistry and biology. Biochim Biophys Acta 811:265–322
4. Marcus RA (1993) Electron transfer reactions in chemistry. theory and experiment. Rev Mod Phys 65:599–610
5. Haken H (1978) Synergetics: an introduction. nonequilibrium phase transitions and self-organization in physics, chemistry, and biology, 2nd edn. Springer, ISBN: 978-3-642-96471-8
6. Haken H (1983) Advanced synergetics: instability hierarchies of self-organizing systems and devices. In: Springer Series in Synergetics, vol 20, ISBN: 978-3-642-45553-7
7. Cave RJ, Newton MD (1996) Generalization of the Mulliken-Hush treatment for the calculation of electron transfer matrix elements. Chem Phys Lett 249:15
8. Cave RJ, Newton MD (1997) Calculation of electronic coupling matrix elements for ground and excited state electron transfer reactions: Comparison of the generalized Mulliken-Hush and block diagonalization methods. J Chem Phys 106:9213–9226
9. Farazdel A, Dupuis M, Clementi E, Aviram A (1990) Electric-field induced intramolecular electron transfer in spiro.pi.-electron systems and their suitability as molecular electronic devices. a theoretical study. J Am Chem Soc 112:4206–4214
10. Dederichs PH, Blügel S, Zeller R, Akai H (1984) Ground states of constrained systems: application to cerium impurities. Phys Rev Lett 53:2512–2515
11. Wu Q, Van Voorhis T (2005) Direct optimization method to study constrained systems within density-functional theory. Phys Rev A **72:**024502–024502-4
12. Wu Q, Van Voorhis T (2006) Direct calculation of electron transfer parameters through constrained density functional theory. J Phys Chem A 110:9212–9218
13. Valiev M, Bylaska EJ, Govind N, Kowalski K, Straatsma TP, van Dam HJJ, Wang D, Nieplocha J, Apra E, Windus TL, de Jong WA (2010) NWChem: a comprehensive and scalable open-source solution for large scale molecular simulations. Comput Phys Commun 181:1477
14. van Hal PA, Meskers SCJ, Janssen RAJ (2004) Photoinduced energy and electron transfer in oligo(p-phenylene vinylene)-fullerene dyads. J Appl Phys A 79:41–46
15. Lemaur V, Steel M, Beljonne D, Brédas J-L, Cornil J (2005) Photoinduced charge generation and recombination dynamics in model donor/acceptor pairs for organic solar cell applications: a full quantum-chemical treatment. J Am Chem Soc 127:6077–6086
16. Kawatsu T, Coropceanu V, Ye A, Brédas J-L (2008) Quantum-chemical approach to electronic coupling: application to charge separation and charge recombination pathways in a model molecular donor–acceptor system for organic solar cells. J Phys Chem C 112:3429

17. Yi Y, Coropceanu V, Bréas J-L (2009) Exciton-dissociation and charge-recombination processes in pentacene/c$_{60}$ solar cells: theoretical insight into the impact of interface geometry. J Am Chem Soc 131:15777–15783
18. Liu T, Cheung DL, Troisi A (2011) Structural variability and dynamics of the P3HT/PCBM interface and its effects on the electronic structure and the charge-transfer rates in solar cells. Phys Chem Chem Phys 13:21461–21470
19. Kanai Y, Grossman JC (2007) Insights on interfacial charge transfer across P3HT/Fullerene photovoltaic heterojunction from Ab initio calculations. Nano Lett 7:1967–1972
20. Li Z, Zhang X, Lu G (2011) Electron structure and dynamics at poly(3-hexylthiophene)/fullerene photovoltaic heterojunctions. Appl Phys Lett 98:083303–083303-3
21. Veldman D, İpek Ö, Meskers SCJ, Sweelssen J, Koetse MM, Veenstra SC, Kroon JM, van Bavel SS, Loos J, Janssen RAJ (2008) Compositional and electric field dependence of the dissociation of charge transfer excitons in alternating polyfluorene copolymer/fullerene blends. J Am Chem Soc 130:7721–7735
22. Guo J, Ohkita H, Benten H, Ito S (2010) Charge generation and recombination dynamics in Poly(3-hexylthiophene)/Fullerene blend films with different regioregularities and morphologies. J Am Chem Soc 132:6154–6164
23. Bakulin AA, Rao A, Pavelyev VG, van Loosdrecht PHM, Pshenichnikov MS, Niedzialek D, Cornil J, Beljonne D, Friend RH (2012) The role of driving energy and delocalized states for charge separation in organic semiconductors. Science 335:1340–1344
24. Fujii M, Yamashita K (2011) Packing effects in organic donor-acceptor molecular heterojunctions. Chem Phys Lett 514:146–150
25. Carmichael HJ (1999) Statistical methods in quantum optics 1: master equations and Fokker-Planck equations. Springer, Berlin. ISBN 978-3-662-03875-8)
26. Meyer H-D, Gatti F, Worth GA (2009) Multidimensional quantum dynamics: MCTDH theory and applications. Wiley, ISBN: 978-3-527-32018-9
27. Tamura H, Burghardt I, Tsukada M (2011) Exciton dissociation at Thiophene/Fullerene interfaces: the electronic structures and quantum dynamic. J Phys Chem C 115:10205–10210
28. Tamura H, Marinazzo R, Ruckenbauer M, Burghardt I (2012) Quantum dynamics of ultrafast charge transfer at a polymer-fullerene interface. J Chem Phys 137:22A540– 22A54-7
29. Jailaubekov AE, Willard AP, Tritsch JR, Chan W-L, Sai N, Gearba R, Kaake LG, Williams KJ, Leung K, Rossky PJ, Zhu X-Y (2013) Nat Mater 12:66
30. Lobaugh J, Rossky PJ (1999) Computer simulation of the excited state dynamics of betaine-30 in acetonitrile. J Phys Chem A 103:9432–9447
31. Sterpone F, Bedard-Hearn MJ, Rossky PJ (2010) Nonadiabatic simulations of exciton dissociation in poly-p-phenylenevinylene oligomers. J Phys Chem A 114:7661–7670
32. Koster LJA, Smits ECP, Mihailetchi VD, Blom PWM (2005) Device model for the operation of polymer/fullerene bulk heterojunction solar cells. Phys Rev B 72:085205
33. Watkins PK, Walker AB, Verschoor GLB (2005) Dynamical monte carlo modelling of organic solar cells: the dependence of internal quantum efficiency on morphology. Nano Lett 5:1814–1818
34. Frost JM, Cheynis F, Tuladhar SM (2006) Influence of polymer-blend morphology on charge transport and photocurrent generation in donor–acceptor polymer blends. J Nelson Nano Lett 6:1674–1681
35. Meng L, Wang D, Li Q, Yi Y, Brédas J-L, Shuai Z (2011) An improved dynamic Monte Carlo model coupled with Poisson equation to simulate the performance of organic photovoltaic devices. J Chem Phys 134:124102–124102-7
36. Ray B, Nair PR, Alam MA (2011) Annealing dependent performance of organic bulk-heterojunction solar cells: a theoretical perspective. Sol Energy Mater Sol Cells 95:3287–3294
37. Hagfeldt A, Boschloo G, Sun L, Kloo L, Pettersson H (2010) Dye-Sensitized solar cells. Chem Rev 110:6595–6663

38. O'Regan B, Grätzel M (1991) A low–cost, high–efficiency solar cell based on dye–sensitized colloidal TiO_2 films. Nature 353:737–740
39. Mathew S, Yella A, Gao P, Humphry-Baker R, Curchod BFE, Ashari-Astani N, Tavernelli I, Rothlisberger U, Nazeeruddin MK, Grätzel M (2014) Dye–sensitized solar cells with 13 % efficiency achieved through the molecular engineering of porphyrin sensitizers. Nat Chem 6:242–247
40. Zhang Z, Yates JT (2012) Band bending in semiconductors: chemical and physical consequences at surfaces and interfaces. Chem Rev 112:5520–5551
41. Kavan L, Grätzel M, Gilbert SE, Klemenz C, Scheel HJ (1996) Electrochemical and photoelectrochemical investigation of single-crystal anatase. J Am Chem Soc 118:6716–6723
42. Peter LM (2007) Dye–sensitized nanocrystalline solar cells. Phys Chem Chem Phys 9:2630–2642
43. Rabi II (1937) Space quantization in a gyrating magnetic field. Phys Rev 51:652–654
44. Robinson GW, Frosch RP (1962) Theory of electronic energy relaxation in the solid phase. J Chem Phys 37:1962–1973
45. Bixon M, Jortner J (1968) Intramolecular radiationless transitions. J Chem Phys 48:715–726
46. Desilvestro J, Grätzel M, Kavan L, Moser J (1985) Highly efficient sensitization of titanium dioxide. J Am Chem Soc 107:2988–2990
47. Anderson S, Constable EC, Dare-Edwards MP, Goodenough JB, Hamnett A, Seddon KR, Wright RD (1979) Chemical modification of a titanium (IV) oxide electrode to give stable dye sensitization without a supersensitiser. Nature 280:572–573
48. O'Regan B, Moser J, Anderson M, Grätzel M (1990) Vectorial electron injection into transparent semiconductor membranes and electric field effects on the dyamics of light-induced charge separation. J Phys Chem 94:8720–8726
49. Nazeeruddin MK, Kay A, Rodicio I, Humphry-Baker R, Muller E, Liska P, Vlachopoulos N, Grätzel M (1993) Conversion of light to electricity by cis–XzBis (2,2'–bipyridyl–4,4'–dicarboxylate) ruthenium (II) charge-transfer sensitizers (X = Cl−, Br−, I−, CN−, and SCN−) on nanocrystalline TiO2 electrodes. J Am Chem Soc 115:6382–6390
50. Wenger B, Grätzel M, Moser J-E (2005) Rationale for kinetic heterogeneity of ultrafast light-induced electron transfer from Ru(II) complex sensitizers to nanocrystalline TiO2. J Am Chem Soc 127:12150–12151
51. Clark WDK, Sutin N (1977) Spectral sensitization of n–Type TiO_2 electrodes by polypyridineruthenium (II) complexes. J Am Chem Soc 99:4676–4682
52. Watson DF, Meyer GJ (2005) Electron injection at dye-sensitized semiconductor electrodes. Annu Rev Phys Chem 56:119
53. Huber R, Moser J-E, Grätzel M, Wachtveitl J (2002) Real-Time observation of photoinduced adiabatic electron transfer in strongly coupled dye/semiconductor colloidal systems with a 6 fs time constant. J Phys Chem B 106:6494–6499
54. Yonehara T, Hanasaki K, Takatsuka K (2012) Fundamental approaches to nonadiabaticity: towards a chemical theory beyond the born-oppenheimer paradigm. Chem Rev 112:499–542
55. Duncan WR, Stier WM, Prezhdo OV (2005) Ab initio nonadiabatic molecular dynamics of the ultrafast electron injection across the alizarin–tio_2 interface. J Am Chem Soc 127:7941–7951
56. Jono R, Fujisawa J, Segawa H, Yamashita K (2011) Theoretical study of the surface complex between TiO_2 and TCNQ showing interfacial charge transfer transitions. J Phys Chem Lett 2:1167–1170
57. Segawa H, Fujisawa J, Kubo T, Uchida S (2009) Composite material for photoelectric conversion material, contains cyano group–containing compound or its reduced form salt exhibiting absorption characteristics at different wavelength region with respect to metal oxide, Japan WO2009110618–A1, EP2249430–A1, US2011011459–A1
58. Jono R, Fujisawa J, Segawa H, Yamashita K (2013) The Origin of the strong interfacial charge-transfer absorption in the surface complex between TiO2 and dicyanomethylene compounds. Phys Chem Chem Phys 5:18584–18588

59. Niu Y, Peng Q, Deng C, Gao X, Shuai Z (2010) Theory of excited state decays and optical spectra: application to polyatomic molecules. J Phys Chem A 114:7817–7831
60. Manzhos S, Segawa H, Yamashita K (2011) A model for recombination in Type II dye–sensitized solar cells: Catechol–thiophene dyes. Chem Phys Lett 504:230–235
61. Manzhos S, Jono R, Yamashita K, Fujisawa J, Nagata M, Segawa H (2011) Study of interfacial charge transfer bands and electron recombination in the surface complexes of TCNE, TCNQ, and TCNAQ with TiO_2. J Phys Chem C 115:21487–21493

Physical Model at the Electrode-Electrolyte Interface

Osamu Sugino

Abstract To understand the mechanism of an electrochemical reaction, it is useful to reveal the atomic-scale details occurring on a model catalyst, such as a well-defined Pt(111) surface. The study using the model catalyst is stronglxy assisted today by the first-principles simulations, which has been originally developed for solid-state physics. As such, the method is not perfectly suited for a realistic electrochemical simulation and thus needs modification. This chapter explains recent advances in electrochemical simulation methods as well as demonstrations performed on noble metal—solution interfaces.

1 Introduction

Electrochemical reactions occurring on a real catalyst are usually too complex to be studied microscopically, and hence for the study of the reaction mechanism, structurally controlled model catalysts are utilized. Some of the model catalysts are found to be surprisingly flat in the atomic scale [1], a fact which has motivated surface scientists to modify their tools for the electrochemical study. In this context, methods for surface sensitive measurement have been developed including infrared and Raman spectroscopy, X-ray scattering and absorption, and scanning tunneling microscopy. The interface has also been studied theoretically by utilizing the first-principles electronic structure calculation, which may be better known as the band-structure calculation. By the collaborative application of these experimental and theoretical tools, understanding of the electrochemical reaction has greatly advanced as reviewed by Koper [2]. Even with such studies, there remains considerable controversy regarding the dominant reaction mechanism even for the

O. Sugino (✉)
The Institute for Solid State Physics, The University of Tokyo,
Kashiwanoha, 5-1-5, Kashiwa, Chiba 277-8581, Japan
e-mail: sugino@issp.u-tokyo.ac.jp

M. Sugiyama et al. (eds.), *Solar to Chemical Energy Conversion*,
Lecture Notes in Energy 32, DOI 10.1007/978-3-319-25400-5_6

Pt(111)—solution interface, which is probably the most studied model catalyst. This indicates that further methodological advance is required.

Most of the band-structure calculation methods utilized for the electrochemical study are based on the density functional theory (DFT) [3] although the accuracy may be improved by using the many-body perturbation theory [4] or else. Owing to the improvements in the computational algorithms, such as the iterative diagonalization method [5], and owing also to the developments of the exchange-correlation energy functional, the DFT-based approach can be applied to increasingly larger systems with improved accuracy. Indeed, O_2 adsorption on a catalyst surface, such as Pt(111) [6], can be now reliably predicted although there still remains accuracy problem, on the order of 0.1 eV, regarding the CO adsorption on Pt(111) [6]. So, in most cases, the problem of the band-structure calculation method is not in the applicability to a surface but in the adaptation to an electrochemical interface (Fig. 1) which is significantly influenced by the solvation effect and the electrode potential effect.

The concept of the electrode potential is essential for an electrochemical study in that the reaction depends very sensitively on the potential. The potential is defined in many textbooks on the basis of difference between the Fermi level of the electrode and the vacuum level near the electrolyte—vacuum interface. Referring to the level-difference under the electrochemical equilibrium condition, the shift of the

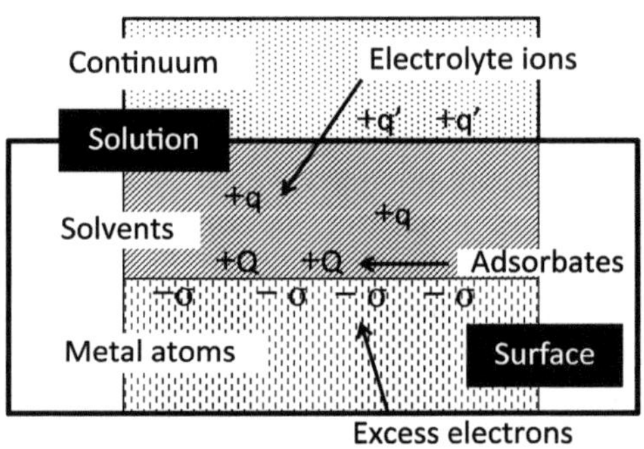

Fig. 1 Typical setup of the physical model of the electrode-electrolyte interface. Metal atoms and solvent molecules are arranged in the form of a slab (applied with the periodic boundary condition in the surface parallel directions) with finite thickness and are interfaced with each other. The interface slab thus constructed (framed with thick line) is the target of the band-structure calculation. The solution away from the interface slab is represented by a continuum, which takes part in the interfacial dielectric screening. The charge induced on the continuum ($+q'$), together with the excess (or deficit) electrons induced on the metal surface ($-\sigma$) and the ionic charges ($+q$ and $+Q$), constitutes the interfacial dipole and causes the electrode potential. By performing the first-principles molecular dynamics simulation, the interface structure is determined. The simulation is mostly done under the condition of fixed number of particles, the electrons and the ions, but is more favorably done under the condition where they are in contact with the reservoir of given chemical potential. Such simulation yields the grand canonical ensemble. Establishing such grand canonical ensemble simulation is the challenge of electrochemical theoreticians

level-difference under a non-equilibrium condition is related to the over- or under-potential of the interface. The potential thus defined reflects the potential drop caused by the interface dipole, which is originated from the dielectric screening property of the interface. Importantly, the dielectric screening accompanies not only the rearrangement of atomic structures at the interface but also the change in the local concentration of the (electrolyte) ions as well as the change in the coverage of adsorbates. This means that the electrochemical interface needs to be treated as an open system and thus the grand potential, or the thermodynamic potential, is the relevant quantity. This is beyond the capability of most of the first-principles simulation codes now, and thus the development of the method is a challenge of electrochemical theoreticians.

2 Modeling of the Electrochemical Interfaces

2.1 Early Stage Modelling

It is important to explain what has been made possible and what has not been in the frontier of the electrochemical simulation. For this purpose, let us begin by explaining the methodology developed in the early stage. At that time, the theory was based on a simple thermodynamic argument, where the potential and the interface dipole were treated only implicitly. Nørskov et al., as reviewed in Ref. [7], used a theory that is based on the chemical equilibrium realized under the standard hydrogen electrode (SHE) condition

$$\mathrm{H_3O^+(aq)} + \mathrm{e^-}(U_0) \leftrightarrow \frac{1}{2}\mathrm{H_2(g)} + \mathrm{H_2O(aq)},$$

where ‘aq’ and ‘g’ parenthesized indicate a molecule in the aqueous and gaseous conditions, respectively, $e^-(U)$ is the electrons in the electrode of potential U, and U_0 is the equilibrium potential (Fig. 2). In this theory, using the energy ΔE required for the (dissociative) adsorption of a hydrogen molecule

$$\frac{1}{2}\mathrm{H_2(g)} \rightarrow \mathrm{H(ads)}$$

under the ultra-high vacuum condition (UHV), the energy required for the Volmer step of the hydrogen evolution reaction

$$\mathrm{H_3O^+(aq)} + \mathrm{e^-}(U_0) \rightarrow \mathrm{H(ads)} + \mathrm{H_2O(aq)}$$

is approximated as ΔE. Simple argument tells that the (local) equilibrium for the Volmer step is realized under a potential U that satisfies $e(U - U_0) = \Delta E$, where e is the elementary charge. Parallel analysis can be applied to other electrochemical reactions as well. Here it is assumed that the energy required for the hydrogen

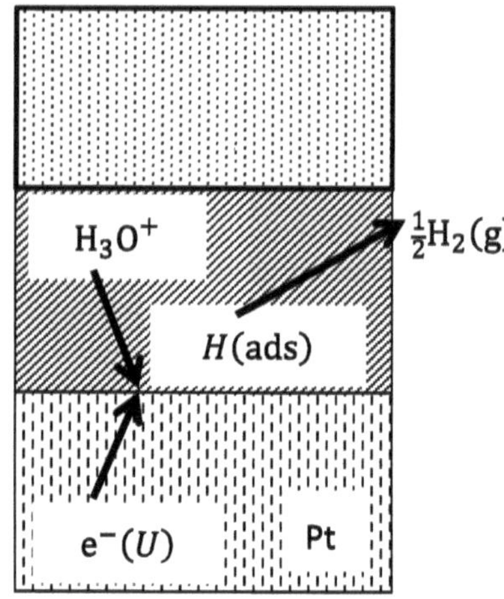

Fig. 2 The hydrogen reduction reaction. This two-step reaction is in equilibrium at U_0, and the first step (Volmer step) is in equilibrium at U when the adsorption energy ΔE satisfies $\Delta E = e(U - U_0)$. Computation of ΔE is the issue of the first-principles calculation

adsorption under UHV is the same as that at the electrochemical interface, of which the justification requires careful comparison with experiment or accurate calculation.

2.2 Solvation Effect

The success of the simple thermodynamic theory relies on the insensitivity of a molecular adsorption to the solvation effect, but the solvation effect will be significantly larger for a charged or a polarized molecule than for a neutral or an unpolarized one. With this in mind, considerable efforts were made to develop a method to incorporate the solvation effect to the first-principles simulation, although the solvent effect had been the central issue of solution chemistry where handy dielectric continuum models, such as polarizable continuum model (PCM) [8], had been developed for a molecular solvation. In the approach of Ref. [9], the solvent is replaced by a dielectric continuum with a cavity, in which the solute is contained. The dielectric constant ϵ_r of the continuum is let smoothly change from one (in the cavity) to the ϵ_r of bulk solvent (in the continuum) depending on the electron density of the solute molecule (that is isolated in the vacuum). This approach was applied to a cavity of infinite diameter, namely the electrochemical interfaces, by other groups [10, 11]. Independently of Ref. [9], different dielectric model was proposed [12], where two parallel dielectric continua are arranged in the form a parallel-plate capacitor (Fig. 3) and let the interface slab be inserted in between. Herein, instead of wrapping the solute (that is the solid surface in this case) with an adaptive cavity as did in Ref. [9], the solute is put into the non-deformable parallel container, but owing to this simplification one can handle much more atoms or molecules and, in addition, follow a finite temperature dynamics, as demonstrated in Ref. [12].

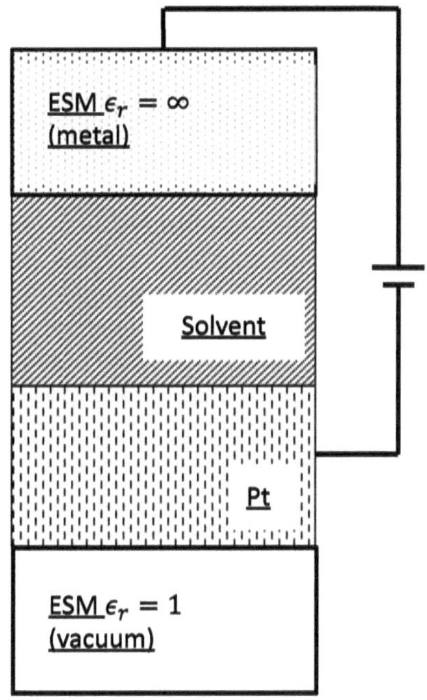

Fig. 3 Typical setup for the electrochemical interface with the effective screening medium (ESM). Metallic ESM is attached *above* the solvent and the vacuum ESM *below* the electrode, by which the electrostatic potential in the electrode and that in the ESM metal can be controlled

2.3 *Effect of Potential*

Another problem of the early thermodynamic model is due to the neglect of the effect of the potential on the reaction environment, which can easily change the reaction mechanism. Its importance can be seen from the potential dependent redistribution of the solute, such as the water reorientation found on the gold surface [13]. Perhaps more importantly, the potential effect is reflected in the shift of the chemical equilibrium backwards or forwards, thereby rearranging the distribution of the reaction intermediates on the surface. Recognizing its importance, efforts were made to develop a method to explicitly apply the potential. Pioneering study was done in Ref. [14] using a small cluster-based model for the interface, and later, the potential was applied to the interface slab models [12, 15]. In applying the potential, however, careful consideration is required on the physics of the electric double layer formation, or the Holmholtz layer formation.

2.3.1 Screeing in the Inner- and Outer-Helmholtz Layer

It is known that the electric double layer is formed by the excess (or deficit) electrons on the electrode on one hand and the compensating electrolyte ions in the solution on the other hand (Fig. 1). When the potential is applied, the electrons are redistributed within the skin depth of the surface, while the ionic redistribution is much more complex: In the outer Helmholtz region, the ions follow the macroscopic

Poisson-Boltzmann equation [15, 16] and screen the electric field taking 1 nm to 1 μm depending on pH of the solution: In the inner Helmholtz region, the ions are bound strongly or weakly on the surface thereby screening the electric field inhomogeneously and, at the same time, disturbing significantly the structure of the solvents, e.g., the hydrogen bond network of liquid water. This indicates that the screening has a microscopic and a macroscopic component, which is considered to require different theoretical treatment such that the latter is treated by introducing excess/deficit electrons and ions explicitly in the model as did in Ref. [17] while the former is treated by a continuum model.

Such method was adopted for a simulation of an acid solution in Ref. [12], where the excess electrons are partly screened by a hydronium ion (H_3O^+) introduced into the slab and partly by the dielectric continuum, which is called the effective screening medium (ESM); no hydronium ions are required under a neutral condition. Applying this model to a Pt(111)—water interface, the value comparable to the experiment was obtained for the magnitude of the interface dipole [12] as well as for that of the interface capacitance [18], indicating that the electric double layer is reasonably well described by such model. It should be mentioned that the simulations were done only near the point of zero charge, where the surface is neutral and, in addition, the surface is neither hydrogenated nor oxidized. At lower or higher potentials, on the contrary, surface is complicated by the adsorbed H or O as well as the electrolyte ions, which has been hampering a first-principles simulation.

2.3.2 Constant Potential Simulation

The simulations mentioned above were done using a given number of ions and electrons, while experiments are done using the potentiostat under the constant potential condition. It is preferable to do the simulation under the constant potential condition although, in principle, the ensemble at constant particle number condition can be transformed to that at constant chemical potential condition. With this in mind, a constant potential scheme was formulated recently [19] using a technique known as the method of the extended Lagrangian. The number of excess electrons, n_e, is taken as a fictitious dynamical variable and is reduced (or increased) when n_e is too high (or low) to keep the potential constant. By this, the potential fluctuates but is kept constant on average. It was shown in Ref. [19] that the excess/deficit electrons follow the grand canonical distribution when equilibrium is reached in the simulation.

In this simulation, although the electrons are let contact with a reservoir of given chemical potential, ions are not. In this sense, the development of the electrochemical simulation is still halfway. In the classical molecular dynamics (MD), the grand canonical MD simulation is conventionally done by introducing stochastic attempts to increase or reduce the number of particles. However, in the first-principles MD, where electrons are also treated within the Born-Oppenheimer approximation, such simulation is hampered by the necessity to remove or add the electrons together with the nuclei, which gives rise to global rearrangement of the electronic wave function. The grand canonical first-principles MD has thus been impractical so far and an algorithmic breakthrough is required.

3 Mechanism Search

Thus far many first-principles calculations have been done to determine the reaction mechanism. Instead of reviewing all such works, however, here we introduce two of ours that illustrate limitation and achievement of the present first-principles calculation.

3.1 Direct First-Principles Md Simulation

Instead of using the modified first-principles scheme mentioned above, Ref. [20] investigated, using a conventional program package, how the oxygen reduction reaction (ORR) will proceed on a M(111) surface, where M = Pt, Pd, Ru, or Au. A metal—water interface was repeated three dimensionally applying the periodic boundary condition, and using the repeated model, the first-principles MD simulation was performed. Initially one O_2 molecule was put on the metal surface and the reaction dynamics was followed by introducing the hydronium ions. The reduction reactions were found to occur through the associative mechanism (see Fig. 4)

$$\begin{aligned} &O_2(\text{ads}) \rightarrow OOH(\text{ads}) \rightarrow OH(\text{ads}) + O(\text{ads}) \\ &\rightarrow H_2O(\text{ads}) + O(\text{ads}) \rightarrow H_2O(\text{ads}) + OH(\text{ads}) \rightarrow H_2O(\text{ads}) \end{aligned}$$

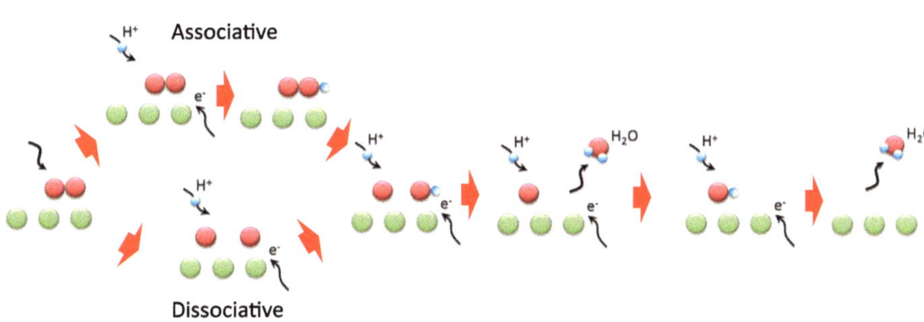

Fig. 4 Reaction paths for the oxygen reduction reaction. *Red*, *white* and *green circles* correspond to oxygen, hydrogen, and metal atoms, respectively. The associative mechanism (*above*) and the dissociative mechanism (*below*) compete on noble metal surfaces. By analyzing the electrochemical measurements and using the first-principles calculations as supplementary information, a crossover in the dominant mechanism was found to occur as the potential is shifted

on Pt and Pd, and the dissociative mechanism

$$\begin{aligned}&O_2(ads) \rightarrow 2O(ads) \rightarrow OH(ads) + O(ads)\\&\rightarrow H_2O(ads) + O(ads) \rightarrow H_2O(ads) + OH(ads) \rightarrow H_2O(ads)\end{aligned}$$

on Ru, while the reaction proceeded only up to a HOOH formation on the Au. This apparently plausible result needs to be accepted with care because of the unrealistically large concentration of the hydronium ions used in that simulation [20]. The concentration corresponds to a negative pH: The potential is lower than the reversible hydrogen electrode, namely the simulation was done under a large overpotential condition, because of the interface dipole formed by the positive charge of the hydronium ion and the compensating negative charge of the electrons on the surface. The fact that the plausible reaction paths were obtained even under such condition might mean that the reaction paths are not so sensitive to the environment except for the case where multiple paths are delicately competing.

3.2 *Implication of Experiment*

Quantitative understanding of the reaction mechanism requires accurate evaluation of the grand potential. Although the first-principles evaluation is not possible now, what can be learned when all existing electrochemical measurements in the literatures are used for analysis? With this in mind, Ref. [21] determined the parameters appearing in the Butler-Volmer theory of the current-voltage curve using available experimental data with some lacking information supplemented by first-principles calculations. The results obtained for the oxygen reduction reaction on Pt(111) are found to support the associative mechanism as the dominant one, and in addition, the dominant mechanism switches from the associative (under the reversible potential condition) to the dissociative (under reduced potential conditions). The analysis shows that the switch in the mechanism relies on the experimental result that the reaction order versus the oxygen pressure is close to one both at low and high current conditions [22, 23]; if it were not for such experiment, conclusion can be different.

The study showed competing associative and dissociative mechanisms and the most dominant one depends rather sensitively on the coverage of the reaction intermediates. In this respect the result obtained in Ref. [20] may be questionable because the coverage effect was neglected. To further investigate this issue it will be important to improve the first-principles simulation to be able to determine the equilibrium coverage by computing the grand potential of atoms and molecules. One may consider if such cumbersome simulation is really necessary because of the recent progress in measuring the reaction intermediates [24]. But, such simulation is undoubtedly valuable to explore, for example, an electrode where experiments are not feasible.

4 Conclusion

Electrochemistry was born in 19th century and, since the beginning of this century, development of the microscopic theory has been accelerated by the surface sensitive measurements and the first-principles calculation, bridging therewith the gap between electrochemistry and surface science or condensed matter physics. Considering the potential that the first-principles calculation has, one can expect further acceleration when the method is adapted to realistic electrochemical environment. The development of the grand canonical simulation is one of the keys, and progress has been made towards this goal, although there are still hurdles to overcome as described in this chapter.

References

1. Itaya K (1998) Prog Surf Sci 58:121–247
2. Koper MTM (2005) J Electrochem Chem 574:375–386
3. Hohenberg P, Kohn W (1964) Phys Rev B 136:864–871
4. Hedin L (1965) Phys Rev A 139:796–823
5. Car R, Parrinello M (1995) Phys Rev Lett 55:2471–2474
6. Eichler A, Mittendorfer F, Hafner J (2000) Phys Rev B 62:4744–4755
7. Nørskov JK, Bligaard T, Rossmeisl J, Christensen CH (2009) Nat Chem 1:37–46
8. Tomasi J, Mennucchi B, Cammi R (2005) Chem Rev 105:2999–3094
9. Scherlis DA et al (2006) J Chem Phys 124:074103 (12 pages)
10. Jinnouchi R, Anderson AB (2008) Phys. Rev. B 77:245417
11. Wang H-F, Liu Z-P (2009) J Phys Chem 113:17502–17508
12. Otani M et al (2008) J Phys Soc Jpn 77:024802
13. Ataka K, Yotsuyanagi T, Osawa M (1996) J Phys Chem 100:10664–10672
14. Anderson AB, Albu TV (1999) J Am Chem Soc 121:11855–11863
15. Gouy G (1910) Compt Rend 149:654
16. Chapman DL (1913) Philos Mag 25:475
17. Skúlaason E et al (2010) J Phys Chem C 114:18182–18197
18. Ando Y, Gohda Y, Tsuneyuki S (2012) Chem Phys Lett 556:9–12
19. Bonnet N, Morishita T, Sugino O, Otani M (2012) Phys Rev Lett 109:266101
20. Okamoto Y, Sugino O (2010) J Phys Chem C 114:4473–4478
21. Bonnet N, Sugino O, Otani M (2014) J Phys Chem C 118:13638–13643
22. Stamenkovic VR et al (2002) J Phys Chem B 106:11970
23. Parthasarathy A et al (1992) J Electrochem Soc 139:2856
24. Jia Q et al (2014) J ElectroChim Soc 161:F1323–F1329

Part III
Chemical, Electrochemical and Photoelectrochemical Approach for Energy Conversion: Necessity of Energy Storage Using Chemical Bonds

Energy Storage in Batteries and Fuel Cells

Tetsuya Kajita and Takashi Itoh

Abstract Batteries and fuel cells (FCs) are the two major types of solar energy storage devices currently in use. Secondary batteries reversibly convert stored chemical energy (e.g., from solar power devices) into electrical energy. FCs generally convert chemical energy from fuel into electricity via chemical reactions with oxygen. Hydrogen is the most common fuel, but hydrocarbons such as natural gas and methanol are sometimes used. FCs are distinct from batteries in that the former require a constant source of fuel and oxygen (e.g., through solar electrolysis of water) to sustain their chemical reactions, and can continuously produce electricity for as long as these inputs are supplied. A system consisting of stored hydrogen in a cylinder tank and an FC system functions as a battery. In this chapter, we discuss batteries and FCs for use with solar power devices, in terms of current practice and future perspectives.

1 Secondary Batteries

Many current applications (e.g., cellular phones, bicycles, mobile computers, vehicles, and large-scale storage) use secondary batteries because they are rechargeable, low in total cost, and flexible, and also because they have a light environmental impact. Batteries can be roughly classified as either physical or chemical. Chemical batteries are further divided into primary (nonrechargeable) and secondary (rechargeable) batteries; we discuss secondary chemical batteries in this section. The electrochemical reactions in secondary batteries are reversible, and such batteries use different combinations of battery elements, such as sodium-sulfur, redox flow, lead-acid, nickel-cadmium, nickel metal hydride, and lithium ion.

T. Kajita · T. Itoh (✉)
Frontier Research Institute for Interdisciplinary Sciences (FRIS),
Tohoku University, 6-3 Aramaki Aoba, Sendai, Miyagi 980-8578, Japan
e-mail: itoh@fris.tohoku.ac.jp

M. Sugiyama et al. (eds.), *Solar to Chemical Energy Conversion*,
Lecture Notes in Energy 32, DOI 10.1007/978-3-319-25400-5_7

A charge-discharge reaction is a fundamental operating mechanism of secondary batteries, which make use of an oxidation-reduction potential based on differential ionization tendencies. A secondary battery consists of cathode and anode active materials, and an electrolyte. The active materials of the cathode (reduction) and anode (oxidation) undergo a redox reaction during battery discharge. The sum of these potentials is the standard battery voltage. Figure 1 shows the electrochemical reaction in a secondary battery.

During charging, the cathode active material (C in Fig. 1) is oxidized, thus producing electrons, and the anode active material (A in Fig. 1) is reduced, thus consuming electrons. In contrast, during discharge the cathode active material is reduced and the anode material is oxidized. These electron flows constitute the current in the external circuit. An electrolyte enables internal ion flow between the anode and cathode, facilitating electrochemical reaction on the surface of the electrodes.

In general, the performance requirements for a secondary battery are high potential (high voltage), high capacity, long charge-discharge cycle life, high output, low cost, and high safety. However, there is no secondary battery that satisfies all of these requirements. The criteria for high-efficacy performance differ by secondary battery type. Figure 2 shows energy density calculations (capacity multiplied by average voltage) for several types of secondary batteries. It is clear that Li ion batteries are the most useful but also the most expensive. In contrast, lead-acid batteries have a lower cost.

We next describe several types of secondary batteries.

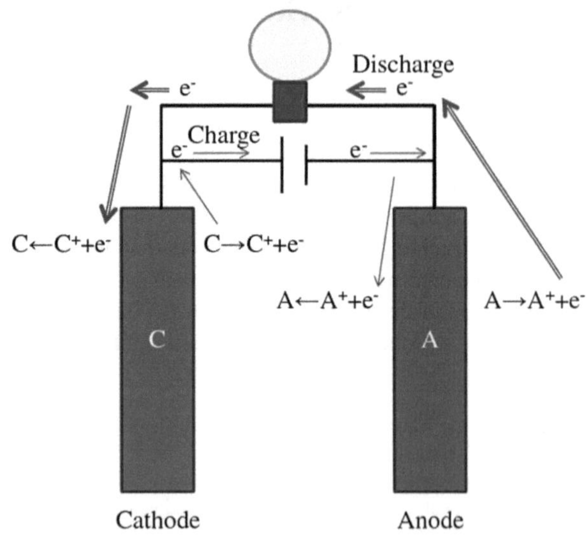

Fig. 1 Schematic of a secondary battery

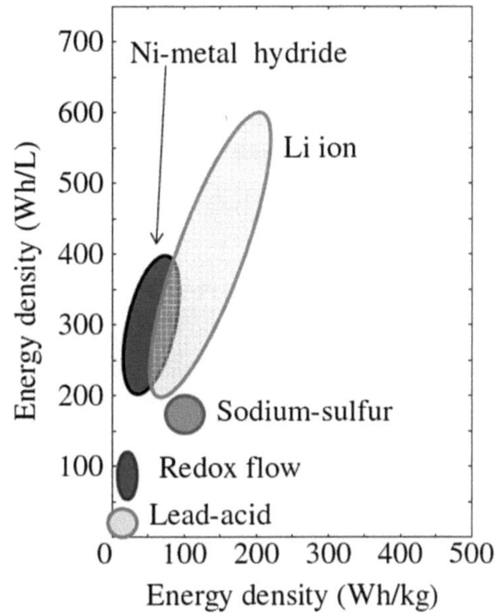

Fig. 2 Energy density in several types of secondary batteries

1.1 Lead-Acid Batteries

Most old types of secondary batteries are lead-acid batteries, which were invented by Gaston Plante (1834–1889) 150 years ago. Lead oxide, lead, and sulfuric acid are the cathode active material, anode active material, and electrolyte, respectively. The electrochemical reactions on each electrode are as follows (Fig. 3).

	Charge state		Discharge state	
Cathode	$PbO_2 + 4H^+ + SO_4^{2-} + 2e$	$\leftrightarrow$	$PbSO_4 + 2H_2O$	(1)
Anode	$Pb + SO_4^{2-}$	$\leftrightarrow$	$PbSO_4 + 2e^-$	(2)

During discharge, both the cathode and anode active materials are converted to lead sulfate ($PbSO_4$). The electrolyte loses its dissolved sulphuric acid and becomes mostly water. The reverse reaction occurs during charging. The battery voltage is 2 V, which is larger than the theoretical potential of water electrolysis.

Lead-acid batteries deteriorate over the course of the charge-discharge cycle. The lead sulfate generated during discharge on the anode readily crystallizes (termed sulfation) and is largely insoluble in the electrolyte, thereby deteriorating battery performance. Additionally, repeated charge-discharge cycles exfoliate lead oxide from the cathode, a situation that can be minimized by optimizing electrode shape. Hydrogen and oxygen generation on the anode and cathode due to water electrolysis via overcharging is a safety problem. However, lead-acid batteries are designed such that the gas is absorbed. This type is called a valve-regulated lead acid battery [1].

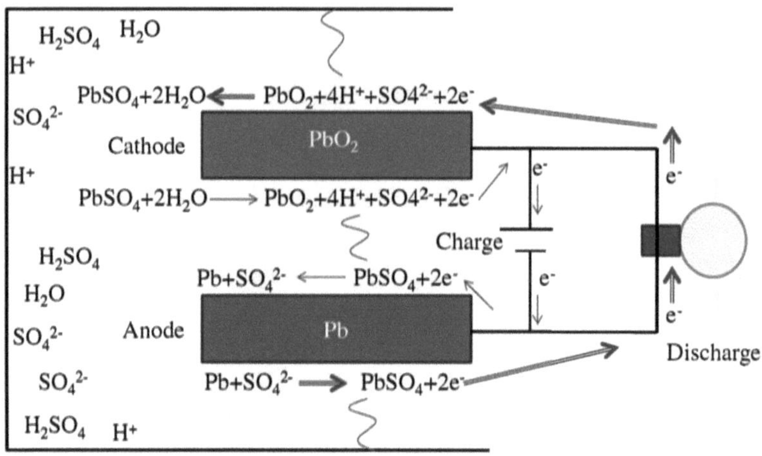

Fig. 3 Schematic of a lead-acid battery

In many applications, lead-acid batteries are used due to their low cost, long cycle life, and high safety, despite having a very low capacity (energy density) compared with other secondary batteries. Lead-acid batteries have recently come into use in large-scale energy storage and in power variation control for large-scale power generation, such as wind and solar.

We next discuss batteries that improve upon the capacity and cycle life of lead-acid batteries.

1.2 Sodium-Sulfur Batteries

Thirty years ago, Ford Motor Company developed sodium-sulfur batteries for electric vehicles. Sulfur and sodium are the cathode and anode active materials, respectively, and a beta alumina is the solid electrolyte. These batteries require a high operating temperature (300–350 °C) because of ionic conductivity in the solid electrolyte and the necessity of keeping the active materials in a molten state [2]. The reactions on the electrodes are as follows (Fig. 4).

$$\begin{array}{lll} & \text{Charge state} & \leftrightarrow \text{Discharge state} \\ \text{Cathode} & 5S + 2Na^{+} + 2e^{-} & \leftrightarrow Na_2S_5 \quad (3) \\ \text{Anode} & Na & \leftrightarrow Na^{+} + e^{-} \quad (4) \end{array}$$

During discharge, sulfur is reduced and becomes Na_2S_5 at the cathode; sodium is oxidized and becomes sodium ion at the anode. The sodium ions migrate toward the cathode through the electrolyte. Although further discharge is possible, solid-state

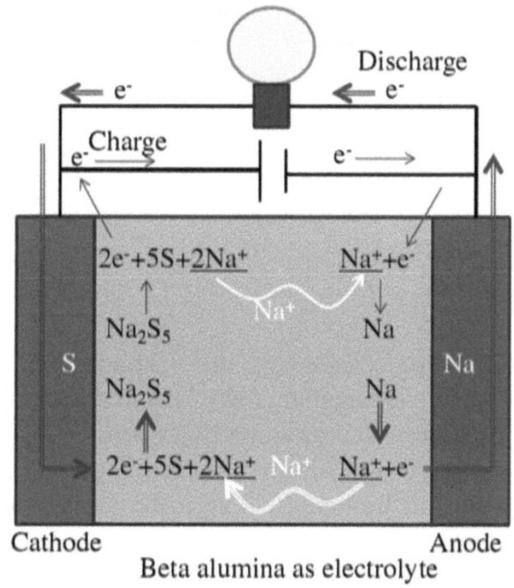

Fig. 4 Schematic of a sodium-sulfur battery

Na_2S_2 forms, which is indicative of poor discharge performance. During charging the reverse electrochemical reaction occurs at the electrodes. The battery voltage is 2 V.

Sodium-sulfur batteries must be handled with care because the active materials are combustible. If the battery burns, extinguishing the fire is difficult because water reacts with sodium. However, the battery can be manufactured at low cost because the component materials are common.

Sodium-sulfur batteries are used in large-scale energy storage due to their low cost, long cycle life, and high capacity, despite the necessity of a very high temperature. Their use is especially common in large urban areas because high energy density enables miniaturization. The high energy density and long cycle life of sodium-sulfur batteries have prompted their use in practical applications such as power load-leveling and power variation control in large-scale power generation. The practical application of sodium-sulfur batteries is expected to advance through the effective use of their high energy density, long cycle life, and low cost.

1.3 Redox-Flow Batteries

The National Aeronautics and Space Administration developed the redox-flow battery in the 1970s [3]. The active materials are in a liquid state and cause an oxidation-reduction reaction at the surface of electrochemically inactive electrodes in the battery. Elements that have multiple ionization states [such as iron (Fe^{3+}/Fe^{2+}), chromium (Cr^{3+}/Cr^{2+}), and vanadium (V^{5+} to V^{2+})] serve as active materials. A proton-exchange membrane separates the active materials. Storage tanks and pumps are needed in order to store and circulate the two liquid active materials from the tanks to the battery. The following discussion focuses on vanadium redox-flow batteries because of their current use in practical applications [4].

Vanadium oxide and vanadium are the cathode and anode active materials, respectively, dissolved in sulfuric acid. The reactions on the electrodes are as follows (Fig. 5).

	Charge state		Discharge state	
Cathode	$VO_2^+ + 2H^+ + e^-$	$\leftrightarrow$	$VO^{2+} + H_2O$	(5)
Anode	V^{2+}	$\leftrightarrow$	$V^{3+} + e^-$	(6)

During discharge, the cathode reduces pentavalent vanadium (VO_2^+) to tetravalent vanadium (VO^{2+}), and the anode oxidizes divalent vanadium to trivalent vanadium. The resulting hydrogen ion (H^+) produced at the cathode migrates to the anode through the proton-exchange membrane and maintains the electrical neutrality of the liquid active materials. During charging, the reverse electrochemical reaction occurs. The battery voltage is 1.4 V. Water electrolysis is problematic

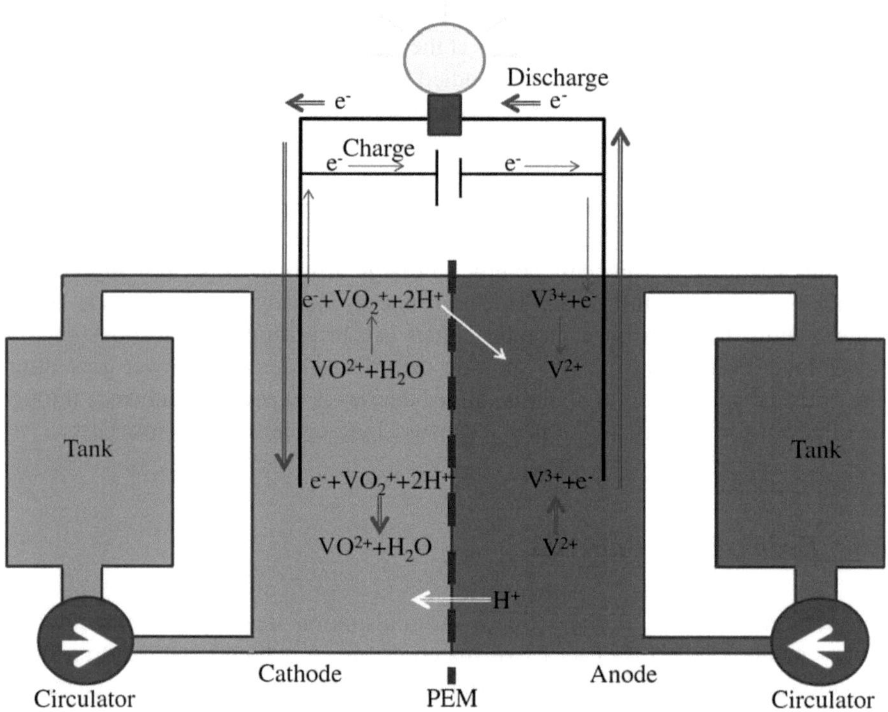

Fig. 5 Schematic of a redox-flow battery

during charging because the active materials are dissolved in acidic water. However, this problem can be addressed by making use of an overvoltage in the inactive electrode materials. For instance, carbon decreases the hydrogen production potential and increases the oxygen production potential. Additional problems such as the battery's low energy density and limitations in large-scale energy storage are due to the battery requiring a tank and circulator.

On the other hand, redox-flow batteries have a longer cycle life and lower cost. Furthermore, the battery has a flexible layout, in which the power-generating section (electrodes and proton-exchange membrane) and power-storing section (tank) are separate. This advantage is particularly effective for large-scale energy storage, which facilitates the use of redox-flow batteries in power-leveling and power variation control. Large-scale demonstrations will be necessary for realizing future applications of redox-flow batteries on a larger scale.

1.4 Ni-Metal Hydride Batteries

Battelle-Geneva Research Center invented the Ni-metal hydride battery in 1967. It uses a hydrogen-absorbing alloy and NiO(OH) as the anode and cathode active materials, respectively. The electrolyte is aqueous potassium hydroxide. The reaction on the electrode is as follows (Fig. 6). M indicates an intermetallic hydrogen-absorbing alloy. In most cases the compounds have the formula AB_5; A is a rare-earth mixture of lanthanum, cerium, neodymium, and praseodymium termed a mischmetal, and B is nickel, cobalt, manganese, and/or aluminum [5, 6].

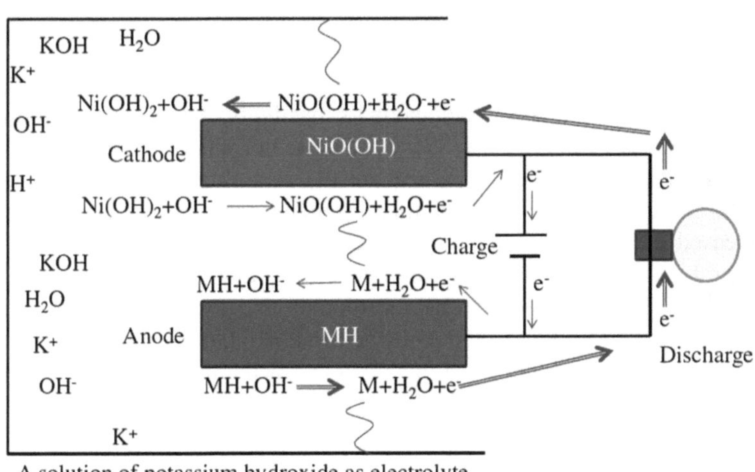

Fig. 6 Schematic of a Ni-metal hydride battery

	Charge state		Discharge state	
Cathode	$NiO(OH) + H_2O + e^-$	$\leftrightarrow$	$Ni(OH)_2 + OH^-$	(7)
Anode	$OH^- + MH$	$\leftrightarrow$	$H_2O + M + e^-$	(8)

During discharge, the cathode reduces NiO(OH) to $Ni(OH)_2$, and the anode oxidizes the hydrogen-absorbing alloy to hydrogen ion. During charging, the reverse electrochemical reaction occurs. The battery voltage is 1.2 V. The battery suppresses gas production during overcharge or overdischarge. The cathode produces oxygen under such conditions. Gas diffuses to the anode, where it reacts to form water, or the anode absorbs the hydrogen. However, hydrogen absorption is not sufficiently fast to prevent excessive hydrogen buildup if the discharge current is too large.

Ni-metal hydride batteries exhibit a well-known memory effect, which decreases their capacity. If the battery is repeatedly recharged after being only partially discharged, the maximum usable capacity in the battery decreases. Repeating the complete charge-discharge cycle several times negates the memory effect.

Ni-metal hydride batteries are used in many applications such as digital cameras, electric razors, and hybrid electric vehicles (HEVs) due to a long cycle life, light environmental load, low internal resistance, and high safety. More than 6 million HEVs equipped with a Ni-metal hydride battery are running worldwide. A demonstration test of a large-scale Ni-metal hydride battery was recently started, with applications in large-scale energy storage, power load-leveling, and power variation control.

1.5 Li Ion Batteries

In the 1970s, M. S. Whittingam proposed a Li ion battery that uses titanium sulfide and lithium metal as the cathode and anode, respectively. A layered transition metal oxide as the cathode active material [7], graphite as the anode active material [8], an ethylene carbonate as the electrolyte solvent [9], and aluminum foil as an electric collector in the cathode [10] are several subsequent innovations. Sony and Asahi Kasei released the first commercial Li ion battery in the 1990s.

In general, Li ion batteries use an organic solvent that has $LiPF_6$ dissolved in it. An organic solvent containing a chain carbonate and cyclic carbonate is almost always used in Li ion batteries, and some additive is included. The reactions at the electrode are as follows. Figure 7 shows the electrochemical reaction that uses the maximum amount of Li (x = 1).

	Charge state		Discharge state	
Cathode	$Li_{(1-x)}CoO_2 + (1 - x)Li^+ + (1 - x)e^-$	$\leftrightarrow$	$LiCoO_2$	(9)
Anode	Li_xC_6	$\leftrightarrow$	$C_6 + xLi^+ + xe^-$	(10)

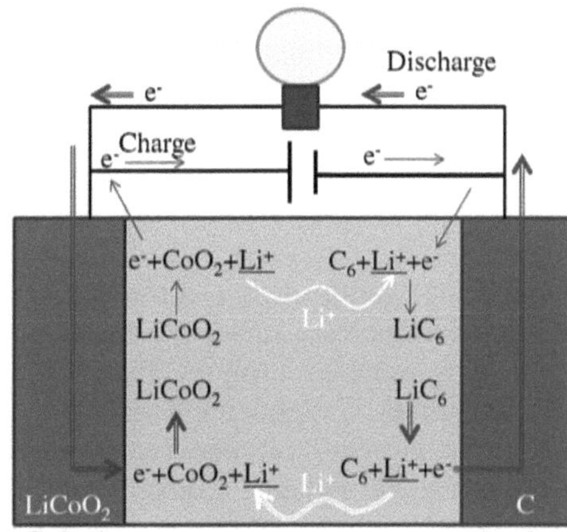

Fig. 7 Schematic of a Li ion battery

During discharge, the cathode reduces $Li_{(1\text{-}x)}CoO_2$, by using intercalating Li ion, to $LiCoO_2$, and the anode oxidizes the Li-intercalated graphite and deintercalates the Li ion. During charging, the reverse electrochemical reaction occurs. The battery voltage ranges from approximately 3–4 V.

During overcharge, the cathode heats up and the electrolyte decomposes, transforming the crystal structure of the cathode active material. This may cause a fire or explosion because the electrolyte is an organic solvent. During overcharge, gas production and Cu foil elusion at the anode destroys the battery's function. As a countermeasure, an electrical circuit in modern batteries prevents overcharge and overdischarge.

Li ion batteries are used in many applications, such as cellular phones, bicycles, mobile computers, HEVs, and electric vehicles, since such batteries have a high capacity (energy density) and long cycle life. The energy density is 2× that of Ni-metal hydride batteries. Improvements in safety and cycle life are needed for large-scale applications.

1.5.1 Future Perspectives for Batteries

Efforts to improve the energy density, safety, cycle life, and cost of secondary batteries are now in progress. An all-solid-state battery is attractive in terms of safety and cycle life. However, low ionic conductivity of the solid electrolyte and of the interface between the electrolyte and active material is a considerable obstacle for practical applications. Metal air secondary batteries that use metal and air (oxygen) as the anode and cathode, respectively, have five times the theoretical energy density of Li ion batteries. However, the basic composition of such an improved battery has been not established yet. Using Na, Mg, and Ca as charge carriers in secondary batteries is drawing attention because of the low material cost. The composition of batteries using these charge carriers is almost the same as that of a Li ion battery, and practical use in the near future is expected.

2 Fuel Cells

Around the world, energy consumption has grown during the last century and continues to increase rapidly to meet growing needs. A possible solution to this problem is water electrolysis. Figure 8 shows a schematic of water electrolysis pertinent to FCs. Upon application of a direct current voltage to the aqueous electrolyte, current flows between the electrodes, and hydrogen and oxygen evolve at the anode and cathode, respectively. The reaction in a FC is the reverse reaction. Supplying hydrogen and oxygen to the anode and cathode, respectively, results in current flow along with the production of water, the only product of an FC. This indicates that FCs are environmentally friendly power-generation systems.

Sir William Robert Grove (1811–1896) invented FCs. They are fundamentally different alternatives to combustion-driven energy-conversion systems in that they directly convert the chemical energy of fuel into electricity [11–15]. Figure 9 shows the principle of the anode and cathode reactions in FCs, which consist of an anode and cathode to which the fuel (hydrogen gas) and oxidant (oxygen gas) flow, respectively. There is an electrolyte between the anode and cathode. Electrochemical reactions occur at both electrodes. The overall reaction in an FC can be obtained by summing both electrode reactions.

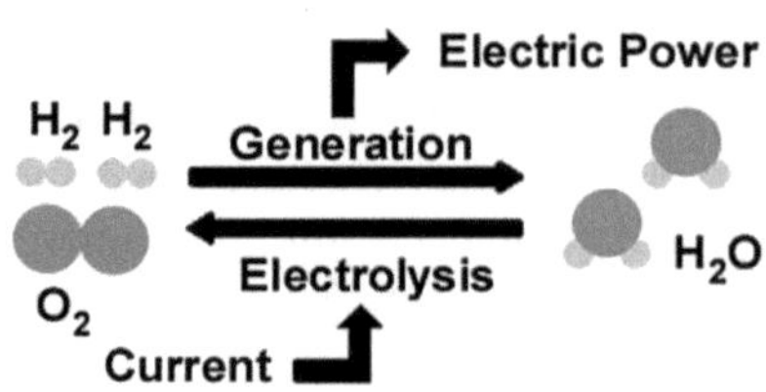

Fig. 8 Schematic for the relationships between water electrolysis and FC reactions

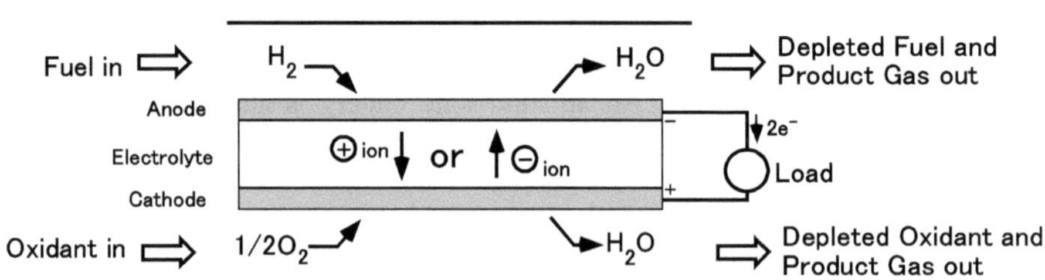

Fig. 9 Principles of the anode and cathode reactions in FCs

$$H_2 + 0.5O_2 \leftrightarrow H_2O \tag{11}$$

There are several types of FCs, which are usually classified by electrolyte and operating temperature (Table 1), as follows.

(a) Polymer electrolyte fuel cell (PEFC)
(b) Alkaline fuel cell (AFC)
(c) Phosphoric acid fuel cell (PAFC)
(d) Molten carbonate fuel cell (MCFC)
(e) Solid oxide fuel cell (SOFC)
(f) Direct methanol fuel cell (DMFC)

MCFCs and SOFCs are high-temperature FCs. PEFCs, AFCs, PAFCs, and DMFCs are low-temperature FCs. In general, low-temperature FCs (except DMFCs) use only H_2 as fuel. MCFCs and SOFCs can also use CO and CH_4 as fuel through internal conversion (12) and reforming (13) reactions, respectively.

$$CO + H_2O \rightarrow H_2 + CO_2 \tag{12}$$

$$CH_4 + 2H_2O \rightarrow 4H_2 + CO_2 \tag{13}$$

Low-temperature FCs are suitable for small-scale applications up to 100 kW (e.g., portable and mobile electric devices, automotive, and house electric power-generation systems). High-temperature FCs are most suitable for mid- and large-scale power-generation systems. The next section introduces various types of FCs in detail.

2.1 Polymer Electrolyte Fuel Cells

PEFCs use thin films of a polymer electrolyte membrane (PEM) as the electrolyte. PEFCs are sometimes also termed proton-exchange membrane fuel cells. Figure 10 shows the principle of the anode and cathode reactions in a PEFC. Hydrogen and oxygen are the anode fuel and cathode oxidant, respectively, and the cell's operating temperature is around 80 °C. PEFCs primarily consist of three elements, which are the PEM, bipolar plates, and catalyst. The future of PEFCs depends on development of these three elements.

The thickness of a PEM is approximately 0.05–0.1 mm. Since a PEM consists of a perfluorosulfonic acid polymer, PEFCs should operate at temperatures less than 80 °C. The PEM is sandwiched between the anode and cathode electrodes. Both the electrodes contain a platinum or platinum alloy catalyst on carbon, and both the catalyst and the carbon have a high specific surface area. The membrane electrode assembly comprises the electrodes and the PEM, and dictates cell performance. The theoretical output voltage is 1.23 V; however, the actual output voltage is

Table 1 Characteristics of various types of FCs

	(a) PEMFC	(b) AFC	(c) PAFC	(d) MCFC	(e) SOFC	(f) DMFC
Fuel	Hydrogen	Hydrogen	Hydrogen	Hydrogen Carbon monoxide (CO) Methane (CH_4)	Hydrogen CO CH_4	Methanol
Oxidant	Oxygen (air)	Oxygen	Oxygen (air)	Oxygen (air)	Oxygen (air)	Oxygen (air)
Electrolyte	Proton-exchange membrane	Potassium hydroxide (KOH)	Phosphoric acid (H_3PO_4)	Molten carbonate [$(Li/Na/K)_2CO_3$]	Zirconia (ZrO_2) doped with Y_2O_3	Proton-exchange membrane
Charge carrier	Proton (H^+)	Hydroxide ion (OH^-)	H^+	Carbonate ion (CO_3^{2-})	Oxide ion (O^{2-})	H^+
Catalyst	Platinum-related metal	Platinum-related metal	Platinum-related metal	None	None	Platinum-related metal
Anode reaction	$H_2 \rightarrow 2H^+ + 2e^-$	$H_2 + 2OH^- \rightarrow 2H_2O + 2e^-$	$H_2 \rightarrow 2H^+ + 2e^-$	$H_2 + CO_3^{2-} \rightarrow H_2O + CO_2 + 2e^-$	$H_2 + O^{2-} \rightarrow H_2O + 2e^-$	$CH_3OH + H_2O \rightarrow CO_2 + 6H^+ + 6e^-$
Cathode reaction	$0.5O_2 + 2H^+ + 2e^- \rightarrow H_2O$	$0.5O_2 + 2H_2O + 2e^- \rightarrow 2OH^-$	$0.5O_2 + 2H^+ + 2e^- \rightarrow H_2O$	$0.5O_2 + CO_2 + 2e^- \rightarrow CO_3^{2-}$	$0.5O_2 + 2e^- \rightarrow O^{2-}$	$1.5O_2 + 6H^+ + 6e^- \rightarrow 3H_2O$
Approx. operating temperature (°C)	30–100	50–260	150–220	600–700	500–1000	20–90
Waste heat utilization	Warm water	Warm water, steam	Warm water, steam	Steam turbine, gas turbine	Steam turbine, gas turbine	Warm water
Approx. efficiency (%)	30–40	45–50	40–45	50–65	55–70	30–40

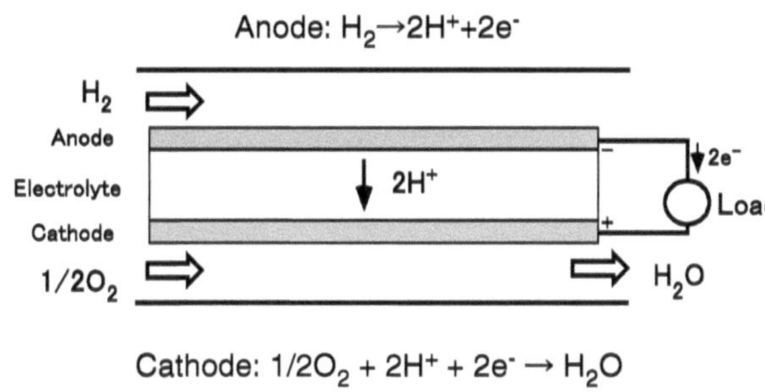

Fig. 10 Principles of the anode and cathode reactions in PEFCs

approximately 0.65–0.75 V because of the anode and cathode overvoltage, and the voltage loss of the cell elements.

Bipolar plates distribute the fuel and oxidant within the cell, separate the individual cells in the stack, carry current and water away from each cell, and keep the cells cool. Nonporous graphite, a variety of coated metals, and a number of composite materials have been suggested for use as bipolar plate materials. The most widely used bipolar plate material is composite materials that contain graphite, which has low corrosion resistance and high conductivity. However, graphite is brittle and lacks mechanical strength, and therefore it is difficult to reduce plate thickness.

Catalysts for PEFCs are also important because PEFCs operate at temperatures less than 100 °C. Since hydrogen does not react with oxygen at such temperatures, PEFCs require a catalyst to lower the activation energy. The catalyst is typically platinum (particles that are several nanometers in diameter with a large specific surface area) or a related metal. Hydrogen and oxygen molecules adsorb on the platinum surface, and dissociate into hydrogen and oxygen atoms at the adsorption site on each electrode.

PEFCs require high-purity hydrogen. At present, steam reforming is the most common industrial hydrogen production method. Ideally, hydrogen production would be solar powered, making it environmentally friendly.

2.2 *Alkaline Fuel Cells*

AFCs use an aqueous solution of potassium hydroxide as the electrolyte. Figure 11 shows the principle of the anode and cathode reactions in an AFC. The reactions at the anode and cathode are as follows.

$$\text{Anode} \quad H_2 + 2OH^- \rightarrow 2H_2O + 2e^- \tag{14}$$

$$\text{Cathode} \quad 0.5O_2 + H_2O + 2e^- \rightarrow 2OH^- \tag{15}$$

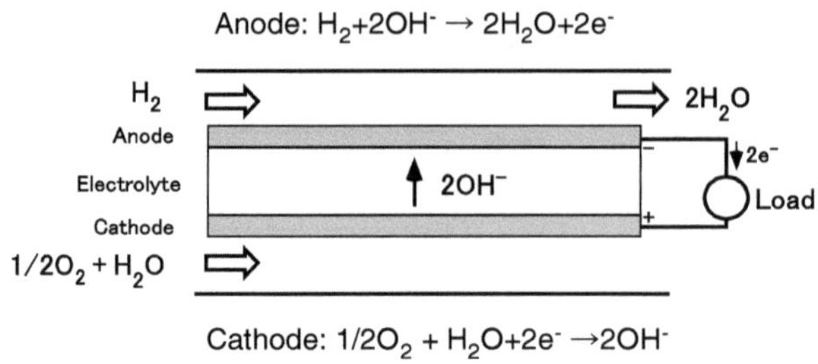

Fig. 11 Principle of the anode and cathode reactions in AFCs

$$\text{Total reaction} \quad H_2 + 0.5O_2 \rightarrow H_2O \tag{16}$$

AFCs can use nickel-related metal as electrodes. The operating temperature is approximately 50–260 °C. Regarding electrolyte durability, only pure hydrogen and oxygen can be used in AFCs because dissolved carbon dioxide impurities decrease cell performance, though development in this area is progressing. AFCs currently have only specialized applications in spacecraft, but use in hydrogen supply infrastructure has been proposed. Technological development may improve the prospects of AFCs.

2.3 Phosphoric Acid Fuel Cells

PAFCs include a phosphoric acid solution impregnated in a separator as the electrolyte. The anode and cathode reactions, as well as the total reactions, of PAFCs are as same as the reactions of PEFCs (Fig. 10), as follows.

$$\text{Anode} \quad H_2 \rightarrow 2H^+ + 2e^- \tag{17}$$

$$\text{Cathode} \quad 0.5O_2 + 2H^+ + 2e^- \rightarrow H_2O \tag{18}$$

$$\text{Total reaction}\, H_2 + 0.5O_2 \rightarrow H_2O \tag{19}$$

The operating temperature and power-generation efficiency are approximately 200 °C and 40 %, respectively. Analogously to PEFCs, the platinum catalyst may deteriorate if there is carbon monoxide in the fuel. Commercialization is progressing and PAFC power-generation plants for onsite 50–200 kW energy production are in development. Many PAFC plants continuously generate electric power for more than 40,000 h, which is the PAFC development lifetime target.

2.4 Molten Carbonate Fuel Cells

MCFCs use a mixture of alkali carbonate (Li_2CO_3, Na_2CO_3, and K_2CO_3) melts as electrolytes at high temperature. Figure 12 shows the principles of the anode and cathode reactions in MCFCs. The anode and cathode reactions are as follows.

$$\text{Anode}\quad H_2 + CO_3^{2-} \rightarrow H_2O + CO_2 + 2e^- \tag{20}$$

$$\text{Cathode}\quad 0.5O_2 + CO_2 + 2e^- \rightarrow CO_3^{2-} \tag{21}$$

$$\text{Total reaction}\; H_2 + 0.5O_2 \rightarrow H_2O \tag{22}$$

MCFCs operate at 600—700 °C. The carbonate mixture forms highly CO_3^{2-} conductive molten salt. The cathode reduces oxygen molecules in the presence of CO_2 to CO_3^{2-}. The carbonate ions modulate electrical conductance inside the cell to the anode. The anode oxidizes hydrogen gas and carbonate ions, producing CO_2 and H_2O as exhaust gas. CO_2 gas can be reused at the cathode. Summing both the anode and cathode reactions gives the overall reaction for MCFCs. Water is the only final product. Although FCs generally use hydrogen fuel, MCFCs can also use CO and CH_4 through internal conversion and internal reforming reactions, respectively.

At MCFC operating temperatures, an anode consisting of nickel alloy and aluminum (and/or chromate) as well as a nickel oxide cathode are sufficient for the necessary electrochemical reactions. The nickel alloy is a catalyst, and reactions (12) and (13) produce CO and CO_2. For this reason, MCFCs can use hydrogen fuel, even if the fuel contains CO from coal gasification gas. Within the cell, it is also possible to reform methane, which is the main component of natural gas. Methane reforming can also be done outside the cell; this is called external reforming and is suitable for large-scale plants. Compared with internal-reforming MCFCs, external-reforming MCFCs are simpler and lighter.

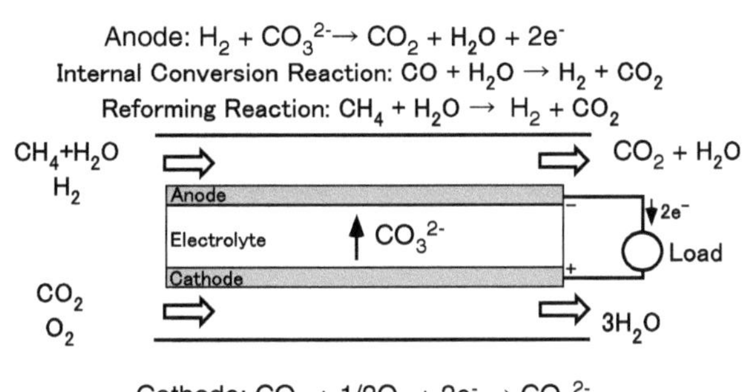

Fig. 12 Principle of the anode and cathode reactions in MCFCs

MCFC melt composition is a key issue pertinent to cell performance and lifetime. Molten carbonate melts upon electrolyte impregnation of the porous $LiAlO_2$, and melt composition of Li–Na-K carbonates depends on melting point, conductivity, and Ni solubility.

2.5 *Solid-Oxide Fuel Cell*

Like MCFCs, SOFCs do not require a platinum catalyst or reformer for internally producing hydrogen fuel. SOFCs also exhibit good power-generation efficiency. The anode and cathode reactions in SOFCs are as follows.

$$\text{Anode} \quad H_2 + O^{2-} \rightarrow H_2O + 2e^- \tag{23}$$

$$\text{Cathode} \quad 0.5O_2 + 2e^- \rightarrow O^{2-} \tag{24}$$

$$\text{Total reaction}\ H_2 + 0.5O_2 \rightarrow H_2O \tag{25}$$

Figure 13 shows the principle of the anode and cathode reactions in SOFCs. They typically operate at approximately 500–1000 °C. O^{2-} is the electrolytic charge carrier, and ZrO_2 (zilconia), doped with 8–10 mol% Y_2O_3 (yttria), is the electrolyte, termed yttria-stabilized zirconia (YSZ). A porous gas-diffusion structure (metallic YSZ and related materials) is the anode. The high porosity and large surface area of the electrodes facilitate rapid mass transport of the reactant and product gases. The cathode is also porous and can be made of a strontium-doped lanthanum manganite $(La_{0.84}Sr_{0.16})MnO_3$ and related materials, as well as lanthanum strontium ferrite and lanthanum strontium cobalite.

SOFCs can attain at least 96 % of their theoretical open-circuit voltage. In the absence of a major internal resistance drop they can operate at high current densities (1000 mA/cm^2), and they have a high fuel-conversion efficiency. Like MCFCs,

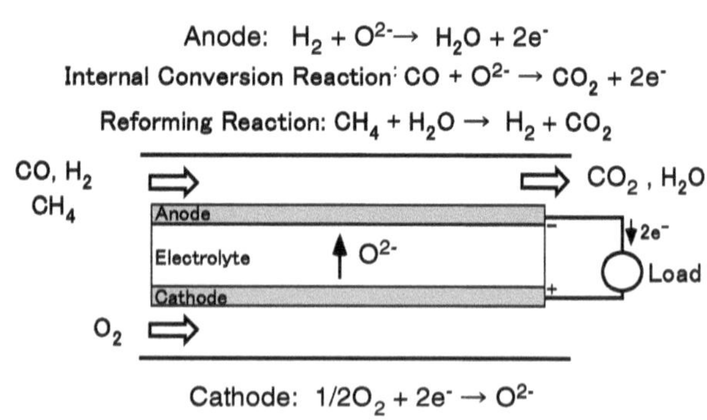

Fig. 13 Principle of the anode and cathode reactions in SOFCs

SOFCs require no CO_2 recycling, leading to further simplification. Their use at alternative thermal power stations, such as thermal electric power plants, is expected.

2.6 *Direct Methanol Fuel Cell*

DMFCs are similar to PEFCs. Instead of hydrogen gas (as in PEFC), DMFC uses liquid methanol directly. Figure 14 shows the principle of the anode and cathode reactions in DMFCs. The chemical reactions at the anode and cathode are as follows.

$$\text{Anode}\quad CH_3OH + H_2O \rightarrow CO_2 + 6H^+ + 6e^- \tag{26}$$

$$\text{Cathode}\quad (3/2)O_2 + 6H^+ + 6e^- \rightarrow 3H_2O \tag{27}$$

$$\text{Total reaction}\; CH_3OH + (3/2)O_2 \rightarrow CO_2 + 2H_2O \tag{28}$$

At the anode, methanol is converted to water, carbon dioxide, protons, and electrons. Oxygen reacts with the proton, and the final products are water and carbon dioxide. The actual reactions in DMFCs are complex. First, methanol adsorbs on the platinum catalyst, is oxidized, and releases a proton.

$$CH_3OH \rightarrow CH_2OH \rightarrow HCHO \rightarrow CHO \rightarrow CO \tag{29}$$

Each oxidation step on the platinum catalyst yields one proton (reaction 29). The final product is carbon monoxide, which poisons the platinum catalyst. Solving this issue will require development of a catalyst that is unaffected by CO (e.g., a platinum-related alloy).

DMFCs have a more serious problem. Methanol molecules permeate the PEM, which is called crossover. The membrane electrode assembly of DMFCs is generally permeable to water and methanol. Unreacted methanol migrates to the cathode and reacts with oxygen, yielding carbon dioxide and water. This process

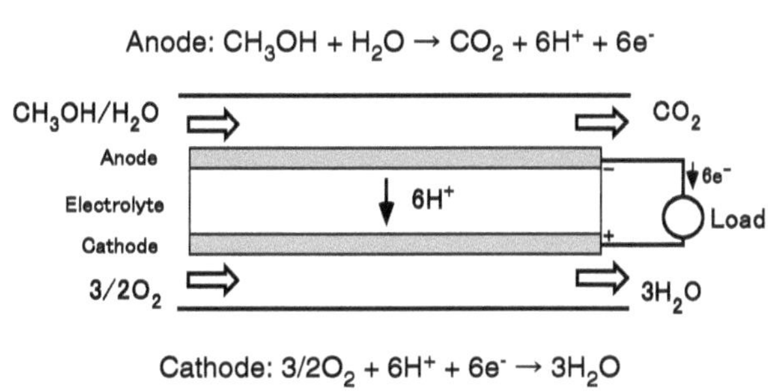

Fig. 14 Principle of the anode and cathode reactions in DMFCs

produces thermal energy not electrical power. One solution may be to decrease the methanol concentration, but doing so decreases the electrical power output. Another solution may be to use a different PEM to decrease methanol permeability.

DMFCs should be useful for mobile electronic devices, laptop computers, and cell phones. These devices require an easy method for methanol refill via a small cartridge. Investigations on using DMFCs in power-assisted bicycles and cars are also progressing.

2.6.1 Future Perspectives for Fuel Cells

Researchers are currently working to improve fuel cells in terms of their energy density, cycle life, and cost. Moreover, many governments are promoting the commercialization of stationary FCs and FC-based electric vehicles. The development of electrolytes, bipolar plates, and catalysts is important for the spread of FCs in the future; also important is new technology for hydrogen generation, transport, and supply. Further development of FCs will be required in order to get a clearer picture of their promise for solving today's energy challenges.

References

1. Misra Sudhan S (2007) J Power Sources 168:40–48
2. Oshima T, Kajita M, Okuno A (2004) Int J Appl Ceram Technol 1(3):269–276
3. Thaller LH (1974) Proceedings of 9th inter-society energy conversion engineering conference. American Society of Mechanical Engineers 924
4. Rychcik M, Skyllas-Kazacos M (1988) J Power Sources 22:59–76
5. Markin TL, Dell RM (1981) J Electroanal Chem 118:217–228
6. Iwakuma C, Kajiya Y, Yoneyama H, Sakai T, Oguro K, Ishikawa H (1989) J Electrochem Soc 136:1351–1355
7. Mizushima K, Jones PC, Wiseman PJ, Goodenough JB (1980) Mater Res Bull 15:783–789
8. Yazami R, Touzain Ph (1983) J Power Sources 9:365–371
9. Fong R, Sacken U, Dahn JR (1990) J Electroanal Chem 137:2009–2013
10. Yoshino A, Sanechika K (1997) Nonaqueous secondary battery. Japan Patent 2128922, B1, 02 May 1997
11. O'Hayre RP, Cha S-W, Colella W, Prinz FB (2006) Fuel cell fundamentals. Wiley, New York
12. Larminie J, Dicks A (2003) Fuel cell systems explained, 2nd edn. Wiley, New York (2003)
13. Carrette L, Friedrich KA, Stimming U (2000) ChemPhys Chem 1:162
14. Kordesch K, Simader G (1996) Fuel cells and their applications. VCH Publishers, New York
15. Acres GJK (2001) J Power Sources 100:60

Energy Storage in C–C, H–H and C–H Bond

Masayuki Otake

Abstract Solar energy is most commonly converted into electricity by employing photovoltaic or solar thermal generation systems. Various technologies for direct storage of electricity have been proposed, although, with the exception of batteries and capacitors, systems based on these technologies are minimally effective for transportation applications. Energy storage in chemical substances is considered much more effective for transportation and long-term storage, especially as fuels for engines, turbines and boilers. In this chapter, chemical energy conversion/storage systems based on C–C, H–H and C–H bonds will be discussed. Advanced catalyst, electrode and process technologies play a key role in these systems.

1 Introduction of Energy Storage in Chemical Compounds

The density of the solar energy flux reaching the surface of the earth is insufficient for direct heating of boilers for power generation. On this account, photovoltaic and other primary solar energy conversion tools have been developed to generate electricity. Electricity can be transported via transmission lines to areas of demand, but it cannot readily be stored, nor transported over very long distances due to the low cost efficiency and significant energy loss. Energy storage in chemical compounds relies on the conversion of energy, especially electricity or heat, into chemicals with high energy densities such as fuels. Hydrogen possesses the highest energy density (highest heat of combustion per unit mass), followed by methane. The properties of these compounds make them suitable for transportation in vessels,

M. Otake (✉)
Mitsubishi Chemical Techno-Research Corp, 16-1 Samoncho,
Shinjuku-ku, Tokyo 160-0017, Japan
e-mail: otake.masayuki@ms.mctr.co.jp

M. Sugiyama et al. (eds.), *Solar to Chemical Energy Conversion*,
Lecture Notes in Energy 32, DOI 10.1007/978-3-319-25400-5_8

and they can typically be produced by adding energy to less energy-rich compounds such as water or CO_2 through chemical, electrochemical, photo electrochemical or photochemical processes. In nature, photosynthetic conversion of H_2O and CO_2 into carbohydrates is the most environment-friendly and efficient chemical reaction, and the resulting products have played a key role in human life. Many researchers have been trying to develop artificial means of photosynthesis, although only a limited conversion of H_2O into H_2 and 1/2 O_2 (Light reaction of photosynthesis) has been reported [1]. A wide variety of technologies is currently available, and there is no doubt that several potential methods of solar energy storage in chemicals are adoptable, especially in C–C, H–H and C–H chemistries.

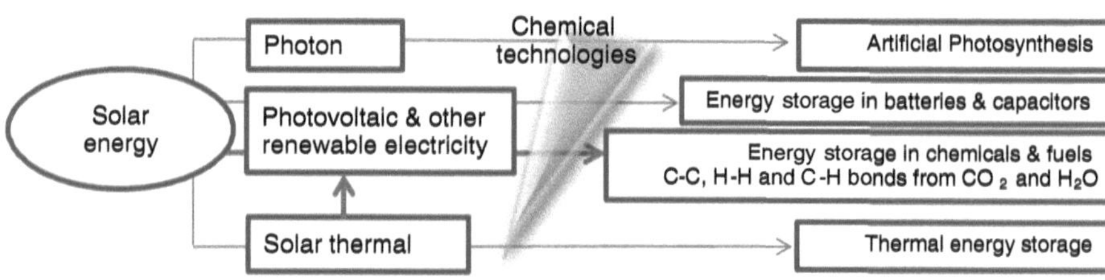

2 Solar Energy Storage by Chemical Means

Various efficient means for storage of electricity have been developed, including mechanical, chemical and electrical storage methods. Solar photon energy can be used not only in photovoltaic systems but also in natural photosynthesis for agricultural, fishery, forestry and horticultural products, as well as photocatalytic processes. Several compounds are known to be reversibly transformed into their structural isomers, and in doing so storing photon energy as heat [2]. Concentrated solar thermal energy has been used for power generation (CSP). 400 MMW and higher electricity CSP facilities have been in operation in US, while application in thermochemical H_2 processes is currently under investigation. Low-level concentrations of solar thermal energy can be used for heating water, drying or regenerating desiccants for air conditioning. Solar energy, however, can be more efficiently stored in C–C, H–H and C–H chemical bonds of fuels, since the energy density of these products are much higher than batteries and other energy storage means. Furthermore, fuels are easy to transport and store, and they boast high energy densities, meaning their relative quantities are smaller.

3 Solar Energy Storage via Chemical Processes

Many chemical processes have been developed or studied for the solar energy conversion and storage. Table 1 classifies and summarizes technologies for solar energy conversion applicable for H_2O and CO_2.

Figure 1 exhibits a classification of nearly 500 papers that appeared in Chemical Abstracts in the 20 years between 1994 and 2014. This figure indicates the predominance of photochemical[1A, 1B], photo electrochemical [2] and solar thermal

Table 1 Classification of solar energy to chemicals conversion technologies

Class No	Classification	Definition	Explanation in more detail
1A	PC-semi	Photochemical H_2O conversion with semiconductor catalysts	Water splitting by semiconductor photocatalysts, Z-scheme, sensitizer
1B	PC-comp	Photochemical H_2O conversion with organometallic catalysts	Water splitting by organometallic photocatalysts, often with dyes and sacrifices
2	PEC	Photo electrochemical H_2O conversion	Water photo electrochemical dissociation with or without external bias potential to the semiconductor
2A	PEC-td	Tandem PEC with monolithic PV system	Smart and energy-saving electrochemical process
3	EC	Electrochemical H_2O conversion in liquid-phase	Water electrolysis with electricity form external, solar or renewable, energy sources
3A	EC-pem	Electrochemical H_2O conversion PEM electrolyte	PEM-type Water electrolysis with electricity from external energy sources
3B	EC-soe	Electrochemical H_2O conversion Solid oxide electrolyte	Solid-electrolyte based water electrolysis with electricity from external energy sources
3C	EC-bio	Microbial Fuel Cell with electro-generating microorganisms	*Shewanella Ionica* (K Hashimoto, Univ Tokyo), *Geobacter metallireducens* (Bruce E Logan, Pennsylvania State Univ)
4	TC	Solar thermal, thermochemical, STEP (solar thermal electrochemical photo) H_2O conversion	Thermochemical water splitting, direct thermal splitting, thermochemical, reforming with solar heat
5	PS	Biological photosynthesis	H_2-generating microorganisms
6	Biomass	Thermochemical or direct biological biomass conversion to H_2	Biomass conversion to syngas, Hydrogen

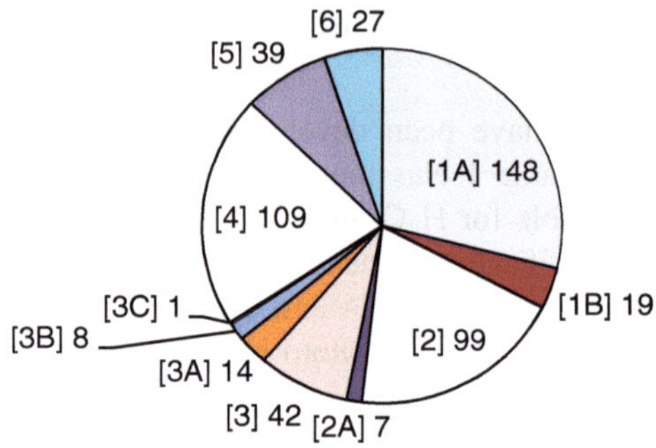

Fig. 1 Numerical distributions of papers based on technical classification in Table 1

[3] H_2O splitting versus other processes. Increases in the annual numbers of papers for each classification [1A], [2, 3] was found during years between 2008 and 2013.

Storage of energy using CO_2 and/or H_2O, chemicals with the lowest energy contents, corresponds to increase in enthalpy of formation of the products. In Fig. 2, the increases in enthalpies of formation per C or H_2 of several compounds with C–C, H–H and C–H bonds are shown. The increments in enthalpy values of hydrocarbon products are larger than oxygen-rich carbohydrates, which are the product of photosynthesis by plants in nature.

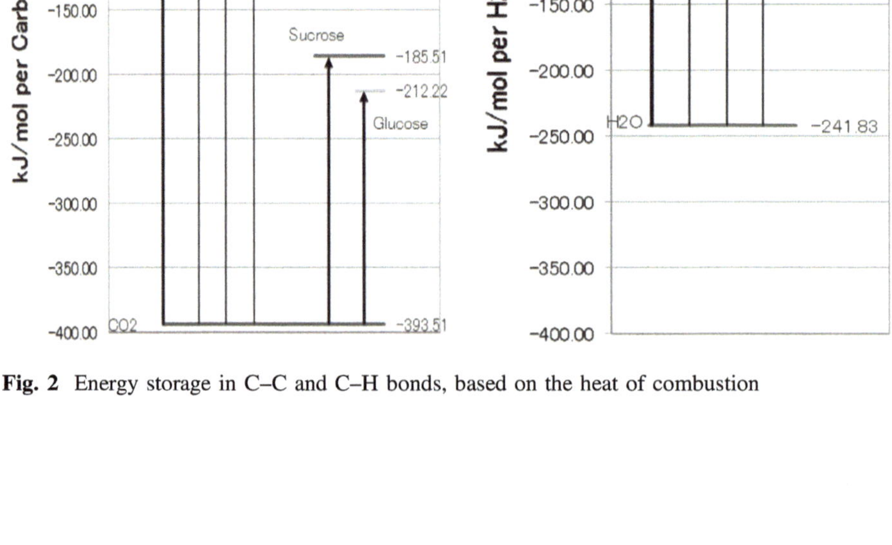

Fig. 2 Energy storage in C–C and C–H bonds, based on the heat of combustion

3.1 C–C Bond Formation as Energy Storage Means

In the photosynthesis by floras, CO_2 and H_2O are converted into carbohydrates through solar photon utilization via the Calvin cycle. Triglyceride fats are produced by oil-producing plants via the TCA cycle. Abiotic and important chemical processes of C–C bond formation for energy storage include C1 chemistry that has been studied extensively in the 1980s. Typical industrial chemical processes [4] include syngas (CO/H_2) conversion via the Fischer-Tropsch reaction to hydrocarbons rich in lower olefins (Fe catalyst) or higher paraffin (Co catalyst) and a combination of syngas conversion to methanol or dimethyl ether and their subsequent conversion to olefins (often referred to as MTO, MTP and DMTP reactions) or gasoline (often referred to as MTG and MTA reactions). In the latter case, syngas is produced by indirect and catalytic reduction of CO_2 with solar energy-based H_2, or its direct electrochemical reduction. Another, but less efficient C–C bond formation mechanism is methane coupling. This C1 compound is in thermodynamic equilibrium with ethylene, acetylene, other C2+ hydrocarbon radicals, H_2 and carbon at temperatures higher than 1000 K. Catalytic oxidative coupling of methane to ethylene was studied extensively in the 1980s, while oxidative or plasma coupling to acetylene is one of the key processes for manufacturing acetylene. There is little information indicating use of solar energy for this process.

C–C bond formation from CO_2 results in typical hydrocarbonaceous products, and their production from CO_2 [C^{4+}] is an efficient process for energy storage. Preliminary or simultaneous multi-electron conversion to such intermediates as CO [C^{2+}], HCOOH[C^{2+}], H_2CO[C^0] and CH_3OH[C^{2-}] is usually required before CO_2 can be effective in C–C [apparently C^{+1}, C^0, C^{-1}, C^{-2} or C^{-3}, assuming H^+ and O^{2-} valence] bond formation. In these CO_2 reductions, chemical, electrochemical, photo electrochemical and microbial means are used, as is summarized in Table 2 and Fig. 3. Matthias Beller [3] reported carbonylation of alkenes with CO_2/H_2. Usually

Table 2 Processes for CO_2 conversion to chemicals with C–C bonds

Hydrocarbons synthesis	Chemical conversion of photosynthetic products or biomass, Fischer-Tropsch conversion of solar syngas (via CO_2 + H_2 RWGS)
Photosynthetic production	CO_2/H_2O + solar irradiation to carbohydrates, lipids, phytols, and other chemicals
Carbohydrates to oxygenates conversion	Ethanol or other oxygenates from solar CO/H_2 via catalytic or biological conversion
Commodity/ fine chemicals from CO_2	Alkene carbonylation with CO_2/H_2, CO_2 fixation or syngas fermentation, Fine chemicals via photochemical reactions Floras and micro-algal photosynthesis, photo-bioreactor Electrochemical CO_2 reduction (Oxalic acid, ethyleneglycol, etc) Electrochemical bioreactor
Oxidative or thermal coupling of lower hydrocarbons	Oxidative coupling of methane to ethylene, Plasma coupling of methane to acetylene, Catalytic conversion of methane to benzene, with CO_2 as a H_2 acceptor

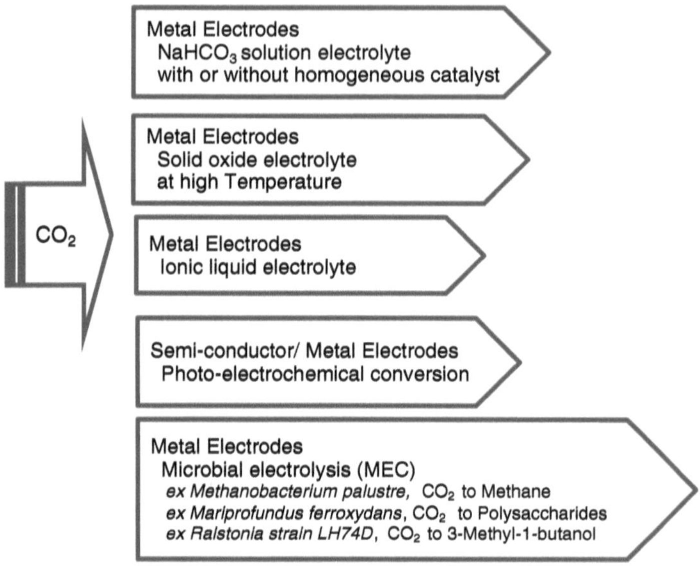

Fig. 3 Various electrochemical CO_2 conversions into organic C–C and C–H bonds

chemical CO_2 reductions use H_2 or CH_4 as economic electron sources. In electrochemical CO_2 reduction, electrolytic reduction in the aqueous or vapor phase was studied over specified electrodes and electrochemical catalyst. As researchers of Brookhaven National Laboratory reported, in imidazolium-based ionic liquids and with $Re(bpy)_2$ catalyst, electrolytic CO_2 reduction proceeded at a substantially low voltage (0.45 V) and high efficiency [5]. Only recently, US-based Liquid Light Inc patented electrochemical reduction of CO_2 to produce oxalic acid, an intermediate to ethylene glycol [6]. In the photochemical or photo electrochemical CO_2 reduction, semiconductor electrodes dipped in the aqueous phase are illuminated, while CO_2 in the aqueous phase is reduced. Product patterns depend strongly on the electrode and photocatalyst, as is shown in Fig. 4 and Table 3.

3.2 H–H Bond Formation as Energy Storage Means

H_2 is the sole compound that possesses an H–H bond. Many routes have already been studied and used for H_2 production. Tremendous amount of H_2, globally more than 40 million tons per year, is produced in industry via steam or partial oxidative reforming of natural gas, petroleum hydrocarbons, coal and other renewable carbonaceous materials. Carbons in the reactant are used to reduce H_2O into H_2, while they are oxidized into CO and CO_2. CO in the syngas is further converted into H_2 via Water Gas Shift (WGS) reaction: the CO-rich gaseous composition is shifted

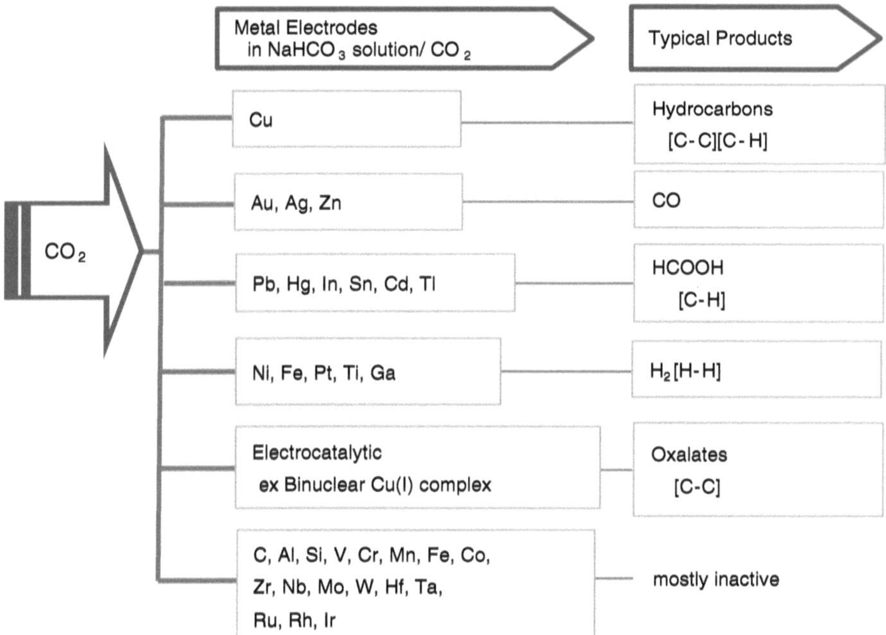

Fig. 4 Electrodes and electrochemical reduction of CO_2 in alkaline aqueous solution

into H_2-rich composition by the following equilibrium with the reverse WGS (RWGS) reaction,

$$CO + H_2O \rightleftarrows CO_2 + H_2$$

The reforming generates CO_2 byproduct, which is increasingly an important source of industrial CO_2, since efficient separation/recovery before atmospheric release is possible. Use of CO_2 sequestration from industrial H_2 plants(CCS) for EOR (enhanced oil recovery) is also increasing in the US. Catalytic or light-driven alkane dehydrogenation to olefin is also reported as an alternative chemical source for H_2.

In Fig. 5, electrochemical routes for H_2O splitting are summarized. Hydro- or geo-powered and electrolysis-based H_2 for ammonia or methanol synthesis is produced in several countries. In the early 1990s, the Euro-Quebec Hydro-Hydrogen project was proposed. A typical commercial electrolyzer for alkaline water is operated at 70–90 °C, up to 3 MPa pressure, 20 A/dm^2 current density at 1.86 V/cell [7]. Improvements in the electrolysis technology have been made in cells with PEM and solid oxide electrolyte membranes, and several national research projects for technology improvements have been executed, mainly in Japan and the US. Several nano-structured electrodes with lower H_2 and O_2 overpotentials have been proposed [8–10]. Assuming massive production of H_2 in

Table 3 Semiconductor catalysts for the photochemical CO_2 reduction

Category	Best system	Co-catalyst	Major product	Rmax	Minor products	Traces
TiO_2-based	TiO_2 (P-25) particles in aq suspension	–	CH_3OH	3.4	CH_4	
	TiO_2 (P-25) particles in Isopropanol suspension	–	HCOOH	1.2		
	TiO_2 anatase, 14 nm crystalline particles	–	CH_4	0.4	CH_3OH	H_2, CO
	Ag/TiO_2 particles	Ag (5.2wt%)	CH_4 + CH_3OH	10.5		
	CuO/TiO_2(P-25) in aq suspension	Cu (2.0wt%)	CH_3OH	443.0		
	Ti-SBA-15	–	CH_4	106.0	CH_3OH	C_2H_4, C_2H_6, O_2
	Glass wool supported sensitized TiO_2-SiO_2	Cu (0.5wt%)	CO	60.0	CH_4	
	Optical fiber coated with sensitized TiO_2	Cu-Fe (0.5wt%)	CH_4, C_2H_4	17.5		
ZnS	ZnS nanoparticles	Cd	HCOOH	75.1	H_2, CO	
	ZnS nanoparticles in silica		HCOOH	7000.0		
CdS	CdS nanoparticles		CO	8.4	H_2	
CuAgInZnS	Mixed sulfide nanoparticles	RuO_2	CH_3OH	34.3		
WO_3	Monoclinic crystals		CH_4	0.3		
	$W_{18}O_{49}$ nanowire		CH_4	666.0		
N-Ta_2O_5	N-Ta_2O_5 nanoparticles	Ru-dcbpy	H_2CO	7700.0	H_2CO	
p-InP	p-InP photocathode	Ru-complex	HCOOH	140.0		

Rmax: maximum formation rate reported for the major product(s), in $\mu molg^{-1}h^{-1}$, see Jacek K Stolarczyk et al. Angew Chem Int Ed, 2013, 52, 7372

the future, tremendous volume of water is required along with the sufficient solar energy. A floating system for renewable electricity generation and sea water electrolysis would be the most practical facility. A chlorine-free electrolytic H_2 and O_2 production system for sea water has been developed in Japan [11].

Several microbes and algae are known to produce H_2 via photobiological water splitting and enzymatic, fermentative or bio-assisted electrolytic pathways.

In artificial photosynthesis, H_2 and O_2 are produced by photochemical or photo electrochemical water splitting.

More than 100 thermochemical H_2O splitting routes to H_2 and O_2 have been proposed after the studies in 1960s. Several of them are still under development today. Thus there are several practical ways for producing H_2 from renewable energy, and all of these can be effective in solar energy storage, as is shown in Table 4.

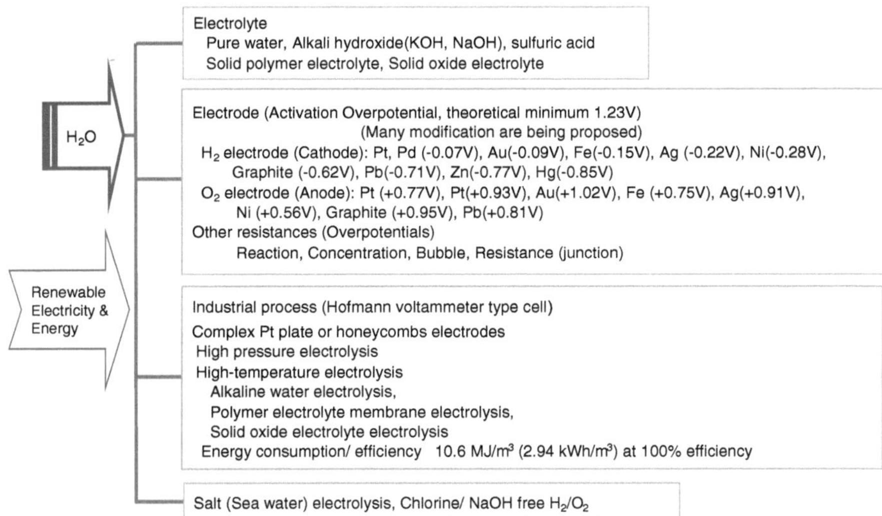

Fig. 5 Electrochemical routes for H_2O splitting

Table 4 Solar energy-based H–H bond formation processes

Hydrogen	Solar-to-hydrogen technologies; H_2O splitting to H_2 and 1/2 O_2, photochemical, electrochemical, photo electrochemical and thermochemical H_2O splitting
	Steam or POX reforming of renewable organic chemicals like methane and biomass, solar thermal conversion
	Photobiological H_2O splitting, biohydrogen via biocatalyzed electrolysis, fermentative or enzymatic H_2 production

3.3 *C–H Bond Formation as Energy Storage Means*

C–H bonds are common to most of organic compounds. The energy storage chemistry in C–H bonds is nearly in parallel to that for organic C–C bonds described in 3.1, since the conversions of inorganic CO_2 to C–C proceed along with its catalytic, electrochemical or photochemical reduction. C–H bond formation for energy storage is also possible in the conversion of H_2-deficient to H_2-rich compounds. Hydroconversion of heavier to lighter hydrocarbons (scission of C–C bond to two $-CH_3$, or hydrocracking) is a typical C–H bond forming reaction in petroleum refining. Hydrodeoxygenation and hydrogenative elimination of N, S and halogen atoms from the starting organic compounds to form C–H bonds are additional examples of an H_2-enrichment chemical process.

Carbohydrates with C–C and C–H bonds are produced in the photosynthesis, while mostly C1 compounds with C–H bond are produced in the photochemical or photo electrochemical reductions of CO_2, as is seen in Fig. 4 and Table 3.

4 Prospects of the Chemical Energy Storage

Although a variety of solar energy storage technologies have been developed, photon, electric and thermal energies are storable in chemicals with C–C, H–H and C–H bonds. In the chemical storage, products bear energy contents higher than CO_2 and H_2O. Hydrogenative conversions of CO_2 to C–C and/or C–H compounds are, however, exothermic. H_2O splitting to H_2 and O_2 is an endothermic process, and hence is the key issue for the chemical energy storage processes. Photochemical,

Table 5 Application of solar or renewable energy storage in chemical bonds

Type	Primary solar energy conversion	Secondary storage process	R&D or industrial status of solar energy storage
[C–C]	CO_2 to syngas via RWGS	Syngas to Fischer-Tropsch or methanol synthesis	No solar H_2 available; Commercial GTL from natural gas but with no solar energy use
	Electrochemical CO_2 conversion	CO_2 to oxalate	Several successful results in labs
	Photocatalytic CO_2 conversion	CO_2 to ethylene	Several successful results in labs
	Photo electrochemical CO_2 conversion		Several successful results in labs
	Solar thermal conversion of biomass	Syngas to Fischer-Tropsch or methanol synthesis	No solar energy-based process
[H–H]	Photocatalytic, Electrochemical, Photo electrochemical & Solar thermal H_2O splitting	Compression of H_2 gas Liquified H_2 H_2 adsorbent Reversible chemical H_2 fixation	H_2 via Hydro- and geo-powered electrolysis of H_2O for commercial NH_3 and methanol production, H_2 storage in high-pressure transportable vessel for FCV (70 MPa) High-pressure H_2 pipeline (currently 21–22 MPa) Various adsorbents, metal hydrides, inorganic hydrides Organic hydride (Methylcyclohexane)
[C–H]	CO_2 to syngas via RWGS	Syngas to Fischer-Tropsch or methanol synthesis	H_2 via Hydro- and geo-powered electrolysis of H_2O for commercial methanol production, Commercial GTL and mega-methanol from natural gas with no solar energy use
	CO_2 conversions as in [C–C]	CO_2 to Formate, methanol, methane	Several successful results in labs

electrochemical and photo electrochemical conversion of CO_2 for energy storage purposes are not mature at present. In Table 5 are summarized the applications of solar or renewable energy in the formation of chemical bonds.

The viability of artificial solar energy storage in chemical substances depends on its competitiveness versus the photosynthetic process in nature and electrical energy storage tools. Various technologies can be applied in the artificial processes; e.g., solar energy concentration to allow compact facilities for energy conversion, various catalysts and promoters already available, and the storage and transportation of primary solar energy to energy-consuming chemical plants. Strategies of chemical energy storage using CO_2 and water as starting materials are increasing in importance. CO_2 is a GHG gas, but can be recovered from chemical plants and boiler tail gas. The US Navy is reportedly [12] studying sea water as a CO_2 and H_2O source, producing fuel oil via Fischer-Tropsch conversion of syngas (H_2/CO). The concentration of CO_2 in sea water is much higher than in pure water, while anodic chlorine formation during the electrochemical H_2 production can be eliminated by optimizing electrodes and electrolytic conditions.

The key chemical for energy storage through C–C, H–H and C–H bonds is hydrogen which can be obtained via splitting of H_2O into H_2 and O_2. If the low cost and abundant renewable H_2 is available, energy storage in C–C and C–H chemical bonds becomes very attractive, and could be economically viable. H_2 is the key intermediate for the chemical energy storage process [13, 14].

References

1. Maeda K, Domen K (2012) In: Annual survey of catalytic science and technologies, 20th anniversary. The Catalysis Society of Japan, p 60
2. Scharf H-D (1979) Angew Chem Int Ed 1979, 18, 652 and 1994, 33 2009
3. Lipang W, Liu Q, Jackstell R, Beller M (2014) Angew Chem Int Ed 53:6310–6320
4. Papers in studies in surface science and catalysis 107 (1997), 119 (1998), 136 (2001) and 147 (2004). In: Natural gas conversion series IV, V, VI and VII. Elsevier
5. Chemical & Engineering News (2014) 26 May 2014, p 41
6. WO2014/100828, WO2014/042783 (Liquid Light Inc)
7. Smolinka T, Garche J, Hebling C, Ehret O (2012) In: Symposium e-water electrolysis and hydrogen as part of the future renewable energy system, Copenhagen/Denmark, 10 May 2012
8. Kuai L, Geng J, Chen C, Kan E, Liu Y, Wang Q, Geng B (2014) Angew Chem Int Ed 53:7547–7551
9. Cheng L, Huang W, Gong Q, Liu C, Li Y, Dai H (2014) Angew Chem Int Ed 53:7860–7863
10. Liu Q, Tian J, Cui W, Jiang P, Cheng N, Asiri AM, Sun X (2014) Angew Chem Int Ed 53:6710–6714
11. Hashimoto K (2000) Hydrogen Energy System HESS, JP4752287, JP5359133 Daiki Ataka Engineering Co, now Hitachi Zosen Corp, 25(1), p 55
12. Willauer HD, DiMascio F, Hardy DR, Williams FW (2014) Ind Eng Chem Res 53:12192–12200
13. Schlogl R (2011) Angew Chem Int Ed 50:6424–6426
14. US DOE (2013) Hydrogen Production Technical Team Roadmap, June 2013

Part IV
Chemical, Electrochemical and Photoelectrochemical Approach for Energy Conversion: Approach Using Chemical Reactions

Thermochemical Water Splitting by Concentrated Solar Power

Hiroki Miyaoka

Abstract Thermochemical water splitting (WS) is an attractive conversion technique of solar energy to hydrogen, which is a type of energy storage and transportation media. The thermochemical WS cycle is composed of several kinds of chemical reactions, and thus they are basically classified by the reaction process and usable solar concentration systems. The cycles operated at more than 1000 °C have a simple reaction process and high theoretical conversion efficiency although large scale solar heat systems with high solar concentration are required. In the case of WS with lower operating temperature than 1000 °C, the reaction process become complicated, and the efficiency is theoretically low. However, various kinds of heat sources can be utilized to operate them. This section discusses the basic properties of typical thermochemical WS cycles and suitable solar heat systems. Especially, alkali metal redox cycles reported in recent years are introduced in detail as a novel hydrogen production technique.

1 Introduction

To establish a sustainable energy system in future, the development of effective utilization techniques of natural energy such as solar, hydro, and wind is important issue. Particularly, energy conversion techniques to secondary energy such as electricity and chemical energy are required to effectively utilize the natural energy because such energy fluctuates with time cycle of earth and is localized to specific areas. In our current society, the infrastructure to use electricity has already been established, suggesting that the conversion from natural energy to electricity is an ideal process. However, it is difficult to store and transport large amount of energy by conventional battery systems. Thus, many researchers recently focus on the

H. Miyaoka (✉)
Institute for Sustainable Sciences and Development, Hiroshima University, 1-3-1 Kagamiyama, Higashi-Hiroshima 739-8530, Japan
e-mail: miyaoka@h2.hiroshima-u.ac.jp

M. Sugiyama et al. (eds.), *Solar to Chemical Energy Conversion*,
Lecture Notes in Energy 32, DOI 10.1007/978-3-319-25400-5_9

conversion process to chemical energy, which has significant advantages for long-time storage and/or long-distance transportation of energy.

The technologies to utilize natural energy strongly depend on environments of areas, suggesting that the suitable techniques should be selected and established for each area. For instance, solar, hydro, and wind power generation systems are suitable for Sun-belt area, Northern Europe, and South America, respectively. In these areas, the generated electric power can be shared with neighbor area and/or countries by electrical grid. For island countries with poor natural resources such as Japan, large amount of energy should be transported from outside countries to sustain and develop the society. In this case, the technologies related to energy storage and carrier as well as power generation are important. Namely, various kinds of technologies should be required for effective utilization of natural energy, and an alternative energy system of current one based on fossil fuels should be realized by combining those technologies.

This section introduces the conversion techniques from solar energy to chemical energy, especially hydrogen, and discusses the characteristic properties of each technique.

2 Solar Heat Technology

The heat energy generated by concentrating sunlight is generally called solar heat. For the thermochemical conversion to chemical energy, solar heat is used because the conversion proceeds via several steps of thermochemical reactions. To establish effective solar heat systems, elemental technologies such as reflector, receiver, heat exchanger, and heat storage media are being investigated. Only typical solar heat systems are introduced below. At present, various types of solar heat systems are considered with the environmental situation and roughly classified by the shape of heliostat and concentration ratio. The heliostat is a mirror to reflect and concentrate sunlight, and it is computationally controlled to chase the sun. The concentration ratio of sunlight is directly related to the possible temperature of the generated heat. Figure 1 shows the schematic images of typical solar heat systems, (a) tower, (b) beam down, (c) trough, (d) fresnel, and (e) dish types, and the properties are listed in Table 1.

In the tower type (a), the sunlight is concentrated to the receiver at the top of the central tower by heliostats placed on ground. In this case, high concentration can be realized. The concentration factor C is 300–1500 [1]. Thus, high temperature around 1000–1500 °C can be generated. The beam down type (b) is equipped with a second reflector and ground receiver with concentrator, leading to the highest concentration of sunlight, $C > 5000$, among them. The available temperature of beam down type is also more than 1000 °C. However, the above systems require large field and enough solar irradiation to exhibit essential performance, suggesting that the place for construction of these plants is limited to particular areas. The trough type system (c) is composed of linear parabolic mirror and collector tube

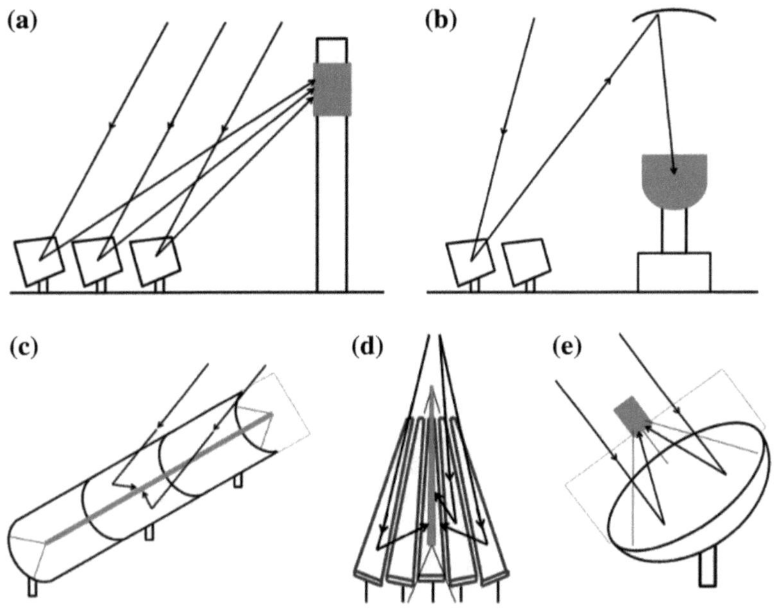

Fig. 1 Schematic images of typical solar heat systems. **a** Tower type. **b** Beam down type. **c** Trough type. **d** Fresnel type. **e** Dish type

Table 1 Performance and characteristics of typical solar heat plants

Type	Temperature T (°C)	Concentration factor C	Scale
Tower	1000–1500	300–1500	Large
Beam down	1000–1500	1000–5000	Middle-large
Trough	400–600	30–100	Middle-large
Dish	>1500	1000–1500	Small

located at the focal line of the concentrator. The solar energy is absorbed and transferred by fluid such as oil. The temperature generated by trough type is 400–600 °C because of low concentration factor. On the other hand, this system is currently utilized as commercial power generation plant due to the simple structure. As a modified system of the trough, the fresnel type (d) is proposed and being developed recently. In this case, the linear mirrors are placed, and the angle of each mirror is controlled with the sun motion, suggesting that it is not necessary to produce the large parabolic mirror with accuracy to realize the high concentration. The parabolic reflector is used for the dish type (e). The sunlight is concentrated to the receiver at the focus point of the dish mirror. Although the dish type is a smaller scale system than others, more than 1500 °C is available because the concentration factor C reaches to 1000–1500.

For the solar heat systems, only direct normal irradiation is available. Thus, the studies on utilization of solar heat are being carried out in mainly "Sun-belt" area such as United State, Spain, and Northern Africa [2–7], which are suitable due to fine and arid climate throughout the year [1]. Recently, the construction of the pilot plants actively proceeds in Asian countries such as Korea and China as well [8, 9]. In Japan, suitable areas for solar heat systems are limited to southern parts, and the disadvantageous climate such as high humidity is problem for the effective utilization. However, the pilot solar heat system of beam down type is constructed at Miyazaki University, and the studies are being achieved by several research groups [10].

So far, solar heat systems have been investigated as a power generation technique. In this case, the concentrated solar energy is used for steam turbine, and then the heat energy is converted to electricity. In large continents such as United State and Europe, the generated electricity can be transferred by the huge power grid structured across the own and/or possibly neighbor countries, and it is effectively utilized by combination with other renewable energy, hydro and wind power [11]. On the other hand, such systems are difficult to be established in island countries such as Japan. Thus, the technologies for the conversion from solar heat to chemical energy, which is medium to store and transport the energy, are required. Typical thermochemical conversion processes using water (H_2O) are water splitting (WS) (1), coal gasification (2), and steam reforming (3) as follows,

$$\text{Water splitting}: H_2O \rightarrow H_2 + 1/2O_2, \tag{1}$$

$$\text{Coal gasification}: H_2O + C \rightarrow H_2 + CO, \tag{2}$$

$$\text{Steam reforming}: H_2O + CH_4 \rightarrow 3H_2 + CO. \tag{3}$$

In reactions (1) and (2)–(3), hydrogen and syngas are generated, respectively. Considering the utilization of hydrogen by fuel cell, H_2 is ideally produced by the water splitting. However, the direct water splitting using solar heat is quite difficult because of the high stability of H_2O. For reactions (2) and (3), the contribution of natural energy is relatively lowered compared with WS due to high energy state of coal and hydrocarbons, which can be burned and generate energy by themselves, although the reaction can be controlled by catalytic process at realizable temperature range. Here, the final product CO_2 is problem because it is considered as one of greenhouse gases, and thus effectual process for the treatment of CO_2 is necessary.

3 Hydrogen Production by Thermochemical Water Splitting

To directly decompose H_2O to H_2 and O_2, more than 4000 °C is required from thermodynamic point of view. However, such extreme high temperature is difficult to generate by using solar heat systems and be handled by existing materials. Thus, thermochemical WS has been considered and investigated so far. In this case, H_2O

is split via several steps of chemical reactions, resulting in the lower operating temperature than 4000 °C. Because the thermochemical WS is composed of chemical reactions, the reaction properties are expected by the thermodynamics of each reaction. Generally, the thermodynamic properties of chemical reaction is expressed by following Gibbs free energy change ΔG,

$$\Delta G = \Delta H - T\Delta S, \tag{4}$$

$$\Delta S = \Delta S^0 + \mathrm{R}\ln\left(\frac{p_0}{p_{\mathrm{pro}}}\right), \tag{5}$$

where ΔH is enthalpy change, ΔS is entropy change, ΔS^0 is standard entropy change, R is gas constant, p_0 is standard pressure, and p_{pro} is partial pressure of gaseous products, respectively. Basically, ΔH and ΔS^0 are more or less determined by the included elements, composition, and molecule or crystal structure of starting materials and products in the reaction. To modify the thermodynamics and reduce the operating temperature, ΔH and/or p_{pro} should be controlled by material and engineering points of view. The substitution of elements and optimization of the composition are generally useful techniques to control ΔH. When the gaseous reaction products are effectively separated and removed from the reaction field, the equilibrium condition is shifted due to decrease in p_{pro}, and then the reaction temperature is lowered like steam reforming using membrane reactor.

The theoretical efficiency of the thermochemical water splitting η_{WS} is calculated by following equation,

$$\eta_{\mathrm{WS}} = \frac{\Delta H_{\mathrm{H_2O}}}{(Q_{\mathrm{endo}} - \alpha Q_{\mathrm{exo}} + W)}, \tag{6}$$

where ΔH_{H2O} is heat of formation of H_2O, 286 kJ/mol, Q_{endo} is total energy required by the endothermic reactions, Q_{exo} is total energy generated by exothermic reactions in the thermochemical WS cycle, α is ratio of heat recovery from Q_{exo}, and W is working energy required to operate the hydrogen production process such as power for pumping of fluid. Since α and W are determined after designing and making the operating systems, the ideal efficiency is estimated under $\alpha = 0$ and $W = 0$. In addition, the efficiency of solar heat systems η_{system} is expressed as follows,

$$\eta_{\mathrm{system}} = \eta_{\mathrm{solar}} \cdot \eta_{\mathrm{Carnot}}, \tag{7}$$

$$\eta_{\mathrm{solar}} = \frac{\alpha_s IC - \epsilon\sigma T^4}{IC}, \tag{8}$$

$$\eta_{\mathrm{Carnot}} = \frac{T - T_{\mathrm{low}}}{T}. \tag{9}$$

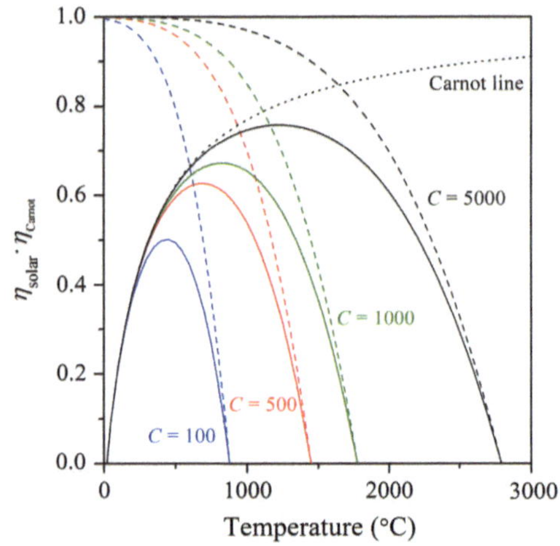

Fig. 2 Conversion efficiency from solar energy to heat energy

The efficiency of receiver for solar energy absorption η_{solar} is defined by the effective absorption of solar radiation α_s, intensity of solar radiation I, concentration factor of the solar heat system C, emittance of the receiver ϵ, the Stefan-Boltzmann constant σ, and operating temperature T, assuming that the heat conduction and convection are negligible [1]. For the utilization of thermochemical cycle, the maximum efficiency is limited by Carnot efficiency η_{Carnot} shown by Eq. (9), where T_{low} is the temperature of cold thermal receiver. Figure 2 shows the conversion efficiency of $\eta_{solar} \cdot \eta_{Carnot}$ as a function of operating temperature T with $\alpha_s = \epsilon = 1$, $I = 1$ kW/m^2, and $T_{low} = 25$ °C for the concentration factor in the range from 100 to 5000. The η_{solar} and η_{Carnot} are shown as broken and dotted lines, respectively. The drop of η_{solar} starts at a lower temperature with smaller concentration factor C [1, 12]. Thus, the maximum efficiency for $C = 100$ is about 50 % around 500 °C and increases with higher C values. If taking into account only energy conversion efficiency, solar heat systems with high concentration factor around 1000–5000 and high operating temperature around 1000–1500 °C are suitable, namely, the tower and beam down types are better than others (Table 1).

4 Thermochemical Water Splitting Cycle

Figure 3 shows the operating temperature of the typical thermochemical WS cycles with heat sources based on solar concentrating systems and unused heat, in which the efficient utilization of unused energy such as exhaust heat from factories is also important issue in future society. As shown in this figure, various thermochemical WS cycles are proposed and investigated so far. The WS cycles are classified into high (>1000 °C) and low (<1000 °C) temperature types. As the typical cycles, the properties of 2-step cycles [13–22], Iodine-sulfur (I-S) cycles [23–25], and sodium

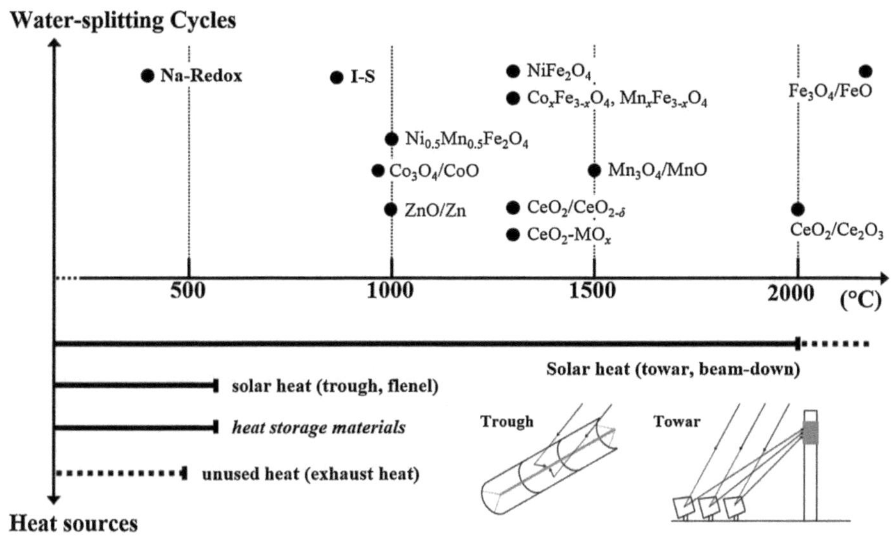

Fig. 3 Operating temperature of typical thermochemical water splitting cycles and temperature range of heat sources based on solar and unused energy

(Na) metal redox cycles [26], which is recently reported as WS of low temperature type, are introduced below.

4.1 *Two-Step Water Splitting Cycles*

The 2-step cycles is generally defined by the following two reactions,

$$M_xO_y \rightarrow M_xO_{y-1} + \frac{1}{2}O_2 \quad \text{(Thermal reduction)}, \tag{10}$$

$$M_xO_{y-1} + H_2O \rightarrow M_xO_y + H_2 \quad \text{(Water splitting)}. \tag{11}$$

where M is transition and rare-earth elements. Basically, x and y are integers. The thermal reduction and water splitting in the cycle proceed via valence fluctuation of M. All the 2-step cycles requires more than 1000 °C to be operated, thus utilization of tower and beam down types is considered. Considering the Carnot efficiency, the potential conversion efficiency is theoretically obtained with the higher temperature range. However, η_{solar} is decreased at the high temperature even with large C value. Thus, the current target temperature of 2-step cycles are set to 1000–1500 °C, and then more than 60 % of efficiency is expected with C = 1000–5000.

In the recent pilot plant, WS by the 2-step cycles is performed by direct irradiation of concentrated sunlight in the reactor placed into receiver part (Fig. 1a, b) because heat storage and transfer materials are not able to withstand such high temperatures. In addition to the high efficiency, another advantage of the cycles is simple reaction process based on a solid-gas reaction. Both reactions can be operated by controlling gas phases, suggesting that the working energy *W* can be suppressed, and the efficient system including heat recovery can be designed. So far, ferrite: Fe_3O_4/FeO [13, 15, 16], cobalt oxide: Co_3O_4/CoO [16], and manganese oxide: Mn_3O_4/MnO [16] cycles have been investigated mainly. Important properties for the 2-step cycles are thermodynamic stability and melting point of the oxides. The oxides have to be reduced below target temperature, 1000–1500 °C, generated by the solar heat systems. Moreover, the melting point is related to the sintering of oxide particles in the high temperature process, which generally causes degeneration of the kinetic properties. To modify the thermodynamic and material properties, the substitution of third elements is useful technique. In fact, various kinds of modified cycles have been proposed [17, 20, 22, 27–33]. In recent years, the 2-step cycle based on ceria CeO_2, which has high melting point, is drawing widespread interest. Although the ceria cycle using valence fluctuation: CeO_2/Ce_2O_3 requires about 2000 °C for the thermal reduction, it was reported that O_2 is released to form nonstoichiometric phase $CeO_{2-\delta}$ below 1500 °C as follows [34],

$$CeO_2 \rightarrow CeO_{2-\delta} + \frac{\delta}{2}O_2, \tag{12}$$

$$CeO_{2-\delta} + \delta H_2O \rightarrow CeO_2 + H_2. \tag{13}$$

In addition to above fundamental studies, more practical researches are also carried out. The scaffold as support of oxides effectively improves the reaction kinetics and suppresses sintering of the particles [20–22]. The circulating fluidized bed reactor is proposed as the efficient reaction system considering the heat recovery [35].

4.2 Iodine-Sulfur (I-S) Cycle

Iodine-sulfur (I-S) cycle is the typical WS cycle of low temperature type. This cycle is composed of three kinds of chemical reactions as follows,

$$SO_2 + I_2 + H_2O \rightarrow 2HI + H_2SO_4 \tag{14}$$

$$2HI \rightarrow I_2 + H_2 \tag{15}$$

$$H_2SO_4 \rightarrow SO_2 + \frac{1}{2}O_2 + H_2O \tag{16}$$

The reaction (12) is generally called Bunsen reaction, which produces HI and sulfuric acid H_2SO_4 from the mixture of SO_2, I_2, and H_2O. The produced HI and H_2SO_4 are thermally decomposed at the processes (13) and (14), respectively.

The most difficult reaction is thermal decomposition of H_2SO_4 and requires about 900 °C. The theoretical conversion efficiency for WS is 67 % without the heat recovery, which is relatively higher value among the reported cycles. The I-S cycle is based on a gas-liquid system unlike the 2-step cycles, which are gas-solid systems mentioned above. Although the reaction yield is essentially limited by thermodynamic equilibrium, this is an advantage from engineering point of view because the fluids are easily transferred by using mechanical pumps in the reaction processes. Historically, the I-S cycle have been investigated as the conversion technique of nuclear energy generated by high temperature gas-cooled reactor (HTGR) to hydrogen via WS [25]. In Japan, the research on this cycle has being carried out in mainly Japan Atomic Energy Agency (JAEA) by using HTTR (High Temperature engineering Test Reactor). The bench-scale plant is constructed in JAEA, and the continuous hydrogen production has been succeeded already [36]. As other energy sources for operating the I-S cycle, solar heat systems of high temperature types such as tower and beam down are available because the enough high temperature than 1000 °C can be produced. While the reactions in I-S cycle proceed under equilibrium conditions as mentioned above, effective catalysts are being investigated in recent years [37].

4.3 Alkali Metal Redox Cycle

Alkali metal redox cycle is proposed in 2012 as a low temperature type. Although studies about the cycles are fundamental research, the sodium (Na) cycle is potentially operated at lower temperature than 600 °C, in which the low temperature heat generated by solar heat system such as trough type is utilizable [26]. The following four kinds of reactions make up the cycle,

$$2\mathrm{MOH(s)} + 2\mathrm{M(l)} \rightarrow 2\mathrm{M_2O(s)} + \mathrm{H_2(g)}, \tag{17}$$

$$2\mathrm{M_2O(s)} \rightarrow \mathrm{M_2O_2(s)} + 2\mathrm{M(g)}, \tag{18}$$

$$2\mathrm{M(g)} \rightarrow 2\mathrm{M(s)} \rightarrow 2\mathrm{M(l)}, \tag{19}$$

$$\mathrm{M_2O_2(s)} + \mathrm{H_2O} \rightarrow 2\mathrm{MOH(s)} + \frac{1}{2}\mathrm{O_2(g)}, \tag{20}$$

where M is alkali metals such as Li, Na, and K. Enthalpy change ΔH and entropy change ΔS of each reaction for all the cycles with p_{pro} = 0.1 MPa are listed in Table 2, where the values are estimated by using NIST database [38].

For the H_2 generation reaction (17), only Li cycle is exothermic. The Na cycle has suitable $\Delta H = 11$ kJ, suggesting that this reaction proceeds below 500 °C under equilibrium condition with p_{pro} = 0.1 MPa. On the other hand, temperatures higher than 500 °C is necessary to operate the reaction of the K cycle because of large ΔH and small ΔS values. To realize the H_2 generation, the partial pressure of

Table 2 Thermodynamic parameters of ΔH (kJ) and ΔS (J/K) for the alkali metal redox cycles

	(17)		(18)		(19)		(20)	
	ΔH	ΔS	ΔH	ΔS	ΔH	ΔS	ΔH	ΔS
Li	−232	53	883	258	−314	−210	−51	62
Na	11	36	540	252	−210	−192	−55	66
K	119	18	409	246	−173	−178	−68	77

H_2 should be decreased to less than 1×10^{-2} Pa. Experimentally, the H_2 generation is clarified at 350 °C with 80 % of reaction yield for the Na cycle [26]. The hydrolysis of all the cycles is exothermic, thus it is expected that the reactions can be controlled below 500 °C. The most difficult reaction is thermal reduction of M by Eq. (18) because of huge ΔH for all the cycles, and thus more than several 1000 °C is necessary under equilibrium state with $p_{pro} = 0.1$ MPa. In other words, this reaction is most important for controlling the cycles at lower temperatures because the thermodynamic properties of other reactions satisfy the conditions to operate the cycle by lower temperature heats than 600 °C. When the reaction (18) is performed at more than 300 °C, the generated metals M are immediately melted, after which some parts of the molten metal vaporize. Therefore, by condensing and removing the generated metal vapor from the reaction field, the entropy value is increased due to the reduction of p_{pro}, which is partial pressure of M vapor near M_2O. As a result, the operating temperature of thermal reduction should be lowered. In fact, the Na generation is experimentally found at 500 °C for the Na cycle.

Figure 4 shows the experimental system. Na_2O is put into the bottom part of the reactor made by Ni-based alloy and is heated up to 500 °C. The generated Na vapor is condensed as a solid phase by the water cooling part equipped at the upper side of the reactor. From the photo of cooling part after the reaction in Fig. 4, it can be clearly seen that the metallic product is generated. The product is assigned to Na metal by X-ray diffraction (XRD) measurements as shown in Fig. 5. These results

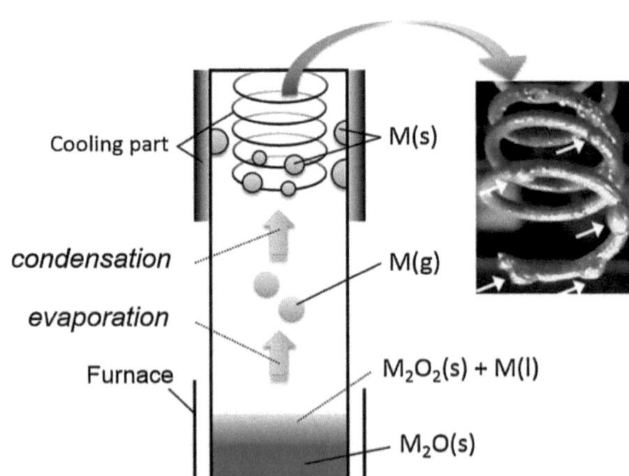

Fig. 4 Schematic image of experimental system and photo of product at cooling part

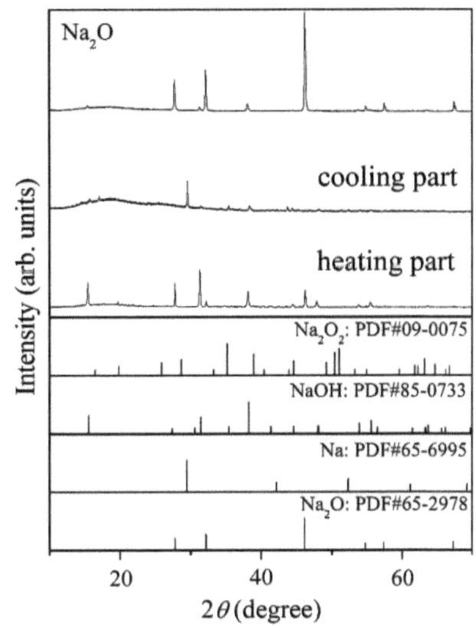

Fig. 5 XRD patterns of Na_2O, the products at cooling, and heating after the reaction at 500 °C

indicate that the temperature of Na generation is significantly decreased by the control of thermodynamic properties using effective material transfer. On the other hand, in the XRD pattern of the product at heating part, only starting material and NaOH as impurity are observed, and peroxide Na_2O_2 as expected product is not found. It is suggested that the phenomena are originated in the decomposition of Na_2O_2 and/or the corrosion of Na_2O_2 with the reactor [39]. Here, the decomposition of peroxide is indicated for the Li cycle as well. Considering the decomposition of peroxide M_2O_2, the alkali metal redox cycles are modified as follows:

$$2MOH(s) + 2M(l) \rightarrow 2M_2O(s) + H_2(g), \quad (21)$$

$$2M_2O(s) \rightarrow M_2O_2(s) + 2M(g), \quad (22)$$

$$M_2O_2(s) \rightarrow 2M(g) + O_2(g), \quad (23)$$

$$4M(g) \rightarrow 4M(s) \rightarrow 4M(l), \quad (24)$$

$$2M(l) + 2H_2O(l) \rightarrow 2MOH(s) + H_2(g). \quad (25)$$

In this cycle, the reactions of (22) and (23) would continuously proceed, indicating that the M_2O is thermally decomposed into M and O_2. The most serious problem is corrosion to utilize the alkali metal cycles safely as practical application. Here, for the K cycle, this problem is mentioned in review [40]. In fact, a strong corrosion is found in the experiments for the K cycle. In the case of Li and Na cycles, although the degree of corrosion is weaker than that of the K cycle, the problem has to be considered. Therefore, the details of reaction process and mechanism including the corrosion have to be investigated in future works. The temperature reduction below 500 °C using entropy control is realized by only alkali metals because of the low melting point. The following Zn/ZnO cycle is similar to the Na cycle because the metal vapor is generated in the process [19]. However, more than 1000 °C is necessary to obtain enough Zn generation due to the higher melting point.

$$ZnO(s) \rightarrow Zn(g) + \frac{1}{2}O_2(g), \tag{26}$$

$$Zn(s) + H_2O \rightarrow ZnO(s) + H_2(g) \tag{27}$$

The cupper-chlorine (Cu-Cl) is another attractive low temperature type WS [41]. The reactions in this cycle are described as follows,

$$2Cu(s) + 2HCl(g) \rightarrow 2CuCl(l) + H_2(g), \tag{28}$$

$$2CuCl(s) \rightarrow 2CuCl(aq) \rightarrow CuCl_2(aq) + Cu(s), \tag{29}$$

$$CuCl_2(aq) \rightarrow CuCl_2(s), \tag{30}$$

$$2CuCl_2(s) + H_2O(g) \rightarrow CuO * CuCl_2(s) + 2HCl(g), \tag{31}$$

$$CuO * CuCl_2(s) \rightarrow 2CuCl(l) + \frac{1}{2}O_2(g). \tag{32}$$

Although the maximum temperature required to operate the cycle is <500 °C, electrochemical process (29) is included. Thus, the design of suitable operation system would be required considering both of thermochemical and electrochemical processes.

For thermochemical WSs such as the Na cycle, the energy conversion efficiency is theoretically lower than that of the high temperature types and the I-S cycle because of the limitation of Carnot efficiency. On the other hand, various kinds of solar heat systems, trough as well as tower types, and heat storage materials, which is possible to be used below 600 °C [42, 43], can be utilized. Therefore, it is expected that the operation time for the hydrogen production via the WS of low temperature type becomes longer than the conventional cycles.

5 Conclusion

Thermochemical water-splitting is one of the key technologies for converting solar energy into hydrogen as energy storage and transportation medium. Currently, various kinds of solar heat systems are investigated, and performance such as realizable temperature strongly depends on the environment of the areas. Thus, the WS cycles should vary according to the environmental conditions as well. The conventional 2-step cycles are composed of simple reactions using valence fluctuation of transition or rare-earth elements in oxides. Although high temperature of

1500 °C is required for water splitting and thermal reduction, the theoretical conversion efficiency is essentially high. The I-S cycle is also a promising WS technique and operated around 900 °C. The bench-scale test has been established to produce hydrogen via continuous three-step reactions. Alkali metal redox cycles have been proposed in recent years. Although their research is still at the laboratory scale, the Na cycle is recognized as a potential WS operated below 600 °C, suggesting that it can be utilized by various kinds of solar heat systems and unused heats as wide practical application.

References

1. Kodama T, Gokon N (2007) Thermochemical cycles for high-temperature solar hydrogen production. Chem Rev 107:4048–4077
2. Sandia National Laboratories. http://energy.sandia.gov/
3. Palumbo R, Keunecke M, Möller S, Steinfeld A (2004) Reflections on the design of solar thermal chemical reactors: thoughts in transformation. Energy 29:727–744
4. Paul Scherrer Institute. http://www.psi.ch/ene/
5. DLR Institute of Solar Research. http://www.dlr.de/sf/en/desktopdefault.aspx/tabid-7066/
6. Weizmann Institute of Science, Solar Research Facilities Unit. http://www.weizmann.ac.il/weizsites/solarenergy/
7. Plataforma solar de Almería. http://www.psa.es/webeng/index.php
8. Korea Institute of Energy Research. http://www.kier.re.kr/eng/05_about/org_solarheat_energy.jsp
9. Solar Energy Laboratory of Chinese Academy of Sciences. http://english.iee.cas.cn/rh/rd/200907/t20090722_24883.html
10. Miyazaki University. http://www.miyazaki-u.ac.jp/topics/20120806
11. White Book -Clean Power from Deserts- (2007) The DESERTEC concept for energy, water and climate security. DESERTEC foundation
12. Fletcher EA, Moen RL (1977) Hydrogen- and oxygen from water. Science 197:1050–1056
13. Nakamura T (1977) Hydrogen production from water utilizing solar heat at high temperatures. Sol Energy 19:467–475
14. Ambriz JJ, Ducarroir M, Sibieude F (1982) Preparation of cadmium by thermal dissociation of cadmium oxide using solar energy. Int J Hydrogen Energy 7:143–153
15. Sibieude F, Ducarroir M, Tofighi A, Ambriz J (1982) High temperature experiments with a solar furnace: The decomposition of Fe_3O_4, Mn_3O_4, CdO. Int J Hydrogen Energy 7:79–88
16. Lundberg M (1993) Model calculations on some feasible two-step water splitting processes. Int J Hydrogen Energy 18:369–376
17. Ehrensberger K, Frei A, Kuhn P, Oswald HR, Hug P (1995) Comparative experimental investigations of the water-splitting reaction with iron oxide $Fe_{1-y}O$ and iron manganese oxides $(Fe_{1-x}Mn_x)_{1-y}O$. Solid State Ionics 78:151–160
18. Steinfeld A, Sanders S, Palumbo R (1999) Design aspects of solar thermochemical engineering- case study: two-step water-splitting cycle using the Fe_3O_4/FeO redox system. Sol Energy 65:43–53
19. Weidenkaff A, Steinfeld A, Wokaun A, Auer PO, Eichler B, Reller A (1999) Direct solar thermal dissociation of zinc oxide: condensation and crystallisation of zinc in the presence of oxygen. Sol Energy 65:59–69
20. Kodama T, Kondoh Y, Yamamoto R, Andou H, Satou N (2005) Thermochemical hydrogen production by a redox system of ZrO_2-supported Co(II)-ferrite. Sol Energy 78:623–631

21. Gokon N, Hasegawa T, Takahashi S, Kodama T (2008) Thermochemical two-step water-splitting for hydrogen production using Fe-YSZ particles and a ceramic foam device. Energy 33:1407–1416
22. Gokon N, Murayama H, Nagasaki A, Kodama T (2009) Thermochemical two-step water splitting cycles by monoclinic ZrO_2-supported $NiFe_2O_4$ and Fe_3O_4 powders and ceramic foam devices. Sol Energy 83:527–537
23. O'Keefe D, Allen C, Besenbruch G, Brown L, Norman J, Sharp R, McCorkle K (1982) Preliminary results from bench-scale testing of a sulfur-iodine thermochemical water-splitting cycle. Int J Hydrogen Energy 7:381–392
24. Sakurai M, Nakajima H, Amir R, Onuki K, Shimizu S (2000) Experimental study on side-reaction occurrence condition in the iodine-sulfur thermochemical hydrogen production process. Int J Hydrogen Energy 25:613–619
25. Kubo S, Nakajima H, Kasahara S, Higashi S, Masaki T, Abe H, Onuki K (2004) A demonstration study on a closed-cycle hydrogen production by the thermochemical water-splitting iodine-sulfur process. Nucl Eng Des 233:347–354
26. Miyaoka H, Ichikawa T, Nakamura N, Kojima Y (2012) Low-temperature water-splitting by sodium redox reaction. Int J Hydrogen Energy 37:17709–17714
27. Ehrensberger K, Kuhn P, Shklover V, Oswald HR (1996) Temporary phase segregation processes during the oxidation of $(Fe_{0.7}Mn_{0.3})_{0.99}O$ in N_2-H_2O atmosphere. Solid State Ionics 90:75–81
28. Tamaura Y, Kojima M, Sano T, Ueda Y, Hasegawa N, Tsuji M (1998) Thermodynamic evaluation of water splitting by a cation-excessive (Ni, Mn) ferrite. Int J Hydrogen Energy 23:1185–1191
29. Steinfeld A, Kuhn P, Reller A, Palumbo R, Murray J, Tamaura Y (1998) Solar-processed metals as clean energy carriers and water-splitters. Int J Hydrogen Energy 23:767–774
30. Kaneko H, Kojima N, Hasegawa N, Inoue M, Uehara R, Gokon N, Tamaura Y, Sano T (2002) Reaction mechanism of H_2 generation for $H_2O/Zn/Fe_3O_4$ system. Int J Hydrogen Energy 27:1023–1028
31. Inoue M, Hasegawa N, Uehara R, Gokon N, Kaneko H, Tamaura Y (2004) Solar hydrogen generation with $H_2O/ZnO/MnFe_2O_4$ system. Sol Energy 76:309–315
32. Kaneko H, Miura T, Ishihara H, Taku S, Yokoyama T, Nakajima H, Tamaura Y (2007) Reactive ceramics of CeO_2-MO_x (M = Mn, Fe, Ni, Cu) for H_2 generation by two-step water splitting using concentrated solar thermal energy. Energy 32:656–663
33. Fresno F, Yoshida T, Gokon N, Fernández-Saavedra R, Kodama T (2010) Comparative study of the activity of nickel ferrites for solar hydrogen production by two-step thermochemical cycles. Int J Hydrogen Energy 35:8503–8510
34. Abanades S, Flamant G (2006) Thermochemical hydrogen production from a two-step solar-driven water-splitting cycle based on cerium oxides. Sol Energy 80:1611–1623
35. Gokon N, Murayama H, Umeda J, Hatamachi T, Kodama T (2009) Monoclinic zirconia-supported Fe_3O_4 for the two-step water-splitting thermochemical cycle at high thermal reduction temperatures of 1400–1600 °C. Int J Hydrogen Energy 34:1208–1217
36. Japan Atomic Energy Agency. http://httr.jaea.go.jp/
37. Machida M, Miyazaki Y, Matsunaga Y, Ikeue K (2011) Efficient catalytic decomposition of sulfuric acid with copper vanadates as an oxygen-generating reaction for solar thermochemical water splitting cycles. Chem Commun 47:9591–9593
38. National Institute of Standards and Technology, NIST Chemistry WebBook. http://webbook.nist.gov/chemistry/
39. Miyaoka H, Ichikawa T, Kojima Y (2014) Thermochemical energy storage by water-splitting via redox reaction of alkali metals. Energy Procedia 49:927–934
40. Abanades S, Charvin P, Flamant G, Neveu P (2006) Screening of water-splitting thermochemical cycles potentially attractive for hydrogen production by concentrated solar energy. Energy 31:2805–2822

41. Naterer G, Suppiah S, Lewis M, Gabriel K, Dincer I, Rosen MA, Fowler M, Rizvi G, Easton EB, Ikeda BM, Kaye MH, Lu L, Pioro I, Spekkens P, Tremaine P, Mostaghimi J, Avsec J, Jiang J (2009) Recent Canadian advances in nuclear-based hydrogen production and the thermochemical Cu-Cl cycle. Int J Hydrogen Energy 34:2901–2917
42. Dincer I, Dost S (1996) A perspective on thermal energy storage systems for solar energy applications. Int J Energy Res 20:547–557
43. Gil A, Medrano M, Martorell I, Lázaro A, Dolado P, Zalba B, Cabeza LF (2010) State of the art on high temperature thermal energy storage for power generation. Part 1—concepts, materials and modellization. Renew Sustain Energy Rev 14:31–55

Photocatalytic Approach for CO_2 Fixation

Kazuhiko Maeda

Abstract In order to address the depletion of fossil fuels and the serious environmental problems accompanying their combustion and the concomitant CO_2 emission, large-scale chemical conversion of CO_2 into energy-rich materials would be an ultimate solution, and several reactions have been proposed. There have been a lot of challenges that have to be addressed in this field of research, but several breakthroughs have been achieved in recent 10 years. In this chapter, photocatalytic CO_2 reduction systems, which are of particular importance, are reviewed, with a focus on both homogeneous and heterogeneous aspects.

Keywords Fossil fuels · CO_2 reduction · Carbon monoxide · Formic acid · Metal complex · Photocatalysis · Photocatalyst · Sunlight · Isotope · Rhenium · Dimethylformamide (DMF) · Triethanolamine (TEOA) · Quantum yield · One electron reduced (OER) species · Turnover number · Electron donor · Supramolecular metal complex · Intramolecular electron transfer · One electron reduced (OER) species · Intramolecular electron transfer · Ruthenium · Manganese · Chlorophylls · Resonance energy transfer · Conjugation · Mesoporous silica · Light-harvesting · Semiconductor · Water oxidation · Langmuir-Hinshelwood type mechanism · Perovskite · Cocatalyst · Silver · Overpotential · Water splitting · Layered double hydroxide (LDH) · Heterogeneous photocatalysis · Acetonitrile · Polymer · Carbon nitride · Visible light · Carbon nitride · Visible light · Action spectrum · Z-scheme · Natural photosynthesis · Sensitizer · Hydrogen evolution · Electron-hole recombination

K. Maeda (✉)
Department of Chemistry, Tokyo Institute of Technology, 2-12-1-NE-2 Ookayama, Meguro-ku, Tokyo 152-8550, Japan
e-mail: maedak@chem.titech.ac.jp

M. Sugiyama et al. (eds.), *Solar to Chemical Energy Conversion*,
Lecture Notes in Energy 32, DOI 10.1007/978-3-319-25400-5_10

1 Introduction

1.1 Research Background

Photocatalytic CO_2 fixation into energy-rich chemicals such as carbon monoxide and formic acid has attracted attention in order to address the depletion of carbon resources and the suppression of global warming as well as to accomplish the CO_2-reduction half cycle in artificial photosynthesis [1, 2]. Because CO_2 is a very stable molecule, the reduction of CO_2 remains a big challenge; it requires a high potential to electrochemically reduce CO_2 by one electron (–1.9 V vs. NHE at pH 7). On the other hand, the required potential for CO_2 reduction can be reduced if one utilizes multi-electron process. For example, the potentials of two-electron reduction of CO_2 into formic acid and CO are –0.61 and –0.53 V (vs. NHE at pH 7), respectively, approximately 1.3 V more positive than the potential required to drive one-electron reduction of CO_2.

$$CO_2 + 2H^+ + 2e^- \rightarrow HCOOH \quad E^\circ = -0.61V$$
$$CO_2 + 2H^+ + 2e^- \rightarrow CO + H_2O \quad E^\circ = -0.53V$$
$$CO_2 + 4H^+ + 4e^- \rightarrow HCHO + H_2O \quad E^\circ = -0.48V$$
$$CO_2 + 6H^+ + 6e^- \rightarrow CH_3OH + H_2O \quad E^\circ = -0.38V$$
$$CO_2 + 8H^+ + 8e^- \rightarrow CH_4 + 2H_2O \quad E^\circ = -0.24V$$

Thus, a catalyst that allows one to promote multi-electron transfer is needed. As described below, certain metal complexes and semiconductors (as well as their composites) work as such (photo)catalysts.

Since the seminal work by Lehn et al. who demonstrated selective CO_2 reduction into CO using Re(I) diimine complexes in 1983 [3], photocatalytic CO_2 reduction has been extensively studied for the purpose of light-to-chemical energy conversion. However, a satisfactory system has yet to be devised to date. As the goal of CO_2 fixation by a photocatalyst is to convert solar energy into chemical energy on a large-scale, a given system has to meet the following requirements.

(1) A given CO_2 fixation system has to be workable under sunlight having low energy density, and to be stable and efficient during long-term operation.
(2) CO_2 reduction has to be promoted while suppressing any side reaction that can occur during the reaction (e.g., H_2 reduction via water reduction).
(3) Water should be used as an electron source.

Unfortunately, no photocatalytic system that satisfies all of the above requirements has been reported to date. Besides, energy conversion scheme via CO_2 reduction where the change in Gibbs energy is positive had not been achieved until very recently. Nevertheless, there were some breakthroughs especially in the last 10 years. This chapter describes recent progress on photocatalytic CO_2 reduction using metal complexes and semiconductors, along with some important early works.

1.2 *Some Important Aspects in Photocatalytic CO_2 Fixation Research*

One should be careful when reading a literature on photocatalytic CO_2 reduction, as some may be problematic. For example, even though a given CO_2 fixation system is claimed to work in water, quantification of O_2, which is the product of water oxidation, is ambiguously or not described. Besides, it is extremely important in any of photocatalytic CO_2 reduction research to investigate the origin of carbon-containing products by isotope tracer experiments with $^{13}CO_2$, because contaminated carbon species may become such materials upon photo-irradiation especially in a heterogeneous system. For example, using a commercially available titania loaded with nanoparticulate metal promoters as a heterogeneous photocatalyst for CO_2 reduction, it has been revealed that both CO and CH_4 were obtained, but the contribution of surface contaminants to the reaction products cannot be neglected [4]. High turnover number with respect to the amount of catalytically active sites or the photocatalyst itself is also an important indicator to judge whether or not a given reaction photocatalyzes CO_2 reduction. In this section, the author would like to focus on representative, but "reliable" systems in this regard, which are considered important milestones in photocatalytic CO_2 reduction research.

2 Metal Complexes

2.1 *Re Diimine Tricarbonyl Type Complexes*

In 1983, Lehn et al. reported that *fac*-$[Re(N\wedge N)(CO)_3X]^+$ (N^N = diimine ligand; X = Cl^-, Br^-) not only works as an efficient CO_2 reduction photocatalyst but also as a catalyst (Scheme 1). These Re(I) complexes selectively produce CO in a dimethylformamide (DMF)/triethanolamine (TEOA) solution. It is noted that even in the presence of water, certain Re(I) complexes are capable of selectively producing CO without noticeable H_2 formation [3].

Follow-up studies have been made to create more active photocatalysts. The chloro ligand on *fac*-$Re(bpy)(CO)_3Cl$ (bpy = 2,2′-bipyridine) undergoes substitution with various phosphorus ligands, PR_3 (R = alkyl, alkoxy, or allyl group),

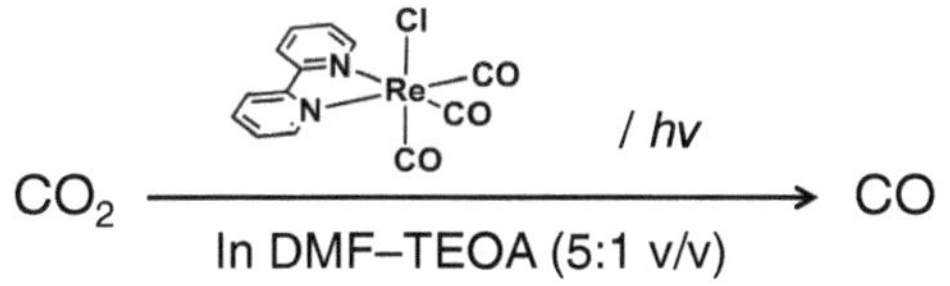

Scheme 1 Photocatalytic CO_2 reduction into CO on *fac*-$[Re(bpy)CO_3Cl]^+$ in a mixed solution of DMF and TEOA

giving *fac*-[Re(bpy)$(CO)_3(PR_3)]^+$ [5]. The photocatalytic activity for CO_2 reduction was found to depend strongly on the phosphorus ligand [6]. For example, *fac*-[Re (bpy)$(CO)_3\{P(OEt)_3\}]^+$ is an efficient photocatalyst for CO formation with a quantum yield (QY) of 38 % at 365 nm [7], which is twice as large compared with that with *fac*-Re(bpy)$(CO)_3$Cl.

The reaction mechanism of photocatalytic CO_2 reduction in a homogeneous system was investigated by Ishitani et al. in detail using three different Re(I) diimine complexes, *fac*-[Re(bpy)$(CO)_3$L] (L = SCN^- (**1–NCS**), Cl^- (**1–Cl**), and CN^- (**1–CN**)) [8]. The reaction scheme is given in Scheme 2. It has been revealed by means of laser spectroscopy that the initial step of the reaction is the reductive quenching of the triplet metal-to-ligand charge transfer (^{3}MLCT) excited-state of the Re complex by TEOA, generating one-electron reduced (OER) species of the rhenium complex $[Re^I(N\wedge N^{\bullet -})(CO)_3X]^-$ [9–11]. The corresponding OER species of these complexes play two important roles of capturing CO_2 after loss of the monodentate ligand (L) and of donating the second electron to CO_2 by another OER species without losing L. In the case of **1–NCS,** the corresponding OER species play these two roles in the photocatalytic reaction, resulting in more efficient CO evolution (30 % QY) than that of **1–Cl** (16 % QY), whose OER species are too short-loved to accumulate during the photocatalytic reaction. On the other hand, **1–CN** showed no photocatalytic ability, because the corresponding OER species does not dissociate the CN- ligand. Based on this mechanistic information, the most efficient photocatalytic system was successfully developed using a mixture of *fac*-[Re(bpy)$(CO)_3(CH_3CN)]^+$ and *fac*-[Re{4,4′-$(MeO)_2$bpy}$(CO)_3\{P(OEt)_3\}]^+$, which respectively work as a catalyst and a redox sensitizer. The QY of this system was 59 % at the optimal condition.

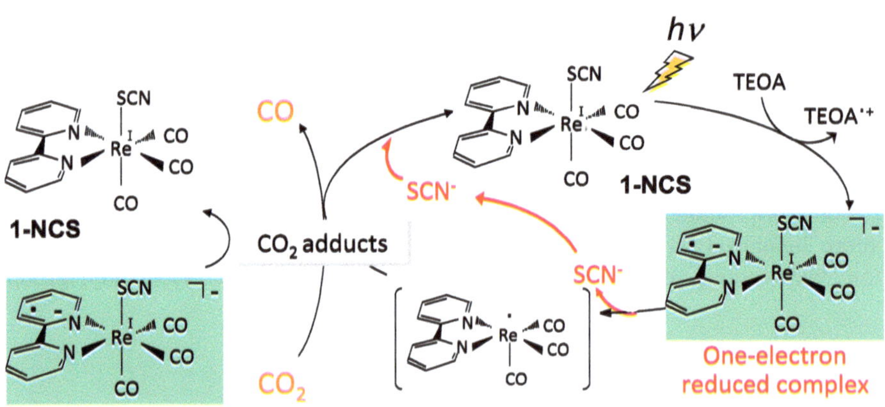

Scheme 2 Photocatalytic reaction mechanism by **1–NCS**. (Reproduced with permission from Ref. [8]. Copyright 2008, American Chemical Society)

Based on the mechanistic study, Ishitani et al. have proposed strategies to develop a highly efficient CO_2 reduction photocatalyst.

(1) Efficient formation of OER species by quenching of ^{3}MLCT excited-state by an electron donor.
(2) Effective production of [Re(LL$^{\bullet-}$)(CO)$_3$] by dissociation of the ligand from the OER species.
(3) Efficient reduction of CO_2 adduct(s) by another OER species.
(4) Prompt recovery of the starting complex by re-coordination of a ligand after CO formation.

A highly efficient CO_2 reduction has thus been established. However, there still remained unclear point(s) on the reaction mechanism, especially a CO_2 addition step to the Re center. So far, 17-electron species, i.e., [Re^0(N^N)(CO)$_3$] and/or [Re^I(N^N$^{\bullet-}$)(CO)$_3$] [3, 12–14] and a Re dimer with CO_2 as a bridge ligand [14, 15] have been proposed as possible intermediates derived from reduced Re complexes. However, no clear evidence had been obtained to clarify the mechanism. Another important aspect in photocatalytic CO_2 reduction on Re complexes is that TEOA works as a special electron donor, which enhances catalytic turnover number and selectivity of CO production, compared with other reductants such as triethylamine [16], suggesting a special action of TEOA during the reaction.

Ishitani et al. pointed out that *fac*-[Re^I(bpy)(CO)$_3${R$_2$N–CH$_2$CH$_2$O–COO}] (R = CH$_2$CH$_2$OH) could be a predominant complex in various photocatalytic CO_2 reduction reactions using [Re^I(N^N)(CO)$_3$X]$^{n+}$ (X = monodentate ligand; $n = 0, 1$) type complexes in a DMF–TEOA mixed solution (Scheme 3) [17]. A DMF-coordinated complex, *fac*-[Re^I(bpy)(CO)$_3$(DMF)]$^+$ (**1**), underwent transformation upon addition of TEOA to generate *fac*-[Re^I(bpy)(CO)$_3$(OCH$_2$CH$_2$NR$_2$)] (**1′**) with an equilibrium constant of 19. Further exposure of **1′** to CO_2 resulted in the generation of **2**, whereas a similar CO_2 treatment of **1** did not. The equilibrium constant between **1′** and **2** (K = [2] /[1′][CO_2]) in a CO_2 atmosphere was estimated to be 1.7×10^3 M^{-1}, which means the exclusive formation of **2** in the CO_2-saturated mixed solution of DMF and TEOA. This also suggests that even a very low concentration of CO_2 is enough to produce the CO_2–TEOA adduct. Electrochemical analysis indicated the superior CO_2 reduction ability of **2** to **1**. On the basis of these observations,

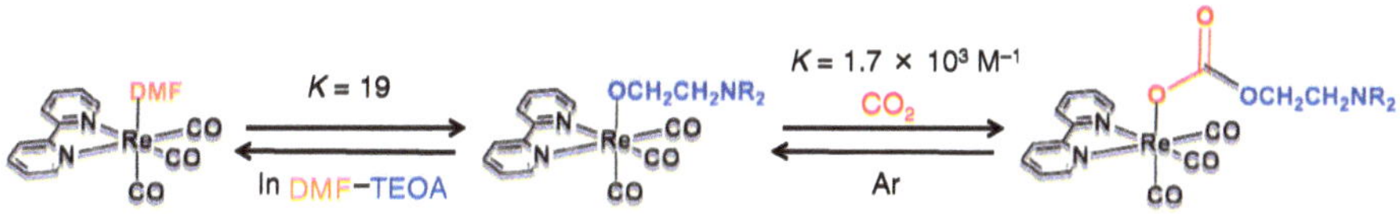

Scheme 3 Structural change of *fac*-[Re(bpy)CO$_3$DMF]$^+$ in the presence of TEOA and CO_2. (Reproduced with permission from Ref. [17]. Copyright 2013, American Chemical Society)

it was claimed that **2** should be the catalytically active species in many photocatalytic CO_2 reduction systems that have been reported so far, because TEOA was used as an electron donor in these reported systems.

2.2 Supramolecular Metal Complexes

As introduced above, Re(I) diimine complexes work as photocatalysts and electrocatalysts for CO_2 reduction. However, there are several problems in these Re-based complexes including (1) insufficient visible-light-absorption, (2) low stability, and (3) reliance on a strong electron donor. A strategy to address the problems (1) and (2) is to couple a catalytic metal complex with a redox photosensitizer so as to improve the electron transfer process from the excited-state sensitizer to the catalytic unit.

Ishitani et al. prepared a series of Ru(II)−Re(I) binuclear complexes (see Fig. 1), and examined their photocatalytic activities [18]. As shown in Fig. 1, improved photocatalytic activity was obtained with the binuclear complex **[d$_2$Ru-Re]$^{2+}$**, compared to either the corresponding Ru or Re unit alone and a physical mixture of the Ru and Re unit. This is attributed to improved intramolecular electron transfer from the OER species, which were produced following the selective excitation and subsequent reductive quenching of the Ru ^{3}MLCT excited state, to the catalytic Re moiety. It is also noted that photocatalytic responses were extended further into the visible region by applying a Ru moiety as the sensitizing unit.

Another important information obtained from this work is that the photocatalytic abilities of the binuclear complexes with a 4-methyl-4′- [1, 10] phenanthroline-[5,6-d]imidazol-2-yl)bipyridine (abbreviated as BL) bridging ligand, i.e., **[Ru-BL-Re]$^{2+}$** and **[Re-BL-Ru]$^{2+}$**, were much lower than that containing a bpyC$_3$bpy bridging ligand (**[d$_2$Ru-Re]$^{2+}$**). When $(CF_3)_2$bpy or bpy were used as peripheral ligands, the binuclear complexes also had poor photocatalytic ability, giving a turnover numbers (TNs) for CO formation of 3 and 50, respectively, for **[tfbRu-Re]$^{2+}$** and **[b$_2$Ru-Re]$^{2+}$** (data not plotted in Fig. 1). In these metal complexes that showed poorer photocatalytic activity, the intramolecular electron transfer was found to be endothermic, hindering the forward electron transfer and leading to inferior performance. In addition to the endothermic character in the intramolecular electron transfer event, it appears that electron localization on the bridging ligand appears to be another decisive factor. One can assume that, electrons in **[Ru-BL-Re]$^{2+}$** are mainly localized on the Ru end of the bridging ligand, because the energy level of the π^* orbital on the phenanthroline-imidazolyl motif of BL is lower than that on the bpy one coordinating to the Re unit. The low electron density on the catalytic Re site in the OER species could account for the low photocatalytic activity of **[Ru-BL-Re]$^{2+}$**.

In the case of **[Re-BL-Ru]$^{2+}$**, in contrast, electron localization must occur on the Re site, which is supposed to contribute to higher photocatalytic activity. It has been reported that photocatalytic CO_2 reduction activities of mononuclear complexes of

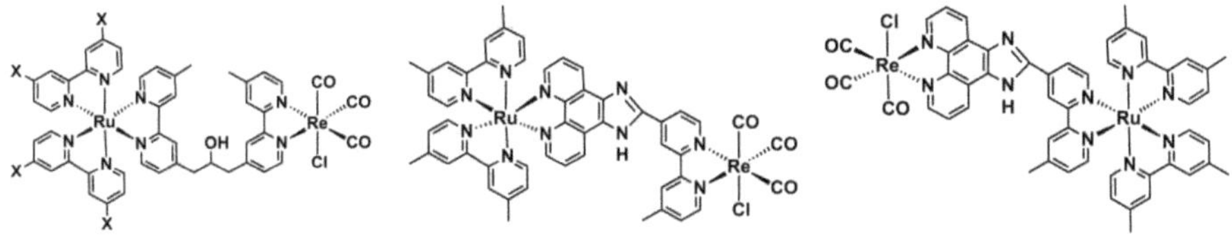

$[b_2Ru\text{-}Re]^{2+}$: X = H
$[d_2Ru\text{-}Re]^{2+}$: X = CH_3
$[tfbRu\text{-}Re]^{2+}$: X = CF_3

$[Ru\text{-}BL\text{-}Re]^{2+}$

$[Re\text{-}BL\text{-}Ru]^{2+}$

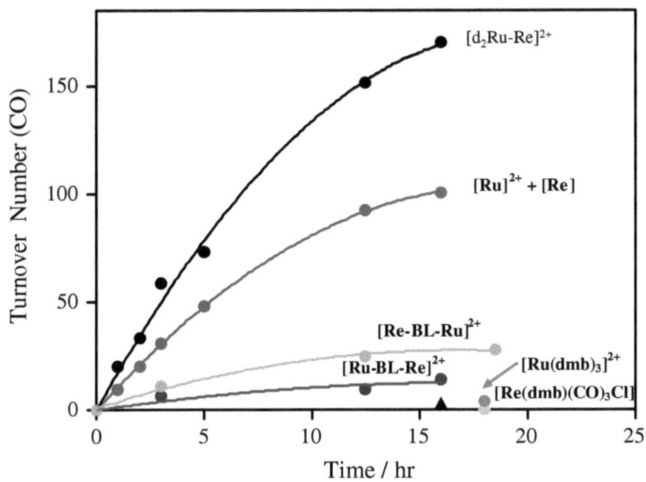

Fig. 1 Photocatalytic activities of supramolecular metal complexes (0.05 mM) for CO formation in a CO_2-saturated DMF–TEOA (5:1 v/v) solution containing 0.1 M 1-Benzyl-1, 4-dihydronicotinamide (BNAH) as a sacrificial electron donor under selective photoexcitation of the Ru moieties (λ > 500 nm). (Reproduced with permission from Ref. [18]. Copyright 2005, American Chemical Society)

fac-$[Re(N\wedge N)(CO)_3(PR_3)]^+$ (R = an alkyl group) depend strongly on their reduction potentials, $E_{1/2}^{red}(LL/LL^{\bullet -})$, and efficient photocatalytic reduction of CO_2 requires $E_{1/2}^{red}(LL/LL^{\bullet -}) < -1.41$ V vs. $Ag/AgNO_3$ [6]. In view of the wide conjugation of BL and the strong electronic interaction across the bridging ligand, the reduction ability of the OER species ($E_{1/2}^{red} = -1.1$ V vs. $Ag/AgNO_3$) should be insufficient for prompt reduction of CO_2 under these conditions.

According to the same strategy, they also recently developed photocatalytic CO_2 reduction systems using various multinuclear Ru complexes, which selectively produce HCOOH with high turnover numbers (~671) under visible light (λ = 480 nm) [19]. This kind of supramolecular metal complexes has been shown to be applicable to a Z-scheme CO_2 reduction system, in combination with a suitable semiconductor material, as will be discussed later.

2.3 Metal Complex Catalysts Based on Earth-Abundant Elements

Most of the metal complexes that have been developed to date for CO_2 reduction are comprised of Re- and Ru-based ones. In order to replace such precious metals from the sustainable chemistry point of view, an earth-abundant material that is sufficiently active, and alternative to precious metals, is highly desirable. Bourrez et al. reported that a manganese-based complex exhibited electrocatalytic activity for CO_2 reduction to CO, although the TON was moderate (TON = 13 for 4 h) [20]. Ishitani et al. used the same Mn complex as a catalyst, in combination with a ruthenium(II) tris-diimine complex as a redox sensitizer, to achieve CO_2 reduction into HCOOH in the presence of BNAH as an electron donor [21]. The QY of HCOOH formation by this Mn complex was 5.9 %, which is comparable to that achieved by a similar Re complex (6.9 %).

2.4 Enhanced Photocatalytic Activity of Rhenium(I) Complex by Light-Harvesting Periodic Mesoporous Organosilica

As briefly mentioned above, sunlight is a very "dilute" energy source (ca. 10 photons nm^{-2} s^{-1}) especially for small molecules such as metal complexes. Therefore, light-harvesting would be highly desirable for efficient photocatalysis by a homogeneous metal complex. In nature, a wheel-like array of chlorophylls in LH1 and LH2 of purple photosynthetic bacteria efficiently absorbs sunlight, funneling the captured energy to a reaction center by resonance energy transfer (RET) with a QY of almost unity [22]. For the construction of an artificial photosynthetic system, the three-dimensional organization of molecular parts, that is, light absorbers and multi-electron catalysts, at appropriate positions is of particular importance, because the RET efficiency is strongly dependent on the distance between the energy donor and acceptor molecules and their orientation [23].

Inagaki et al. employed a mesoporous biphenyl-silica (Bp-PMO) anchoring *fac*-$[Re^{I}(bpy)(CO)_3(PPh_3)]^{+}(OTf)^{-}$ (OTf = CF_3SO_3) in the mesochannels for CO_2 reduction [24]. The structure of this hybrid material is depicted in Fig. 2. The incident photons (λ = 280 nm) were effectively absorbed by the biphenyl groups in Bp-PMO, and the excited energy was funneled into the Re complex by RET. As the result, photocatalytic CO evolution from CO_2 was enhanced by a factor of 4.4, compared with direct excitation of the Re complex. In addition, Bp-PMO helped to protect the Re complex against photo-decomposition. These results demonstrate the potential of PMOs as a light-harvesting antenna for designing various photoreaction systems, mimicking the natural photosynthesis.

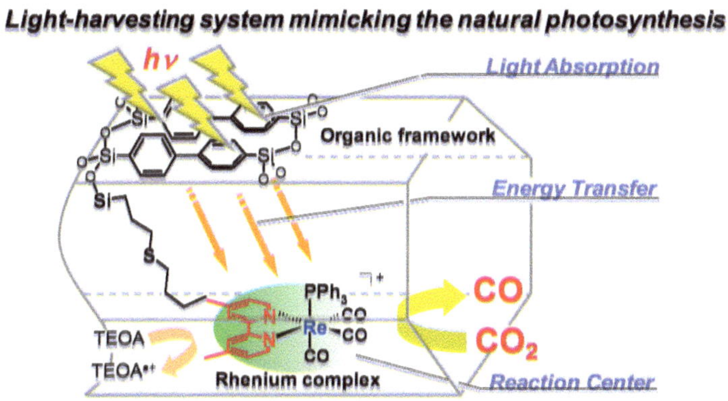

Fig. 2 Schematic representation of light-harvesting by PMO and enhancing of photocatalysis of Re complex. (Reprinted with permission from Ref. [24]. Copyright 2010, American Chemical Society)

3 Semiconductors

As described above, certain metal complexes catalyze CO_2 reduction to CO or HCOOH photocatalytically or electrochemically with high selectivity and QYs in a homogeneous system. From the viewpoint of large-scale application and efficient solar energy utilization, however, semiconductor-based heterogeneous photocatalysts would be more advantageous over molecular-based homogeneous catalysts, considering their superior oxidation ability to utilize a mild reductant (ideally, water) and potential recyclability [25]. In this section, several kinds of heterogeneous photocatalysts based on inorganic semiconductors for CO_2 reduction are described.

3.1 CO_2 Reduction on Wide Gap Semiconductors in the Presence of CH_4 or H_2

Tanaka et al. have developed several wide-gap semiconductors (some of them may be mentioned as insulators) such as ZrO_2 and MgO for CO_2 reduction into CO in the presence of CH_4 or H_2 gas as a reductant [26–28]. Their results indicated that CO_2 adsorbed on ZrO_2 and MgO, which are solid base oxides, is reduced to formate species by gas phase H_2. Importantly, the stable linear form of CO_2 transforms into a reactive species upon adsorption, which are more susceptible to reduction than the linear form.

Their group also reported that β-Ga_2O_3 having a band gap of ca. 4.5 eV showed higher activity for photocatalytic CO_2 reduction into CO in the presence of H_2 than other solid bases such as MgO [29]. This reaction is subject to a Langmuir-Hinshelwood type mechanism, where the monodentate bicarbonate species was

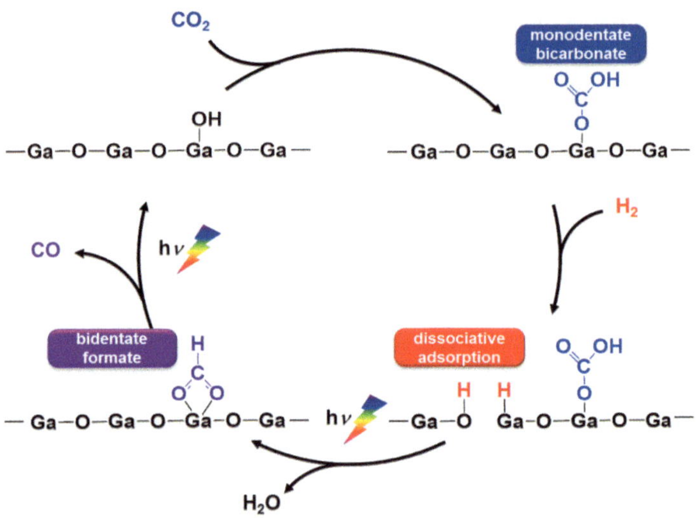

Scheme 4 Mechanism of photocatalytic reduction of CO_2 over β-Ga_2O_3 in the presence of H_2. (Reprinted with permission from Ref. [29]. Copyright 2010, American Chemical Society)

reduced by the dissociatively adsorbed hydrogen, thereby generating bidentate formate on β-Ga_2O_3 that was finally decomposed into CO under photo-irradiation (Scheme 4). Despite the inferior affinity of β-Ga_2O_3 with CO_2, the dissociatively adsorbed H_2 on β-Ga_2O_3 was responsible for the higher photocatalytic activity.

3.2 *Semiconductor Photocatalysts Workable in Water*

As described above, some of wide-gap semiconductor oxides (or insulators) have been shown to exhibit activity for CO_2 reduction into CO in the presence of CH_4 or H_2 as an electron donor. However, a semiconductor photocatalyst that is capable of reducing CO_2 using water as an electron source and of producing stoichiometric amount of O_2 had not been reported until recently.

In 2011, Kudo et al. reported that $ALa_4Ti_4O_{15}$ (A = Ca, Sr, and Ba) layered perovskites having 3.79–3.85 eV band gaps showed photocatalytic activity for CO_2 reduction to form CO and water oxidation into O_2 [30]. $ALa_4Ti_4O_{15}$ (A = Ca, Sr, and Ba) has been originally developed as a photocatalyst for overall water splitting into H_2 and O_2 under UV irradiation [31]. Table 1 summarizes the photocatalytic activities of $ALa_4Ti_4O_{15}$ (A = Ca, Sr, and Ba) for CO_2 reduction in water. While the CO_2 reduction activity was negligible in the absence of a cocatalyst, the activity could be enhanced by modification with Ag cocatalysts. Among three semiconductors tested, the Ba derivative was found to exhibit the highest performance. Under Ar bubbling condition, water splitting reaction proceeds on Ag/$BaLa_4Ti_4O_{15}$,

Table 1 Photocatalytic activities for CO_2 reduction over $ALa_4Ti_4O_{15}$ (A = Ca, Sr, and Ba) modified with Ag cocatalyst under UV irradiation ($\lambda > 200$ nm)[a]

Entry	Photocatalyst	Cocatalyst (wt%)	Loading method	Activity/μmol h^{-1}			
				H_2	O_2	CO	HCOOH
1	$CaLa_4Ti_4O_{15}$	0	–	1.3	0.6	0.07	0
2	$CaLa_4Ti_4O_{15}$	Ag (1.0)	Liquid-phase reduction	3.2	6.6	9.3	0.4
3	$SrLa_4Ti_4O_{15}$	0	–	0.8	0.5	0.06	0
4	$SrLa_4Ti_4O_{15}$	Ag (1.0)	Liquid-phase reduction	4.8	5.8	7.1	0.8
5	$BaLa_4Ti_4O_{15}$	0	–	5.3	2.4	0	0
6	$BaLa_4Ti_4O_{15}$	Ag (1.0)	Liquid-phase reduction	5.6	12	19	0.4
7	$BaLa_4Ti_4O_{15}$	Ag (1.0)	Impregnation[b]	5.6	8.7	8.9	0.3
8	$BaLa_4Ti_4O_{15}$	Ag (1.0)	Photodeposition	10	7.0	4.3	0.3
9[c]	$BaLa_4Ti_4O_{15}$	Ag (1.0)	Liquid-phase reduction	20	11	0	0

[a]Catalyst 0.3 g, water 360 mL, CO_2 flow system (15 mL min^{-1}), a 400 W high-pressure mercury lamp, an inner irradiation quartz cell
[b]Calcination at 723 K for 1 h in air, followed by reduction with H_2 at 473 K for 2 h
[c]Ar flow

but continuous CO_2 bubbling of the reactant solution allowed one to proceed CO_2 reduction into CO and HCOOH as the major and minor product, respectively. Although H_2 evolution via overall water splitting could not be suppressed completely, modification of $BaLa_4Ti_4O_{15}$ with ~10 nm Ag nanoparticles, which was achieved by employing the liquid-phase reduction method, improved the selectivity for CO_2 reduction while suppressing the undesirable H_2 evolution.

Even under CO_2 bubbling, cocatalysts of NiO_x, Ru, Cu, and Au did not achieve any appreciable CO_2 conversion, but promoted overall water splitting. Ag is known to be an efficient electrocatalyst for CO_2 reduction, but has relatively large overpotential for H_2 evolution [32]. This is a preferable feature for use as a cocatalyst for CO_2 reduction on a semiconductor photocatalyst, and appears to contribute to the superior performance of Ag-loaded material for CO_2 fixation. Kudo et al. also pointed out that continuous bubbling of the reactant suspension with CO_2 was important to get more CO, suggesting the occurrence of some backward reactions. It should be noted that in the optimal condition, the ratio of reduction/oxidation products $\{(CO + HCOOH + H_2)/O_2\}$ was almost equal to 2/1, consistent with the reaction stoichiometry. The stoichiometric evolution of O_2 clearly indicated that water was consumed as a reducing reagent (an electron donor) for the CO_2 reduction. Thus, an uphill reaction of CO_2 reduction accompanied with water oxidation was achieved using the $Ag/BaLa_4Ti_4O_{15}$ photocatalyst.

Teramura et al. have developed layered double hydroxides (LDHs; $[M^{2+}_{1-x}M^{3+}_x(OH)_2]^{x+}(A^{n-})_{x/n}\ mH_2O$) as new heterogeneous photocatalysts for CO_2 reduction

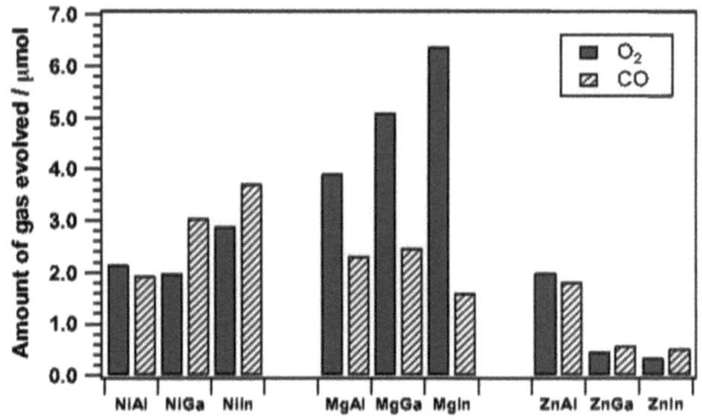

Fig. 3 Yields of O_2 and CO for the photocatalytic conversion of CO_2 in water, over various LDHs ($M^{2+}/M^{3+} = 3$) after 10 h of photo-irradiation ($\lambda > 200$ nm). (Reprinted with permission from Ref. [33]. Copyright 2011, Wiley–VCH Verlag GmbH and Co. KGaA)

workable in water under UV irradiation ($\lambda > 200$ nm) [33]. LDHs are natural or synthetic clays that consist of brucite ($Mg(OH)_2$)-like positively charged two-dimensional sheets interleaved with anionic species (A^{n-}) such as CO_3^{2-} for charge compensation, in which some divalent cations such as Mg^{2+} are substituted by trivalent cations. Although the reaction products did not meet the stoichiometry (in most cases, excess O_2 evolution was observed) and the reason still remains unclear, several kinds of LDHs showed activity for CO and O_2 evolution in water, as shown in Fig. 3. As exemplified by Mg-In LDH, interestingly, neither Mg- nor In-hydroxide gave CO or O_2 from water in the presence of CO_2, but the combination of the two metals to make a Mg-In LDH resulted in clearly observable CO and O_2 evolution. It indicates the importance of the formation of LDH structure for driving CO_2 reduction in water. They also conducted isotope tracer experiments with $^{13}CO_2$, which indicated that the main source of CO generated was CO_2 molecules in gas phase, but some residual CO_3^{2-} ions in the interlayer became the source of CO.

4 Metal-Complex/Semiconductor Hybrid Photocatalysts

4.1 The Proof-of-Concept

As introduced above, certain metal complexes based on rhenium or ruthenium catalyze CO_2 reduction to CO or HCOOH (photo)catalytically with high selectivity and QYs. However, the oxidation ability of these metal complexes is in general too low to oxidize water. By contrast, while the selectivity for CO_2 reduction is not very high, the stability of semiconductors for oxidation reactions is attractive.

On the basis of these backgrounds, one can simply draw a composite material, as shown in Scheme 5, consisting of a light-absorbing semiconductor and a catalytic

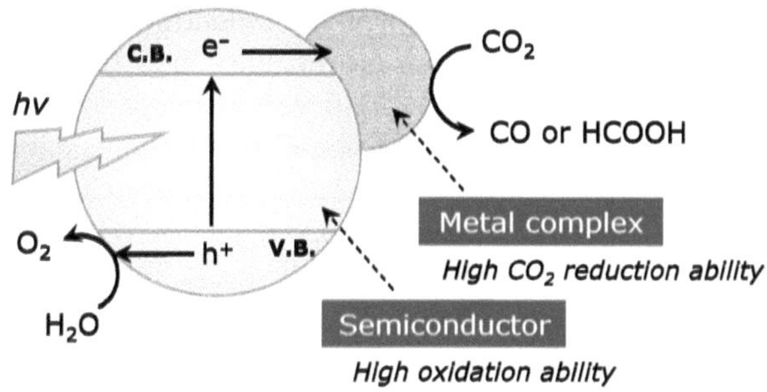

Scheme 5 Photocatalytic CO_2 reduction and water oxidation on a metal complex/semiconductor hybrid. V.B. and C.B. indicate the valence and conduction band of a semiconductor, respectively

metal complex, which shows high performance both for water oxidation and CO_2 reduction. In 2010, Morikawa et al. reported a proof-of-concept of this kind of a metal-complex/semiconductor hybrid photocatalyst to reduce CO_2 into formic acid [34]. They used a p-type semiconductor, N-doped Ta_2O_5 (energy gap = ca. 2.6 eV), and ruthenium(II) complexes $[Ru(bpy)_2(CO)_2]^{2+}$, $[Ru(dcbpy)(bpy)(CO)_2]^{2+}$ or $[Ru(dcbpy)_2(CO)_2]^{2+}$ (dcbpy: 4,4′-dicarboxy-2,2′-bipyridine; bpy: 2,2′-bipyridine). $[Ru(dcbpy)_2(CO)_2]^{2+}$ is an electrocatalyst for CO_2 reduction into HCOOH [35]. Under irradiation of the composite with visible light in a mixed solution of acetonitrile (MeCN) and TEOA, electrons and holes are generated in the conduction and valence band of N-Ta_2O_5, respectively. Here, the adsorbed metal complexes do not essentially harvest the incident photons. The conduction band electrons move to the adsorbed Ru complex, thereby reducing CO_2 into HCOOH. On the other hand, holes left behind the valence band are consumed by oxidation of TEOA. The selectivity of HCOOH production was more than 75 % before the turnover number underwent saturation. Unfortunately, however, N-Ta_2O_5 is unable to oxidize water into molecular O_2 because the valence band potential is more negative than the water oxidation potential. Therefore, this system is not applicable to an artificial photosynthetic assembly using water as an electron source.

4.2 Metal-Complex/Polymeric Semiconductor Hybrid

As an alternative semiconductor material that has stronger oxidation ability, Maeda et al. focused on carbon nitride polymers [36, 37]. Carbon nitride is an earth-abundant polymer semiconductor photocatalyst, which has recently been developed for water splitting with visible light by the same group [38, 39]. It has several proposed allotropes with diverse properties, but the graphitic phase is regarded as the most stable under ambient conditions. Graphitic carbon nitride is yellow powder with high chemical stability both in acid and base, exhibiting a steep absorption edge at around 450 nm and a tail extending to 600 nm. Importantly, the material shows photocatalytic activity for water oxidation, in contrast to nitrogen doped Ta_2O_5. However, there had been no reliable report on photocatalytic CO_2 reduction using C_3N_4 as a photocatalyst until very recently.

Mesoporous graphitic carbon nitride (mpg-C_3N_4) polymers with a ruthenium complex, *cis, trans*-[Ru{4,4′-($CH_2PO_3H_2$)$_2$-2,2′-bipyridine}$(CO)_2Cl_2$] (abbreviated **Ru** for simplicity), that works a catalyst for CO_2 reduction were combined together to create a new CO_2 reduction photo-assembly [36]. Table 2 summarizes CO_2 reduction activities, which were tested in a MeCN–TEOA mixture (4:1 v/v) under >400 nm irradiation. Mpg-C_3N_4 alone did not show any activity for CO_2 reduction. However, combining mpg-C_3N_4 with **Ru** resulted in the production of HCOOH and CO as CO_2 reduction products, with H_2 as a byproduct. Under optimal condition, turnover number with respect to the adsorbed **Ru** exceeded 200 after 20 h of visible light irradiation, with selectivity of formic acid production of higher than 80 %. These numbers are higher than those recorded by the previous report using nitrogen doped Ta_2O_5. Control experiments showed that using an insulator, alumina, instead of carbon nitride, did not give any products. Nothing happened without catalyst sample as well. When the reaction was conducted under argon atmosphere, no carbon-containing product was obtained, evolving H_2 alternatively. Without TEOA, the amounts of produced HCOOH and CO became very low.

Figure 4 shows an action spectrum of formic acid production on **Ru**/mpg-C_3N_4. The apparent quantum yield (AQY) decreased with increasing the wavelength of incident light, and reached zero at 550 nm. This change in AQY corresponds to the light-absorption profile of carbon nitride, which is shown by red curve. It clearly means that the formic acid production originates from light absorption by carbon nitride. Because **Ru**/mpg-C_3N_4 photocatalyst consists of large amount of carbon, isotope tracer experiments were conducted using $^{13}CO_2$ as the reactant. Interestingly, the main product of HCOOH was found to originate solemnly from

Table 2 Photocatalytic activities for CO_2 reduction over **Ru**/mpg-C_3N_4 in a CO_2-saturated MeCN$^-$–TEOA mixed solution under visible light (λ > 400 nm)[a]

Entry	Photocatalyst	Solution	Amount of products (5 h)/nmol		
			H_2	CO	HCOOH
1	mpg-C_3N_4	MeCN–TEOA	354	0	0
2	**Ru**/mpg-C_3N_4	MeCN–TEOA	1267	580	5455
3	**Ru**/Al_2O_3	MeCN–TEOA	7	0	0
4	None	MeCN–TEOA	0	0	0
5[b]	**Ru**/mpg-C_3N_4	MeCN–TEOA	3772	0	0
6	**Ru**/mpg-C_3N_4	MeCN	0	160	110

[a]Catalyst 8.0 mg (**Ru** 7.8 μmol g^{-1} adsorbed), MeCN–TEOA mixed solution (4:1 v/v) 4 mL, a 450 W high-pressure mercury lamp with a $NaNO_2$ solution filter, a Pyrex test tube cell (11 mL capacity)

[b]Ar atomsphere. (Reproduced with permission from Ref. [36]. Copyright 2013, The Royal Society of Chemistry)

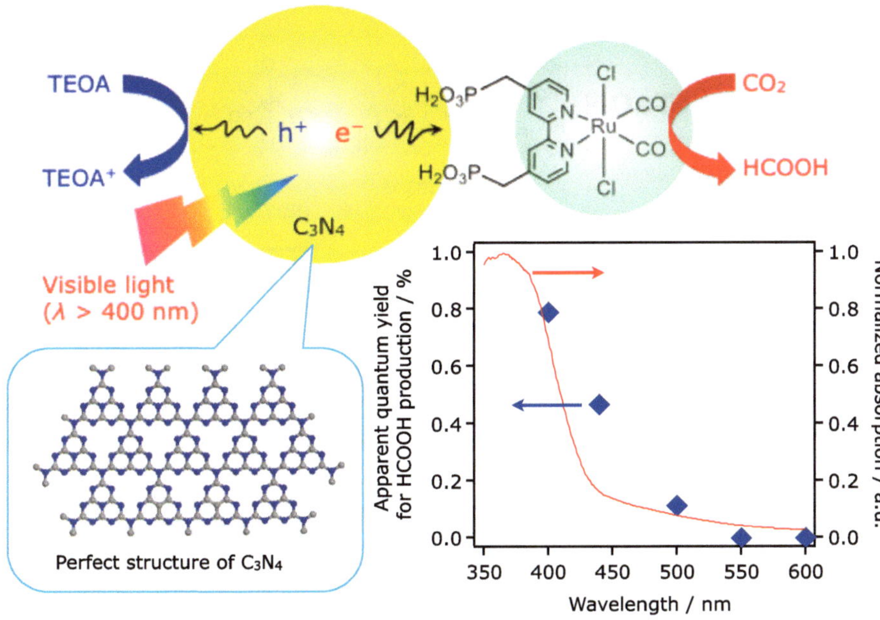

Fig. 4 Schematic illustration of photocatalytic CO_2 reduction into HCOOH on **Ru**/mpg-C_3N_4 under visible light, along with the action spectrum. (Reproduced with permission from Ref. [36]. Copyright 2013, The Royal Society of Chemistry)

CO_2, not the decomposition of C_3N_4 component. Based on these results, it was concluded that **Ru** and carbon nitride work as CO_2 reduction catalyst and light-absorber, respectively. Also, TEOA is an electron donor to scavenge holes in the valence band of carbon nitride, and a proton source.

In contrast to inorganic semiconductors, it is possible to control both bulk and surface properties of C_3N_4 based on an organic chemistry protocol [40], thereby modulating the band-gap structure and introducing a desired organic moiety that anchors a metal-complex catalyst. Besides, C_3N_4 is structurally flexible, exhibiting various shapes with the aid of a hard template such as silica during the synthesis [41]. Maeda et al. have also reported that the activity of **Ru**/C_3N_4 for the CO_2 reduction reaction is sensitive to specific surface area and crystallinity of carbon nitride, but is largely insensitive to the pore size and the volume [37].

4.3 Artificial Z-Scheme

Very recently, Ishitani et al. developed a new type of photocatalytic CO_2 fixation system using Ag-loaded TaON semiconductor and a Ru(II) binuclear complex, which works according to Z-scheme principle somewhat similar to natural

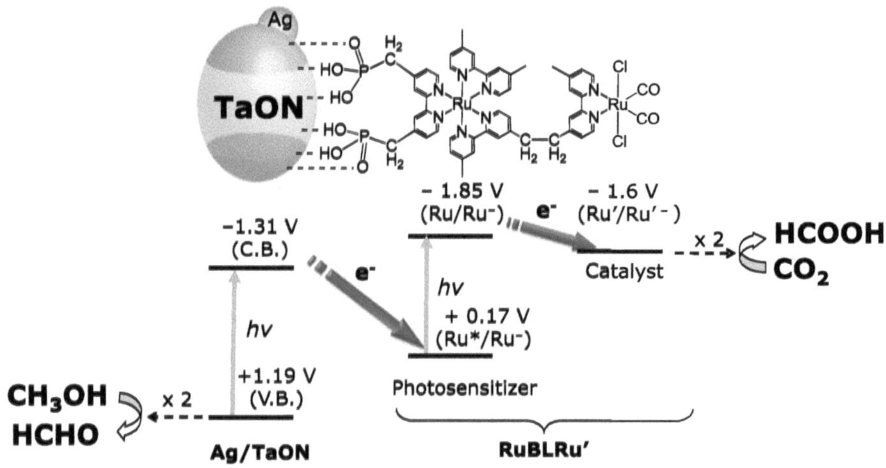

Scheme 6 Artificial Z-scheme for photocatalytic CO_2 reduction using Ag-loaded TaON semiconductor and a Ru(II) binuclear complex (RuBLRu'). (Reproduced with permission from Ref. [42]. Copyright 2013, American Chemical Society)

photosynthesis in green plants [42]. As illustrated in Scheme 6, two components of TaON and light-harvesting Ru unit both undergo photoexcitation upon visible light in the initial step. The photogenerated hole in the TaON valence band oxidizes methanol, and the conduction band electrons move to the excited or oxidized photosensitizer unit, but cannot be transferred to the ground state. An OER species generated as a result of the interfacial electron transfer is consumed by intramolecular electron transfer, which is thermodynamically down-hill, finally reducing CO_2 into HCOOH on the catalytic Ru unit. Because formic acid production from CO_2 involves a two-electron reduction, the stepwise two-photon absorption and subsequent electron transfer processes would occur twice during the reduction of CO_2 to give one HCOOH molecule. It should be noted that the whole reaction is energetically up-hill, involving a positive change in the Gibbs energy of 83.0 kJ mol^{-1}. Isotope tracer experiments indicated that this hybrid material photocatalytically produced formic acid as the major reduction product and formaldehyde as the oxidation product from CO_2 and methanol, respectively. Under visible light ($\lambda > 400$ nm), both Ag/TaON and the sensitizer unit in the supramolecular complex undergo photoexcitation. The conduction band electrons in Ag/TaON migrate to the excited state or oxidized sensitizer unit, producing a one-electron reduced species. Subsequent intermolecular electron transfer occurs from the one-electron- reduced species in the photosensitizer unit to the catalyst unit, as it is a thermodynamically downhill process. Finally, holes left in the valence band of Ag/TaON oxidize methanol to give formaldehyde, whereas electrons transferred to the catalyst unit reduce CO_2 into formic acid. Since the CO_2 reduction to give formic acid is a two-electron process, the stepwise two-photon absorption and the subsequent electron transfer events are likely to occur twice during the reaction to give one HCOOH molecule.

Currently, the main problem of this Z-scheme system includes competitive H_2 evolution that lowers the selectivity of CO_2 reduction, significant electron–hole recombination in the TaON component [43], and possible back electron transfer from the excited-state photosensitizing unit to Ag/TaON, which is thermodynamically a down-hill process. To address these problems, refinement of preparation condition of TaON as well as proper design of the metal complex component to maximize the forward electron transfer rate is required. In addition, role(s) of Ag deposits on TaON need to be clarified, as the efficiency of this system becomes very low in the absence of the Ag modification.

5 Summary and Future Outlook

In this chapter, photocatalytic CO_2 reduction both in homogeneous and heterogeneous systems are reviewed. Certain metal complexes consisting rhenium, ruthenium, or manganese are shown to work as efficient (photo)catalysts for CO_2 reduction into CO or HCOOH with high quantum yields and selectivity. Unfortunately, however, no metal complex that is capable of oxidizing water to drive CO_2 reduction has been reported so far. On the other hand, semiconductor photocatalysts having high photooxidation ability could be applicable to a CO_2 reduction system that should utilize water as an electron source. For example, $ALa_4Ti_4O_{15}$ (A = Ca, Sr, and Ba) modified with Ag nanoparticles exhibit activity for CO_2 reduction and water oxidation under band-gap irradiation. A suitable combination between a metal complex and a semiconductor led to the development of new visible-light CO_2 reduction systems. Importantly, some of them could work non-sacrificially, converting visible-light energy into chemical energy.

References

1. Morris AJ, Meyer GJ, Fujita E (2009) Molecular approaches to the photocatalytic reduction of carbon dioxide for solar fuels. Acc Chem Res 42:1983–1994
2. Takeda H, Ishitani O (2010) Development of efficient photocatalytic systems for CO_2 reduction using mononuclear and multinuclear metal complexes based on mechanistic studies. Coord Chem Rev 254:346–354
3. Hawecker J, Lehn J-M, Ziessel R (1983) Efficient photochemical reduction of CO_2 to CO by visible light irradiation of systems containing $Re(bipy)(CO)_3X$ or $Ru(bipy)_3^{2+}$–Co^{2+} combinations as homogeneous catalysts. J Chem Soc Chem Commun 536–538
4. Yui T, Kan A, Saitoh C, Koike K, Ibusuki T, Ishitani O (2011) Photochemical reduction of CO_2 using TiO_2: effects of organic adsorbates on TiO_2 and deposition of Pd onto TiO_2. ACS Appl Mater Interfaces 3:2594–2600
5. Hori H, Koike K, Ishizuka M, Takeuchi K, Ibusuki T, Ishitani O (1997) Preparation and characterization of $[Re(bpy)(CO)_3L][SbF_6]$ (L = phosphine, phosphite). J Organomet Chem 530:169–176

6. Koike K, Hori H, Ishizuka M, Westwell JR, Takeuchi K, Ibusuki T, Enjouji K, Konno H, Sakamoto K, Ishitani O (1997) Key process of the photocatalytic reduction of CO_2 using [Re (4,4-X_2-bipyridine)$(CO)_3PR_3]^+$ (X = CH_3, H, CF_3; PR_3 = Phosphorus Ligands): dark reaction of the one-electron-reduced complexes with CO_2. Organometallics 16:5724–5729
7. Hori H, Johnson FPA, Koike K, Ishitani O, Ibusuki T (1996) Efficient photocatalytic CO_2 reduction using $[Re(bpy)(CO)_3\{P(OEt)_3\}]^+$. J Photochem Photobiol A Chem 96:171–174
8. Takeda H, Koike K, Inoue H, Ishitani O (2008) Development of an efficient photocatalytic system for CO_2 reduction using rhenium(I) complexes based on mechanistic studies. J Am Chem Soc 130:2023–2031
9. Kutal C, Weber MA, Ferraudi G, Geiger D (1985) A mechanistic investigation of the photoinduced reduction of carbon dioxide mediated by tricarbonylbromo(2,2'-bipyridine) rhenium(I). Organometallics 4:2161–2166
10. Kalyanasundaram K (1986) Luminescence and redox reactions of the metal-to-ligand charge-transfer excited state of tricarbonylchloro-(polypyridyl)rhenium(I) complexes. J Chem Soc Faraday Trans 2(82):2401–2415
11. Kutal C, Corbin AJ, Ferraudi G (1987) Further studies of the photoinduced reduction of carbon dioxide mediated by tricarbonylbromo(2,2′-bipyridine)rhenium(I). Organometallics 6:553–557
12. Smieja JM, Benson EE, Kumar B, Grice KA, Seu CS, Miller AJM, Mayer JM, Kubiak CP (2012) Kinetic and structural studies, origins of selectivity, and interfacial charge transfer in the artificial photosynthesis of CO. Proc Natl Acad Sci 109:15646–15650
13. Smieja JM, Kubiak CP (2010) Re(bipy-tBu)$(CO)_3Cl^-$improved catalytic activity for reduction of carbon dioxide: IR-spectroelectrochemical and mechanistic studies. Inorg Chem 49:9283–9289
14. Hayashi Y, Kita S, Brunschwig BS, Fujita E (2003) Involvement of a binuclear species with the Re−C(O)O−Re moiety in CO_2 reduction catalyzed by tricarbonyl rhenium(I) complexes with diimine ligands: strikingly slow formation of the Re−Re and Re−C(O)O−Re species from Re(dmb)(CO)3S (dmb = 4,4′-Dimethyl-2,2′-bipyridine, S = Solvent). J Am Chem Soc 125:11976–11987
15. Agarwal J, Fujita E, Schaefer HF III, Muckerman JT (2012) Mechanisms for CO Production from CO_2 using reduced rhenium tricarbonyl catalysts. J Am Chem Soc 134:5180–5186
16. Lehn J-M, Ziessel R (1982) Photochemical generation of carbon monoxide and hydrogen by reduction of carbon dioxide and water under visible light irradiation. Proc Natl Acad Sci 79:701–704
17. Morimoto T, Nakajima T, Sawa S, Nakanishi R, Imori D, Ishitani O (2013) CO_2 capture by a rhenium(I) complex with the aid of triethanolamine. J Am Chem Soc 135:16825–16828
18. Gholamkhass B, Mametsuka H, Koike K, Tanabe T, Furue M, Ishitani O (2005) Architecture of supramolecular metal complexes for photocatalytic CO_2 reduction: Rutheniu−rhenium Bi-and tetranuclear complexes. Inorg Chem 44:2326–2336
19. Tamaki Y, Morimoto T, Koike K, Ishitani O (2012) Photocatalytic CO_2 reduction with high turnover frequency and selectivity of formic acid formation using Ru(II) multinuclear complexes. Proc Natl Acad Sci 109:15673–15678
20. Bourrez M, Molton F, Chardon-Noblat S, Deronzier A (2012) [Mn(bipyridyl)$(CO)_3$Br]: an abundant metal carbonyl complex as efficient electrocatalyst for CO_2 reduction. Angew Chem Int Ed 50:9903–9906
21. Takeda H, Koizumi H, Okamoto K, Ishitani O (2014) Photocatalytic CO_2 reduction using a Mn complex as a catalyst. Chem Commun 50:1491–1493
22. Pullerits T, Sundström V (1996) Photosynthetic light-harvesting pigment-protein complexes: toward understanding how and why. Acc Chem Res 29:381–389
23. Alstrum-Acevedo JH, Brennaman MK, Meyer TJ (2005) Chemical approaches to artificial photosynthesis. 2. Inorg Chem 44:6802–6827
24. Takeda H, Ohashi M, Tani T, Ishitani O, Inagaki S (2010) Enhanced photocatalysis of rhenium(I) complex by light-harvesting periodic mesoporous organosilica. Inorg Chem 49:4554–4559

25. Maeda K (2011) Photocatalytic water splitting using semiconductor particles: History and recent developments. J Photochem Photobiol C: Reviews 12:237–268
26. Kohno Y, Tanaka T, Funabiki T, Yoshida S (1997) Photoreduction of carbon dioxide with methane over ZrO_2. Chem Lett 993–994
27. Kohno Y, Ishikawa H, Tanaka T, Funabiki T, Yoshida S (2001) Photoreduction of carbon dioxide by hydrogen over magnesium oxide. Phys Chem Chem Phys 3:1108–1113
28. Teramura K, Tanaka T, Ishikawa H, Kohno Y, Funabiki T (2004) Photocatalytic reduction of CO_2 to CO in the presence of H_2 or CH_4 as a reductant over MgO. J Phys Chem B 108:346–354
29. Tsuneoka H, Teramura K, Shishido T, Tanaka T (2010) Adsorbed species of CO_2 and H_2 on Ga_2O_3 for the photocatalytic reduction of CO_2. J Phys Chem C 114:8892–8898
30. Iizuka K, Wato T, Miseki Y, Saito K, Kudo A (2011) Photocatalytic reduction of carbon dioxide over Ag cocatalyst-loaded $ALa_4Ti_4O_{15}$ (A = Ca, Sr, and Ba) using water as a reducing reagent. J Am Chem Soc 133:20863–20868
31. Miseki Y, Kato H, Kudo A (2009) Water splitting into H_2 and O_2 over niobate and titanate photocatalysts with (111) plane-type layered perovskite structure. Energy Environ Sci 2:306–314
32. Hori Y, Wakabe H, Tsukamoto T, Koga O (1994) Electrocatalytic process of CO selectivity in electrochemical reduction of CO_2 at metal electrodes in aqueous media. Electrochim Acta 39:1833–1839
33. Teramura K, Iguchi S, Mizuno Y, Shishido T, Tanaka T (2012) Photocatalytic conversion of CO_2 in water over layered double hydroxides. Angew Chem Int Ed 51:8008–8011
34. Sato S, Morikawa T, Saeki S, Kajino T, Motohiro T (2010) Visible-light-induced selective CO_2 reduction utilizing a ruthenium complex electrocatalyst linked to a p-type nitrogen-doped Ta_2O_5 semiconductor. Angew Chem Int Ed 49:5101–5105
35. Ishida H, Tanaka K, Tanaka T (1987) Electrochemical CO_2 reduction catalyzed by ruthenium complexes $[Ru(bpy)_2(CO)_2]^{2+}$ and $[Ru(bpy)_2(CO)Cl]^+$. Effect of pH on the formation of CO and $HCOO^-$. Organometallics 6:181–186
36. Maeda K, Sekizawa K, Ishitani O (2013) A polymeric-semiconductor–metal-complex hybrid photocatalyst for visible-light CO_2 reduction. Chem Commun 49:10127–10129
37. Maeda K, Kuriki R, Zhang M, Wang X, Ishitani O (2014) The effect of the pore-wall structure of carbon nitride on photocatalytic CO_2 reduction under visible light. J Mater Chem A 2:15146–15151
38. Wang X, Maeda K, Thomas A, Takanabe K, Xin G, Carlsson JM, Domen K, Antonietti M (2009) A metal-free polymeric photocatalyst for hydrogen production from water under visible light. Nat Mater 8:76–80
39. Maeda K, Wang X, Nishihara Y, Lu D, Antonietti M, Domen K (2009) Photocatalytic activities of graphitic carbon nitride powder for water reduction and oxidation under visible light. J Phys Chem C 113:4940–4947
40. Zhang J, Chen X, Takanabe K, Maeda K, Domen K, Fu X, Antonietti M, Wang X (2010) Synthesis of a carbon nitride structure for visible-light catalysis by copolymerization. Angew Chem Int Ed 49:441–444
41. Goettmann F, Fischer A, Antonietti M, Thomas A (2006) Chemical synthesis of mesoporous carbon nitrides using hard templates and their use as a metal-free catalyst for friedel-crafts reaction of benzene. Angew Chem Int Ed 45:4467–4471
42. Sekizawa K, Maeda K, Koike K, Domen K, Ishitani O (2013) Artificial Z-scheme constructed with a supramolecular metal complex and semiconductor for the photocatalytic reduction of CO_2. J Am Chem Soc 135:4596–4599
43. Maeda K, Higashi M, Lu D, Abe R, Domen K (2010) Efficient nonsacrificial water splitting through two-step photoexcitation by visible light using a modified oxynitride as a hydrogen evolution photocatalyst. J Am Chem Soc 132:5858–5868

Part V
Chemical, Electrochemical and Photoelectrochemical Approach for Energy Conversion: Approach Using Electrochemical Reactions

Water Splitting Using Electrochemical Approach

Akira Yamaguchi, Toshihiro Takashima, Kazuhito Hashimoto and Ryuhei Nakamura

Abstract For electrochemical water splitting, a number of bioinspired and biomimetic Mn-based materials have been developed; however, the catalytic performances markedly differ between natural and synthetic Mn catalysts. Based on the recent in situ detection of surface intermediates for the oxygen evolution reaction (OER) by MnO_2, this chapter introduces the design rationale for the efficient OER catalysts, and discusses the evolutional origin of natural Mn_4-clusters to provide a better understanding of the differences in activity between natural and man-made OER catalysts.

1 Introduction

Water is by far the most abundant source of electrons in nature. In biological systems, plants, algae, and cyanobacteria utilize water as the primary electron source to maintain cellular function and growth, thereby contributing to homeostasis and sustainable ecosystems. For humans to mimic such sustainable natural systems, the development of artificial water splitting systems that efficiently utilize water as an electron source is needed for generating renewable fuels [1–4]. In particular, relatively clean water at neutral pH is the most desirable resource, as it is abundant and safe for handling. However, artificially extracting electrons from water requires excess energy owing to the inherent difficulty of catalyzing the

A. Yamaguchi · R. Nakamura (✉)
Biofunctional Catalyst Research Team, RIKEN Center for Sustainable Resource Science,
2-1 Hirosawa, Wako, Saitama 351-0198, Japan
e-mail: ryuhei.nakamura@riken.jp

T. Takashima
Clean Energy Research Center, University of Yamanashi, Kofu, Yamanashi 400-8511, Japan

K. Hashimoto
Department of Applied Chemistry, The University of Tokyo, 7-3-1 Hongo, Bunkyo-ku,
Tokyo 113-8656, Japan

M. Sugiyama et al. (eds.), *Solar to Chemical Energy Conversion*,
Lecture Notes in Energy 32, DOI 10.1007/978-3-319-25400-5_11

overall four-electron/four-proton reaction, known as the oxygen evolution reaction (OER) ($2H_2O \rightarrow 4H^+ + 4e^- + O_2$). In contrast, the tetrameric Mn cluster of photosystem II (PSII) in photosynthetic organisms catalyzes the OER with remarkably high activity and selectivity, as demonstrated by the overpotential values of 160–300 mV [5–7] and maximum turnover frequencies of $\sim 10^3\ s^{-1}$ [8].

Inspired by natural water-splitting catalysts, OER catalysts composed of Mn, which is an inexpensive and abundant element, have attracted considerable research interest. A number of bioinspired and biomimetic Mn-based materials have been developed as water-splitting catalysts; however, the catalytic performances markedly differ between natural and synthetic Mn catalysts, particularly under neutral pH conditions. Despite intensive studies aiming at understanding the catalytic process of natural and artificial catalysts, the reasons underlying the differences in ability are unclear. Furthermore, the fundamental question as to why Mn is the only element found in nature that is capable of efficient water splitting remains unanswered. This question is particularly relevant to the evolution of PSII, as oxygen reduction reactions in respiration ($O_2 + 4H^+ + 4e^- \rightarrow 2H_2O$), which is the reverse reaction of the OER and operates at essentially the same potential region with OER (1.23 V vs. RHE), exploit several metal elements, such as Fe and Cu, but not Mn, as catalytic centers.

To help resolve the above questions, this chapter describes recent development in OER catalysts designed based on electrochemical systems, particularly focusing on MnO_2-based catalysts. First, a brief overview of electrochemical OER catalysts, including a comparison of Mn-, noble metal-, and Co-based catalysts, is presented in Sects. 2 and 3. In Sect. 4, the results of spectroscopic examination of MnO_2-based OER catalysts are presented, and the importance of Mn^{3+} in the overall water-splitting reaction is discussed. Based on these findings, the origin of the differences in activity between natural and man-made MnO_2 OER catalysts is discussed in Sect. 5. In Sect. 6, a strategy for controlling proton transfer and improving the catalytic activity of MnO_2 for OER under neutral pH conditions is proposed. In the final section, perspectives related to the evolutional origin of natural Mn_4-clusters are described.

2 Electrochemical Water Oxidation Catalysts with 3d-Block Elements

To achieve high catalytic activity for the electrochemical OER, a number of noble metals, such as Ir and Ru, have been adopted as catalysts. For example, in 1971, Trasatti et al. [9] reported that a RuO_2 film electrode prepared on titanium exhibited high OER activity, with an overpotential at 10 mA cm^{-2} of 200 mV in 1 M $HClO_4$. This finding inspired the development of a number of colloidal and heterogeneous RuO_2-based OER catalysts [10, 11]. However, despite their high catalytic activity, RuO_2-based catalysts exhibit anodic corrosion at high anodic potential, thereby limiting their practical application.

IrO_x are potentially suitable candidates for OER catalysts as they exhibit both high catalytic activity and stability [10, 12, 13]. Yagi et al. [14] reported that a monolayer of citrate-stabilized IrO_2 nanoparticles (50–100 nm) deposited on an ITO electrode had catalytic water oxidation activity with an onset potential of 1.2 V vs. SHE at pH 5.3. Furthermore, by decreasing the particle size (1.6 ± 0.6 nm), the catalytic activity of IrO_2 drastically increased, with an overpotential of 150–250 mV at 0.5 mA cm^{-2} over a pH range of 1.5–13, and showed a current efficiency of 100 % at an overpotential of 250 mV at 0.5 mA cm^{-2} [15, 16]. Although these catalysts displayed high catalytic activity and durability over a wide range of pH, the high cost and scarcity of noble metals severely limit their widespread use as water-splitting catalysts. Thus, a distinct need exists for more abundant materials that can serve as efficient catalysts in such systems.

For the utilization of abundant elements, particularly 3d-block transition metal elements, as OER catalysts, it is necessary to understand the mechanisms underlying OER activity from a theoretical perspective [17–19]. One possible determinant of OER activity is metal ion (M)–OH bond strength. In the early 1950s, Rüetschi et al. [17] theoretically investigated the OER under alkaline conditions with several types of noble metal and 3d-block transition metal electrodes, including electrodes composed of Ag, Au, Cd, Co, Cu, Fe, Ni, Pb, Pd, and Pt, and found that the overpotential for OER linearly decreases with increasing M–OH bond energy.

Another potential determinant is the occupancy of the e_g orbital in active metal sites. Iwakura et al. [18] reported that OER activity of transition metal oxides with spinel structures ($Co_xFe_{3-x}O_4$, $Mn_xFe_{3-x}O_4$, and $Ni_xFe_{3-x}O_4$) increases with increasing number of unpaired d-electrons when the rate-determining step is the generation of OH radical via oxidation of OH^- ($HO^- \rightarrow HO^\bullet + e^-$), but decreases with increasing number of unpaired d-electron if the rate-determining step is the reaction between OH radical and OH^-. Bockris et al. [19] investigated the relationship between OER activity and multiple factors, such as the effective magnetic moment, formation enthalpy of the corresponding hydroxides, M–OH bond strength, and d-electron numbers, of 18 perovskite-type oxides. The authors concluded that OER activity is related to the occupancy of the antibonding orbitals of M–OH and predicted a volcano plot for OER efficiency as a function of M–OH bond strength. Motivated by these theoretical studies and the success of d-band modeling to predict catalytic activity, in 2011, Shao-Horn et al. [20] investigated the OER activity of $Ba_{0.5}Sr_{0.5}Co_{0.8}Fe_{0.2}O_{3-\delta}$, which has a perovskite structure, and found this compound catalyzes the water oxidation reaction under alkaline conditions with an overpotential of 250 mV at current density of 50 μA cm^{-2}. In addition, these researchers investigated the OER activities of more than 10 metal oxides and found that the activity showed a volcano-shaped dependence on the electron occupancy of an e_g orbital of surface transition metal cations [20]. The highest predicted catalytic activity based on volcano plots was for the $Ba_{0.5}Sr_{0.5}Co_{0.8}Fe_{0.2}O_{3-\delta}$-based catalyst, which possesses one electron in its 3d e_g orbital (Fig. 1). Taken together, these studies demonstrate that the d-band model is effective for the rational design of OER catalysts comprised of 3d-block transition metal elements under alkaline conditions [17–20].

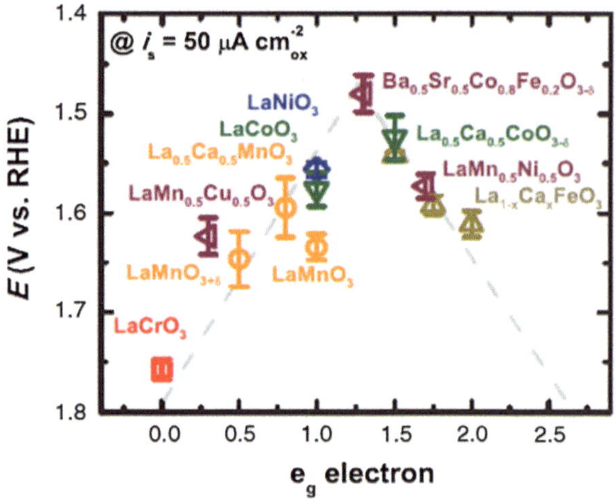

Fig. 1 The volcanic relation between the OER activity, defined by the overpotentials at 50 μA $cm^{-2}{}_{ox}$ of the OER current, and the occupancy of the e_g-symmetry electron of the transition metal (B in ABO_3) under alkaline condition [20]

It is important to note, however, that predicted catalytic efficiencies based on d-band models are not always applicable for the neutral pH conditions. For example, Raabe et al. [21] examined the surface conditions of $Pr_{1-x}Ca_xMnO_3$ (PCMO) during the OER at pH 7 by in situ transmission electron microscopy (TEM) and revealed that an amorphous PCMO layer formed with a concomitant decrease of OER activity. Furthermore, in situ electron energy-loss spectroscopy (EELS) analysis revealed that the oxidation state of the Mn site was reduced to 2.0 from the initial value of 3.2 during the reaction. This tendency for the deterioration of OER activity at neutral pH cannot be directly explained by the d-band model and is most likely because the overall rate-determining step is the storing process of oxidative power, not the downstream reactions, such as H_2O adsorption and O-O bond formation, as discussed in Sect. 4.

Among 3d-block transition metal elements, the OER catalytic efficiency of soluble aquo and hydroxo complexes of CoO_x has been widely investigated under both heterogeneous and homogeneous conditions at neutral pH since the early 1980s [22–24]. In 2008, Nocera et al. [23] reported the in situ formation of a catalytic film composed of cobalt phosphates, hydroxide, and oxides from a neutral solution of Co(II) and phosphate ions upon application of a positive potential to an ITO electrode. This film catalyst exhibited OER activity with an overpotential of 410 mV at 1 mA cm^{-2} at pH 7. In 2009, a preparation of cubic Co_3O_4 nanoparticles (5.9 ± 1.0 nm) showed an electrochemical OER current density of 10 mA cm^{-2}, with an overpotential of 330 mV, under alkaline conditions [24]. Although these Co-based materials display high catalytic activity even at neutral pH, as is the case for IrO_2, which are active over a wide pH range, Mn-based OER catalysts exhibit low activity under neutral pH conditions, as described in the following section.

3 Manganese-Based Catalysts Inspired by the Natural Oxygen-Evolving Center

Inspired by the natural O_2 evolution center, large numbers of Mn-based water oxidation catalysts, including metal complexes and bulk oxides such as MnO_2, Mn_2O_3, and MnOOH, have been investigated [25–43]. Mn-based catalysts demonstrate high OER activity under alkaline conditions. For example, a complex metal oxide $MnFe_2O_4$ catalyzes electrochemical OER in 1 M KOH at an overpotential of 400 mV at 100 mA cm^{-2} [42], and similarly, nanostructured MnOOH electrodeposited on a gold electrode at pH 14 with an overpotential of 390 mV at 50 mA cm^{-2} [27]. In addition, efficient OER catalytic activity was demonstrated for Mn_2O_3 with an overpotential of 290 mV at 1 mA cm^{-2} at pH 14 [28].

Despite the high catalytic OER activity of Mn-based catalysts under alkaline conditions, their activity dramatically decreases at neutral pH. This response was clearly demonstrated by Tamura et al. [29], who examined the OER activity of MnO_2 electrocatalysts that were synthesized on Ti or Pt electrodes by the thermal decomposition of $Mn(NO_3)_2$ under highly acidic (1 N H_2SO_4), highly alkaline (1 N KOH), and neutral (1 N K_2SO_4) conditions. Although the electrodes exhibited high catalytic activity under alkaline and acidic conditions, with an overpotential of 300 mV at 1 mA cm^{-2}, the activity was markedly lower under neutral conditions, with overpotential of 610 mV. This tendency has also been observed with other Mn-based materials, including MnOOH and Mn_2O_3 [26, 28]. Therefore, the catalytic OER activity of Mn-based catalysts varies drastically between alkaline and neutral pH.

4 Electrospectroscopic Examination of Manganese Oxide Water Oxidation Catalyst

To determine why synthetic Mn catalysts show a sharp decline in catalytic activity under neutral conditions, we recently performed a spectroelectrochemical characterization of intermediate species of the OER for an MnO_2 electrode [44–46]. During the OER at neutral pH, a new absorption band with a maximum around 510 nm was observed at the potential slightly negative than the onset for OER. Utilizing pyrophosphate as a probe molecule, the new absorption peak was assigned to the d-d transition of Mn^{3+} that was generated via the injection of electrons from water to MnO_2. We also examined the pH dependence of the onset potential of the water oxidation current on the MnO_2 electrode and found that under neutral conditions (pH 4–8), the onset potential for OER remained constant at approximately 1.5 V vs. SHE, whereas it sharply shifted in the negative direction and showed

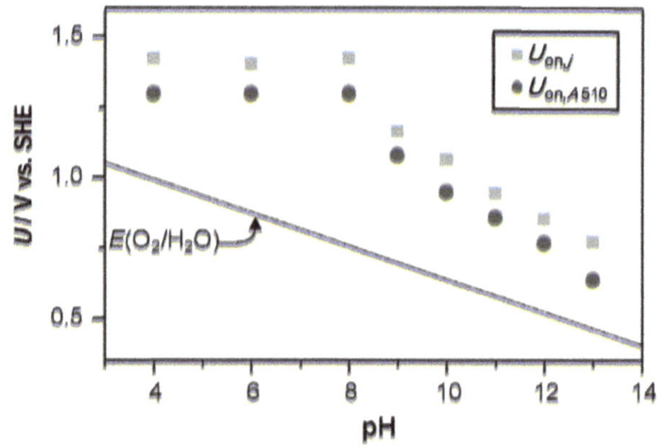

Fig. 2 pH dependences of the onset potential for oxidation current ($U_{on,j}$, *squares*) and optical absorption at 510 nm ($U_{on,A510}$, *circles*). The *solid line* represents the standard potential for OER [44]

linear pH-dependent behavior under alkaline conditions (Fig. 2). Notably, the OER current was accompanied by the appearance of the d-d transition absorption of Mn^{3+}, indicating that Mn^{3+} acts as an intermediate for the OER. However, Mn^{3+} is unstable at neutral pH due to charge disproportionation ($2Mn^{3+} \rightarrow Mn^{2+} + Mn^{4+}$), but exists stably under alkaline conditions. Based on these experimental observations and the pH-dependent redox behavior of Mn^{3+}, we proposed a new model to explain the low activity of MnO_2 OER catalysts under neutral conditions (Fig. 3). In this model, electrons are injected from water to MnO_2 prior to the evolution of O_2, resulting in the formation of Mn^{3+} at the MnO_2 electrode surface. Because Mn^{3+} is unstable due to the disproportionation reaction, Mn^{2+} must be electrochemically oxidized to Mn^{3+} to sustain the catalytic turnover. The regeneration of Mn^{3+} from

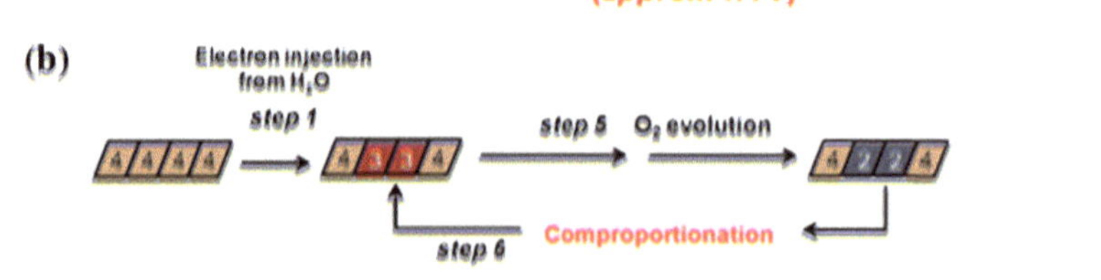

Fig. 3 Oxidation states of Mn ions involved in OER on a δ-MnO_2 electrode at **a** pH 4–8 and **b** pH ≥ 9 [44]

Mn^{2+} serves as the rate-determining step of the OER under neutral conditions. This model is supported by the experimental observation that the onset potential for OER under neutral conditions remains constant at approximately 1.5 V, as the oxidation potential of Mn^{2+} to Mn^{3+} ($Mn^{2+} \rightarrow Mn^{3+} + e^-$) is approximately 1.4 V regardless of the solution pH (Fig. 3). Thus, the origin of the large overpotential for the OER with MnO_2 catalysts is attributable to the inherent instability of Mn^{3+} under neutral conditions. Then the new question arises: does Mn^{3+} stably exist in the tetrameric Mn cluster of PSII during the reaction cycle, known as Kok cycle?

5 Natural Versus Man-Made MnO_2

A key process in natural photosynthesis is the four-electron/four-proton water oxidation reaction that produces O_2. This reaction occurs in the oxygen-evolving complex (OEC) of PSII. Among 3d-block transition metal elements, only Mn is utilized as a photosynthetic center in OEC [47, 48]. Similarly, in almost all photosynthetic organisms, the μ-oxo-bridged tetrameric Mn cluster $CaMn_4O_5$, functions as an OEC [47]. This Mn_4-cluster catalyzes the oxidation of water to molecular oxygen with a low overpotential (160–300 mV) [5–7] and high turnover frequency (maximum 10^3 s^{-1}) under mild conditions [8]. The catalytic cycle of the Mn_4-cluster has been extensively investigated both spectroscopically and computationally [49–56]. In the Kok cycle, the OEC is sequentially oxidized by four photogenerated oxidizing equivalents. Although the oxidation states of the intermediates, which are denoted as S_n states (n = 0 − 4), are still debated, they are commonly assigned from S_0, $3Mn^{III}$ Mn^{IV}, to S_4, which is a putative $3Mn^{IV}$, Mn^{V}, or $4Mn^{IV}$-ligand radical that promotes O−O bond formation and O_2 release. In the Kok cycle, Mn^{III} exists stably in spite of the fact that Mn^{3+} is vulnerable to charge disproportionation in man-made Mn catalysts (Fig. 3). Therefore, the regulation between charge comproportionation ($Mn^{2+} + Mn^{4+} \rightarrow 2Mn^{3+}$) and charge disproportionation ($2Mn^{3+} \rightarrow Mn^{2+} + Mn^{4+}$) is of key importance for enhancing the OER activity of MnO_2-based materials at neutral pH. Recently, we enhanced OER activity by the formation of an N-Mn bond via coordination of the amine groups of poly(allylamine hydrochloride) to surface Mn sites of MnO_2 electrodes [45]. Using this approach, charge disproportionation is suppressed by the introduction of asymmetry into the Mn-centered crystal field, thereby eliminating the orbital degeneracy of Mn^{3+} ($t_{2g}^3 e_g^1$). Nevertheless, the molecular mechanisms behind the stabilization of Mn^{3+} relative to charge disproportionation in OEC is largely unknown. The knowledge developed in charge ordering of perovskite materials may provide new insight in this subject [57, 58].

6 Strategy for Efficient Manganese Oxide-Based Water Oxidation Under Neutral Conditions

In Sect. 4, we described a reaction model for the OER mechanism of MnO_2 catalyst, and proposed a strategy to increase the catalytic activity based on this model in Sect. 5. In the reaction model, the onset potential for OER of MnO_2 catalyst under neural conditions is determined by the regeneration of Mn^{3+} from the Mn^{2+} generated in the disproportionation reaction [44–46]. On examination of the pH dependence of the onset potential, it is clear that the reaction mechanism changed at a pH of approximately 10 (Fig. 2). It is important to note that the pK_a of the coordinated water ligand of Mn^{2+} is 10.6 [59], indicating that the protonation state of the ligand changes at around pH 10; i.e., the ligand is protonated at neutral pH and deprotonated at alkaline pH. However, because the pK_a of the coordinated water ligand of Mn^{3+} is reported to be 0.7 [59], the water ligand exists in the deprotonated state over a wide pH range. As the onset potential for OER was shown experimentally to be pH-independent under neutral conditions (Fig. 2), the regeneration of Mn^{3+} from Mn^{2+} proceeds via Mn^{3+}-OH_2, because proton transfer is not involved in the rate-determining step. Thus, the regeneration of Mn^{3+} is not a proton-coupled process, rather, it involves the sequential transfer of electrons and protons.

In contrast to the sequential electron/proton transfer step in man-made MnO_2 OER catalysts, the amino acid residues coordinating to the Mn_4-cluster in PSII construct a complex hydrogen-bonding network and regulate the transfer of protons generated in the water oxidation reaction [47, 60, 61]. Moreover, proton transfer is coupled to electron transfer in each oxidizing step of the Kok cycle [56]. Therefore, proton transfer is involved in the catalytic cycle of the natural enzyme, but it is absent in the rate-determining step of the OER by man-made MnO_2 catalysts. Based on this difference, the control of proton transfer appears to be one of the reasons for the higher catalytic activity of Mn_4-clusters. If this is the case, the OER catalytic activity of man-made catalysts can be enhanced by induction of concerted proton-coupled electron transfer (CPET) [62], as occurs in the OEC of natural systems.

As a strategy to induce CPET in artificial MnO_2 catalysts, we focused on the libido rule of acid-base catalysis [63–65], which was proposed by Jencks in 1972 [64] and states that *general acid-base catalysis of reactions coupled to proton transfer (PT) in aqueous solution can occur only at sites that undergo a large change in pK_a in the course of the reaction and when this change in pK_a converts an unfavorable to a favorable PT with respect to the catalyst; i.e., the pK_a of the catalyst is intermediate between the initial and final pK_a values of the substrate site.* Based on this rule governing acid-base catalytic reactions, we anticipated that CPET would be induced by the addition of base reagents with a pK_a between those of Mn^{2+}–OH_2 (10.6) and Mn^{3+}–OH_2 (0.7) (Fig. 4).

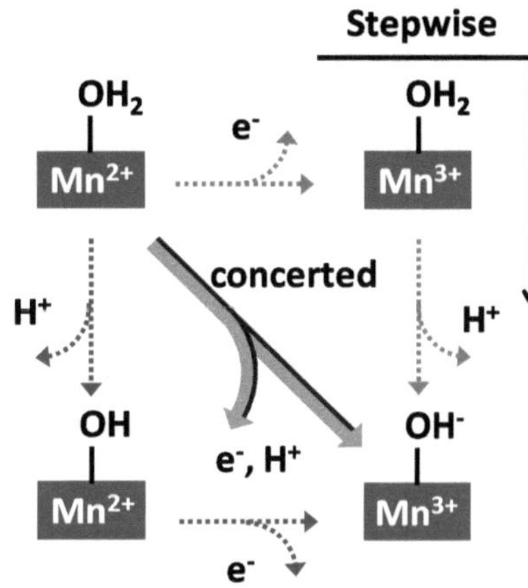

Fig. 4 Electrooxidation of Mn^{2+} to Mn^{3+} on the surface of MnO_2 electrodes

As possible CPET inducers, we selected pyridine because of its relatively high durability for electrochemical oxidation. Introducing pyridine into the electrolyte solution enhanced the electrochemical OER activity of the MnO_2 electrode, as demonstrated by the 200 mV decrease in an overpotential and markedly enhanced current density. Importantly, the onset potential for the OER showed clear pH dependence under neutral conditions, but were nearly constant in the absence of pyridine. This behavior indicates that introduction of the base reagent induced proton transfer in the rate-determining step, as was predicted from the libido rule. The induction of simultaneous electron and proton transfer reactions is also supported by the experimental observation that the kinetic isotope effect, which is determined by the ratio of current density under H_2O to that under D_2O and is defined at 1.29 V vs. SHE and the predicted degree of proton transfer that occurs in the rate-determining step, were both increased by the addition of pyridine to the MnO_2-based catalytic system.

To achieve even higher OER activity, we selected several candidate pyridine derivatives with pK_a's higher than that of pyridine with the expectation that proton transfer would be accelerated due to the stronger proton-extracting ability of the derivatives. Upon addition of the methyl group-substituted pyridines 3-methylpyridine (β-picolline), 4-methylpyridine (γ-picolline), 2,6-dimethylpyridine (2,6-lutidine), and 2,4,6-trimethylpyridine (γ-collidine), into the electrolyte solution of the MnO_2 catalyst, the OER activity increased with increasing pK_a value (Fig. 5) and was found to approximately half of that observed under alkaline conditions in the presence of γ-collidine. In addition, the KIE also increased with increasing pK_a value. Taken together, these results indicate that bases with higher proton-extracting ability promote greater proton transfer in the rate-determining step and thereby increase the OER activity of the catalyst. Yet it should be noted that the base has the possibility to affect several elementary steps other than the oxidation of Mn from 2+ to 3+ in a course of electrocatalytic water oxidation.

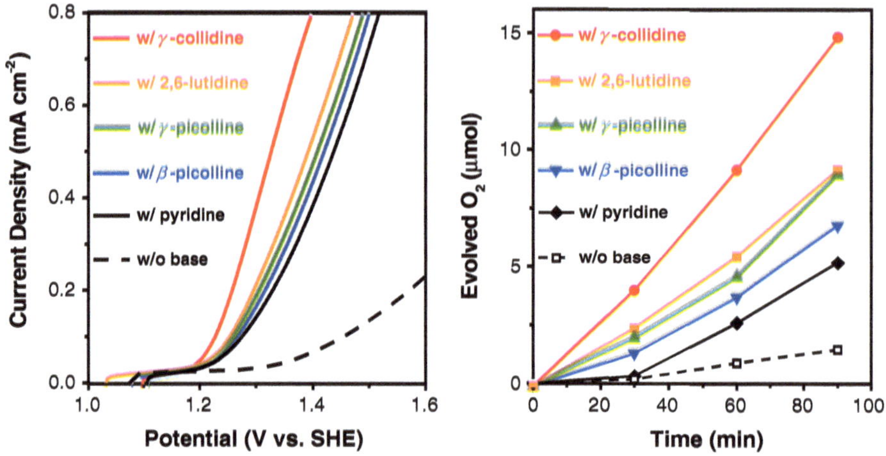

Fig. 5 *Left* Potential dependence of current density for MnO_2 electrodes in the presence of pyridine derivatives and *Right* time course of O_2 production observed for MnO_2 electrodes at an applied potential of +1.39 V versus SHE at pH = 7.5 in the presence of pyridine derivatives

7 Elemental-Based Strategy for Biological Water Splitting

In the previous sections, we described several reasons for the differences in OER activity between natural and man-made Mn-based catalysts, and introduced a strategy for enhancing the activity of the latter. In this section, we will attempt to answer why Mn was adopted for OEC in natural systems. The possible evolutionary origin of natural Mn_4-clusters has been discussed in the literature [66–68]. Russell et al. [66] suggested that the Mn_4-cluster originated from the colloidal cluster of $CaMn_4O_9 \cdot 3H_2O$ and was generated by the super-ultraviolet irradiation-induced reaction between Mn^{2+} and Ca^{2+} [66]. Sauer and Yachandra [67] compared the bond lengths of Mn–Mn in the Mn_4-cluster (2.7–2.8 and 3.3 Å) with those of several Mn-based minerals and speculated that hollandite (mainly composed of α-MnO_2) was possibly the evolutionary origin of the Mn_4-cluster due to the structural similarity between hollandite and the OEC.

As mentioned in the Introduction, the OEC does not use 3d-block elements other than Mn, unlike other enzymatic reaction centers, such as those used for O_2 reduction, even though these elements, particularly Fe, are much more abundant in nature [69]. Armstrong [68] considered that Mn was selected of natural catalytic systems because of its various oxidation states [68] and hypothesized together with Dismukes et al. [70] that the precursor for natural OEC was the Mn(II)-bicarbonate cluster in the Archean sea.

Photosynthesis is thought to have evolved 2.7 billion years ago with the emergence of cyanobacteria [71]. As a result of this activity, the O_2 concentration in ancient earth atmosphere dramatically increased 2.4 billion years ago, promoting the evolution of multicellular organisms. In contrast, the CO_2 concentration of the earth's atmosphere gradually decreased and was not strongly correlated to the O_2 concentration. In high CO_2 environments, Mn ions predominantly exist as $MnCO_3$, but form MnO_x through reaction with O_2. Geological examination of the Kalahari Desert revealed that a critical change in mineral composition from carbonate to oxides occurred 2.3 billion years ago, [72]. The geologic stratum of the Kalahari Desert in southern Africa retains the metamorphosed record for the past ~1.6–4 billion years ago, and the desert's manganese field reflects a major transition of the oxidation state of the Earth' surface. Specifically, a stratum of FeO_x was deposited over 2.7 billion years ago underneath a $MnCO_3$ layer. These findings indicate that the concentration of dissolved Mn^{2+} in shallow water could be much larger than that of Fe^{2+}, and that this difference in element availability may have led to the evolutionary selection of Mn over Fe in photosynthetic OEC systems. This constraint in element availability may have been intensified by the positive feedback manner of the photosynthetic process, as CO_2 is consumed and O_2 is produced, which would promote precipitation of Fe ions as Fe oxides.

Despite of this availability difference between dissolved Mn and Fe ions, it is possible that early photosynthetic centers adopted Fe. To examine the outcome of this selection, the bioenergetics of such systems were considered. The OER is one of the strongest oxidative reactions ($E° = 1.23$ V) in biological systems. Due to this property, other reactions, such as the oxidation of amino acids especially histidine, may also occur if the OER proceeds. Therefore, for PSII to have evolved, the energy gained from photosynthesis must be larger than the energy required to restore or replace the decomposed proteins. In other words, the reaction selectivity towards OER over amino acid oxidation should be sufficiently high to maintain the catalytic cycle for PSII evolution. One strategy to avoid the self-decomposition of amino acids is ligand adoption. The Mn_4-cluster is coordinated by 6 carboxylate ligands [47]. Such the high number of carboxylate ligands is unique for PSII, as other enzymatic active centers are typically surrounded by thiol or histidine groups, which are less tolerant toward oxidative decomposition than carboxylate ligands. In addition to the adaptation of carboxylate ligands, it is speculated that the autocatalytic electron transfer under protein coordination environments is an important consideration for the selective oxidation of water by OEC. Mn can induce chemical autocatalysis in the potential range of the OER of water [73, 74] and this property could be advantageous to induce multi-electron transfer for realizing highly selective OER [75]. OER is at the heart of the natural photosynthetic process, and also the efficiency-limiting reaction of artificial photosynthetic process designed to produce hydrogen through proton reduction or convert carbon dioxide to liquid fuel. Therefore, seeking the elemental-based strategy in biological water splitting will bring a new design rationale for functional analogues of the OEC for efficient and robust water splitting at neutral pH.

References

1. Fujishima A, Honda K (1972) Electrochemical photolysis of water at a semiconductor electrode. Nature 238:37–38
2. Nozik AJ (1978) PHOTOELECTROCHEMISTRY: APPLICATIONS TO SOLAR ENERGY CONVERSION. Annu Rev Phys Chem 29:189–222
3. Bard AJ, Fox MA (1995) Artificial photosynthesis: solar splitting of water to hydrogen and oxygen. Acc Chem Res 28:141–145
4. Lewis NS, Nocera DG (2006) Powering the planet: chemical challenges in solar energy utilization. Proc Natl Acad Sci USA 103:15729–15735
5. Vass I, Styring S (1991) pH-dependent charge equilibria between Tyrosine-D and the S states in photosystem ii. Estimation of Relative Midpoint Redox Potentials. Biochemistry 30:830–839
6. Geijer P, Morvaridi F, Styring S (2001) The S_3 state of the oxygen-evolving complex in photosystem ii is converted to the $S_2Y_Z^{\bullet}$ state at alkaline pH. Biochemistry 40:10881–10891
7. Metz JG, Nixon PJ, Rogner M, Brudvig GW, Diner BA (1989) Directed alteration of the D1 polypeptide of photosystem II: evidence that Tyrosine-161 is the redox component, Z, connecting the oxygen-evolving complex to the primary electron donor, P680. Biochemistry 28:6960–6969
8. Ananyev G, Dismukes GC (2005) How fast can photosystem ii split water? Kinetic performance at high and low frequencies. Photosynth Res 84:355–365
9. Trasatti S, Buzzanca G (1971) Ruthenium dioxide: a new interesting electrode material. Solid state structure and electrochemical behaviour. J Electroanal Chem 29:A1–A5
10. Harriman A, Richoux M, Christensen PA, Moseri S, Neta P (1987) Redox reactions with colloidal metal oxides. Comparison of radiation-generated and chemically generated $RuO_2 \cdot 2H_2O$. J Chem Soc Faraday Trans 1(83):3001–3014
11. Kiwi J, Grätzel M (1979) Colloidal redox catalysts for evolution of oxygen and for light-induced evolution of hydrogen from water. Angew Chem Int Ed 18:624–626
12. Harriman A, Pickering IJ, Thomas JM, Christensen PA (1988) Metal oxides as heterogeneous catalysts for oxygen evolution under photochemical conditions. J Chem Soc Faraday Trans 1 (84):2795–2806
13. Mills A, Russell T (1991) Comparative study of new and established heterogeneous oxygen catalysts. J Chem Soc Faraday Trans 87:1245–1250
14. Yagi M, Tomita E, Sakita S, Kuwabara T, Nagai K (2005) Self-assembly of active IrO_2 colloid catalyst on an ITO electrode for efficient electrochemical water oxidation. J Phys Chem B 109:21489–21491
15. Nakagawa T, Bjorge NS, Murray RW (2009) Electrogenerated IrO_x nanoparticles as dissolved redox catalysts for water oxidation. J Am Chem Soc 131:15578–15579
16. Nakagawa T, Beasley CA, Murray RW (2009) Efficient Electro-oxidation of water near its reversible potential by a mesoporous IrO_x nanoparticle film. J Phys Chem C 113:12958–12961
17. Rüetschi P, Delahay P (1955) Influence of electrode material on oxygen overvoltage: a theoretical analysis. J. Chem. Phys. 23:556–560
18. Iwakura C, Nashioka M, Tamura H (1982) Nihon Kagaku Kaishi 8:1294–1298
19. Bockris JO, Otagawa T (1984) The electrocatalysis of oxygen evolution on perovskites. J Electrochem Soc 131:290–301
20. Suntivich J, May KJ, Gasteiger HA, Goodenough JB, Shao-Horn Y (2011) A perovskite oxide optimized for oxygen evolution catalysis from molecular orbital principles. Science 334:1383–1385
21. Raabe S, Mierwaldt D, Ciston J, Uijttewaal M, Stein H, Hoffmann J, Zhu Y, Blöchl P, Jooss C (2012) In situ electrochemical electron microscopy study of oxygen evolution activity of doped manganite perovskites. Adv Funct Mater 22:3378–3388
22. Brunschwig BS, Chou MH, Creutz C, Ghosh P, Sutin N (1983) Mechanisms of water oxidation to oxygen: cobalt(iv) as an intermediate in the aquocobalt(ii)-catalyzed reaction. J Am Chem Soc 105:4832–4833

23. Kanan MW, Nocera DG (2008) In Situ Formation of an Oxygen-Evolving Catalyst in Neutral Water Containing Phosphate and Co^{2+}. Science 321:1072–1075
24. Esswein AJ, McMurdo MJ, Ross PN, Bell AT, Tilley TD (2009) Size-dependent activity of Co_3O_4 nanoparticle anodes for alkaline water electrolysis. J Phys Chem C 113:15068–15072
25. Gorlin Y, Jaramillo TF (2010) A bifunctional nonprecious metal catalyst for oxygen reduction and water oxidation. J Am Chem Soc 132:13612–13614
26. Mohammad AM, Awad MI, El-Deab MS, Okajima T, Ohsaka T (2008) Electrocatalysis by nanoparticles: optimization of the loading level and operating pH for the oxygen evolution at crystallographically oriented manganese oxide nanorods modified electrodes. Electrochim Acta 53:4351–4358
27. El-Deab MS, Awad MI, Mohammad AM, Ohsaka T (2007) Enhanced water electrolysis: electrocatalytic generation of oxygen gas at manganese oxide nanorods modified electrodes. Electrochem Commun 9:2082–2087
28. Morita M, Iwakura C, Tamura H (1979) The anodic characteristics of massive manganese oxide electrode. Electrochim Acta 24:357–362
29. Morita M, Iwakura C, Tamura H (1977) The anodic characteristics of manganese dioxide electrodes prepared by thermal decomposition of manganese nitrate. Electrochim Acta 22:325–328
30. Morita M, Iwakura C, Tamura H (1978) The anodic characteristics of modified Mn oxide electrode: $Ti/RuO_x/MnO_x$. Electrochim Acta 23:331–335
31. Robinson DM, Go YB, Greenblatt M, Dismukes GC (2010) Water oxidation by λ-MnO_2: catalysis by the cubical Mn_4O_4 subcluster obtained by delithiation of spinel $LiMn_2O_4$. J Am Chem Soc 132:11467–11469
32. Najafpour MM, Ehrenberg T, Wiechen M, Kurz P (2010) Calcium manganese(III) oxides ($CaMn_2O_4{\cdot}xH_2O$) as biomimetic oxygen-evolving catalysts. Angew Chem Int Ed 49:2233–2237
33. Jiao F, Frei H (2010) Nanostructured manganese oxide clusters supported on mesoporous silica as efficient oxygen-evolving catalysts. Chem Commun 46:2920–2922
34. Yagi M, Narita K (2004) Catalytic O_2 Evolution from water induced by adsorption of $[(OH_2)(terpy)Mn(\mu\text{-}O)_2Mn(terpy)(OH_2)]^{3+}$ complex onto clay compounds. J Am Chem Soc 126:8084–8085
35. Narita K, Kuwabara T, Sone K, Shimizu K, Yagi M (2006) Characterization and activity analysis of catalytic water oxidation induced by hybridization of $[(OH_2)(terpy)Mn(\mu\text{-}O)_2Mn(terpy)(OH_2)]^{3+}$ and clay compounds. J Phys Chem B 110:23107–23114
36. Brimblecombe R, Swiegers GF, Dismukes GC, Spiccia L (2008) Sustained water oxidation photocatalysis by a bioinspired manganese cluster. Angew Chem Int Ed 47:7335–7338
37. Limburg J, Vrettos JS, Liable-Sands LM, Rheingold AL, Crabtree RH, Brudvig GW (1999) A functional model for O–O bond formation by the O_2-evolving complex in photosystem II. Science 283:1524–1527
38. Naruta Y, Sasayama M, Sasaki T (1994) Oxygen evolution by oxidation of water with manganese porphyrin dimers. Angew Chem Int Ed Engl 33:1839–1841
39. Gao Y, Åkermark T, Liu J, Sun L, Åkermark B (2009) Nucleophilic attack of hydroxide on a Mn^V Oxo complex: a model of the O–O bond formation in the oxygen evolving complex of photosystem II. J Am Chem Soc 131:8726–8727
40. Najafpour MM, Haghighi B, Ghobadi MZ, Sedigh DJ (2013) Nanolayered manganese oxide/poly(4-vinylpyridine) as a biomimetic and very efficient water oxidizing catalyst: toward an artificial enzyme in artificial photosynthesis. Chem Comm 49:8824–8826
41. Cheng F, Shen J, Peng B, Pan Y, Tao Z, Chen J (2011) Rapid room-temperature synthesis of nanocrystalline spinels as oxygen reduction and evolution electrocatalysts. Nat Chem 3:79–84
42. Singh RN, Singh JP, Cong HN, Chartier P (2006) Effect of partial substitution of Cr on electrocatalytic properties of $MnFe_2O_4$ towards O_2-evolution in alkaline medium. Int J Hydrogen Energy 31:1372–1378

43. Robinson DM, Go YB, Mui M, Gardner G, Zhang Z, Mastrogiovanni D, Garfunkel E, Li J, Greenblatt M, Dismukes GC (2013) Photochemical water oxidation by crystalline polymorphs of manganese oxides: structural requirements for catalysis. J Am Chem Soc 135:3494–3501
44. Takashima T, Hashimoto K, Nakamura R (2012) Mechanisms of pH-dependent activity for water oxidation to molecular oxygen by MnO_2 electrocatalysts. J Am Chem Soc 134:1519–1527
45. Takashima T, Hashimoto K, Nakamura R (2012) Inhibition of charge disproportionation of MnO_2 electrocatalysts for efficient water oxidation under neutral conditions. J Am Chem Soc 134:18153–18156
46. Takashima T, Yamaguchi A, Hashimoto K, Irie H, Nakamura R (2014) In situ UV-vis absorption spectra of intermediate species for oxygen-evolution reaction on the surface of MnO_2 in neutral and alkaline media. Electrochemistry 82:325–327
47. Umena Y, Kawakami K, Shen J-R, Kamiya N (2011) Crystal structure of oxygen-evolving photosystem II at a resolution of 1.9 Å. Nature 473:55–61
48. Yano J, Kern J, Sauer K, Latimer MJ, Puskar Y, Biesiadka J, Loll B, Saenger W, Messinger J, Zouni A, Yachandra VK (2006) Where water is oxidized to dioxygen: structure of the photosynthetic Mn_4Ca cluster. Science 314:821–825
49. Haumann M, Muller C, Liebisch P, Iuzzolino L, Dittmer J, Grabolle M, Neisius T, Meyer-Klaucke W, Dau H (2005) Structural and oxidation state changes of the photosystem II manganese complex in four transitions of the water oxidation Cycle ($S_0 \rightarrow S_1$, $S_1 \rightarrow S_2$, $S_2 \rightarrow S_3$, and $S_{3,4} \rightarrow S_0$) characterized by X-ray absorption spectroscopy at 20 K and room temperature. Biochemistry 44:1894–1908
50. Siegbahn PE (1827) Water oxidation mechanism in photosystem II, including oxidations, proton release pathways, O-O bond formation and O2 release. Biochim Biophys Acta 1003–1019:2013
51. Gatt P, Stranger R, Pace RJ (2011) Application of computational chemistry to understanding the structure and mechanism of the Mn catalytic site in photosystem II—a review. J Photochem Photobiol, B 104:80–93
52. Galstyan A, Robertazzi A, Knapp EW (2012) Oxygen-evolving Mn cluster in photosystem II: the protonation pattern and oxidation state in the high-resolution crystal structure. J Am Chem Soc 134:7442–7449
53. Kanda K, Yamanaka S, Saito T, Umena Y, Kawakami K, Shen J-R, Kamiya N, Okumura M, Nakamura H, Yamaguchi K (2011) Labile electronic and spin states of the $CaMn_4O_5$ cluster in the PSII system refined to the 1.9 Å X-ray resolution. UB3LYP computational results. Chem Phys Lett 506:98–103
54. Gatt P, Petrie S, Stranger R, Pace RJ (2012) Rationalizing the 1.9 Å crystal structure of photosystem II—a remarkable Jahn–Teller balancing act induced by a single proton transfer. Angew Chem Int Ed 51:12025–12028
55. Roelofs TA, Liang WC, Latimer MJ, Cinco RM, Rompel A, Andrews JC, Sauer K, Yachandra VK, Klein MP (1996) Oxidation states of the manganese cluster during the flash-induced S-state cycle of the photosynthetic oxygen-evolving complex. Proc Natl Acad Sci USA 93:3335–3340
56. Sproviero EM, Gascon JA, McEvoy JP, Brudvig GW, Batista VS (2008) A model of the oxygen-evolving center of photosystem II predicted by structural Refinement based on EXAFS Simulations. J Am Chem Soc 130:6728–6730
57. Mazin II, Khomskii DI, Lengsdorf R, Alonso JA, Marshall WG, Ibbreson RM, Podlesnyak A, Martínez-Lope MJ, Abd-Elmeguid MM (2007) Charge ordering as alternative to Jahn-Teller distortion. Phys Rev Lett 98:176406
58. Mizokawa T, Khomskii DI, Sawatzky GA (2000) Spin and charge ordering in self-doped Mott insulators. Phys Rev B 61:11263–11266
59. Amin M, Vogt L, Vassiliev S, Rivalta I, Sultan MM, Bruce D, Brudvig GW, Batista VS, Gunner MR (2013) Electrostatic effects on proton-coupled electron transfer in oxomanganese complexes inspired by the oxygen-evolving complex of photosystem II. J Phys Chem B 117:6217–6226

60. Meyer TJ, Huynh MH, Thorp HH (2007) The possible role of proton-coupled electron transfer (PCET) in water oxidation by photosystem II. Angew Chem Int Ed 46:5284–5304
61. Ogata K, Yuki T, Hatakeyama M, Uchida W, Nakamura S (2013) All-atom molecular dynamics simulation of photosystem II embedded in thylakoid membrane. J Am Chem Soc 135:15670–15673
62. Yamaguchi A, Inuzuka R, Takashima T, Hayashi T, Hashimoto K, Nakamura R (2014) Regulating proton-coupled electron transfer for efficient water splitting by manganese oxides at neutral pH. Nat Commun doi:10.1038/ncomms5256
63. Medina-Ramos J, Oyesanya O, Alvarez JC (2013) Buffer effects in the kinetics of concerted proton-coupled electron transfer: the electrochemical oxidation of glutathione mediated by $[IrCl_6]^{2-}$ at variable buffer pK_a and concentration. J Phys Chem C 117:902–912
64. Jencks WP (1972) Requirements for general acid-base catalysis of complex reactions. J Am Chem Soc 94:4731–4732
65. Jencks WP (1972) General acid-base catalysis of complex reactions in water. Chem Rev 72:705–718
66. Russell MJ, Hall AJ (2002) Chemiosmotic coupling and transition element clusters in the onset of life and photosynthesis. The Geochemical News 113:6–12
67. Sauer K, Yachandra VK (2002) A possible evolutionary origin for the Mn_4 cluster of the photosynthetic water oxidation complex from natural MnO_2 precipitates in the early ocean. Proc Natl Acad Sci USA 99:8631–8636
68. Armstrong FA (2008) Why did nature choose manganese to make oxygen? Phil Trans R Soc B 363:1263–1270
69. http://pubs.usgs.gov/fs/2002/fs087-02/
70. Dismukes GC, Klimov VV, Baranov SV, Kozlov YN, Dasgupta J, Tyryshkin A (2001) The origin of atmospheric oxygen on earth: the innovation of oxygenic photosynthesis. Proc Natl Acad Sci USA 98:2170–2175
71. Lane N (2010) The rollercoaster ride to an oxygen-rich world. New Scientist 205:2746
72. Kirschvink JL, Gaidos EJ, Bertani LE, Beukes NJ, Gutzmer J, Maepa LN, Steinberger RE (2000) Paleoproterozoic snowball earth: extreme climatic and geochemical global change and its biological consequences. Proc Natl Acad Sci USA 97:1400–1405
73. Fan X, Hou J, Sun D, Xi S, Liu Z, Du J, Luo J, Tao C (2013) Mn-oxides catalyzed periodic current oscillation on the anode. Electrochim Acta 102:466–471
74. Olexová A, Melicherčík M, Treindl L (1997) Oscillatory oxidation of Mn(II) ions by hexacyanoferrates(III) and bistability in the reduction of MnO_2 by hexacyanoferrates(II) in a CSTR. Chem Phys Lett 268:505–509
75. Tributsch H (1994) The challenge of non-linear and co-operative mechanisms for electrocatalysis. Electrochim Acta 39:1495–1502

CO_2 Reduction Using Electrochemical Approach

Yoshio Hori

Abstract Electrochemical Reduction of CO_2 may contribute to energy storage process. It has been a challenging subject, since it competes with hydrogen evolution in aqueous electrolytes. The studies started in the beginning of 20th century using metal electrodes of high hydrogen overvoltage, and the main product remained formate for long time. A breakthrough appeared in 1980s; Cu electrode gives rise to hydrocarbons, and Au and Ag form CO. This chapter provides an overview of the electrocatalysis of metal electrodes in CO_2 reduction, and briefly discusses some problems to be solved.

1 Introduction

Carbon dioxide, the most important greenhouse gas as well as a potential carbon resource abundant on the earth, is rapidly increasing in the atmospheric concentration for the recent two centuries. Chemical fixation of CO_2 is an attractive technique for utilization of carbon resource and for possible reduction of atmospheric concentration. CO_2 reduction might practically contribute to a new energy storage process in the future, which could support intermittent renewable natural energy resources. From the chemical point of view, CO_2 is the stablest among carbon based substances under the environmental conditions. Thus, CO_2 can be electrochemically reduced to useful materials under mild conditions, but with remarkably low energy conversion efficiency roughly 30–40 % [1, 2]. It may discourage practical application of CO_2 reduction in very near future. However, the significance of the CO_2 reduction should be reevaluated in various aspects of science, since many interesting facts in basic electrochemistry have been revealed by the recent studies.

Y. Hori (✉)
Chiba University, Shirakawa 4-9-13-1201, Koto-ku, Tokyo 135-0021, Japan
e-mail: y-hori@white.plala.or.jp

M. Sugiyama et al. (eds.), *Solar to Chemical Energy Conversion*,
Lecture Notes in Energy 32, DOI 10.1007/978-3-319-25400-5_12

This section attempts to provide a brief description of electrochemical reduction of CO_2 at metal electrodes in aqueous solutions, stressing basic problems involved with experimental aspects. Some review papers are introduced, which contain the history and classical works of CO_2 reduction [3–10].

2 Fundamental Problems in CO_2 Reduction and Experimental Procedures

2.1 *Electrochemical Equilibria in Aqueous Media*

The equilibrium potentials of CO_2 reduction are in the same range as hydrogen evolution reaction (HER) in aqueous electrolyte solutions. For example, electrochemical reduction of CO_2 to $HCOO^-$ in aqueous solution is calculated on the basis of thermodynamic data, as shown below with respect to the standard hydrogen electrode (SHE) at pH 7.0 at 25 °C.

$$CO_2 + H_2O + 2e^- = HCOO^- + OH^- (-0.43\ \text{V vs. SHE at pH 7.0}). \quad (1)$$

$HCOO^-$ is predominantly produced in neutral pH region instead of HCOOH from CO_2. The standard electrode potential of HER at pH 7.0 is also given.

$$2H_2O + 2e^- = H_2 + 2OH^- (-0.414\ \text{V vs. SHE at pH 7.0}) \quad (2)$$

Nevertheless, CO_2 reduction does not take place easily in aqueous media, and the actual electrolysis potentials for CO_2 reduction are much more negative than the equilibrium ones. The reason is that the intermediate species CO_2^- is formed by an electron transfer to a CO_2 molecule as the first step at highly negative potential as discussed later.

The standard potentials for CO_2 reduction and HER, estimated from thermodynamic data, decrease similarly with pH. However, the actual potential of CO_2 reduction does not depend on pH, whereas that of HER decreases in accordance with proton activity. Thus HER prevails over CO_2 reduction in acidic solutions, and most studies of CO_2 reduction in aqueous media have been made in neutral pH region.

The standard electrode potentials of CO_2 reduction are given below for the formation of CO, CH_4, C_2H_4, C_2H_5OH, and C_3H_7OH. The values are estimated from thermodynamic data in aqueous media at 25 °C with respect to SHE. The standard potentials are conveniently given at pH 7.0, where most of actual CO_2 reductions are carried out.

$$CO_2 + H_2O + 2e^- = CO + 2OH^- (-0.52\ V) \tag{3}$$

$$CO_2 + 6H_2O + 8e^- = CH_4 + 8OH^- (-0.25\ V) \tag{4}$$

$$2CO_2 + 8H_2O + 12e^- = C_2H_4 + 12OH^- (-0.34\ V) \tag{5}$$

$$2CO_2 + 9H_2O + 12e^- = C_2H_5OH + 12OH^- (-0.33\ V) \tag{6}$$

$$3CO_2 + 13H_2O + 18e^- = C_3H_7OH + 18OH^- (-0.32\ V) \tag{7}$$

2.2 *Equilibria of CO_2 Related Species in Aqueous Solution*

Thermodynamic equilibria of CO_2 related species in aqueous media are important in CO_2 chemistry. Thermodynamic considerations are summarized briefly below. An equilibrium of gaseous CO_2 with HCO_3^- is

$$CO_2 + H_2O = HCO_3^- + H^+. \tag{8}$$

pH is related to the activities of other substances, $[HCO_3^-]$ and $[CO_2]$.

$$pH = pK_{a1} + \log[HCO_3^-] - \log[CO_2], \quad pK_{a1} = 6.35\ (25\,°C) \tag{9}$$

where K_{a1} is the equilibrium constant of the reaction Eq. (8).

Taking into account of solubility of gaseous CO_2 in the solution and another aqueous equilibrium,

$$HCO_3^- = CO_3^{2-} + H^+ \tag{10}$$

pH values for equilibrium Eq. (8) in given $[HCO_3^-]$ solutions under $P(CO_2)$ 1 atm are presented in Table 1.

Various electrolyte solutions are employed for CO_2 electrolysis, not only HCO_3^- solutions. Since CO_2 can not be present practically in basic aqueous solutions, CO_2 reduction has been studied with neutral to acidic solutions. It is apparent that aqueous 0.1 M KOH is converted to 0.1 M $KHCO_3$ after saturation and equilibration with CO_2 gas, and that aqueous 0.1 M K_2CO_3 to 0.2 M $KHCO_3$. If a solution with buffer ability, such as phosphate buffer solution, is employed, the problem is more complicated. The pH of a phosphate buffer solution changes from the value of the originally prepared one by equilibration with CO_2, forming HCO_3^-.

Table 1 pH of the solution of given activity $[HCO_3^-]$ under $P(CO_2)$ = 1 atm

$[HCO_3^-]$/mol dm^{-3}	1	0.3	0.1	0.03	0.01
pH of the solution	7.82	7.30	6.82	6.30	5.82

In addition to the equilibrium between CO_2 related species, the buffer reaction equilibrium must be taken into account [9].

2.3 *Variation of pH at the Electrode/Electrolyte Interface During CO_2 Reduction*

Equilibrium values of pH in the electrolyte can be estimated as shown above. Nevertheless, the electrolyte solution close to the electrode is not necessarily in equilibrium during electrolysis. OH^- is generated at the electrode, when H_2O molecule is involved with a cathodic reduction. For example,

$$CO_2 + 6H_2O + 8e^- = CH_4 + 8OH^-. \tag{11}$$

The rate of neutralization between OH^- and CO_2 is slow in aqueous solution at the ambient temperature, as is well known, [11]

$$CO_2 + H_2O = H_2CO_3 \text{ (slow)} \tag{12}$$

$$CO_2 + OH^- = HCO_3^- \text{ (slow)} \tag{13}$$

whereas the following reactions proceed rapidly.

$$H_2CO_3 + OH^- = HCO_3^- + H_2O \text{ (instantaneous)} \tag{14}$$

$$HCO_3^- + OH^- = CO_3^{2-} + H_2O \text{ (instantaneous)} \tag{15}$$

Thus the pH adjacent to the electrode rises much above that of the bulk solution. CO_2 molecules are present in more or less nonequilibrium high pH region at the electrode/electrolyte interface. The enhancement of pH is more significant in solutions, such as K_2SO_4, KCl and $NaClO_4$, which do not release protons. Anions with buffer ability can neutralize OH^-, mitigating the pH enhancement at the electrode.

$$OH^- + HCO_3^- = H_2O + CO_3^{2-} \tag{16}$$

$$OH^- + HPO_4^{2-} = H_2O + PO_4^{3-} \tag{17}$$

Thus pH at the electrode/electrolyte interface in a phosphate buffer solution is actually lower than in a KCl one, leading to anionic effect in the product distribution in CO_2 reduction.

2.4 Electroactive Species in CO_2 Reduction

Teeter and Rysselberghe showed that CO_2 is reduced to HCOOH at a Hg cathode with faradaic efficiency nearly 100 % [12]. They concluded on the basis of their polarographic study that the electroactive species of CO_2 reduction is CO_2 molecule, neither HCO_3^- nor CO_3^{2-} [13]. Several articles were published later, arguing that HCO_3^- can be reduced at a Hg coated Cu electrode, [14] Pd impregnated polymer electrodes, [15] or metallic Pd electrodes [16]. However, the partial current density of CO_2 reduction was below 0.1 mA cm^{-2}, [15] or the data were not confirmed by quantitative chemical analysis [16].

The solubility of CO_2 is as low as 30 mM in water under 1 atm at the ambient temperature. Thus the rate of electrochemical reduction of molecular CO_2 in aqueous media must be limited by the transport process to a low value. If HCO_3^- or CO_3^{2-} is electrochemically reduced, CO_2 reduction will be highly promoted without limitation of transport process.

Hori and Suzuki studied electrochemical reduction of $NaHCO_3$ on the basis of chemical analysis of the product $HCOO^-$ [17]. They showed that HCO_3^- is apparently reduced to $HCOO^-$ at a Hg electrode in aqueous $NaHCO_3$ solutions. The partial current density of $HCOO^-$ formation is proportional to the concentration of $NaHCO_3$ at a constant potential, if CO_3^{2-} and HCO_3^- activity ratio is kept constant at a constant pH (Eq. 10). It increases with the increase of the negative potential, reaching a limiting value (Fig. 1). The limiting current is below 1.0 mA cm^{-2} for a mixture of 0.05 M CO_3^{2-} and 1.0 M HCO_3^-; the value is too low under usual

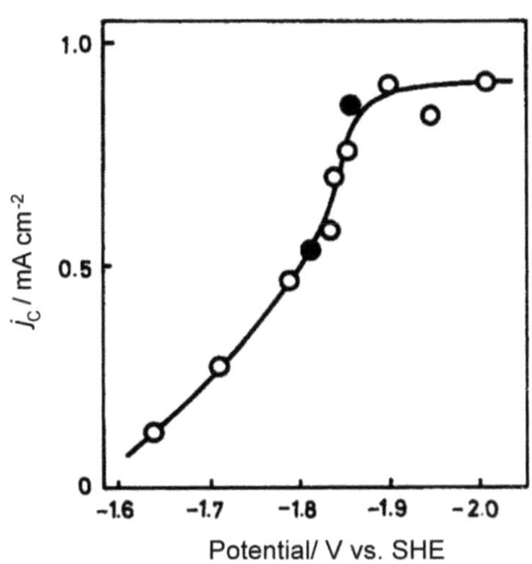

Fig. 1 Partial current density of $HCOO^-$ formation j_c versus electrode potential from reduction of HCO_3^- with a Hg pool electrode. Electrolyte: $NaHCO_3$ (m = 1.0 mol kg^{-1})$^-$ Na_2CO_3 (m = 0.05 mol kg^{-1}). The potential axis negative to the *right*. (Reproduced from Ref. [17] with permission of The Electrochemical Society)

experimental conditions, if HCO_3^- is assumed to be the electroactive species and directly reduced to $HCOO^-$ at the electrode. In addition, the partial current density of $HCOO^-$ formation is not affected by the convective motion of the electrolyte solution. A detailed analysis of the results showed that the CO_2 molecules formed in the decomposition of HCO_3^- are electrochemically reduced to $HCOO^-$ at the electrode. Eggins et al. later examined reduction of HCO_3^- at a glassy carbon electrode by cyclic voltammetry. They applied a similar analytical procedure as above, and confirmed that the decomposition of HCO_3^- to CO_2 is the rate determining step in the reduction of HCO_3^- [18].

2.5 Deactivation of Electrodes and Purity of the Electrolyte Solution

Trace amount of impurities interferes with surface process on electrodes as well as any heterogeneous interfacial reactions. Many workers studied HER on various metal electrodes. They attempted to obtain any correlation between the hydrogen overvoltage and the properties of the electrode material. During the course of these works, widely scattered voltage current relations were published by different workers. Kuhn et al. reviewed the previous works, and decided to lay down criteria to select the experimental data for their discussions [19]. They chose the data by only those workers who had used preelectrolysis for the purification of the electrolyte solution. Bockris described the importance of preelectrolysis of the electrolyte solutions in the study of electrode process in detail on the basis of his own studies for long years as well as the preceding studies [20, 21]. The preelectrolysis is to scavenge the solution with an cathodically polarized large area electrode, usually platinum black.

The electrochemical reduction of CO_2 is also sensitive to the cleanliness of the electrode, similar with the hydrogen electrode process. Any chemical reagents contain impurities of ppm level according to manufacturers. For example, $KHCO_3$ of the best quality contains 5 ppm of heavy metals as Pb and 5 ppm of Fe, respectively as maximum. 0.1 M $KHCO_3$, prepared from this reagent, will contain ca. 1 μM of heavy metals or Fe ion. These metal ions are reduced to metals and deposited on the electrode surface with one monolayer or more in 10 min or less during the CO_2 reduction under most experimental electrolysis conditions. The deposited metal will severely affect the electrocatalytic property of the electrode. Hydrogen overpotential drops, leading to HER prevalent in the total current. Thus, an effective purification of the electrolyte solution is required to obtain reliable results of the electrocatalytic property of metal electrodes.

Deactivation of metal electrodes in CO_2 reduction has been reported by many workers, and regarded as an unavoidable defect irrespective of the potential importance of CO_2 reduction. The "deactivation" was described; the formation of CH_4 and C_2H_4 at Cu electrode decays rapidly and hydrogen evolution prevails in short period. Au and Ag were also reported to suffer from the "deactivation". Various reasons of the deactivation were suggested, such as carbonaceous or organic substances on the electrode surface intermediately formed during the CO_2 reduction, which eventually poison the reaction [22–27].

Hori et al. revealed that poisoning of the reaction is derived from surface contamination caused by deposition of impurity heavy metals in the electrolyte solution originally contained in the chemical reagent. They showed that the deactivation of the electrode does not take place in electrolyte solutions properly purified by the preelectrolysis [28].

Another possibility of the deactivation may be derived from water from which electrolytic solutions are prepared. Very tiny amount of organic substances, such as surface active reagent, is contained in water [29]. These organic substances are sometimes very hard to remove by distillation. Adsorption of surface active reagents will easily prevent electrocatalytic activity of any electrode. Some organic substances suppress CO_2 reduction at Cu electrode. Slight amount of trimethylamine, occasionally contained in anion exchange membrane, prevents CO_2 reduction with enhanced HER [28].

3 Electrocatalysis of Metals in CO_2 Reduction

3.1 Various Metals

HER readily takes place in aqueous electrolytes by cathodic polarization, competing with CO_2 reduction. The rate is proportional to the proton activity in the electrolyte at a constant potential at various metal electrodes [30]. HER is prevalent particularly in acidic solutions, whereas CO_2 molecules do not exist in a basic solution. Thus most of CO_2 reduction have been studied with neutral electrolyte solutions.

Many studies of electrochemical reduction of CO_2 in early years were carried out in aqueous media with metal electrodes of high hydrogen overvoltage such as mercury and lead, aiming at suppression of HER. Eyring and his coworkers studied CO_2 reduction at a Hg electrode in detail; they showed that $HCOO^-$ is exclusively produced with the faradaic efficiency 100 % in neutral aqueous electrolytes [31]. Hori and Suzuki revealed that the partial current of $HCOO^-$ formation at a Hg electrode does not depend on pH at a constant potential, whereas HER is proportional to proton activity [32].

CO_2 reduction proceeds at various metal electrodes not only at those of high hydrogen overvoltage. The product selectivity depends greatly on the metal electrodes, [33] as tabulated in terms of faradaic efficiency with the current densities in Table 2. The product selectivity is greatly affected by the purity of the electrode metals as well as that of the electrolyte solution.

Table 2 Faradaic efficiencies of products in CO_2 reduction at various metal electrodes[a]

Electrode	Potential versus SHE/V	Current density/mA cm^{-2}	Faradaic efficiency (%)					
			CH_4	C_2H_4	CO	$HCOO^-$	H_2	Total
Pb	−1.63	5.0	0.0	0.0	0.0	97.4	5.0	102.4
Hg	−1.51	0.5	0.0	0.0	0.0	99.5	0.0	99.5
Tl	−1.60	5.0	0.0	0.0	0.0	95.1	6.2	101.3
In	−1.55	5.0	0.0	0.0	2.1	94.9	3.3	100.3
Sn	−1.48	5.0	0.0	0.0	7.1	88.4	4.6	100.1
Cd	−1.63	5.0	1.3	0.0	13.9	78.4	9.4	103.0
Au	−1.14	5.0	0.0	0.0	87.1	0.7	10.2	98.0
Ag	−1.37	5.0	0.0	0.0	81.5	0.8	12.4	94.6
Zn	−1.54	5.0	0.0	0.0	79.4	6.1	9.9	95.4
Pd	−1.20	5.0	2.9	0.0	28.3	2.8	26.2	60.2
Ga	−1.24	5.0	0.0	0.0	23.2	0.0	79.0	102.0
Cu	−1.44	5.0	33.3	25.5	1.3	9.4	20.5	103.5[b]
Ni	−1.48	5.0	1.8	0.1	0.0	1.4	88.9	92.4[c]
Fe	−0.91	5.0	0.0	0.0	0.0	0.0	94.8	94.8
Pt	−1.07	5.0	0.0	0.0	0.0	0.1	95.7	95.8
Ti	−1.60	5.0	0.0	0.0	tr.	0.0	99.7	99.7

Electrolyte 0.1 M $KHCO_3$, Temperature: 18.5 ± 0.5 °C
[a]The data are taken from Hori et al. [34]
[b]The total value contains C_2H_5OH (5.7 %), C_3H_7OH (3.0 %), C_3H_5OH (1.4 %), CH_3CHO (1.1 %), C_2H_5CHO (2.3 %) in addition to the tabulated substances
[c]The total value contains C_2H_6 (0.2 %)

Metal electrodes are divided into 4 groups in accordance with the product selectivity as presented in Table 2. Pb, Hg, In, Sn, Cd, Tl, and Bi give formate ion as the major product. Au, Ag, Zn, Pd, and Ga, the 2nd group metals, form CO as the major product. Cu electrode produces CH_4, C_2H_4, alcohols and aldehydes in quantitatively reproducible amounts. The 4th metals, Ni, Fe, Pt, and Ti, do not practically give product from CO_2 reduction continuously, but hydrogen evolution occurs. The classification of metals appears loosely related with that in the periodic table. However, the correlation is not very strong, and the classification such as *d* metals and *sp* metals as revealed for HER does not appear relevant.

These electrode metals may be regrouped broadly into 2 categories, i.e. CO formation metals and formate formation metals [34]. Cu electrode yields CH_4, C_2H_4 and alcohols from CO_2, in which process CO is formed as an intermediate species as described later. Ni, Pt and other platinum family metals give CO strongly adsorbed on the electrode surface, which practically prevents further reduction of CO_2 under atmospheric pressure, forming H_2 in aqueous systems. Au, Ag, Zn, Pd, and Ga give CO as a major product in CO_2 reduction. These metals (Cu, Au, Ag, Zn, Pd, Ga, Ni and Pt) belong to CO formation metals. Formate formation metals

(Pb, Hg, In, Sn, Cd, and Tl) give formate as a major product in neutral aqueous electrolytes in CO_2 reduction.

Such a classification obtained from the selectivity in aqueous systems may be combined with the results in nonaqueous systems. CO formation metals in aqueous electrolyte yield CO in propylene carbonate (PC). Some formate formation metals in aqueous electrolyte, Hg, Tl and Pb, give oxalic acid in PC. The intermediate metals, Cd, Sn, and In, give CO in PC solutions as the major product, whereas formic acid in aqueous solutions. Extending above mentioned concept, with use of another technique, adatom modified electrode, one can obtain the order of CO selectivity of metals; Au > Ag > Cu > Zn > > Cd > Sn > In > Pb > Tl ≈ Hg. The details may be referred elsewhere [34].

3.2 *Copper Electrode*

Copper metal electrode can reduce CO_2 to CH_4, C_2H_4 and alcohols in aqueous electrolytes. The product distribution depends on electrolyte, potential and temperature. Figure 2 shows results of controlled potential reduction of CO_2 at a Cu electrode in 0.1 M $KHCO_3$. The faradaic yields of CO and $HCOO^-$ rise at –0.9 V,

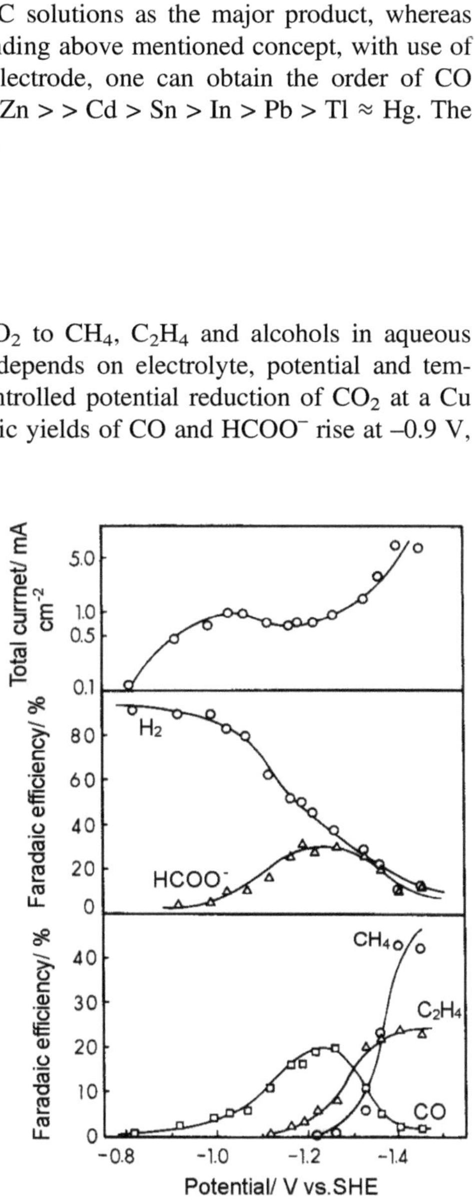

Fig. 2 Variation of the faradaic efficiencies of various products in electrochemical reduction of CO_2 at a Cu electrode in controlled potential electrolysis, 0.1 mol kg^{-1} $KHCO_3$, 1 atm CO_2, 18 °C. The potential axis negative to the *right*. (Reproduced from Ref. [35] with permission of The Royal Society of Chemistry)

Table 3 Electrochemical reduction of CO in 0.1 M $KHCO_3$ at various metal electrodes[a]

Electrode	Potential/V versus SHE	Faradaic efficiency/%				Heat of adsorption of CO/kcal mol^{-1}
		CH_4	C_2H_4	Other HCs and alcohols[b]	H_2	
Au	−1.49	0.0	0.0	0.0	101.6	9.2
Cu	−1.40	16.3	21.2	12.5	45.5	17.7
Ni	−1.46	2.6	0.3	0.7	94.2	40.8
Pt	−1.29	0.1	0.0	0.0	96.8	46.6

[a]Current density 2.5 mA cm^{-2}
[b]C_2H_6, C_2H_5OH, n-C_3H_7OH, HCHO

reaching maxima at −1.20 to −1.25 V, and drop with the increase of the cathodic potential. C_2H_4 begins to increase at −1.1 V; CH_4 starts at −1.2 V, rising steeply with the potential.

CO can also be electrochemically reduced to hydrocarbons and alcohols at a copper electrode with a similar product distribution with CO_2 reduction (Table 3) [35, 36], whereas $HCOO^-$ is not reduced at all at a Cu electrode. Infrared spectroscopic measurements showed that CO is adsorbed at Cu electrodes in CO_2 saturated electrolyte solution in on-top mode on polycrystal surface [37] and in bridged mode on Cu(111) surface [38]. These facts evidence that CO is intermediately formed at copper electrode during CO_2 reduction and further reduced to hydrocarbons and alcohols. The surface of Cu electrode is covered with CO with the coverage more than 90 % with HER severely suppressed [39].

Figure 2 shows that CH_4 formation starts at a more negative potential than C_2H_4, and rises more steeply with the increase of the negative potential. C_2H_4 formation is more favorable than CH_4 in high pH solutions. Such different features strongly suggest that CH_4 and C_2H_4 are produced via different reaction paths from the common starting substance, CO. The partial current of C_2H_4 formation from reduction of CO was analyzed in relation with the electrode potential; a linear Tafel relationship was obtained regardless of pH of the electrolyte [40]. CH_4 formation is different; the partial current is proportional to proton activity, giving a linear Tafel relationship. These discussions lead to a molecular reaction pathway diagram in which CH_4 formation path is separated from C_2H_4 in an early stage of the reaction, shown in Fig. 3, as also confirmed in a study with on-line mass spectroscopy [41]. Adsorbed CHO on Cu electrode was discussed in favor of hydrocarbon formation over COH as an intermediate in a study of density function theory [42, 43].

The reason has not yet been fully elucidated why only Cu can yield hydrocarbons and alcohols from CO_2 with high current density. An explanation was published in terms of CO adsorption. As described in a previous section, Au, Cu, Ni and Pt electrodes yield CO from CO_2. The electrocatalytic activities of these metals are compared for electroreduction of CO in reference to the heat of adsorption of

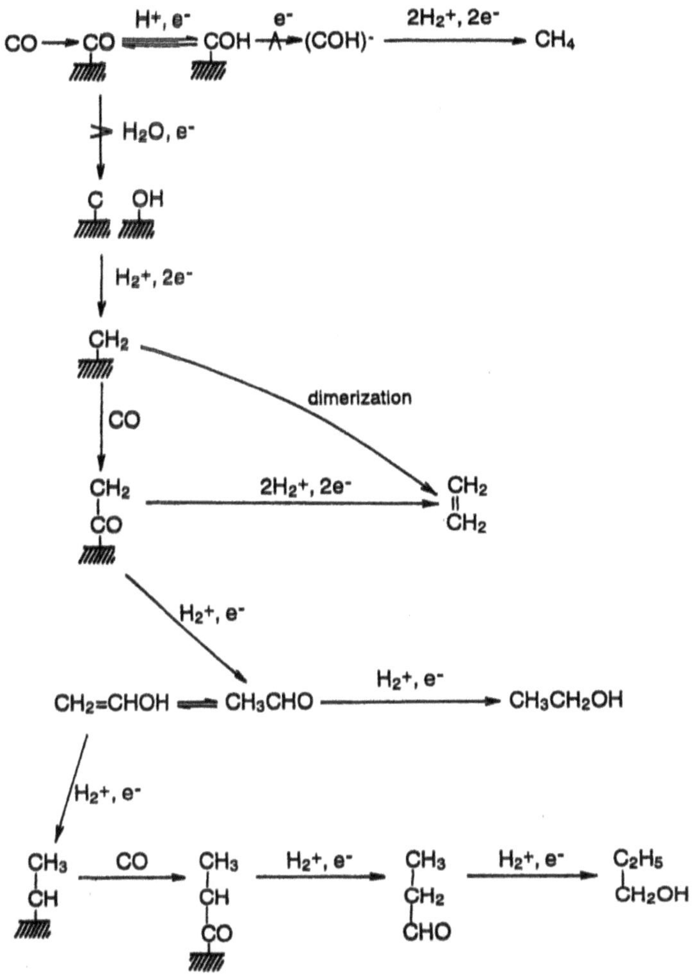

Fig. 3 Molecular reaction pathway diagram of CO reduction. H_2^+ represents H_2^+ or a combined H (ads) + H^+. The → with a ∧ through it denotes a rate-determining step. (Reproduced from Ref. [40], Copyright (1997) with permission of The American Chemical Society)

CO (Table 3) [44]. On the basis of the heats of adsorption, CO will scarcely interact with Au surface; thus CO will not be reduced at Au. Ni and Pt have high heats of adsorption, and adsorb CO strongly. CO, highly stabilized on the surface, cannot be further reduced. The heat of adsorption of Cu is intermediate among these metals, leading to CO adsorption with moderate strength; CO is effectively reduced to hydrocarbons at Cu. Watanabe et al. studied a molecular orbital calculation of CO adsorption on Cu and Ag. They showed that higher adsorption strength of CO on Cu than Ag leads to more effective formation of hydrocarbons [45, 46]. Recently

Peterson and Norskov basically supported the concept on the basis of a density functional theory approach [47].

CO is weakly adsorbed on Cu electrode, in equilibrium with gaseous CO dissolved in the electrolyte solution close to the electrode. When the electrolyte is stirred, CO desorbs easily from the surface [35]. Sakata et al. reported that the yield of CO from a Cu electrode is greatly enhanced when the electrolyte solution is stirred, as is compatible with the weak adsorption of CO on the electrode surface [48].

3.3 Copper Electrode; Effect of Anionic Species

The product distribution of Cu electrode is greatly affected by anionic and cationic species of the electrolyte solution. Table 4 shows the results of electrochemical reduction of CO_2 at a constant current density 5 mA cm^{-2} in various 0.1 M electrolyte solutions [35]. $KHCO_3$, KCl, $KClO_4$, and K_2SO_4 solutions favor the CO_2 reduction. K_2HPO_4 solutions highly promote HER rather than CO_2 reduction at less negative potential; 0.5 M K_2HPO_4 gives much higher H_2 yields than 0.1 M K_2HPO_4.

Such an enhancement of HER is attributed to lower pH value at the electrode/electrolyte interface caused by the buffer action of anions, mentioned previously. pH would rise locally at the interface due to OH^- generation in cathodic reactions in aqueous media. Nevertheless, the buffer action of HPO_4^{2-} neutralizes the OH^-, keeping the pH at a lower value. HER goes up with the concentration of K_2HPO_4 owing to its buffer action. A similar process proceeds for $KHCO_3$ solution. KCl, $KClO_4$, and K_2SO_4 do not have buffer ability, and thus the pH at the electrode/electrolyte interface rises. No significant change in the product distribution is observed with the variation of KCl concentration as shown in Table 4.

In addition to the competition between CO_2 reduction and HER, a product selectivity is also observed in accordance with anionic species. KCl, $KClO_4$, and K_2SO_4 solutions provide higher yields of C_2H_4 and alcohols than CH_4 in comparison with 0.1 M $KHCO_3$. Locally high pH region, formed at the electrode in these electrolytes, favors reduction of the intermediate CO to C_2H_4 and alcohols rather than to CH_4 as discussed previously.

3.4 Copper Electrode; Effects of Cationic Species

The cationic species also affects CO_2 reduction at metal electrodes. Paik et al. studied electroreduction of CO_2 to $HCOO^-$ at a Hg pool electrode in Li^+, Na^+ and $(C_2H_5)_4N^+$ hydrogencarbonate solutions [31]. The electrode potential at a constant current increases in the positive direction with the sequence of $Li^+ < Na^+ < (C_2H_5)_4N^+$. The results were discussed in connection with the variation

Table 4 Faradaic efficiencies of products from the electroreduction of CO_2 at a Cu electrode at 5 mA cm^{-2} in various solution at 19 °C

Solution	Conc./mol dm^{-3}	pH[a]	Potential/V versus SHE	Faradaic efficiency (%)							
				CH_4	C_2H_4	EtOH	PrOH	CO	$HCOO^-$	H_2	Total
$KHCO_3$	0.1	6.8	−1.41	29.4	30.1	6.9	3.0	2.0	9.7	10.9	92.0
KCl	0.1	5.9	−1.44	11.5	47.8	21.9	3.6	2.5	6.6	5.9	99.8
KCl	0.5		−1.39	14.5	38.2	[b]	[b]	3.0	17.9	12.5	
$KClO_4$	0.1	5.9	−1.40	10.2	48.1	15.5	4.2	2.4	8.9	6.7	96.0
K_2SO_4	0.1	5.8	−1.40	12.3	46.0	18.2	4.0	2.1	8.1	8.7	99.4
K_2HPO_4	0.1	6.5	−1.23	17.0	1.8	0.7	tr	1.3	5.3	72.4	98.5
K_2HPO_4	0.5	7.0	−1.17	6.6	1.0	0.6	0.0	1.0	4.2	83.3	96.7

[a]pH values were measured for bulk solution after electrolyses
[b] Not analysed
Reproduced from Ref. [35] with permission of The Royal Society of Chemistry

Table 5 Faradaic efficiencies of the products in the electrochemical reduction of CO_2 in various 0.1 M hydrogencarbonate solutions[a]

Electrolyte	Potential/ V versus SHE	Faradaic efficiency (%)								C_1/ C_2
		CH_4	C_2H_4	CO	EtOH	PrOH	$HCOO^-$	H_2	Total	
$LiHCO_3$	−1.45	32.2	5.2	tr	1.6	tr	4.7	60.5	104.2	6.19
$NaHCO_3$	−1.45	55.1	12.9	1.0	4.2	0.6	7.0	25.1	105.9	4.27
$KHCO_3$	−1.39	32.0	30.3	0.5	10.9	1.6	8.3	14.5	98.1	1.06
$CsHCO_3$	−1.38	16.3	30.5	2.4	7.2	4.4	15.8	24.4	101.0	0.53

[a]Current density: 5.0 mA cm^{-2}. Temperature: 18.5 °C

Reproduced from Ref. [49] with permission of The Chemical Society of Japan

of the potential at the outer Helmholz plane (OHP), and qualitatively ascribed to the greater specific adsorption of the larger cations. Hori and Suzuki later confirmed Paik et al.s' observation at a Hg electrode [32].

The features of CO_2 reduction at a Cu electrode also depends remarkably on cationic species. Table 5 shows faradaic efficiencies of CO_2 reduction at a Cu electrode in 0.1 M various alkali hydrogencarbonate solutions at a constant current density 5.0 mA cm^{-2} [49]. Hydrogen formation is prevalent in the Li^+ electrolyte. The formation of C_2H_4 and alcohols increases relatively in the order $Li^+ < Na^+ < K^+ < Cs^+$. The CH_4 and C_2H_4 ratio in the faradaic efficiency $C1/C_2$ obviously shows a trend $Li^+ > Na^+ > K^+ > Cs^+$. The potential during the electrolyses is in the order $Li^+ < Na^+ < K^+ < Cs^+$. An identical feature appears in electrochemical reduction of CO, an intermediate species in CO_2 reduction [49]. Such a cationic effect in CO_2 reduction at a Cu electrode was confirmed by other workers [50, 51]. These cationic effect may be derived from the extent of the specific adsorption of the cations to the electrode [49].

3.5 *Surface Treatment, Alloying and Modification of Metal Electrodes*

Surface treatments also affect CO_2 reduction. Acid treatment and thermal treatment are often applied in order to remove surface contamination as well as surface stress caused by mechanical polishing. The product selectivity of Cu electrode depends greatly on these surface treatments [22, 52, 53]. Frese reported formation of CH_3OH at intentionally preoxidized Cu electrodes. The maximum partial current of CH_3OH production reached 15 mA cm^{-2} [54]. However, Koga et al. did not detect CH_3OH from Cu electrodes oxidized in various manners [53]. Steady formation of CH_3OH from Cu electrodes at a high current density has not yet been confirmed by other workers to date. Tang et al. showed that nanoparticle covered Cu surface is favored for C_2H_4 formation and discussed in terms of density function theory [43].

Nanoparticle covered Au electrode was prepared by reducing amorphous surface oxide of Au electrode. Electrolysis current as high as 6 mA cm^{-2} was observed with high selectivity of CO. The overvoltage was reduced, and the electrode was not deactivated for more than 8 h [55].

Cu based alloys have been used for CO_2 reduction. Cu–Ni and Cu–Fe alloys, formed by in situ deposition during CO_2 reduction, gradually lose CH_4 and C_2H_4 yields simultaneously with increased H_2 evolution with the increase of Ni or Fe coverage on the Cu surface [56]. A Cu–Cd electrode, also prepared by in situ deposition, produced CH_4 and C_2H_4. The yields gradually dropped with the increase of Cd coverage, whereas CO formation rose [57]. Watanabe et al. studied various Cu based alloys; Cu–Ni, Cu–Sn, Cu–Pb, Cu–Zn, Cu–Cd. They reported that the major products are CO and $HCOO^-$ [58]. Kyriacou and Anagnostopoulos prepared surface alloy electrodes Cu–Au. They showed that the surface alloying severely suppresses the formation of hydrocarbons and alcohols, leading to the increase of CO formation [59].

Modifications of Cu electrode have been reported with introduction of oxygen, sulfur, or chlorine atoms to the surface in order to improve the product selectivity [22, 54, 60–63]. It is not confirmed whether or not these atoms still remain on the electrode surface during electrolytic reduction of CO_2. These atoms would be removed from the surface by cathodic polarization during the electrolysis, and newly reduced Cu surface may be composed of crystallites with favorable atomic configurations for preferred selectivity.

4 Problems with Regard to Efficient Utilization of CO_2 Reduction

4.1 Energy Utilization Efficiency and Overpotential

CO_2 can be electrochemically reduced to useful materials under mild conditions. However, energy conversion efficiency, defined as the ratio of the free energy of the products obtained in electrochemical CO_2 reduction and that consumed in the reduction, would be roughly 30–40 % [1, 2]. Such low energy efficiencies prevent the electrochemical reduction of CO_2 from practical process.

The standard potential of the reduction of CO_2, calculated from the thermodynamic data, depends on the products, −0.52 V versus SHE at pH 7 at 25 °C for CO for example. The reduction of CO_2 actually takes place at much more negative potentials than the values estimated from the thermodynamic data. Such high overpotentials result from involvement of intermediate species which requires highly negative potential for formation; $CO_2{\cdot}^-$ anion radical is the presumed intermediate species. The standard potential of $CO_2^{\cdot-}$ formation is −2.21 V versus

SCE in dimethyl formamide (DMF), [64] −1.90 [65] or −1.85 V versus SHE [66] in aqueous media.

Some ideas have been proposed to circumvent the formation of high potential intermediate, $CO_2\cdot^-$. Bocarsly showed pyridine and pyridinium redox system can mediate CO_2 reduction in aqueous media as an electron shuttle, preventing hydrogen evolution. The redox system promotes the reduction of CO_2 to formate, formaldehyde and methanol subsequently with hydrogenated Pd electrode. The redox system contributes to stabilization of $CO_2\cdot^-$ radical by complexation, working as a multielectron carrier. The overvoltage was greatly reduced by several hundreds mV [67, 68].

An ionic liquid electrolyte, 1-ethyl-3-methylimidazolium tetrafluoroborate (EMIM-BF4), is also effective for reduction of overvoltage in CO_2 reduction. Using an Ag electrode, Masel et al. succeeded in CO_2 reduction to CO with overvoltage as low as 0.17 V. They showed a selective formation of CO with an current efficiency higher than 96 % for 7 h. This process is also derived from stabilization of $CO_2\cdot^-$ radical by complexation with the electrolyte solution. Energy conversion efficiency reached 87 % [69]. In an extension, Kenis et al. synthesized nitrogen containing Ag complexes, and utilized them as the catalysts in gas diffusion electrode in a flow cell reactor. The presence of these nitrogen containing complexes reduced overvoltage of CO_2 reduction to CO, and enhanced the product selectivity. They suggest the cocatalysis effect of the ligand in the reaction [70].

4.2 *Enhancement of CO_2 Transport to the Electrode Interface*

Low solubility of CO_2 in water restricts the transport of CO_2 to the electrode in aqueous system. Under moderate experimental conditions, the highest current density will be limited to 20 mA cm^{-2} or so for formation of CO or $HCOO^-$ for example. Many trials have been attempted to solve the problem.

4.2.1 Elevated Pressure

Ito and his coworkers applied elevated pressure up to 20 atm to CO_2 reduction in aqueous and nonaqueous electrolyte solutions with Zn, In, Sn and Pb electrodes [71–73]. They showed enhanced current density of 30 mA cm^{-2} at –1.7 V versus SCE under 20 atm. The faradaic efficiency of HCOOH formation increased with the increase of CO_2 pressure.

Sakata and his coworkers studied CO_2 reduction under elevated pressure extensively [74, 75]. They could enhance the current density to 163 mA cm^{-2} with a Cu electrode at –1.64 V versus Ag/AgCl under 30 atm with major products of

CH_4, C_2H_4 and CO [48]. The maximum partial current density of HCOOH formation amounted to 560 mA cm^{-2} with an In electrode under 60 atm [76].

The solubility of CO_2 is high in methanol. Fujishima and his coworkers employed CO_2 methanol mixtures under elevated pressure as the electrolyte solution. Tetraalkyl ammonium salts were used for the supporting electrolytes. They showed that CO_2 reduction can proceed with a Cu electrode at 200–500 mA cm^{-2} under 40–60 atm [77].

Later Mazin et al., and Li and Prentice also published studies on CO_2 reduction at Cu electrodes in methanol based electrolyte and ethanol based one under elevated pressure [78, 79].

4.2.2 Gas Diffusion Electrode

Gas diffusion electrode (GDE), a porous composite electrode developed for fuel cell technology, has been applied to CO_2 reduction by many workers.

Mahmood et al. reported GDE with Pb particles as the catalyst which can reduce CO_2 to HCOOH. The highest current density was 100 mA cm^{-2} or more at –1.8 V versus SCE [80]. In and Sn showed lower catalytic activity. Furuya et al. employed Ru-Pd alloy particles as the catalyst, giving HCOOH at the current density 80 mA cm^{-2} at –1.1 V versus SHE [81]. They studied simultaneous electrolysis of CO_2 and nitrite or nitrate ions using GDEs with Cu catalyst particles and various metal phthalocyanines. They obtained urea as the reaction product other than CO and NH_3 [82–84]. Ni−, Co− and Sn− phthalocyanines were tested for CO_2 reduction as the catalyst particle. These catalysts gave CO (for Ni− and Co− phthalocyanines) or HCOOH (Sn−, Cu− and Zn− phthalocyanine) as the reduction products at high current density such as ca. 100 mA cm^{-2} [85–87].

GDEs with Cu particles as the catalyst produce C_2H_4 and C_2H_5OH from CO_2 at high current densities with high faradaic efficiencies. Sammells et al. communicated the current density 667 mA cm^{-2} in KOH electrolyte solution with the current yield of 53 % for C_2H_4 formation at –4.75 V versus Ag/AgCl [88]. Ikeda et al. showed that C_2H_4 and C_2H_5OH are the main products with the current density 300 mA cm^{-2} at –1.25 V versus SHE [89].

Schwarz et al. incorporated Perovskite-type electrocatalysts $A_{1.8}A'_{0.2}CuO_4$ (A = La, Pr, and Gd; A′ = Sr and Th) in GDEs with the electrolyte solution 1 M KOH. They reported that CO_2 was reduced to CH_3OH, C_2H_5OH and n-C_3H_7OH with the total current yield of alcohols 40 % of the current density 180 mA cm^{-2} [90]. They studied various perovskite type oxides, and argued that such oxides without Cu are not active for CO_2 reduction.

Sakata and his coworkers employed GDEs, loaded with Pt catalyst, and reduced CO_2 at elevated pressure higher than 5 atm in 0.5 M $KHCO_3$, whereas the GDE scarcely reduced CO_2 at 1 atm. Under 20 atm, CO_2 was reduced mainly to CH_4 with the current density of 600 mA cm^{-2} with the faradaic efficiency of ca. 50 % for total CO_2 reduction at –1.93 V versus Ag/AgCl [91, 92]. They further studied Fe, [93]

Pd and Ag [94] for the catalyst particles for GDEs, and demonstrated that the current density of CO_2 reduction to CO amounted to 3.05 A cm^{-2} with an Ag loaded GDE.

Tryk et al. studied GDEs composed of active carbon fiber and loaded with catalysts Ni, Fe, Pd metals, porphyrins, and phthalocyanines [95–97]. The GDEs gave partial current density of CO_2 reduction up to 80 mA cm^{-2} with production of CO under atmospheric pressure. Kenis et al. studied flow electrolytic cells equipped with GDEs loaded with nitrogen containing Ag complexes as mentioned previously [70].

4.2.3 Solid Polymer Electrolytes

Ion exchange membrane coated with porous catalyst metal (Solid Polymer Electrolyte, SPE) as well as GDE can provide gas phase electrolysis of CO_2. DeWulf et al. applied an SPE with Cu as the catalyst layer on a cation exchange membrane (CEM) Nafion 115. The SPE reduced CO_2 to CH_4 and C_2H_4 for a while, but the current density for CO_2 reduction dropped below 1 mA cm^{-2} after 70 min electrolysis [98].

Sammells et al. prepared a Cu coated SPE electrode using a CEM Nafion 117 [99, 100]. They reduced CO_2 to C_2H_4 and C_2H_6 with the current density of 10–30 mA cm^{-2} with the terminal voltage 1.5–3.5 V. The faradaic efficiency for CO_2 reduction remained less than 10 %.

Kunugi et al. studied Cu coated SPEs prepared from a CEM (Nafion 117) and an anion exchange membrane (AEM) Selemion AMV, attempting to reduce CO_2 from combustion flue gas [101]. The SPE from Nafion 117 formed C_2H_4 as the product with the maximum partial current 2 mA cm^{-2} at –1.8 V versus SCE. Both types of SPE retained stable activity in CO_2 reduction for 5 h.

Hori et al. prepared porous silver coated SPEs with an AEM (Selemion AMV). Their silver coated SPE reduced CO_2 mainly to CO with the partial current density 60 mA cm^{-2} at –2.8 V versus SHE [102]. They pointed out that a CEM based SPE is deteriorated in a short period of CO_2 reduction, when the SPE is combined with electrolyte solution other than strong acid. If $KHCO_3$ solution is used for the electrolyte solution for example, K^+ will be accumulated at the electrode metal/CEM interface in the cathode gas chamber during the electrolysis, since cations (K^+) are transported across CEM. OH^- is simultaneously generated at the interface by the cathodic CO_2 reduction. KOH, thus formed in this process, is neutralized by CO_2 to form $KHCO_3$ at the interface. The electrode metal/CEM interface is easily destroyed by the accumulation of $KHCO_3$ [102].

A modification of SPE was reported, using a porous membrane made of polyvinylidene fluoride polymer deposited with thin porous Au film [103]. CO_2 was reduced to CO with the faradaic efficiency 75 % at the partial current density 20 mA cm^{-2} as estimated from the difference of the two currents measured under argon and CO_2 atmospheres separately.

4.2.4 Three Phase Electrodes

Three phase electrodes were devised aiming at enhancement of the transport process in CO_2 reduction. Ogura et al. reported Cu and Ag net cathodes with CO_2 bubbled from the bottom of the electrolysis cell. They showed that the cathodic current is enhanced [27, 104]. Koleli et al. used fixed bed reactor packed with Pb or Sn granules. CO_2 was supplied from the bottom of the reactor with K_2CO_3 or $KHCO_3$ electrolyte solution [105, 106]. Their reactor reduced CO_2 mainly to HCOOH with high faradaic efficiency. Akahori et al. examined another three phase electrode device, composed of In impregnated Pb coated carbon fiber or Pb coated glass fiber electrode [107]. Chandrasekaran et al. reported a Pb coated stainless steel mesh electrode in a flow reactor cell which formed mainly formate with the current efficiency of 93 % [108].

References

1. Russell PG, Kovac N, Srinivasan S, Steinberg M (1977) J Electrochem Soc 124:1329
2. Hori Y (1993) In: Tomkiewicz M et al, Proceedings of the symposium on environmental aspects of electrochemistry and photoelectrochemistry, vol 93–18. The Electrochemical Society, p 1
3. Jitaru M, Lowy DA, Toma M, Toma BC, Oniciu L (1997) J Appl Electrochem 27:875
4. Halmann MM, Steinberg M (1999) Greenhous gas, Science and Technology. Lewis Publishers, Carbon Dioxide Mitigation
5. Hori Y (2003) In: Handbook of fuel cells, fundamentals, technology and applications, vol 2. Wiley, p 720
6. Chaplin RPS, Wragg AA (2003) J Appl Electrochem 33:1107
7. Scibioh MA, Viswanathan B (2004) Proceedings of the Indian national science academy, part A: physical sciences, vol 70, pp 407–462
8. Gattrell M (2006) J Electroanal Chem 594:1
9. Hori Y (2008) Modern aspects of electrochemistry, vol 42. In: Vayenas CG et al, Springer, p 89
10. Jhong H-RM, Ma S, Kenis PJA (2013) Cur Opin Chem Eng 2:191
11. Pinsent BRW, Pearson L, Roughton FJW (1956) Trans Faraday Soc 52:1512
12. Teeter TE, Van Rysselberghe P (1954) J Chem Phys 22:1759
13. Van Rysselberghe P, Alkire GJ, McGee JM (1946) J Amer Chem Soc 68:2050
14. Udupa KS, Subramanian GS, Udupa HVK (1971) Electrochim Act 16:1593
15. Stalder CJ, Chao S, Wrighton MS (1984) J Am Chem Soc 106:3673
16. Spichiger-Ulmann M, Augustynski J (1985) J Chem Soc, Faraday Trans 1(81):713
17. Hori Y, Suzuki S (1983) J Electrochem Soc 130:2387
18. Eggins BR, Bennett EM, McMullan EA (1996) J Electroanal Chem 408:165
19. Kuhn AT (1997) Electrochemistry. In: Bloom H, Gutmann F (eds) The past thirty and the next thirty years. Plenum Press p 139
20. Bockris JO'M, Khan SUM (1993) Surface electrochemistry, a molecular level approach. Plenum Press, New York, p. 9
21. Bockris JO'M, Reddy AKN (1970) Modern electrochemistry, vol 2. Plenum Press, New York
22. Kyriacou G, Anagnostopoulos A (1992) J Electroanal Chem 322:233
23. Jermann B, Augustynski J (1994) Electrochim Act 39:1891

24. Friebe P, Bogdanoff P, Alonso-Vante N, Tributsch H (1997) J. Catal. 168:374
25. Kedzierzawski P, Augustynski J (1994) J Electrochem Soc 141:L58
26. Kostecki R, Augustynski J (1994) Ber Bunsen-Ges Phys Chem 98:1510
27. Yano H, Shirai F, Ogura K (2002) J Electroanal Chem 533:113
28. Hori Y, Konishi H, Futamura T, Murata A, Koga O, Sakurai H, Oguma K (2005) Electrochim Acta 50:5354
29. Sawyer DT. Sobkowiak A, Roberts JL, Jr (1995) Electrochemistry for Chemists, 2nd edn. Wiley
30. Vetter KJ (1967) Electrochemical kinetics, theoretical and experimental aspects. Academic Press, New York, p 556
31. Paik W, Andersen TN, Eyring H (1969) Electrochim Act 14:1217
32. Hori Y, Suzuki S (1982) Bull Chem Soc Jpn 55:660
33. Hori Y, Kikuchi K, Suzuki S (1985) Chem Lett 1695
34. Hori Y, Wakebe H, Tsukamoto T, Koga O (1994) Electrochim Act 39:1833
35. Hori Y, Murata A, Takahashi R (1989) J Chem Soc Faraday Trans I 85:2309
36. Hori Y, Murata A, Takahashi R, Suzuki S (1987) J Amer Chem Soc 109:5022
37. Hori Y, Koga O, Yamazaki H, Matsuo T (1995) Electrochim Acta 40:2617
38. Shaw SK, Berna A, Feliu JM, Nichols RJ, Jacob T, Schiffrin DJ (2011) Phys Chem Chem Phys 13:5242
39. Hori Y, Murata A, Yoshinami Y (1991) J Chem Soc, Faraday Trans 87:125
40. Hori Y, Takahashi R, Yoshinami Y, Murata A (1997) J Phys Chem B 101:7075
41. Schouten Y, van der Kwon CJM, Ham Z, Qin MTM (2011) Koper. Chem Sci 2:1902
42. Durand WJ, Peterson AA, Studt F, Abild-Pedersen F, Norskov JK (2011) Sur Sc 605:1354
43. Tang W, Peterson AA, Varela AS, Jovanov ZP, Bech L, Durand WJ, Dahl S, Norskov JK, Chorkendorff I (2012) Phys Chem Chem Phys 14:76
44. Hori Y, Murata A, Takahashi R, Suzuki S (1987) Chem Lett 1665
45. Watanabe K, Nagashima U, Hosoya H (1993) Chem Phys Lett 209:109
46. Watanabe K, Nagashima U, Hosoya H (1994) Appl Surf Sci 75:121
47. Peterson AA, Norskov JK (2012) J Phys Chem Lett 3:251
48. Hara K, Tsuneto A, Kudo A, Sakata T (1994) J Electrochem Soc 141:2097
49. Murata A, Hori Y (1991) Bull Chem Soc Jpn 64:123
50. Kyriacou GZ, Anagnostopoulos AK (1993) J Appl Electrochem 23:483
51. Kaneco S, Katsumataa H, Suzuki T, Ohta K (2006) Electrochim Act 51:3316
52. Kim JJ, Summers DP, Frese KW Jr (1988) J Electroanal Chem 245:223
53. Koga O, Nakama K, Murata A, Hori Y (1989) Denki Kagaku (Electrochem) 57:1137
54. Frese KW Jr (1991) J Electrochem Soc 138:3338
55. Chen YH, Li CW, Kanan MW (2012) J Am Chem Soc 134:19969
56. Hori Y, Murata A, Ito S, Yoshinami Y, Koga O (1989) Chem Lett 1567
57. Hori Y, Murata A (1990) Chem Lett 1231
58. Watanabe M, Shibata M, Kato A, Azuma M, Sakata T (1991) J Electrochem Soc 138:3382
59. Kyriacou G, Anagnostopoulos A (1992) J Electroanal Chem 328:233
60. Hara K, Tsuneto A, Kudo A, Sakata T (1997) J Electroanal Chem 434:239
61. Momose Y, Sato K, Ohno O (2002) Surf Interface Anal 34:615
62. Ogura K, Yano H, Shirai F (2003) J Electrochem Soc 150:D163
63. Yano H, Tanaka T, Nakayama M, Ogura K (2004) J Electroanal Chem 565:287
64. Lamy E, Nadjo L, Savéant J-M (1977) J Electroanal Chem 78:403
65. Schwarz HA, Dodson RW (1989) J Phys Chem 93:409
66. Surdhar PS, Mezyk SP, Armstrong DA (1989) J Phys Chem 93:3360
67. Seshadri G, Lin C, Bocarsly AB (1994) J Electroanal Chem 372:145
68. Cole EB, Lakkaraju PS, Rampulla DM, Morris AJ, Bocarsly AB (2010) J Amer Chem Soc 132:11539
69. Rosen BA, Salehi-Khojin A, Thorson MR, Zhu W, Whipple DT, Kenis PJA, Masel RI (2011) Science 334:643
70. Tornow CE, Thorson MR, Ma S, Gewirth AA, Kenis PJA (2012) J Am Chem Soc 134:19520

71. Ito K, Ikeda S, Iida T, Niwa H (1981) Denki Kagaku 49:106
72. Ito K, Ikeda S, Iida T, Nomura A (1982) Denki Kagaku 50:463
73. Ito K, Ikeda S, Okabe M (1980) Denki Kagaku 48:247
74. Hara K, Kudo A, Sakata T (1995) J Electroanal Chem 391:141
75. Nakagawa S, Kudo A, Azuma M, Sakata T (1991) J Electroanal Chem 308:339
76. Todoroki M, Hara K, Kudo A, Sakata T (1995) J Electroanal Chem 394:199
77. Saeki T, Hashimoto K, Kimura N, Omata K, Fujishima A (1995) J Electroanal Chem 390:77
78. Mazin VM, Mysov EI, Grinberg VA (1997) Russ J Electrochem 33:779
79. Li J, Prentice G (1997) J Electrochem Soc 144:4284
80. Mahmood MN, Masheder D, Harty CJ (1987) J Appl Electrochem 17:1159
81. Furuya N, Yamazaki T, Shibata M (1997) J Electroanal Chem 431:39
82. Shibata M, Yoshida K, Furuya N (1998) J Electrochem Soc 145:595
83. Shibata M, Furuya N (2001) J Electroanal Chem 507:177
84. Shibata M, Furuya N (2003) Electrochim Act 48:3953
85. Mahmood MN, Masheder D, Harty CJ (1987) J Applied Electrochem 17:1223
86. Furuya N, Matsui K, Motoo S (1988) Denki Kagaku 56:288
87. Savinova ER, Yashnik SA, Savinov EN, Parmon VN (1992) React Kinet Catal Lett 46:249
88. Cook RL, MacDuff RC, Sammells AF (1990) J Electrochem Soc 137:607
89. Ikeda S, Ito T, Azuma K, Ito K, Noda H (1995) Denki Kagaku 63:303
90. Schwartz M, Cook RL, Kehoe VM, MacDuff RC, Patel J, Sammells AF (1993) J Electrochem Soc 140:614
91. Hara K, Kudo A, Sakata T, Watanebe M (1995) J Electrochem Soc 142:L57
92. Hara K, Sakata T (1997) J Electrochem Soc 144:539
93. Hara K, Sakata T (1995) Anal Sci Technol 8:683
94. Hara K, Sakata T (1997) Bull Chem Soc Jpn 70:571
95. Yamamoto T, Hirota K, Tryk DA, Hashimoto K, Fujishima A, Okawa M (1998) Chem Lett 825
96. Tryk DA, Yamamoto T, Kokubun M, Hirota K, Hashimoto K, Okawa M, Fujishima A (2001) Appl Organomet Chem 15:113
97. Magdesieva TV, Yamamoto T, Tryk DA, Fujishima A (2002) J Electrochem Soc 149:D89
98. DeWulf DW, Bard AJ (1988) Catal Lett 1:73
99. Cook RL, MacDuff RC, Sammells AF (1988) J Electrochem Soc 135:1470
100. Cook RL, MacDuff RC, Sammells AF (1990) J Electrochem Soc 137:187
101. Komatsu S, Tanaka M, Okumura A, Kunugi A (1995) Electrochim Act 40:745
102. Hori Y, Ito H, Okano K, Nagasu K, Sato S (2003) Electrochim Act 48:2651
103. Stevens GB, Reda T, Raguse B (2002) J Electroanal Chem 526:125
104. Ogura K, Yano H, Tanaka T (2004) Catal Today 98:515
105. Koleli F, Balun D (2004) Appl Catal A 274:237
106. Koleli F, Atilan T, Palamut N, Gizir AM, Aydin R, Hamann CH (2003) J Appl Electrochem 33:447
107. Akahori Y, Iwanaga N, Kato Y, Hamamoto O, Ishii M (2004) Electrochemistry (Tokyo, Japan) 72:266
108. Subramanian K, Asokan K, Jeevarathinam D, Chandrasekaran M (2007) J Appl Electrochem 37:255

CO_2 Reduction Using an Electrochemical Approach from Chemical, Biological, and Geological Aspects in the Ancient and Modern Earth

Akira Yamaguchi, Yamei Li, Toshihiro Takashima, Kazuhito Hashimoto and Ryuhei Nakamura

Abstract The past few decades have witnessed significant advances in our understanding of the synthesis routes and development of electrocatalysts for the reduction of CO_2 to hydrocarbons. It is also notable that the research field related to the Origin of Life theory starts to recognize the significance of electrochemical CO_2 reduction by metal-sulfide minerals as the primary step for organic carbon synthesis. In this chapter, we describe recent progress in the development of catalysts for CO_2 reduction in electrochemical systems, particularly from the viewpoint of the Origin of Life theory, and discuss the perspectives related to the evolutional origin of carbon monoxide dehydrogenases (CODHs), known as one of the most active natural enzymes for CO_2 reduction.

1 Introduction

In an attempt to mitigate the adverse effects of global warming, there has been considerable efforts to produce sustainable electricity from renewable energy sources, such as hydro, solar, wind, and geothermal power. Despite these efforts, renewable energy has inherent limitations: intermittency and fluctuation in the amount of generated energy. One possible solution for overcoming this limitation is to convert generated electricity into H_2 via electrochemical water-splitting reactions. Although electrolysis appears to be a promising solution for producing

A. Yamaguchi · Y. Li · R. Nakamura (✉)
Biofunctional Catalyst Research Team, RIKEN Center for Sustainable Resource Science, 2-1 Hirosawa, Wako, Saitama 351-0198, Japan
e-mail: ryuhei.nakamura@riken.jp

T. Takashima
Clean Energy Research Center, University of Yamanashi, Kofu, Yamanashi 400-8511, Japan

K. Hashimoto
Department of Applied Chemistry, The University of Tokyo, 7-3-1 Hongo, Bunkyo-ku, Tokyo 113-8656, Japan

M. Sugiyama et al. (eds.), *Solar to Chemical Energy Conversion*, Lecture Notes in Energy 32, DOI 10.1007/978-3-319-25400-5_13

reliable and sustainable energy, it is highly desired that the generated H_2 is converted into liquid form for transport, a process that requires considerable energy. Although the commercialization of H_2 as an energy fuel has increased in recent years, the processing and transport of H_2, and the associated infrastructure, remain a major restriction for the widespread use of H_2 as an energy fuel. Another potential solution for the effective use of electricity generated by renewable energy would be to electrochemically convert CO_2 into hydrocarbons. Compared to H_2, hydrocarbons are more easily transformed to liquid form, and further, the infrastructure for hydrocarbon distribution, storage and utilization has been firmly established.

The past few decades have witnessed significant advances in our understanding of the synthesis routes and development of electrocatalysts for the reduction of CO_2 to hydrocarbons. CO_2 reduction catalysts are typically metals or metal alloys, such as pure metallic Cu, which is considered to be an ideal catalytic material for the production of hydrocarbons with high Faradaic efficiency (FE). However, the selective production of hydrocarbons requires an overpotential (η) on the order of 1 V to avoid the formation of H_2 gas. Furthermore, if water is oxidized to O_2 at the anode, the η for the electrochemical conversion of CO_2 to hydrocarbons increases to as high as 2 V. Such high η values are a major obstacle for improving the energy efficiency of hydrocarbon synthesis, which is a necessary prerequisite for the practical application of heterogeneous electrocatalysts in CO_2 fixation systems.

Biological systems convert CO_2 into hydrocarbons with remarkably low η and high FE through diverse enzymatic reactions. The most outstanding example is carbon monoxide dehydrogenases (CODHs), which are found in certain chemolithoautotrophic microorganisms, including acetogens and methanogens, convert CO_2 into CO with a turnover frequency as high as 45 s^{-1} [1, 2]. CODHs can also electrochemically reduce CO_2 to CO at potentials near the thermodynamic potential with the FE of nearly 100 % [1, 2]. The remarkable efficiency of CODHs for CO_2 reduction has inspired numerous scientists to determine the underlying multi-step mechanisms for developing bioinspired catalysts for carbon utilization.

The family of CODHs has evolved over billions of years, and they have used Fe, Ni, Cu and Mo as active sites. Thus, one of the fundamental issues to consider is how the active cores of CODHs, namely Fe_4S_4Ni and MoSCu, have evolved from environmentally abundant rock-forming elements and been functionalized specific for CO_2 reduction with the aid of N- and S-containing peptides. As proposed in the Origin of Life theory, the surface of bisulfide-bearing hydrothermal precipitates with trace metals, such as Ni, Cu, W and Mo, is suggested to have played a critical role in prebiotic organic synthesis by CO_2 reduction on ancient Earth [3–5]. Given those facts, seeking the evolutional transition of inactive FeS minerals to active inorganic analogs of natural enzymes is what scientists can learn from nature's trails in order to acquire the effective catalysts. In other words, in addition to determining the multi-step mechanisms by which modern-day enzymes mediate CO_2 reduction, study of the evolutionary transition of prebiotic catalysts to biotic ones and the geochemical conditions that fostered the transition may provide a new design rationale for the functional analogues of biological catalysts.

Based on the above considerations, this chapter describes recent progress in the development of catalysts for CO_2 reduction in electrochemical systems, particularly from the viewpoint of the Origin of Life theory, which concerns the early evolution of energy metabolism on ancient Earth. After briefly overviewing electrochemical CO_2 reduction by metal electrodes in Sects. 2 and 3, bioinspired approaches for catalytic CO_2 reduction are presented, and the importance of the "frustrated Lewis acid-base pair motif" for CO_2 activation is discussed in Sects. 4 and 5. In Sects. 6 and 7, recent progress on the Origin of Life theory is introduced, and the interaction between electrocatalysts and biogeochemical communities is discussed as a platform for exploring functional analogues of biological catalysts for CO_2 reduction. In Sects. 8, perspectives related to the bioenergetic evolution of CO_2 reduction and carbon assimilation are described.

2 CO_2 Reduction by Metal Electrodes

The first report in the late 19th century on the electrochemical conversion of CO_2 to organic chemicals described the reduction of CO_2 to formic acid [6]. Early electrochemical conversion studies utilized mercury and amalgam, such as zinc-amalgam and sodium-amalgam, as electrocatalysts and demonstrated the high FEs for the production of formic acid [6–12].

Since the late 1970s, researchers have searched for metal electrodes capable of acting as catalysts for CO_2 reduction. For example, Ito et al. investigated CO_2 reduction by Zn, Pb, Sn, In, and Cd metal electrodes using chromotropic acid testing and infrared spectroscopy [13] and revealed that only formate was produced by all of these catalysts during the reduction of CO_2. In 1985, Hori et al. conducted galvanostatic electrolysis at a constant current density of 5 mA cm^{-2} on various metal electrodes, including Cu, Au, Ag, Zn, Pd, Ga, Pb, Hg, In, Sn, Cd, Tl, Ni, Fe, Pt, and Ti [14, 15], and found that formate was predominantly formed at Cd, In, Sn, and Pb cathodes, whereas carbon monoxide was the primary product at Ag and Au cathodes (Table 1). Furthermore, among the examined metals, the Cu cathode was specifically active for the eight-electron reduction of CO_2 to form methane. Based on the work by Ito and Hori et al., electrode metals can be categorized into four groups: (i) metal electrodes that selectively produce formate (Hg, In, Pb, and Sn electrodes); (ii) those that mainly produce carbon monoxide (Au, Ag, Zn, and Pd electrodes), (iii) electrodes that yield hydrocarbons and alcohols (Cu electrode), and (iv) those that show no or little activity toward CO_2 reduction (Ni, Fe, Pt and Ti electrodes). Although the factors controlling the product dependence on the metals remain unknown [15, 16], the results reported by Ito et al. and Hori et al. clearly indicate that the nature of the metal electrodes influence the products formed during electrochemical CO_2 reduction.

Table 1 CO_2 reduction products on metal electrodes in aqueous solution reported by Hori et al. [14]

Metal electrode	Electrode potential	Faradaic efficiency/%		Lower limit/Upper limit		
	(V vs. SHE)	$HCOO^-$	CO	CH_4	H_2	Total
Cd	−1.66 ± 0.02	65.3/67.2	6.2/11.1	0.2	14.9/22.2	93/100
Sn	−1.40 ± 0.04	65.5/79.5	2.4/4.1	0.1/0.2	13.4/40.8	94/110
Pb	−1.62 ± 0.03	72.5/88.8	0.3/0.6	0.1/0.2	3.8/30.9	94/100
In	−1.51 ± 0.05	92.7/97.6	0.9/2.2	0.0	1.6/4.5	93/102
Zn	−1.56 ± 0.08	17.6/85.0	3.3/63.3	0.0	2.2/17.6	90/98
Cu	−1.39 ± 0.02	15.4/16.5	1.5/3.1	37.1/40.0	32.8/33.0	87/92
Ag	−1.45 ± 0.02	1.6/4.6	61.4/89.9	0.0	10.4/35.3	99/106
Au	−1.14 ± 0.01	0.4/1.0	81.2/93.0	0.0	6.7/23.2	100/105
Ni	−1.39	0.3	0.0	1.2	96.3	98
Fe	−1.42	2.1	1.4	0.0	97.5	101

3 Nitrogen-Containing Molecules as Cocatalysts for Electrochemical CO_2 Reduction

One approach for improving the catalytic activity and FF for electrochemical CO_2 reduction on metal electrodes is the utilization of nitrogen (N)-containing compounds as catalysts. In 1994, Seshadri et al. [17] reported that hydrogenated Pd electrodes electrochemically reduced CO_2 to methanol with an FE of 30 % and η of less than 200 mV in the presence of pyridine. Morris et al. [18] further studied pyridine-catalyzed CO_2 reduction with Pt electrodes by changing the catalyst concentration, temperature, and pressure, and confirmed that the reaction kinetics follow the first order with respect to both CO_2 and pyridine. Based on these experimental results, combined with density functional theory (DFT) calculations, the authors proposed the reduced activation energy for CO_2 reduction by pyridine stabilization of carbamate species, which are possible intermediates for CO_2 reduction, by the Lewis basicity of the pyridyl N and metal electrode surface. Further calculations by Keith et al. [19] indicated that the active species was a "surface-bound" dihydropyridine (DHP), rather than a pyridinyl radical in solution.

The use of ionic liquids as catalysts for CO_2 reduction has also been demonstrated as an effective approach to lower η and increase FEs. For example, Rosen et al. [20] electrochemically reduced CO_2 with an Ag cathode in an ionic liquid electrolyte and achieved an η of less than 200 mV and an FEs of more than 96 % for CO production. Notably, the authors utilized 1-ethyl-3-methylimidazolium fluoroborate (EMIM-BF_4) with the expectation that this compound would complex with CO_2 to stabilize the intermediate CO_2^- radical species. The discovery of ionic liquid-based CO_2 reduction has significantly expanded the range of materials that can function as electrocatalysts for CO_2 reduction [21–26].

Although these observations clearly indicate that N-containing molecular catalysts are effective for electrochemical CO_2 reduction in terms of the stabilization of intermediate species, the properties of the metal electrodes and their contribution to catalytic activity have not been examined in detail.

4 Cooperative Activation of CO_2 by Metals and Amino Acids in Biological Systems

The former section described the effectiveness of using homogeneous molecular catalysts for achieving high FE and low η for electrochemical CO_2 reduction. Because the electrochemical reaction proceeds at the metal electrode surface, the interaction between the metal electrodes and CO_2 and/or other reaction intermediates, in addition to the presence of N-containing compounds, should be taken into consideration when designing CO_2 reduction catalysts. One of the most sophisticated examples of the cooperative effect of metals and N-containing compounds for CO_2 reduction is the natural enzymatic system CODH [27, 28]. There are two main types of CODHs: Mo-containing CODHs, which are used by aerobic bacteria [29], and Ni-containing CODHs, which are found in anaerobic bacteria [30]. The crystal structure of Mo–CODH was first reported by Dobbek et al. in 1999 [31] at a resolution of 2.2 Å (Fig. 1, left). The active site of Mo–CODH is composed of molybdopterin-cytosine dinucleotide and S-selanylcysteine connected to a [2Fe–2S] cluster and flavin-adenine dinucleotide. The same research group determined the crystal structure of Ni–CODH in 2001 [32] at a resolution of 1.63 Å (Fig. 1, right).

To better understand the catalytic mechanisms of CO_2 reduction, a series of spectroscopic, biochemical, and structural experiments have been conducted for Ni–CODH as summarized in a review paper [33], because it catalyzes the interconversion between CO_2 and CO [2], while Mo–CODH can only catalyze the

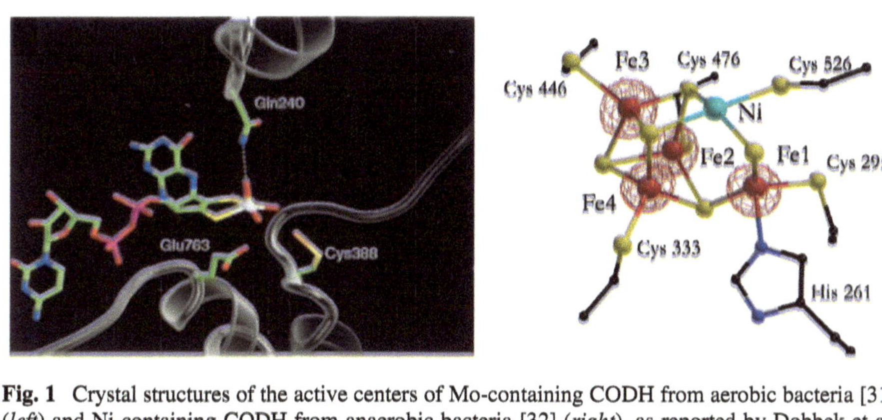

Fig. 1 Crystal structures of the active centers of Mo-containing CODH from aerobic bacteria [31] (*left*) and Ni-containing CODH from anaerobic bacteria [32] (*right*), as reported by Dobbek et al.

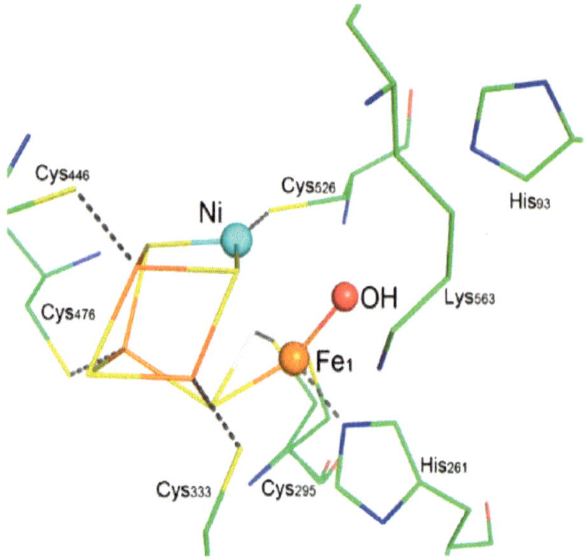

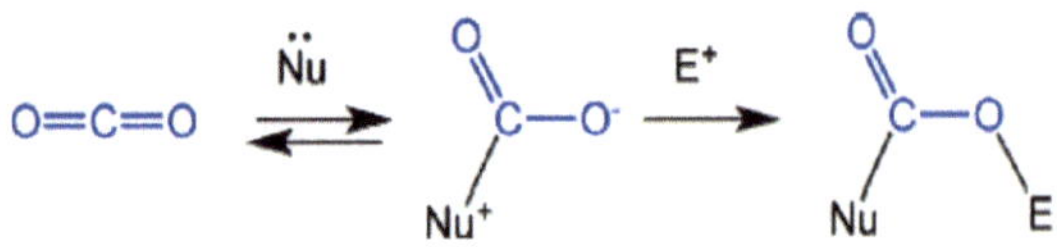

Fig. 2 *Ball-and-stick* drawing of the active site of Ni-containing CODH. A combination of nucleophilic (Nu and Ni sites in the CODH active center) and electrophilic (E, histidine 93 in CODH) interactions facilitates the reduction of CO_2 [27]

oxidation of CO to CO_2. The findings from these studies highlight the importance of nucleophilic and electrophilic interactions for the cooperative activation of CO_2 [27]. For Ni–CODH, the cooperative activation of CO_2 occurs through the nucleophilic interaction of a Ni site to a carboxylate carbon atom, and also involves electrophilic interactions of a lysine residue and Fe site with one of the carboxylate oxygen atoms [27]. These combined interactions, which is a so-called "frustrated Lewis acid-base pair motif" [34], are essential for decreasing the C–O bond order of CO_2, enabling Ni–CODH to effectively reduce CO_2 to CO without the use of strong electrophiles or nucleophiles (Fig. 2). The concept of a "frustrated Lewis acid-base pair" has been proposed for a number of enzymatic reactions in biological systems; however, few examples are applicable for the rational design of electrocatalysts.

5 Metal Sulfides as Biomimetic Catalysts

As the active sites of CODHs are composed of metal-sulfur clusters, metal sulfides are possible candidates for CO_2 reduction catalysts; however, metal sulfides have received relatively little attention as electrocatalysts for CO_2 reduction compared to electrocatalysis with metals and metal alloys. Inspired by the active sites for enzymatic CO_2 reduction in natural systems, Hidai et al. synthesized Fe–S cubane clusters, $[Fe_4S_4(SR)_4]^{2-}$ (R = $CH_2C_6H_5$ or C_6H_5), and evaluated the potential of these clusters for electrochemical CO_2 reduction in *N, N*-dimethylformamide (DMF) [35]. The addition of $[Fe_4S_4(SR)_4]^{2-}$ clusters on a mercury pool electrode resulted in an approx. 700 mV decrease in η. Although the main reaction product was oxalate, the addition of small amount of water increased the production rate of both formate and CO. Tanaka et al. synthesized a metal-sulfur cluster [36] composed of Mo, Fe and S ($[Mo_2Fe_6S_8(SEt)_9]^3$ to mimic the enzymes that found in the reductive carboxylic cycle in photosynthetic bacteria. The model reaction of α-keto acid synthesis proceeded in the presence of the cluster under acetonitrile conditions with an applied electrode potential of −1.55 V versus SCE: $RC(O)SEt + CO_2 + 2e^- \rightarrow RC(O)COO^- + EtS^-$ (R = CH_3, C_2H_5, C_6H_5). In addition to α-keto acid synthesis, β-keto acid synthesis was also promoted by $[Fe_4S_4(SPh)_4]^{2-}$ and $[Mo_2Fe_6S_8(SPh)_9]^{3-}$; $8PhC(O)CH_3 + 8CO_2 + 2NO_2^- + 6e^- \rightarrow 8PhC(O)CH_2COO^- + N_2 + 4H_2O$ [37]. In 2011, Kanatzidis et al. reported [38] that a chalcogel-type material, in which Fe–S cubane clusters and transition metals (Pt, Co, Ni, Sn, or Zn) were connected by Sn–S clusters, was capable of electrochemical CO_2 reduction. Notably, the CO_2 reduction efficiency in DMF tended to be higher when Ni or Co was utilized as transition metal sites. In all of these reports, the electrochemical reduction of CO_2 was conducted in an organic solvent, and the use of metal-sulfide clusters in aqueous solution has not been extensively examined, even though CODHs can catalyze the reduction of CO_2 to CO with high FE and low η under aqueous conditions [2]. To functionalize metal sulfides as CO_2 reduction catalysts, the cooperative effects of the metal centers and the surrounding N-containing molecules are an important consideration, as demonstrated in natural systems, as described in Sect. 4.

In 2014, having the idea that $FeNi_2S_4$ (violarite) has the structural similarity to the Fe_4S_4Ni active site in CODH [39], the electroreduction of CO_2 by $FeNi_2S_4$ (violarite) in aqueous solution was reported by our group [40]. Although Fe_3S_4 (greigite) inefficiently reduces CO_2, the efficiency of CO and CH_4 production was substantially improved by the substitution of Fe with Ni to form violarite. The electroreduction of CO_2 by $FeNi_2S_4$ in aqueous solution was further enhanced by modifying the surface of violarite with amine compounds, such as triethylamine and poly-allylamine hydrochloride (Fig. 3). As CO_2 reduction by CODH involves cooperative activation via both nucleophilic and electrophilic interactions [27], the enhanced activation of CO_2 reduction by substitution of Fe with Ni and the surface modification by amine compounds is of particular interest for the rational design of the efficient CO_2 reduction catalysts.

To clarify the role of Ni in the active site of CODH for the electroreduction of CO_2, Nørskov et al. [41] performed DFT calculations for a $[Fe_4S_4]$ cubane cluster with and without the substitution of Fe with Ni. The substitution of Ni in place of Fe was found to markedly decrease the activation energy for the reduction of CO_2 to CO, but the activation energy for COOH production was not affected. The authors attributed this decrease to a change in the coordination environment of the adsorbed COOH (COOH*) intermediate resulting from the Ni substitution (Fig. 4). Specifically, COOH* only binds to S sites in $[NiFe_3S_4]$ clusters, but binds to both S

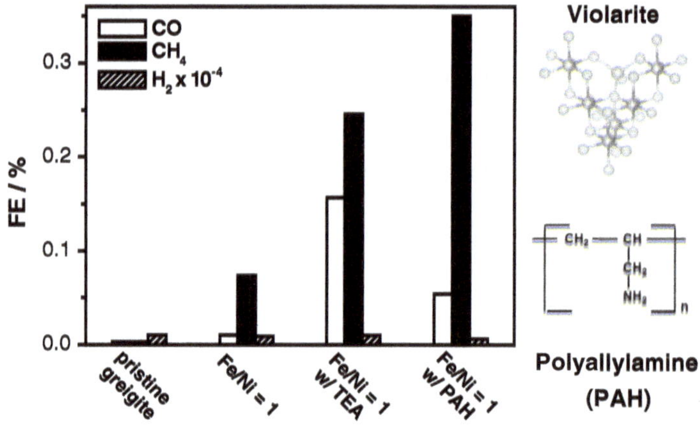

Fig. 3 Faradaic efficiencies (FEs) for CO, CH_4 and H_2 production by greigite, violarite (Fe/Ni = 1) and violarite (Fe/Ni = 1) modified with triethylamine (TEA) or polyallylamine (PAH) under a potentiostatic condition of −1.3 V versus SHE for 30 min [40]. The electrolyte pH was 5.5

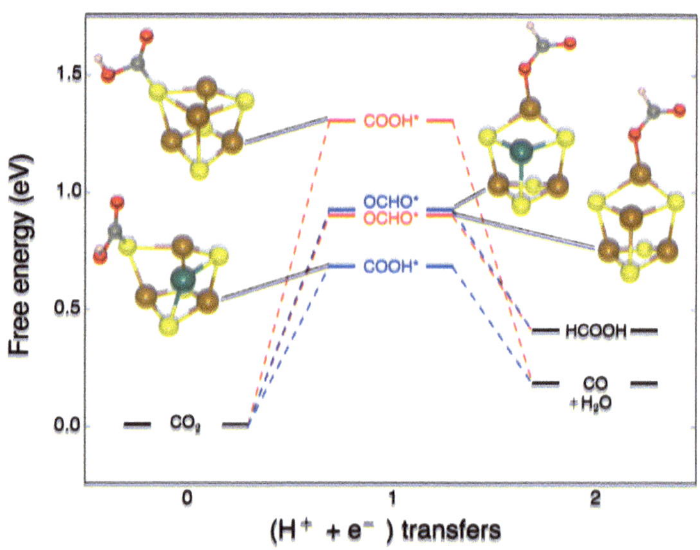

Fig. 4 Free energy diagram for CO_2 reduction on $[Fe_4S_4]^{2-}$ (*left top and right bottom*) and $[NiFe_3S_4]^{2-}$ (*left bottom and right top*) cubane complexes based on DFT calculations reported by Nørskov et al. [41]

and Fe in $[Fe_4S_4]$ clusters. Interestingly, a further decrease in the activation energy was predicted by additional Ni substitution, but only one or zero Ni-substituted cluster was predicted to be stable during the catalytic cycle. Notably, however, the computational study by Nørskov et al. did not consider the function of amino acid residues in the activation process of CO_2 reduction. As we reported previously, amine compounds play a cooperative role with Ni sites for electrochemical CO_2 reduction [40]. Therefore, a DFT study on the role of amino acids in this process is expected to provide new insights into the differences between CODH and abiotic catalysts comprised of $NiFe_3S_4$ clusters, and aid in incorporating the function of the 'frustrated Lewis acid-base pair motif' into man-made electrocatalysts.

An interesting and unanswered question is how prebiotic catalysts for CO_2 reduction evolved from readily available elements in the environments, and further, how these catalysts were functionalized to achieve low η and high FE. Concerning these points, numerous hypotheses related to chemical evolution in the prebiotic ocean and the early evolution of energy metabolism in ancient Earth have been proposed [3–5, 42]. In Sects. 6, 7 and 8, we describe the intimate relationship between processes related to the Origin of Life and electrochemistry, and discuss how these processes can help shape the design rationale for heterogeneous electrocatalysts for efficient CO_2 reduction.

6 Prebiotic Organic Synthesis by Metal Sulfides

Hydrothermal vents in the deep sea floors are surrounded by diverse ecosystems that are sustained by the chemical energy within hydrothermal fluid [43]. It is generally considered that the hydrothermal environment played a crucial role in the emergence of life on primordial earth. Although it remains a topic of debate as to whether primordial life was autotrophic or heterotrophic, the discovery of hydrothermal vents in the 1970s led researches to consider that the first species were autotrophic in nature [44]. One of the main problems of the autotrophic origin theory is the process of carbon fixation; namely, how were organic compounds continuously synthesized at hydrothermal vent? In an attempt to answer this question, Wächtershäuser proposed that CO_2 reduction proceeded at the surface of pyrite (FeS_2) [45, 46]. In this model, organic molecules synthesized on pyrite self-organized to form the first metabolizing cell. The driving force of CO_2 reduction in this system is the oxidative formation of pyrite ($4CO_2 + 7H_2S + 7FeS \rightarrow (CH_2\text{–}COOH)_2 + 7FeS_2 + 4H_2O$, $G^O = -420$ kJ/mol). This hypothesis is supported by the experimental observations that thiol (CH_3S^-) is formed from CO_2 in the presence of FeS [47]; acetic acid is formed from CH_3SH and CO on iron-nickel sulfide [5]; and peptides are formed from amino acids with CO and iron-nickel sulfide [48].

In contrast to Wächtershäuser's hypothesis, Russell proposed that CO_2 in ocean water reacts with H_2 contained in alkaline hydrothermal fluid in three-dimensional nano- or micro-sized cavities composed of iron sulfides [49]. The main difference between the hypotheses of Wächtershäuser and Russell is that the former considers FeS as an energy source for CO_2 reduction, whereas the latter treats FeS as a component for the reactors, with H_2, rather than FeS, acting as the electron donor for CO_2 reduction. Of note, iron sulfides, particularly Nickelian mackinawite

($FeNiS_2$), Ni-containing greigite ($NiFe_5S_8$), and violarite ($FeNi_2S_4$), are speculated to act as the "catalysts" for CO_2 reduction in the cavities proposed in Russell's model, as these materials share structural similarity to [Ni–Fe] hydrogenase, the C-cluster in CODH, and the ACS active site A-cluster, respectively [39]. Russell et al. further hypothesized that H_2, when adsorbed onto and/or absorbed into the mineral cubane sites, is dissociated into a proton, electron, and reactive hydrogen atom ($H^{\cdot}$), which then reduces CO_2 to CO [42]. Subsequently, the generated CO reacts with methane thiol (CH_3SH) produced in the Earth's crust [45] to form thioester acetyl methylsulphide, which is then hydrolyzed to acetate [5].

7 Bioelectrochemical CO_2 Reduction at Deep-Sea Hydrothermal Vents

To explain how processes related to the Origin of Life and electrochemistry are being merged, we introduce our recent findings regarding the energetics of CO_2 reduction at deep-sea hydrothermal vents. In 2010, we reported that deep-sea hydrothermal vents function as naturally occurring electrochemical fuel cells, which are powered by the reductive chemicals stored in the Earth's crust [50]. We found that natural minerals dominated by $CuFeS_2$ and FeS_2 from Black smoker chimneys in the Mariner hydrothermal field display excellent metallic electrical conduction over a 10 cm distance. In addition, we also demonstrated that deep-sea minerals promote the electrocatalytic oxidation of H_2S and H_2, as well as the electroreduction of O_2 and ferric ions. These findings led us to propose that a new type of energy propagation for CO_2 reduction and carbon assimilation occurs in the deep floor. In addition to the reductive chemicals, such as H_2 and H_2S, emitted from hydrothermal vents, we speculate that high-energy electrons from hot reduced hydrothermal fluid are directly delivered to cold oxidative seawater in the form of electrical current, with the same operation principals as electrochemical fuel cells (Fig. 5).

Based on our experimental findings, together with the hypothesis of Russell et al. that Ni/greigite may be the evolutionary origin for the $NiFe_3S_4$ cluster in CODH [39, 42], we also investigated the energetics for electrocatalytic CO_2 reduction by $FeNi_2S_4$ [40]. Through the examination of potential-dependent CO_2 reduction activity, it was found that CO_2 reduction by H_2 in hydrothermal fluids involves a strong endergonic electron transfer reaction. Therefore, a naturally occurring proton-motive force (PMF) as high as 200 mV must be established across the hydrothermal vent chimney wall. These findings are consistent with the speculation that highly alkaline hydrothermal vents in the Hadean ocean are the most likely system that supported prebiotic organic synthesis [42], as the chimney interior of these hydrothermal vents has a pH of 9–10 and the outer walls are bathed in ocean water with a pH of 5–6, generating a PMF of 180–300 mV. Of note, in addition to the PMF, steep temperature difference between hydrothermal fluid and sea water

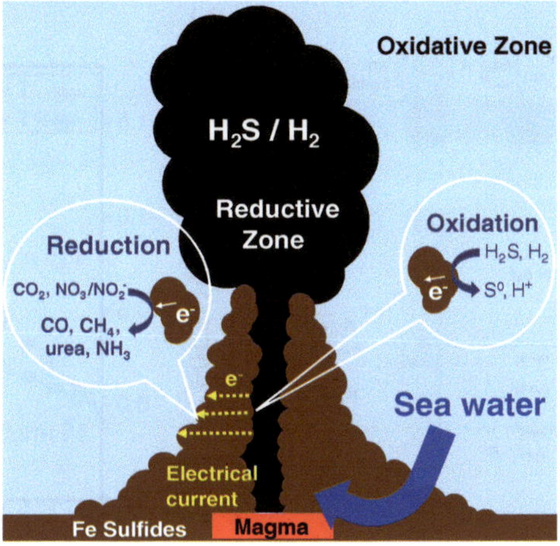

Fig. 5 Proposed model for electrical current generation in the chimneys of deep-sea hydrothermal vents [50]. Electrons generated via the oxidation of H_2S in hydrothermal fluid (*Reductive Zone*) are transported through the conductive chimney wall to sea water (*Oxidative Zone*) where they reduce Fe^{3+}, O_2, and CO_2

realized by low thermal conductivity of natural minerals resulted in the generation of high reduction energy, which is comparable to the photosynthetic CO_2 fixation center (Fig. 6) [51]. Therefore, the geochemical 'settings' of electrochemical reactors in the Hadean Ocean satisfy the thermodynamic and kinetic preconditions that are necessary to initiate the electroreduction of CO_2 to hydrocarbons. However, it is interesting to consider how prebiotic catalysts for CO_2 reduction were functionalized to achieve low η and high FE in these naturally occurring electrochemical fuel cells.

8 Evolution of Active Cores of CODHs and Challenges for the Design of Efficient CO_2 Reduction Catalysts

Finally, to help solve the questions related to the functionalization of prebiotic CO_2 reduction catalysts, here, we discuss the importance of N-containing molecules in CO_2 conversion, particularly focusing on the role of C–N bonding and N-assisted CO_2 reduction in the evolution of chemical reactions underlying the emergence of primitive life forms, and the implications of these processes in the design of electrocatalysts for the reduction of CO_2 to hydrocarbons.

As discussed in Sects. 3, 4 and 5, N-containing species functions as cocatalysts for electrochemical CO_2 reduction by either stabilizing the reduced CO_2 intermediates or directly activating CO_2 molecules by weakening the C=O bond via the concept of frustrated Lewis acid-base pair motif. We therefore speculate that

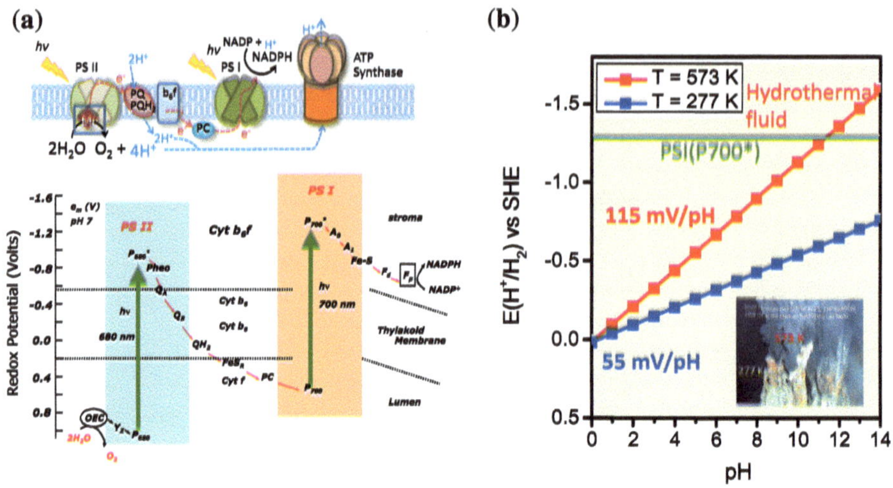

Fig. 6 **a** (*top*) The schematic image of the components of natural photosynthetic system, and (*bottom*) the energy diagram of each component; photosynthetic electron transfer chain. **b** Temperature and pH dependence of the standard redox potential of H^+/H_2 couple. Based on our results for electrochemical CO_2 reduction on violarite, a pH gradient of larger than 4 between hydrothermal fluid and ocean water is required to promote electrochemical CO_2 reduction on the surface of iron sulfide. Considering the large temperature difference between hydrothermal fluid (573 K) and ocean water (277 K), the driving force of the reaction should be high. The pH and temperature gradient between hydrothermal fluid and sea water can provide alkaline hydrothermal vent with the reduction power as high as that of photosynthetic CO_2 fixation center (PSI)

N-containing compounds in the deep-sea hydrothermal environment could aid in the conversion of oceanic CO_2 into organics or even the formation of C–N bonded compounds. From a thermodynamic perspective, the potential gradient generated from the pH, temperature, and redox disequilibrium between hydrothermal vents and acidic ocean is favorable for the reduction of nitrite/nitrate, which might exist in the oxidative ocean environment, to ammonia (E^0 = 0.897–0.875 V vs. RHE) (Fig. 6). During the reduction, in situ-formed amine-like species have the potential to coordinate with CO_2 to reduce the overpotential for CO_2 reduction and/or form C–N bonded compounds, such as urea. Of particular note is that such a C–N bond forming reaction has already been demonstrated by Shibata et al., in which urea was synthesized via simultaneous electrochemical reduction of CO_2 and nitrite/nitrate using Cu as an electrode [52, 53].

Although deep-sea electrochemical reactors are theoretically feasible to drive CO_2 reduction and C–N bond formation, determining how these inorganic minerals functioned, at least with activity comparable to their enzymatic counterparts, remains challenging. In addition, a large gap exists in our understanding of the related function of generally inactive artificial catalysts and robust bioenzymes. Russell et al. [3, 54] assumed (Fig. 7) that the mineral counterparts were the original precursors to various present-day enzymes, including [NiFe]-hydrogenases, CODH,

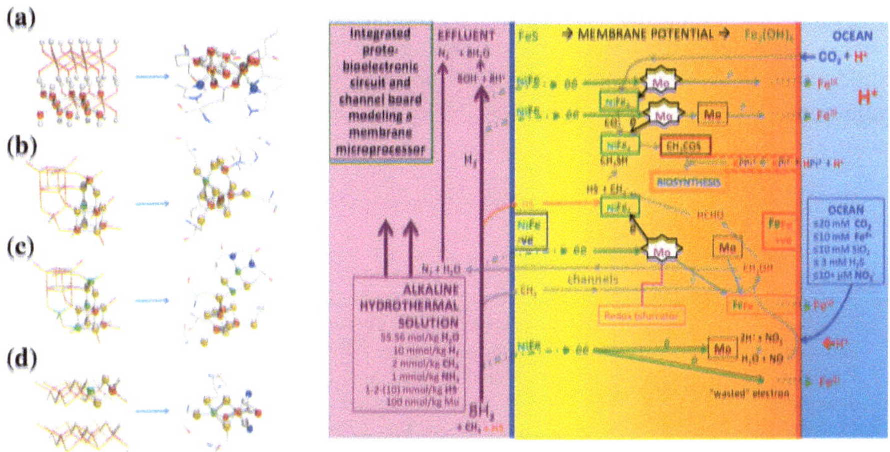

Fig. 7 *Left* Structural comparisons between transition element sulfides and oxides found in the precipitate membranes with the active sites of metalloenzymes present in the Last Universal Common Ancestor (LUCA) [39]: **a** $[Fe(OH)_2]_n$ (cf. brucite) with methane monooxygenase, **b** greigite with acetyl CoA synthase, **c** violarite with CODH, **d** nickelian mackinawite with [NiFe]-hydrogenase. *Right* The hypothetical proto-bioelectronic and proto-bioprotonic circuitry and channels in the inorganic membrane that feed the redox and proton disequilibria to the putative carbon-fixing and pyrophosphatase nano engines [54]

nitrate reductase, methane monooxygenase, methanol dehydrogenase, and pyrophosphate synthetase, acting as molecular engines. Russell et al. [3, 54] also assumed that the mineral counterparts convert redox and proton gradients to drive the endergonic reactions required for the initial stages of metabolism and ultimately life itself. This proposed theory inspired us to not only study the possible abiotic formation of C–N bonded compounds, but also to obtain insight into the design of CO_2 reduction systems from the catalytic functions of extant enzyme active centers. There should be some evolutionary reasons for the elemental strategy that nature has selected, such as energetic and kinetic priority towards some critical reactions. We speculate that C–N bonding reactions via simultaneous reduction of CO_2 and nitrite/nitrate [52, 53] may be one such critical reaction and therefore deserve detailed analysis to test the feasibility of the deep-sea emergence of primitive life theory.

The electrocatalytic formation of C–N bonds may allow for the synthesis of complex chemicals, such as urea, urethane derivatives, amides, carbamates, and isocyanate [55], by the insertion of CO_2 into other N-containing species. Such C–N bonded compounds comprise a large family of basic organic chemicals and fuels. It is expected that the coupling of different (reduced) N-compounds with CO_2-derived species may lead to the formation of various kinds of C–N bonded compounds. Although the utilization of CO_2 to synthesize such C–N bonded compounds through thermal catalysis has already been commercialized, high temperature/pressure conditions are typically required, which are energy intensive to maintain [56, 57]. We expect that better catalysts based on earth-abundant elements could be developed by understanding the bio-enzymatic functions, which could achieve the CO_2 reduction or C–N bonding with high efficiency under mild conditions.

References

1. Svetlitchnyi V, Peschel C, Acker G, Meyer O (2001) Two membrane-associated NiFeS-carbon monoxide dehydrogenases from the anaerobic carbon-monoxide-utilizing eubacterium *Carboxydothermus hydrogenoformans*. J Bacteriol 183:5134–5144
2. Parkin A, Seravalli J, Vincent KA, Ragsdale SW, Armstrong FA (2007) Rapid and efficient electrocatalytic CO_2/CO Interconversions by *Carboxydothermus hydrogenoformans* CO Dehydrogenase I on an Electrode. J Am Chem Soc 129:10328–10329
3. Nitschke W, Russell MJ (2009) Hydrothermal focusing of chemical and chemiosmotic energy, supported by delivery of catalytic Fe, Ni, Mo/W Co, S and Se, forced life to emerge. J Mol Evol 69:481–496
4. Koonin EV, Martin W (2005) On the origin of genomes and cells within inorganic compartments. Trends Genet 21:647–654
5. Huber C, Wächtershäuser G (1997) Activated acetic acid by carbon fixation on (Fe, Ni)S under primordial conditions. Science 276:245–247
6. Royer E (1870) Reduction of carbonic acid into formic acid. C R Acad Sci 70:731–735
7. Coehn A, Jahn S (1904) Über elektrolytische Reduction der Kohlensäure. Ber dt Chem Ges 37:2836–2842
8. Ehrenfeld R (1905) Zur elektrolytischen Reduction Kohlensäure. Ber dt Chem Ges 38:4138–4143
9. Fischer F, Prziza O (1914) Über die elektrolytische Reduktion von unter Druck gelöstem Kohlendioxyd und Kohlenoxyd. Ber dt Chem Ges 47:256–260
10. Rabinowitsch M, Maschowetz A (1930) Elektrochemische Gewinnung von Formiaten aus Kohlensäure. Z Elektrochem Angew 36:846–850
11. Van Rysselberghe P, Alkire GJ (1944) Polarographic reduction of Carbon Dioxide. J Am Chem Soc 66:1801
12. Chechel PS, Antropov LI (1958) Electrochemical method for production of Sodium Formate from Carbon Dioxide and Sodium Amalgam. J Appl Chem 31:1846–1850
13. Ito K, Murata T, Ikeda S (1975) Electrochemical Reduction of Carbon Dioxide to Organic Compounds. Bull. Nagoya Inst. Tech. 27:209–214
14. Hori Y, Kikuchi K, Suzuki S (1985) Production of CO and CH_4 in electrochemical reduction of CO_2 at metal electrodes in aqueous Hydrogen Carbonate Solution. Chem Lett 1695–1698
15. Hori Y, Wakabe H, Tsukamoto T, Koga O (1994) Electrocatalytic process of CO selectivity in electrochemical reduction of CO_2 at metal electrodes in aqueous media. Electrochim Acta 39:1833–1839
16. Nakato Y, Yano S, Yamaguchi T, Tsubomura H (1991) Reactions and mechanism of the electrochemical reduction of Carbon Dioxide on Alloyed Copper-Silver Electrodes. Denki Kagaku 59:491–498
17. Seshadri G, Lin C, Bocarsly AB (1994) A new homogeneous electrocatalyst for the reduction of carbon dioxide to methanol at low overpotential. J Electroanal Chem 372:145–150
18. Morris AJ, McGibbon RT, Bocarsly AB (2011) Electrocatalytic Carbon Dioxide activation: the rate-determining step of Pyridinium-Catalyzed CO_2 reduction. Chem Sus Chem 4:191–196
19. Keith JA, Carter EA (2013) Electrochemical reactivities of Pyridinium in solution: consequences for CO_2 reduction Mechanisms. Chem Sci 4:1490–1496

20. Rosen BA, Salehi-Khojin A, Thorson MR, Zhu W, Whipple DT, Kenis PJA, Masel RI (2011) Ionic liquid-mediated selective conversion of CO_2 to CO at low overpotentials. Science 334:643–644
21. Rosen BA, Zhu W, Kaul G, Salehi-Khojin A, Masel RI (2013) Water enhancement of CO_2 conversion on silver in 1-Ethyl-3-Methylimidazolium tetrafluoroborate. J Electrochem Soc 160:H138–H141
22. Watkins JD, Bocarsly AB (2014) Direct Reduction of Carbon Dioxide to formate in high-gas-capacity ionic liquids at Post-Transition-Metal Electrodes. Chem Sus Chem 7:284–290
23. Sun L, Ramesha GK, Kamat PV, Brennecke JF (2014) Switching the reaction course of electrochemical CO_2 reduction with ionic liquids. Langmuir 30:6302–6308
24. Grills DC, Matsubara Y, Kuwahara Y, Golisz SR, Kurtz DA, Mello BA (2014) Electrocatalytic CO_2 reduction with a homogeneous catalyst in ionic liquid: high catalytic activity at low over potential. J Phys Chem Lett 5:2033–2038
25. Quezada D, Honores J, García M, Armijo F, Isaacs M (2014) Electrocatalytic reduction of carbon dioxide on a cobalt tetrakis(4-aminophenyl)porphyrin modified electrode in $BMImBF_4$. New J Chem 38:3606–3612
26. Zhou F, Liu S, Yang B, Wang P, Alshammari AS, Deng Y (2014) Highly selective electrocatalytic reduction of carbon dioxide to carbon monoxide on silver electrode with aqueous ionic liquids. Electrochem Commun 46:103–106
27. Appel AM, Bercaw JE, Bocarsly AB, Dobbek H, DuBois DL, Dupuis M, Ferry JG, Fujita E, Hille R, Kenis PJA, Kerfeld CA, Morris RH, Peden CHF, Portis AR, Ragsdale SW, Rauchfuss TB, Reek JNH, Seefeldt LC, Thauer RK, Waldrop GL (2013) Frontiers, opportunities, and challenges in biochemical and chemical catalysis of CO_2 fixation. Chem Rev 113:6621–6658
28. Doukov TI, Iverson TM, Seravalli J, Ragsdale SW, Drennan CL (2002) A Ni-Fe-Cu center in a bifunctional Carbon Monoxide Dehydrogenase/Acetyl-CoA Synthase. Science 298:567–572
29. Meyer O, Gremer L, Ferner R, Ferner M, Dobbek H, Gnida M, Meyer-Klaucke W, Huber R (2000) The role of Se, Mo and Fe in the structure and function of Carbon Monoxide Dehydrogenase. Biol Chem 381:865–876
30. Ferry JG (1995) CO Dehydrogenase. Annu Rev Microbiol 49:305–333
31. Dobbek H, Gremer L, Meyer O, Huber R (1999) Crystal structure and mechanism of CO dehydrogenase, a molybdo iron-sulfur flavoprotein containing S-selanylcysteine. Proc Natl Acad Sci USA 96:8884–8889
32. Dobbek H, Svetlitchnyi V, Gremer L, Huber R, Meyer O (2001) Crystal structure of a Carbon Monoxide Dehydrogenase reveals a [Ni–4Fe–5S] cluster. Science 293:1281–1285
33. Ragsdale SW, Kumar M (1996) Nickel-Containing Carbon Monoxide Dehydrogenase/ Acetyl-CoA synthase. Chem Rev 96:2515–2539
34. Stephan DW (2008) “Frustrated Lewis pairs”: a concept for new reactivity and catalysis. Org Biomol Chem 6:1535–1539
35. Tezuka M, Yajima T, Tsuchiya A, Matsumoto Y, Uchida Y, Hidai M (1982) Electroreduction of Carbon Dioxide Catalyzed by Iron-Sulfur Clusters $[Fe_4S_4(SR)_4]^{2-}$. J Am Chem Soc 104:6834–6836
36. Komeda N, Nagao H, Matsui T, Adachi G, Tanaka K (1992) Electrochemical Carbon Dioxide Fixation to Thioesters Catalyzed by $[Mo_2Fe_6S_8(SEt)_9]^{3-}$. J Am Chem Soc 114:3625–3630
37. Tanaka K, Wakita R, Tanaka T (1989) Electrochemical Carboxylation Coupled with Nitrite Reduction Catalyzed by $[Fe_4S_4(SPh)_4]^{2-}$ and $[Mo_2Fe_6S_8(SPh)_9]^{3-}$. J Am Chem Soc 111:2428–2433
38. Yuhas BD, Prasittichai C, Hupp JT, Kanatzidis MG (2011) Enhanced electrocatalytic reduction of CO_2 with Ternary Ni–Fe_4S_4 and Co–Fe_4S_4–Based Biomimetic Chalcogels. J Am Chem Soc 133:15854–15857
39. Nitschke W, McGlynn SE, Milner-White EJ, Russell MJ (1827) On the antiquity of metalloenzymes and their substrates in bioenergetics. Biochim Biophys Acta 871–881:2013

40. Yamaguchi A, Yamamoto M, Takai K, Ishii T, Hashimoto K, Nakamura R (2014) Electrochemical CO_2 Reduction by Ni-containing Iron sulfides: how is CO_2 electrochemically reduced at bisulfide-bearing deep-sea hydrothermal precipitates? Electrochim Acta 141:311–318
41. Varley JB, Hansen HA, Ammitzbøll NL, Grabow LC, Peterson AA, Rossmeisl J, Nørskov JK (2013) Ni–Fe–S Cubanes in CO_2 reduction electrocatalysis: a DFT study. ACS Catal 3:2640–2643
42. Russell MJ (2007) The alkaline solution to the emergence of life: energy, entropy and early evolution. Acta Biotheor 55:133–179
43. Reysenbach A-L, Shock E (2002) Merging Genomes with Geochemistry in Hydrothermal Ecosystems. Science 296:1077–1082
44. Weiss RF, Lonsdale P, Lupton JE, Bainbridge AE, Craig H (1977) Hydrothermal plumes in the galapagos rift. Nature 267:600–603
45. Wächtershäuser G (1988) Pyrite formation, the first energy source for life: a hypothesis. Syst Appl Microb 10:207–210
46. Wächtershäuser G (1990) Evolution of the first metabolic cycles. Proc Natl Acad Sci USA 87:200–204
47. Heinen W, Lauwers AM (1996) Organic sulfur compounds resulting from the interaction of iron sulfide, hydrogen sulfide and carbon dioxide in an anaerobic aqueous environment. Orig Life Evol Biosph 26:131–150
48. Huber C, Wächtershäuser G (1998) Peptides by activation of Amino Acids with CO on (Ni, Fe)S surfaces: implications for the origin of life. Science 281:670–672
49. Russell MJ, Daniel RM, Hall AJ, Sherringham JA (1994) A hydrothermally precipitated catalytic iron sulphide membrane as a first step toward life. J Mol Evol 39:231–243
50. Nakamura R, Takashima T, Kato S, Takai K, Yamamoto M, Hashimoto K (2010) Electrical current generation across a black smoker chimney. Angew Chem Int Ed 49:7692–7694
51. Ang R, Khan, AU, Tsujii N, Takai K, Nakamura R, Mori T, Thermoelectricity generation and electron-magnon scattering in natural chalcopyrite mineral from a Deep-Sea Hydrothermal Vent, Angew Chem Int Edn., in press
52. Shibata M, Yoshida K, Furuya N (1995) Electrochemical synthesis of urea on reduction of carbon dioxide with nitrate and nitrite ions using Cu-loaded gas-diffusion electrode. Electroanal Chem 387:143–145
53. Shibata M, Yoshida K, Furuya N (1995) Electrochemical synthesis of urea at gas-diffusion electrodes, III. Simultaneous reduction of carbon dioxide and nitrite ions with various metal catalysts. J Electrochem Soc 145:595–600
54. Russell MJ, Barge LM, Bhartia R, Bocanegra D, Bracher PJ, Branscomb E, Kidd R, McGlynn S, Meier DH, Nitschke W, Shibuya T, Vance S, White L, Kanik I (2014) The drive to life on wet and icy worlds. Astrobiology 14:308–343
55. Tsipis CA, Karipidis PA (2005) Mechanistic Insights into the Bazarov synthesis of Urea from NH_3 and CO_2 using electronic structure calculation methods. J Phys Chem A 109:8560–8567
56. Aresta M, Dibenedetto A, Angelini A (2014) Catalysis for the valorization of exhaust carbon: from CO_2 to chemicals, materials, and fuels. Technological use of CO_2. Chem Rev 114:1709–1742
57. Sakakura T, Choi JC, Yasuda H (2007) Transformation of carbon dioxide. Chem Rev 107:2365–2387

Electrochemical Water Splitting Coupled with Solar Cells

Katsushi Fujii

Abstract The simplest method for solar to chemical energy conversion is water splitting using electrochemical cells operated by solar cells. Both two devices have been established and well-studied, however, their coupling is difficult due to the variability of the solar radiation. In this chapter, the coupling technique and efficiency improvement of the combination of the electrochemical cell and solar cell are discussed based on the basic principles of both devices.

1 Introduction

Electrochemical water splitting has a long history, starting with William Nicholson and Anthony Carlisle who in 1800 used the voltaic pile (battery) invented a few weeks before by Alessandro Volta [1]. The industrial water splitting method to hydrogen and oxygen was developed by Dmitry Lachinov in 1888 [2]. Today, electrochemical water splitting is one of the major water splitting methods alongside photoelectrochemical, photochemical, photobiological, and thermochemical water splitting.

Solar cells also have a long history. The first demonstration of photovoltaic effect was performed by Alexandre Edmond Becquerel via an electrode in a conductive solution exposed to light. In 1887, James Moser introduced the dye sensitized photoelectrochemical cell. The first solid state photovoltaic cell was built by Charles Fritts in 1883, which consisted of semiconductor selenium with an extremely thin layer of gold to form junctions [3]. The first practical photovoltaic cell with a diffused silicon p-n junction was developed by Daryl Chapin, Calvin Souther Fuller, and Gerald Pearson in 1954 at Bell Laboratories [4]. After that solar cells

K. Fujii (✉)
Global Solar Plus Initiative, The University of Tokyo, 4-6-1 Komaba, Meguro-Ku, Tokyo 153-8904, Japan
e-mail: k.fujii@rcast.u-tokyo.ac.jp

M. Sugiyama et al. (eds.), *Solar to Chemical Energy Conversion*,
Lecture Notes in Energy 32, DOI 10.1007/978-3-319-25400-5_14

based on the p-n junction concept spread widely to become a representative means of creating renewable energy.

The combined use of electrochemical water splitting and solar cells began in 1970s in order to produce hydrogen energy in a sustainable way [5]. The system is the combination of two well-known technologies, however, electrochemical cell operated by solar cells was accompanied with many technical difficulties. This chapter discusses electrochemical water splitting coupled with solar cells.

2 Electrochemical Water Splitting

Electrochemical water splitting is the method of decomposing water (H_2O) into hydrogen (H_2) and oxygen (O_2) using a current path through the water (electrolyte) [6]. The most simple electrochemical water splitting can be performed as shown in Fig. 1. In this case, the water includes some acid or salt, which is not oxidized/reduced by the water splitting (adding NaCl usually produces Cl_2 instead of O_2 from the anode.) The electrode must also not be oxidized/reduced by water splitting like carbon.

In water splitting, the chemical reactions in acidic solutions are basically;

Cathode (reduction):

$$2H^+(aq) + 2e^- \rightarrow H_2(g), \tag{1}$$

Anode (oxidation):

$$4H_2O(\ell) \rightarrow O_2(g) + 4H^+(aq) + 4e^-, \tag{2}$$

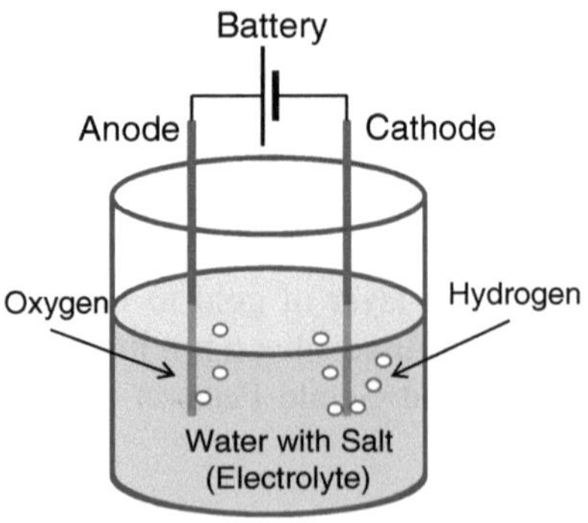

Fig. 1 Schematic electrochemical water splitting

in basic solution are;

Cathode (reduction):

$$2H_2O(\ell) + 2e^- \rightarrow H_2(g) + 2OH^-, \tag{3}$$

Anode (oxidation):

$$4OH^-(aq) \rightarrow O_2(g) + 2H_2O(\ell) + 4e^-. \tag{4}$$

The reduction occurs at the cathode surface to produce hydrogen gas with the electrons in the cathode supplied to the hydrogen ions (protons) in the electrolyte. At the anode surface, water is oxidized to produce oxygen with electrons transferred from water molecules to the anode. Thus, the total reaction is;

$$2H_2O(\ell) \rightarrow 2H_2(g) + O_2(g). \tag{5}$$

In neutral solution, both half reactions occur.

The thermodynamics of the electrochemical reactions is defined by the Gibbs energy difference (ΔG) as shown in Fig. 2. The standard formation of Gibbs energy (under 25 °C, 1 atm) changes in the Eq. (5) as follows;

$$\Delta_f G^0 = +237.13\,[\mathrm{kJ/mol}]. \tag{6}$$

This Gibbs energy change shows the form of water is the stable state, i.e., hydrogen can storage energy. The Gibbs energy change includes not only the enthalpy change (ΔH: Enthalpy changes with chemical reaction and phase change for example.) but also the entropy change ($-\Delta TS$: Entropy changes with the state of the materials using the S of the entropy); $\Delta G = \Delta H - T\Delta S$. In Eq. (5), the state of materials change (the change of the entropy) at 25 °C is;

$$\Delta S^0 = 163.34[\mathrm{J/molK}], -T\Delta S^0 = -48.70\,[\mathrm{kJ/molK}], \tag{7}$$

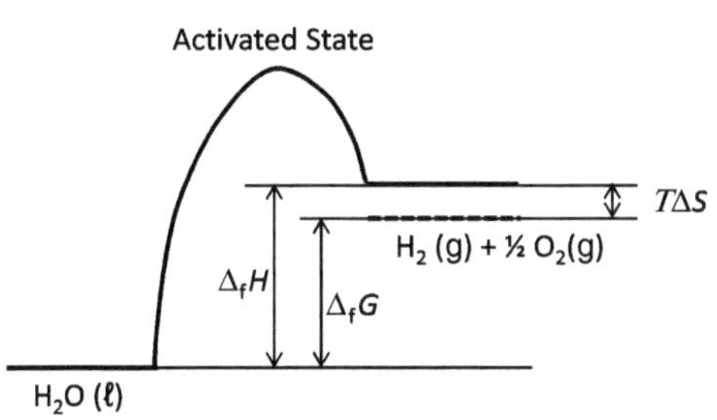

Fig. 2 Schematic thermodynamics for water splitting

and the enthalpy change is;

$$\Delta_f H^0 = +285.83[\text{kJ/mol}]. \quad (8)$$

When the closed cell is used for the electrolysis (the total volume; V and temperature; T are constant), the Helmholtz energy ($\Delta F = \Delta H - T\Delta S - p\Delta V = \Delta U - T\Delta S$) has to be used because the pressure changes. For the open cell (the pressure; P and temperature; T are constant), the Gibbs energy change is enough for the consideration because the pressure does not change.

For the electrolysis, the $\Delta_f G$ of the reaction;

$$2H^+(aq) + 2e^- = H_2(g), \quad (9)$$

is zero at pH 0 with standard condition. The required voltage for water splitting when using Faraday constant F; $\Delta G = -nF\Delta E$ (n: the number of electrons transferred, F: Faraday constant 96485 C/mol, ΔE: potential difference) from $\Delta_f G$ in Eq. (6);

$$\Delta_f E = -1.23\,[\text{V}] \quad (10)$$

The negative value of the required voltage shows that energy must be added for water splitting.

The current—voltage (I-V) characteristics are shown in Fig. 3 schematically. The turn on voltage is higher than 1.23 V even using open cell due to the existence of the activation energy for the chemical reaction of water splitting. The slope after the turn on the current shows the series resistance of the overall of the electrolysis circuit. The reductions of the overpotential and resistance improve the electrochemical performance of water splitting.

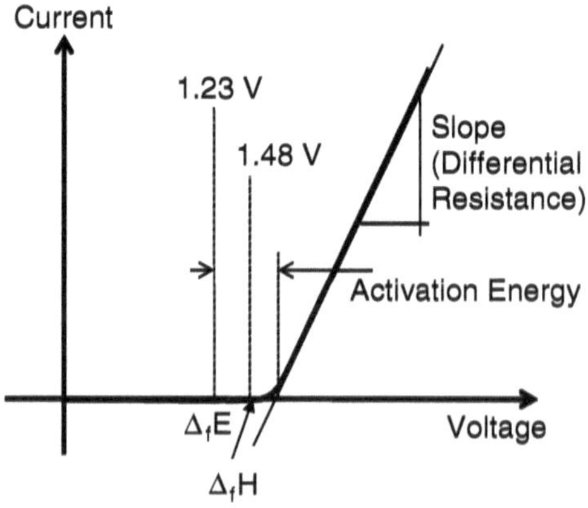

Fig. 3 Schematic current–voltage characteristics for water splitting. Δ_fE and Δ_fH show the required voltage calculated from Gibbs free energy change and enthalpy change

3 Solar Cells

In order to perform electrochemical water splitting by solar power, solar energy has to be converted to electric energy using solar cells. Solar cell is a device which converts light to electricity usually using semiconductor p-n junction [7]. When a photon hits a piece of semiconductor, one of three things happens;

1. the photon passes through the semiconductor when the photon energy is lower than that of semiconductor band gap,
2. the photon reflects the semiconductor surface,
3. the photon is absorbed by the semiconductor when the energy is higher than that of the semiconductor band gap. In this case, the electron-hole pair (and heat) generates in the semiconductor.

This generated electron-hole pair is the source of the electricity. The generated electron-hole pair usually recombines and emits light or heat at the position. For carrier separation, there are two main mechanisms;

1. drift of carriers driven by an electric field established across the device,
2. diffusion of carriers due to their random thermal motion.

In thick solar cells, there are no electric fields in the active region, thus the dominant mode of carrier separation is diffusion. In order to realize effective carrier separation, the diffusion length of minority carrier must be larger than the cell thickness. This condition is achieved only when the semiconductor is almost perfect and very few defects exist in the semiconductor. Usually, the diffusion length of minority carrier is very short due to the existence of defects and high probability of the recombination with majority carrier. Thus, the semiconductor p-n junction, which extends to the whole thickness of the cell, is used in order to separate the electron-hole pair. The schematic view of solar cell is shown in Fig. 4.

In order to understand the electronic behavior of solar cell, the electrically equivalent model is important. The equivalent circuit model of a solar cell is shown in Fig. 5. An ideal solar cell consists of a current source and a diode (due to the existence of p-n junction or Schottky junction), which is parallel to the current source. In practice, a solar cell has a shunt resistance (R_{SH}) and a series resistance (R_S) with the components of current source and diode.

Using this equivalent circuit, the output current produced by the solar cell (I) is defined as;

$$I = I_L - I_D - I_{SH}, \tag{11}$$

where, I_L is photogenerated current, I_D is diode current, and I_{SH} is shunt current. The output voltage (V) is defined using the voltage across the diode and shunt resistance (V_j) and the series resistance;

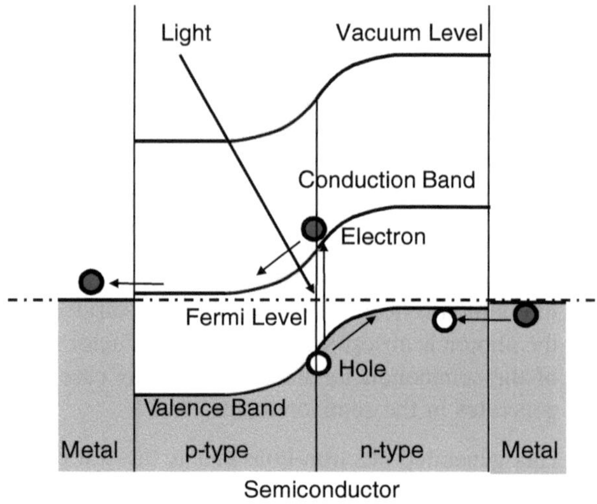

Fig. 4 Band diagram of a semiconductor solar cell under short circuit condition

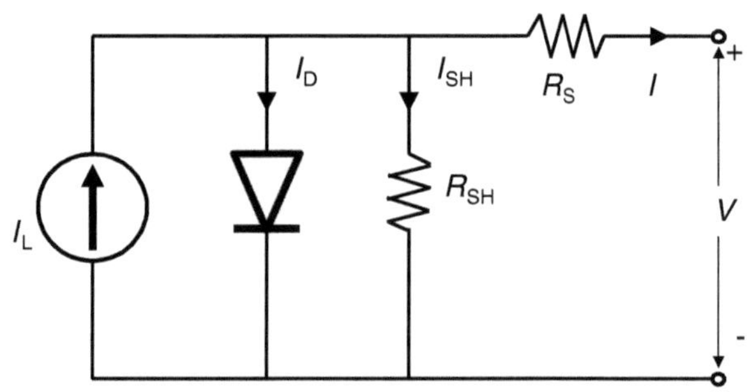

Fig. 5 Equivalent circuit of a solar cell

$$V_j = V + IR_s. \tag{12}$$

The current path through the diode is defined by the Shockley diode equation;

$$I_D = I_0\left\{\exp\left[\frac{qV_j}{nkT}\right] - 1\right\}, \tag{13}$$

where, I_0 is the reverse saturation current, n is diode ideality factor (usually n is in-between 1 to 2, and 1 for an ideal diode), q is elementary charge, k is Boltzmann's constant, and T is absolute temperature. The current passing through the shunt diode is defined by Ohm's law;

$$I_{SH} = \frac{V_j}{R_{SH}}. \tag{14}$$

Substituting Eq. (12–14) into Eq. (11), the characteristic equation of a solar cell is obtained;

$$I = I_L - I_0\left\{\exp\left[\frac{q(V+IR_s)}{nkT}\right] - 1\right\} - \frac{V+IR_s}{R_{SH}}. \tag{15}$$

From Eq. (15), open circuit voltage (V_{OC}) at the output terminals which is defined as $I = 0$ with the assumption of the shunt resistance being high enough to neglect the final term of Eq. (15);

$$V_{OC} \approx \frac{nkT}{q}\ln\left(\frac{I_L}{I_0} + 1\right). \tag{16}$$

Similarly, short circuit current (I_{SC}) at $V = 0$ for high-quality solar cell (low R_s and I_0 and high R_{SH});

$$I_{SC} \approx I_L. \tag{17}$$

Typical I–V characteristics are shown in Fig. 6. The maximum power conversion point is shown as P_{max} with the current and voltage being I_{max} and V_{max}, respectively. The fill factor (FF) is defined as;

$$FF = \frac{P_{max}}{V_{OC}I_{SC}}. \tag{18}$$

This FF is the key parameter of the performance of solar cell. Solar cell with high FF shows low R_s and high R_{SH}, that is, high-quality solar cell.

Temperature affects the solar cell characteristics Eq. (15) directly via T in the exponential term and I_0 of the reverse saturation current. The net effect is linear

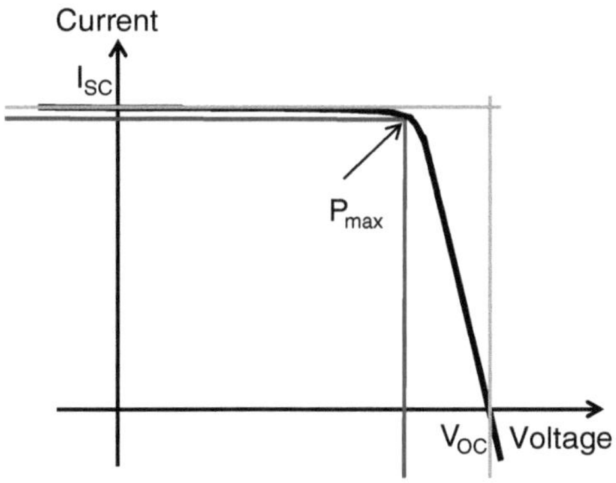

Fig. 6 Schematic current –voltage characteristics for solar cell. I_{SC}, V_{OC}, and P_{max} are short–circuit currents, open-circuit voltage, and maximum power point with current I_{max} and voltage V_{max}, respectively

reduction of V_{OC} with increasing temperature. In addition, I_{SC} increases slightly with temperature due to the bandgap shrinkage in the solar cell material induced by temperature.

4 Electrochemical Cells

Electrochemical cell is a device for the water splitting reaction. Although the wider meaning of electrochemical cell is a device which either derives electrical energy from chemical reaction or facilitates chemical reaction by electrical energy, the device discussed here is the latter definition [8]. The typical structure of the electrochemical cell for water splitting is shown in Fig. 7. It consists of an ion exchange materials in-between an anodic and a cathodic electrode. Proton exchange polymer is usually used as the ion exchange materials for water splitting. The smallest required applied voltage to operate this device for water splitting is the $\Delta_f E$ as defined in Eq. (10) due to the requirement of the chemical reaction.

The simplest equivalent circuit for an electrochemical cell is usually shown as Fig. 8a. The R_{ser}, C_p, and R_p are series resistance of the cell, capacitance and parallel resistance at the electrode-electrolyte interface, respectively. This equivalent circuit is useful for the impedance analysis at a certain voltage V, however, this model is not appropriate for the current–voltage characteristics in DC operations. The equivalent circuit for electrochemical cells in DC operations can be described as Fig. 8b when we discuss the optimal connection between solar cells and electrochemical cells. An electrochemical cell can be used as a fuel cell at its reverse bias region under the existence of hydrogen and oxygen in the electrolyte. For this regime of fuel-cell operation, the equivalent circuit is completely different from the one for electrochemical-cell operation and this is not discussed here.

Using this equivalent circuit, the current flowing the electrochemical cell (I) is defined as;

$$I = I_{Dfwd} + I_{prl}, \tag{19}$$

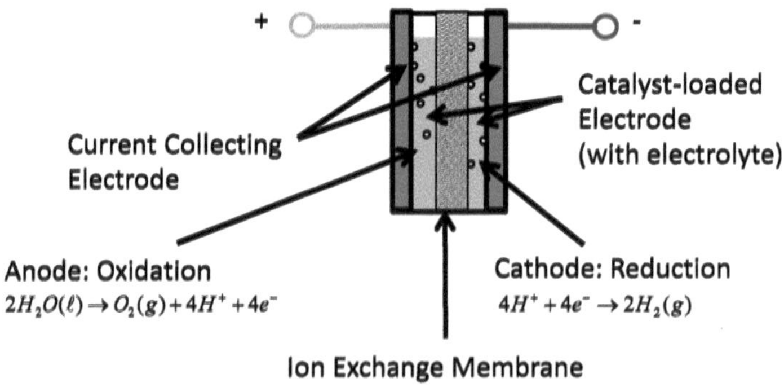

Fig. 7 Schematic electrochemical cell structure for water splitting

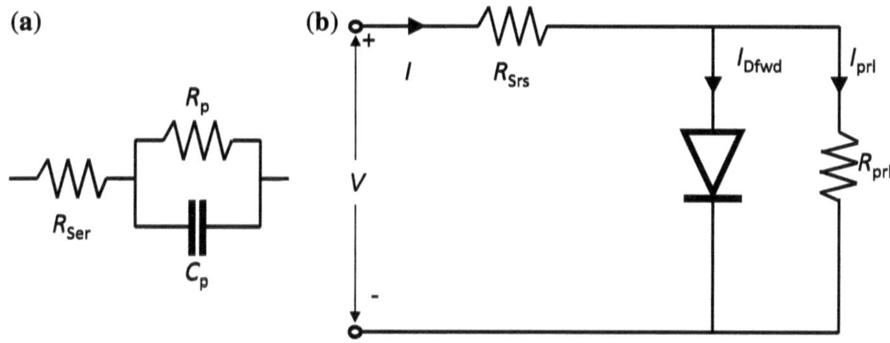

Fig. 8 Equivalent circuit for electrochemical cell **a** for impedance analysis and **b** for DC current circuit

where I_{Dfwd} is the forward diode current and I_{prl} is the parallel resistance current. The current path through the diodes are defined by the Shockley diode equation;

$$I_{Dfwd} = I_{0fwd}\left\{\exp\left[\frac{q(V - IR_{srs})}{nkT}\right] - 1\right\}, \tag{20}$$

where, I_{0fwd} is the reverse saturation current for forward diode and I_{0rev} is the reverse saturation current for the reverse diode. The current passing through the parallel diode is defined by Ohm's law;

$$I_{prl} = \frac{(V - IR_{srs})}{R_{prl}}. \tag{21}$$

Substituting Eq. (20) and Eq. (21) into Eq. (19), the characteristic equation of an electrochemical cell is obtained;

$$I = I_{0fwd}\left\{\exp\left[\frac{q(V - IR_{srs})}{nkT}\right] - 1\right\} + \frac{V - IR_{srs}}{R_{prl}}. \tag{22}$$

5 Electrochemical Cell Operated by Solar Cells

In order to convert solar energy to chemical energy like hydrogen generated from water splitting, electrochemical cells have to be operated by solar cells. In order to maximize the energy conversion efficiency in both solar cells and electrochemical cells, it is mandatory to employ an appropriate operation point (voltage and current) for both devices. When a solar cell and an electrochemical cell are connected directly in a series, the electric circuit requires the current and voltage under operation to be the same for the two devices. Therefore, the crossing point between

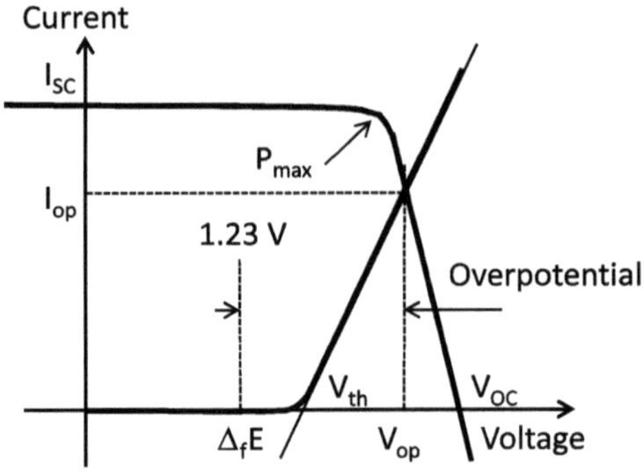

Fig. 9 Schematic current–voltage characteristics of the electrochemical cell operated by solar cell. Iopand Vopindicatethe operating current and voltage, respectively. Overvoltage at the operating current is also shown in the graph

the current – voltage curves for the two devices determines the current (I_{op}) and the voltage (V_{op}) under operation as shown in Fig. 9.

The energy transfer efficiency from sunlight to the chemical free energy of hydrogen is usually defined as solar to hydrogen (*STH*) efficiency, which is the ratio of the free energy of the produced hydrogen to the input energy of the sun during a unit time;

$$STH = \frac{n_r m F \Delta_f E}{P_{sun}}, \tag{23}$$

where, n_r is the number of electrons transferred for the chemical reaction (2 for hydrogen generation as shown in Eq. (1) or (3)), m is the production rate of the hydrogen (mol/s), F is the Faraday constant (C/mol), $\Delta_f E$ is the minimum required voltage to produce hydrogen from water calculated from standard formation Gibbs energy (1.23 V for the standard condition), and P_{sun} is the energy of the sun light. *STH* can be decomposed into several elements to analyze the origin of the loss.

1. Conversion efficiency of solar cell (η_{sc}) at the power maximum point

$$\eta_{sc} = \frac{P_{max}}{P_{sun}} = \frac{I_{max} V_{max}}{P_{sun}}. \tag{24}$$

2. Matching efficiency (η_{op}) between the operation point and the power maximum point of solar cell

$$\eta_{op} = \frac{I_{op} V_{op}}{I_{max} V_{max}}. \tag{25}$$

3. The efficiency due to overpotential (η_{ov}), which is the ratio between the operation point voltage and the free energy stored as hydrogen($\Delta_f E$)

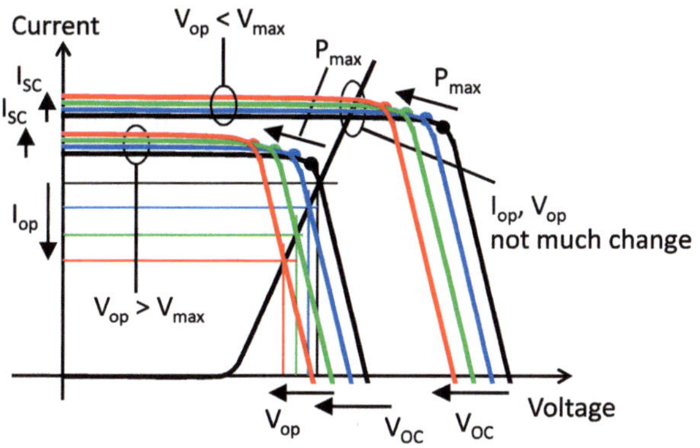

Fig. 10 Schematic current–voltage characteristics of the electrochemical cell operated by solar cell. I_{op} and V_{op} decrease with increasing solar cell operating temperature when $V_{max} < V_{op} < V_{oc}$

$$\eta_{ov} = \frac{\Delta_f E}{V_{op}} = \frac{\Delta_f E}{\Delta_f E + V_{ov}}, \tag{26}$$

where, V_{ov} is the overpotential. In order to improve η_{ov}, overvoltage must be reduced.

4. Faradic efficiency (η_{Fa})

$$\eta_{Fa} = \frac{n_r m F}{I_{op}}. \tag{27}$$

This Faradic efficiency describes the efficiency with the charge transferred into a system where an electrochemical reaction is facilitated. That is the ratio of the produced hydrogen and transferred electron. For the *STH* of Eq. (23) is described with using Eqs. (24)–(27);

$$STH = \eta_{sc}\eta_{op}\eta_{ov}\eta_{Fa}. \tag{28}$$

Each efficiency has to be improved in order to obtain the high *STH* from Eq. (28).

The temperature dependence of the solar cell affects the operation for the direct connection of the solar cell and electrochemical cell. Since the V_{oc} reduces linearly with increasing operating temperature, the operation efficiency is drastically influenced by the operating temperature of solar cell when the operating voltage V_{op} between V_{max} and V_{oc} is as shown in Fig. 10. The operation is relatively stable when the operating voltage V_{op} is smaller than V_{max}. This indicates that the operating voltage V_{op} should be smaller than V_{max} when the electrochemical cell is operated by solar cells with direct connection.

6 Water Electrolysis by Conventional Solar Cells

The practical way of water electrolysis by solar cell is the connection of the planer-type conventional solar cell and polymer-electrolyte electrochemical cell (PEEC). This connection was first reported in 1977 [5], and the usage of the concentrated photo voltaic cell (CPV) to improve the performance was proposed in 1982 [9]. The idea at the beginning was to directly couple the solar cell and electrochemical cell. Voltage and current matching was considered by controlling the number of solar cells and electrochemical cells [5, 9, 10]. The next proposal was the usage of maximum power point tracking (MPPT) to operate solar cells in order to increase hydrogen production [11–13]. The solar cell output voltage was adjusted to the electrochemical cell using a DC/DC converter after that [14]. This connection of the solar cell and electrochemical cell was applied to the polymer electrolyte electrochemical cell and fuel cell system to store and use the renewable energy as chemical energy [15].

Due to the variability of solar radiation, efficient coupling is not easy. The solar cell output voltage adjustment by DC/DC converter with a specific algorithm was proposed in 2008 [16]. The efficiency of electricity to hydrogen with DC/DC converter was greater than 90 % for most voltages [17]. The exception was 87 % when solar cell input power was low because the converter operated near the open-circuit voltage. When a DC/DC converter is used, the efficiency of the converter is also an important parameter for enhancing the total efficiency of the system. Thus, direct coupling of solar cell and electrochemical cell has been also studied [18]. The relative sizing between the solar cell and electrochemical cell was proposed based on the simple modeling of both current–voltage curves [19]. The estimated overall annual energy loss due to direct coupling compared with that using MPPT was estimated to be only a few % [20]. Direct coupling by controlling the number of solar cells and electrochemical cells was also demonstrated [21].

Problems faced in the improvement of the connection of the planer-type solar cell and electrochemical cell are system integration and control. Although the optimized system of the coupling increased the hydrogen generation efficiency, the solar to hydrogen energy conversion efficiency (*STH*) was reported to be 12.4 % [22], which is close to the efficiency of solar cells.

7 Water Electrolysys by Concentrated Solar Cells

In order to improve solar to hydrogen energy conversion efficiency using water splitting, the efficiency of solar cells is the largest bottleneck. The approach of using highly efficient solar cells was gradually applied to the concentrator-array design of solar cell in 1982 [9]. The efficiency of the concentrated-array solar cell made by Si was 11 %, thus, the overall solar to hydrogen conversion efficiency (*STH*) was reported to be 7 %.

The first approach from photoelectrochemical water splitting was to be used for multiple-junction type photoelectrochemical electrodes [23]. The electrode was similar to the multiple-junction solar cell and had the structure of p-$In_{0.5}Ga_{0.5}P$/n-GaAs/p-GaAs connecting to a metal electrode. The light to hydrogen conversion efficiency under tungsten-halogen white light with 3.0 mol/L H_2SO_4 aqueous solution was reported to be 12.4 %. A photohydrolysis system was proposed in 2000 to be an $Al_xGa_{1-x}As$/Si double-junction solar cell directly attached to Pt black and RuO_2 electrodes dipped into 1.0 mo/L $HClO_4$ aqueous solution [24]. The energy conversion efficiency from sunlight to hydrogen was 18.3 % under AM0 (130 mW/cm^2).

These two ideas of concentrated solar cell (CPV) and multiple-junction solar cell were combined and used for water splitting as reported in 2007 [25]. The hydrogen generation was performed by a $Ga_xIn_{1-x}P/Ga_yIn_{1-y}As$ double-junction solar cell system and a polymer electrolyte electrochemical cell (PEEC) using Pt-cathode and Ir-anode under 500-times solar light concentration. Unique characteristics include direct attachment of the electrochemical cell under the double junction solar cell and the usage of PEEC. The electrolyte for PEEC is a proton exchanging polymer membrane, thus the system does not require conductive aqueous electrolytes and pure water is used as the source of hydrogen. The energy conversion efficiency from real sunlight to hydrogen was 15.1 % from the recalculation used with the hydrogen Gibbs energy of 1.23 V, which is commonly reported value. The efficiency was reported to be 18.1 % with hydrogen Helmholtz energy of 1.48 V. Since the electrochemical water splitting structure was located just underneath of the tandem type solar cell, the device was specially designed for solar to hydrogen energy conversion.

To establish a "stand-alone total natural energy" system for not only energy generation but also energy usage, a high-efficiency and simple device is suitable for converting sunlight into storage energy. In addition, the separation of electricity and hydrogen generation is indispensable because the energy storage device must be used in conjunction with other energy sources like water and/or wind power generation. Therefore, a much simpler water-splitting hydrogen generation system, combining a PEEC and a concentrated photovoltaic cell (CPV) [26] was proposed. Since the PEEC and CPV can be set separately, the system provides much more flexibility from the design point of view. The energy conversion efficiency under 10-times solar simulator light to hydrogen was 15.3 % with the operating voltage matching EC and CPV.

The typical systems of highly efficient solar cells and electrochemical cell for water splitting are shown in Fig. 11 schematically. The structures, especially those for the solar cells, are complicated. Even though, it is clear that applying high efficiency solar cell to water splitting hydrogen generation is effective for improving conversion efficiency compared to the combination of the planer solar cell and electrochemical cell.

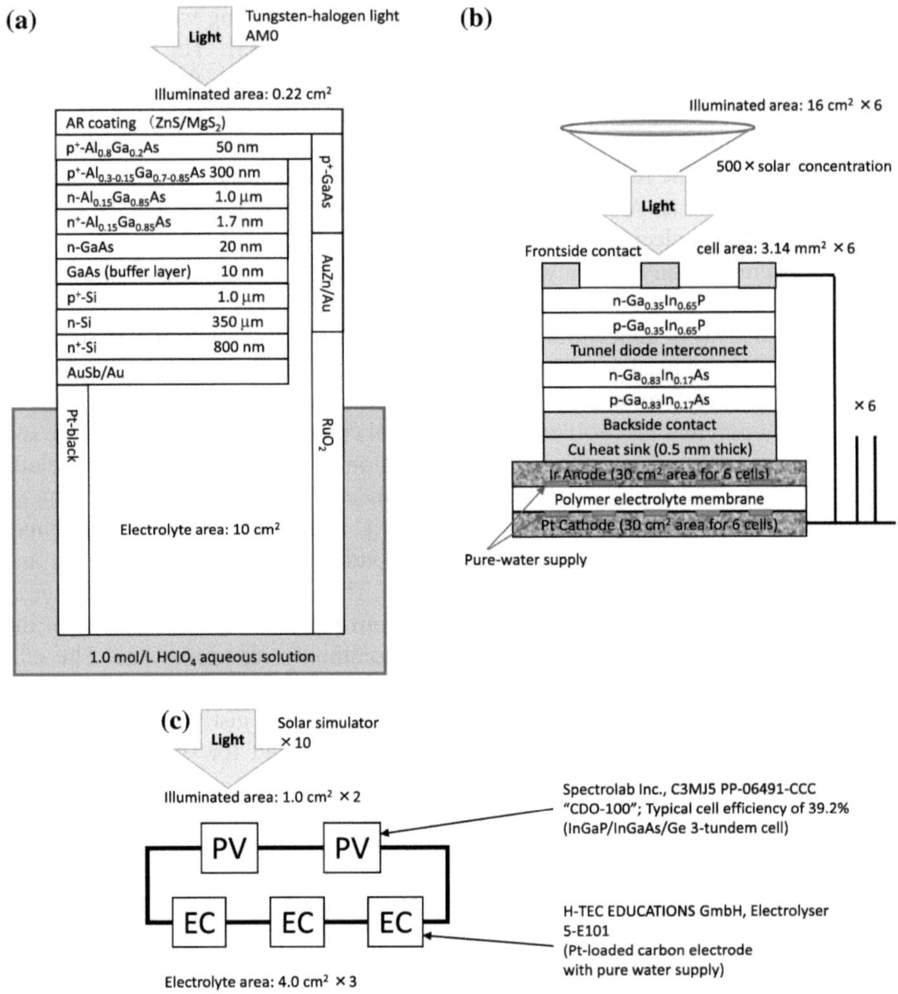

Fig. 11 Schematic diagrams for water electrolysis by concentrated solar cells. **a** Proposed by Licht et al. [24]. **b** Proposed by Peharz et al. [25], and **c** Proposed by Fujii et al. [26]

8 Future Improvements

Keypoints in the improvement of solar to hydrogen conversion efficiency (*STH*) include conversion efficiency improvement of solar cell, voltage and current of electrochemical cell matching the maximum power point of solar cell, electrochemical cell overpotential reduction, and Faradic efficiency improvement. They are discussed with Eq. (28).

The improvement of the conversion efficiency of solar cell affects the increase of *STH* as discussed in the previous section. Another important aspect for the design of the solar cell is the open circuit voltage or the voltage for the maximum power point. It is clear from the discussion of current matching of Eq. (25), that the operating point for an electrochemical cell should match the maximum power point of a solar cell to reduce loss. This means that the design of the solar cell, when it is coupled with a water electrolyzer, must take into account not only the efficiency of the solar cell itself at its power maximum point but also how the voltage at the power maximum match the operation voltage of an electrolyzer.

For the purpose of water splitting, it is therefore necessary to employ a solar cell whose maximum power point voltage locates around 1.7 V, which is a typical operating voltage of an electrochemical cell at a typical current density such as 0.1 to 1.0 A/cm . An alternative way is to connect plural solar cells and electrochemical cells in series so that the sum of the operation voltages is controlled to well match the solar cell and the electrochemical cells.

The reduction of the overvoltage of electrochemical cell is also important. The hydrogen evolution overvoltage is not so high (<0.05 V for 1 mA/cm^2) when appropriate water-reduced electrochemical catalyst is used, however, the oxygen evolution overvoltage is still high (about 0.15 V for 1 mA/cm^2) even when a water-oxidized electrochemical catalyst is used [27]. This shows that the development of oxygen evolution electrochemical catalyst is important to improve electrochemical cell performance.

Faradic efficiency was reported to be affected by the current density of the electrochemical cell [26], that is, there exists an optimized current density. This can be explained by the fact that parasitic reaction becomes dominant at low current densities, and the product blocks the reaction at high current density. The keypoint for the improvement is to find the factor for increasing the optimized current density.

9 Conclusion

Electrochemical water splitting by solar cells is the most practical solar to chemical energy conversion technique. Both solar cells and electrochemical cells used for generating hydrogen with splitting water have a long history and are well-studied. However, the optimized coupling of the two devices is not easy and is still under discussion. Conversion efficiency is defined by the conversion efficiency of the solar cell, voltage and current matching of solar cell and electrochemical cell, overpotential of the electrochemical cell, and Faradic efficiency. The usage of high efficiency solar cell like concentrated photo voltaic cell improves solar to hydrogen conversion efficiency. Since both solar cells and electrochemical cells are not suitable for solar to chemical energy conversion, there are many problems even at the device level that have to be resolved.

References

1. Trasatti S (1999) 1799–1999: Alessandro Volta's 'Electric Pile' Two hundred years, but it doesn't seem like it. J Electroanal Chem 460:1–4; de Levie R (1999) The electrolysis of water. J Electroanal Chem 476:92–93. (Erratum: Trasatti S (2000) Erratum to "Water electrolysis: who first?", 481:112)
2. http://en.wikipedia.org/wiki/Dmitry_Lachinov, http://en.wikipedia.org/wiki/Electrolysis_of_water. (2013.11.18)
3. Fritts CE (1883) On a new form of selenium photocell. Am J Sci **26:**465, http://en.wikipedia.org/wiki/Timeline_of_solar_cells (2013.11.18)
4. Chapin DM, Fuller CS, Pearson GL (1954) A new silicon pn junction photocell for converting solar radiation into electrical power. J Appl Phys 25:676–677
5. Costogue EN, Yasui RK (1977) Performance data for a terrestrial solar photovoltaic/water electrolysis experiment. Sol Energy 19:205–210
6. Reference for water electrolysis. For example, Grimes CA, Varghese OK, Ranjan S (2008) Light, water, hydrogen, the solar generation of hydrogen by water photoelectrolysis. Springer, ISBN: 978-0-387-6828-9, pp 1–111, Chap 1–2
7. See solar cell device principles. For example, Haeberlin H (2012) Photovoltaics system design and practice, English Version. Wiley, ISBN: 978-1-119-99285-1, pp 79–125, Chap 3
8. Reference for electrochemical cell. For example, Bird AJ, Faulkner LR (2001) Electrochemical methods, fundamentals and applications, 2nd edn. Wiley, ISBN: 0-471-04372-9, pp 1–43, Chap 1
9. Esteve D, Ganibal C, Steinmetz D, Vialaron A (1982) Performance of a photovoltaic electrolysis system. Int J Hydrogen Energy 7:711–716
10. Fischer M (1986) Review of hydrogen production with photovoltaic electrolysis systems. Int J Hydrogen Energy 11:495–501
11. Siegel A, Schott T (1988) Optimization of photovoltaic hydrogen production. Int J Hydrogen Energy 13:659–675
12. Garcia-Conde AG, Rosa F (1993) Solar hydrogen production: a Spanish experience. Int J Hydrogen Energy 18:995–1000
13. Arriaga LG, Martinez W, Cano U, Blud H (2007) Direct coupling of a solar-hydrogen system in Mexico 32:2247–2252
14. Solmecke H, Just O, Hackstein D (2000) Comparison of solar hydrogen storage systems with and without power electric DC-DC-converters. Renewable Energy 19:333–338
15. Shapiro D, Duffy J, Kimble M, Pien M (2005) Solar-powered regenerative PEM electrolyzer/fuel cell system. Sol Energy 79:544–550
16. Garcia-Valverde R, Miguel C, Martinez-Bejar R, Urbina A (2008) Optimized photovoltaic generator-water electrolyzer coupling through a controlled DC-DC converter. Int J Hydrogen Energy 33:5352–5362
17. Garrigos A, Blanes JM, Carrasco JA, Lizan JL, Beneito R, Molina JA (2010) 5 kW DC/DC converter for hydrogen generation from photovoltaic sources. Int J Hydrogen Energy 35:6123–6130
18. Clarke RE, Giddey S, Ciacchi FT, Badwal SPS, Paul B, Andrews J (2009) Direct coupling of an electrolyser to a solar PV system for generating hydrogen. Int J Hydrogen Energy 34:2531–2542
19. Garcia-Valverde R, Espinosa N, Urbina A (2011) Optimized method for photovoltaic-water electrolyzer direct coupling. Int J Hydrogen Energy 36:10574–10586
20. Paul B, Andrews J (2008) Optimal coupling of PV arrays to PEM electrolysers in solar-hydrogen systems for remote area power supply. Int J Hydrogen Energy 33:490–498
21. Maeda T, Ito H, Hasegawa Y, Zhou Z, Ishida M (2012) Study on control method of the stand-alone direct-coupling photovoltaic—water electrolyzer. Int J Hydrogen Energy 37:4819–4828

22. Gibson TL, Kelly NA (2010) Predicting efficiency of solar powered hydrogen generation using photovoltaic-electrolysis devices. Int J Hydrogen Energy 35:900–911
23. Khaselev O, Turner JA (1998) Production via water splitting a monolithic photovoltaic-photoelectrochemical device for hydrogen. Science 280:425–427
24. Licht S, Wang B, Mukerji S, Soga T, Umeno M, Tributsch H (2000) Efficient solar water splitting, exemplified by RuO2-catalyzed AlGaAs/Si photoelectrolysis. J Phys Chem B 104:8920–8924
25. Peharz G, Dimroth F, Wittstadt U (2007) Solar hydrogen production by water splitting with a conversion efficiency of 18 %. Int J Hydrogen Energy 32:3248–3252
26. Fujii K, Nakamura S, Sugiyama M, Watanabe K, Bagheri B, Nakano Y (2013) Characteristics of hydrogen generation from water splitting by polymer electrolyte electrochemical cell directly connected with concentrated photovoltaic cell. J. Hydrogen Energy 38:14424–14432
27. Cook TR, Dogutan DK, Reece SY, Surendranath Y, Teets TS, Nocera DG (2010) Solar energy supply and storage for the legacy and nonlegacy worlds. Chem Rev 110:6474–6502

Part VI
Chemical, Electrochemical and Photoelectrochemical Approach for Energy Conversion: Approach Using Photoelectrochemical Reactions

Photoelectrochemical Approach for Water Splitting

Joel W. Ager

Abstract The splitting of liquid water into hydrogen and water via photoelectrochemical (PEC) approaches is described. If sunlight is used as the illumination source, the overall process provides a means to convert solar power into chemical energy. PEC water splitting is the direct coupling of the following processes: (1) absorption of solar light in a material and the creation of electrons and holes, (2) transport of electrons and holes to the absorber/water interface, and (3) evolution of hydrogen from the electrons and oxygen from the holes, often with the assistance of catalysts. The distinctions between this process and the related approach of coupling photovoltaic (PV) elements to hydrogen evolution (HER) and oxygen evolution (OER) catalysts will be discussed. The history of research on PEC water splitting dating back to its discovery in the early 1970s is summarized. The basic design principles of PEC water splitting device, with an emphasis on the type and number of PV absorbing elements are discussed and state of the art demonstrations are summarized.

1 Introduction

A practical method to use sunlight to generate chemical fuels would be a carbon-neutral energy source which could dramatically change the landscape of global energy generation and storage [1–3]. Here, approaches which use sunlight to split water into hydrogen and oxygen will be discussed. The hydrogen proceeded by such a process would represent a sustainably-generated, storable chemical fuel [4], and could be used for powering transportation [5–7] and, potentially, for grid scale energy storage [8].

J.W. Ager (✉)
Joint Center for Artificial Photosynthesis and Materials Sciences Division, Lawrence Berkeley National Laboratory and Materials Science and Engineering, University of California, Berkeley, CA, USA
e-mail: JWAger@lbl.gov

M. Sugiyama et al. (eds.), *Solar to Chemical Energy Conversion*,
Lecture Notes in Energy 32, DOI 10.1007/978-3-319-25400-5_15

The electrochemical water splitting half reactions are as follows [9, 10]:

$$2H^{+} + 2e^{-} \rightarrow H_2 \quad (1)$$

$$H_2O \rightarrow 2e^{-} + 2H^{+} + 1/2\,O_2 \quad (2)$$

or

$$H_2O + 2h^{+} \rightarrow 2H^{+} + 1/2\,O_2, \quad (2a)$$

where reaction (1) is the hydrogen evolution reaction (HER) and reactions (2) and (2a) are equivalent ways of depicting the oxygen evolution reaction (OER). The free energy change for water splitting to hydrogen (and oxygen) by reactions (1) and (2) under standard-state conditions is $\Delta G° = +237$ kJ/mol of H_2 or $\Delta E° = -1.23$ V. At pH0 the redox potentials for reactions (1) and (2) are +0 and +1.23 V versus NHE (normal hydrogen electrode), respectively. The redox potentials of (1) and (2) shift at the Nerstian rate of 0.059 V/pH unit; it has thus become typical in the PEC water splitting field to report potentials versus RHE (reversible hydrogen electrode) to take this effect into account.

The force required to drive reactions (1) and (2) in the fuel-generating direction can be provided by a simple DC electrical bias, which optionally can be provided by a photovoltaic cell. When driven by solar light, these types of water splitting approaches are often termed "PV+electrolyzer". Here, we focus on photoelectrochemical (PEC) water splitting, which is the direct coupling of the following processes: (1) absorption of solar light in a material and the creation of electrons and holes, (2) transport of electrons and holes to the absorber/water interface, and (3) evolution of hydrogen from the electrons and oxygen from the holes, often with the assistance of catalysts.

We follow the taxonomy proposed by Nielander et al. [11] in delineating the types of PEC approaches and devices discussed in this book chapter. PV-coupled systems in which the photovoltage is independent of the solid/electrolyte interface are termed *PV-biased electrosynthetic* devices and are homologous to the "PV +electrolyzer" terminology used above. The key distinguishing feature of a PEC device is that the charge separation required for the voltage generation is provided by the nature of the solid-electrolyte interface. We further distinguish *photocatalytic* approaches, in which reactions (1) and (2) on the surface of a material, usually nanostrucured, which is both light absorbing and catalytic. These approaches have been reviewed extensively [12–17].

PEC water splitting cells are often evaluated by measuring the photocurrent density. In this case, the solar to hydrogen (STH) conversion efficiency, η, for spontaneous solar water splitting at standard temperature and pressure of H_2 and O_2 is given by [18, 19]:

$$\eta = \frac{(1.23\text{V})\left(J_{\text{op}}\right)}{P_{in}} \quad (3)$$

where 1.23 V represents the free energy of the products, J_{op} is the operational photocurrent density, or the rate of hydrogen production converted to a current density, and P_{in} is the incident irradiance (100 mW cm^{-2} at 1 sun, AM 1.5G). This method assumes 100 % conversion of the photocurrent to the desired hydrogen product; measurement of the products to verify the expected 2:1 H_2:O_2 ratio is clearly more direct.

One set of obviously crucial components of a PEC water splitting system are the materials that capture and convert sunlight. The process imposes unique demands

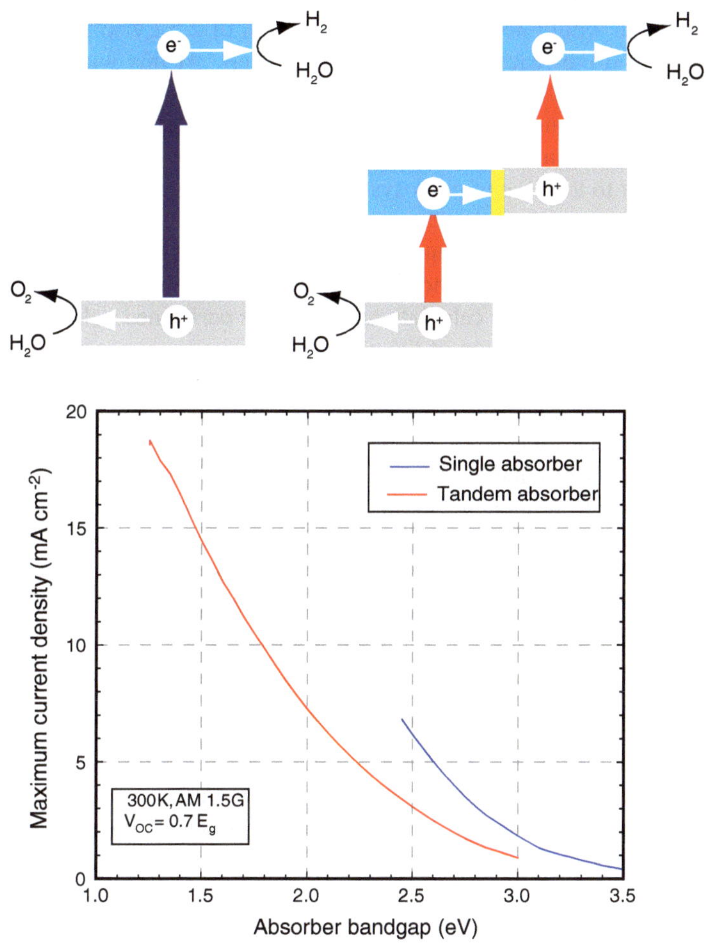

Fig. 1 Schematic (*top*) and maximum current densities (*below*) of 1 photon and 2 photon (*tandem*) methods for photoelectrochemical fuel production. The maximum current density as a function of bandgap was calculated assuming a minimum voltage of 1.7 V and a V_{OC} of 70 % of the bandgap; for the tandem cell, two cells with equal bandgaps were assumed (the maximum current is half that of a single cell at the same gap). Even higher currents are possible if the two absorbers have unequal gaps in a spectrally splitting approach

on the light absorber(s), relative to photovoltaics. If it is to proceed spontaneously, without added bias, either a single material [12, 20, 21], or two or more absorbers arranged in a voltage-additive tandem cell format connected electrically in series and current-matched spectrally [22–24], must at minimum provide the thermodynamically required voltage of 1.23 V to split water, as depicted schematically in Fig. 1. In practice, some overpotential will always be present, with a minimal value likely involving operation at a total required voltage of 1.4–1.5 V. Because this operating voltage should be at the maximum power point, the open-circuit voltage developed by the system therefore needs to be 1.7–1.8 V. Some insight into the number absorbers required can be gained by the simple analysis shown in Fig. 1. A single material would need a band gap in excess of 2.5 eV to provide sufficient photovoltage to drive reactions (1) and (2) by itself. At this wide band gap, the amount of solar light is limited and the maximum photocurrent is less than 5 mA cm^{-2}, corresponding to an STH efficiency limit of 6 %. Clearly, tandem absorber systems, even if equal band gaps and a side by side configuration are assumed (similar to the case of photosystems I and II in nature [25, 26]), are capable of much larger current densities.

It is interesting to consider what the maximum achievable STH conversion is, by analogy to the well-known "Schockley-Queisser limit" in power generating PV cells [27]. A number of such calculations have been performed, taking into account key differences in the operation of the two types of energy conversion devices. For example, in contrast to solar cells, where the operating current can be rapidly adjusted through use of power electronics, spontaneous water splitting devices operate at a fixed current based on the reaction thermodynamics, kinetic overpotential requirements of catalysts, series resistance losses, and current matching for all series-connected components. While the assumptions regarding catalyst overpotentials and device architectures vary, the consensus of these studies is that a STH conversion efficiency of >25 % is possible with a two coupled absorbers in an integrated systems in which the catalyst and absorber areas are equivalent [28–32].

2 Absorber Electronic Structure

For reaction (1) to be energetically favored at the surface of light absorber, the conduction band of the material has to be higher in energy than the H^+/H_2 redox potential. Similarly, for reaction (2) to be favored, the valence band must lie lower than the O_2/H_2O potential (n.b. this is not a requirement for PV-biased electrosynthetic systems). Figure 2 depicts the band alignments for a number of elemental, III–V, II–V, and oxide semiconductors. Similar analyses have been provided by Grätzel [33], Bak et al. [12], and Chen and Wang [34].

If a single absorbing material is used for PEC water splitting, its conduction and valence bands must "straddle" the potentials for the fuel producing redox reactions, with its conduction band edge lying higher than the HER redox potential and its

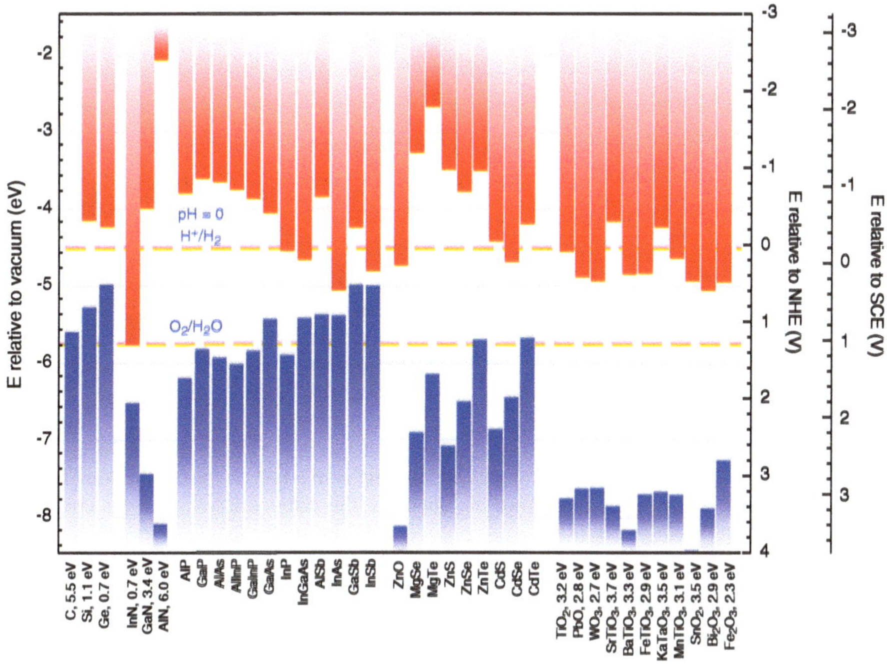

Fig. 2 Valence band (*blue*) and conduction band (*red*) positions of a elemental, III–V, II–V, and oxide semiconductors compared to the water splitting redox potentials at pH 0. Band positions with respect to vacuum are obtained from Walukiewicz [35] and Van de Walle and Neugebauer [36] Oxide data is adapted from Bak et al. [12]

valence band edge lying lower than the OER redox potential. Examination of Fig. 2 reveals that some wide band gap III–V semiconductors (e.g. GaN) and a number of oxides (e.g. $SrTiO_3$, $KaTaO_3$) meet this requirement. However, the band gaps of these materials, and others that meet this band alignment criterion, are >3 eV, limiting the STH efficiency which can be obtained with this approach (Fig. 1).

A general schematic of a tandem PEC system is shown in Fig. 3. In this device, a p-type semiconductor is used as a photocathode for HER and an n-type semiconductor is used for OER. The two light absorbers are electrically connected in series for voltage additivity. The charge separation and transport is created by a combination of the depletion layers formed by the solid-electrolyte interface at the photocathode and photoanode and by the charge selective nature of the interface, which can be enhanced by HER and OER catalysts, respectively. The choice of possible materials in this type of configuration is greatly expanded. For example, many III–V semiconductors have conduction band edges higher than the HER redox potential and most oxides have a suitably situated valence band edge for OER.

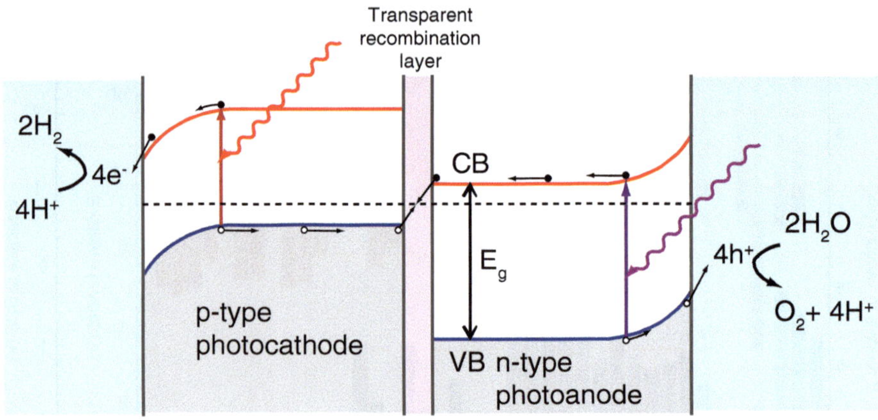

Fig. 3 Schematic of an idealized tandem-junction photoanode and photocathode device during steady-state operation. The process of solar water splitting is overlaid on the equilibrium diagram. Proton conduction in the electrolyte from the anode to the cathode is required for continuous operation. (Adapted from Nozik [22])

3 Historical Perspective and Current State of the Art

In the 1972 discovery of PEC water splitting, Fujishima and Honda employed a single absorber approach with a TiO_2 serving as a photoanode performed the OER and an electrically connected Pt cathode performing the HER [20]. Soon after, a number of oxides including $SrTiO_3$ were also found to be active for water splitting [21], and a large scale demonstration of sustained H_2 generation by sunlight was performed using TiO_2 photoanodes [37]. However, the wide band gaps of these materials, which necessitate UV illumination, have limited the STH efficiency which can be achieve this type of one-absorber approach.

Soon after the Fujishima and Honda discovery, Nozik [22] developed an energy analysis that, by analogy to tandem-junction solar cells, showed that a tandem-junction architecture consisting of a p-type photocathode and an n-type photoanode could achieve a higher STH efficiency than a single photoelectrode.. An idealized version of this type of approach is shown in Fig. 3. Initial experimental verification of this concept was achieved with a number of multiple-electrode tandem architectures including p-GaP/n-TiO_2, p-CdTe/n-TiO_2, p-CdTe/n-$SrTiO_3$, and p-GaP/n-$SrTiO_3$ [38, 39]. However, the stabilities of the active components, especially the photoanode, emerged as a critical issue [40]. In fact, many of the materials shown in Fig. 2 including all of the III–V and II–VI semiconductors are thermodynamically unstable against corrosion under illuminated OER conditions in water [34, 41]. This greatly reduces the number of materials which can be used in the type of design shown in Fig. 2.

In the late 1980s through the early 2000s groups explored triple-junction (3 J) amorphous silicon (*a*-Si) cells to drive solar water splitting [42–46]. Because the open-circuit photovoltage of 3-junction (3 J) *a*-Si solar cells under solar cell conditions can be over 2 V, they possess adequate overpotential to drive solar water splitting. These types of demonstrations, as well as more recent examples which use earth-abundant catalysts [47], employ a PV-biased electrosynthetic geometry and do not have a solid-liquid junction at either the cathode or anode. The highest STH efficiencies obtained with this approach are close to 8 % [45, 46].

Beginning in the mid-1980s, development of high-efficiency 1 and 2 J solar cells led to a rapid rise in reported efficiencies for solar water splitting. These PEC cells utilized multiple-electrode architectures consisting of InP and GaAs [23], and stacked 2 J architectures of *a*-Si [48, 49]. In the late 1990s and early 2000s, high-efficiency approaches based on III–V and Si/III–V 2 J architectures were developed [24, 50]. Reported STH efficiencies for these PV-biased electrosynthetic systems are in the 12–18 % range.

Since 2010, there have been significant efforts to replace noble-metal electrocatalysts with those made from less expensive elements, and to use metal-oxide light absorbers that may be more stable. Such efforts resulted in renewed interest in using PEC water splitting approaches which use at least one charge separating solid-liquid junction (SLJ), usually at the photoanode. Overall water splitting without added bias has been demonstrated with metal oxide photoanodes coupled to dye-sensitized solar cells [51], 2 and 3 J silicon based solar cells [52, 53], and to

Table 1 Reported solar to hydrogen efficiencies for overall PEC water splitting by devices which have at least one SLJ

Publication institute(s)	Device structure (cathode//anode) and catalysts electrolyte pH	Reported STH efficiency
Khaselev and Turner (1998) [24] NREL	SLJ $GaInP_2$(p)//GaAs(pn) integrated Pt HER catalyst wired Pt anode pH 0	12.4 % at 11 suns
Kainthla et al. (1987) [23] Texas A&M	SLJ InP(p)//GaAs(n) SLJ integrated Pt HER catalyst integrated MnO OER catalyst pH 14	8.2 %
Abdi et al. (2014) [53] HZB Berlin	a-Si:H(pin)//nc-Si:H(pin)//$BiVO_4$(n) SLJ wired Pt cathode integrated CoPi OER catalyst pH 7	5.2 %
Abdi et al. (2013) [52] Delft University, HZB Berlin	*a*-Si:H(pin)//*a*-Si:H(pin)//$BiVO_4$(n) SLJ *a*-Si:H (pin)//$BiVO_4$(n) SLJ wired Pt cathode integrated CoPi OER catalyst pH 7.3	4.9 %, 3.6 %
Brillet et al. (2012) [51] EPFL	DSSC//WO_3(n) SLJ, pH 0 DSSC//Fe_2O_3(n) SLJ, pH 13.6 wired Pt cathode integrated ALD Al_2O_3 + Co catalyst	3.1 %, 1.17 %
Gaillard et al. (2010) [60] U. Hawaii	a-Si(pin)//a-Si(pin)//WO_3(n) SLJ wired Pt cathode no OER catalyst 0.3 M H_3PO_4.	3.0 %
Gurudayal et al. (2015) [54] Nanyang Tech. U	$CH_3NH_3PbI_3$//Fe_2O_3 SLJ wired Pt cathode No OER catalyst pH 13.6	2.4 %

DSSC dye-sensitized solar cell; *a-Si* amorphous silicon; *nc-Si* nanocrystalline silicon
Illumination was simulated AM1.5G unless otherwise stated. Values are listed in decreasing order of reported STH efficiency

halide perovskite solar cells [54]. All oxide-based approaches have been demonstrated as well using p-Cu_2O as the photocathode and $BiVO_4$ as the photoanode [55]. Since 2013, demonstrations using Si nanowires in combination with metal oxide photoanode components photoelectrodes have been reported [56, 57].

Efficiency trends in PEC water splitting have been reviewed recently by Rongé et al. [58] and Ager et al. [59]. Here, we compare reported STH efficiencies of selected demonstrations which used at least one SLJ for charge separation in Table 1 in decreasing order of claimed efficiency. One demonstration, with 12.4 % reported efficiency under 11x solar concentration, uses a SLJ at the III–V based photocathode [24]. Another, notably, uses a III–V-based photoanode [23]. The rest employ metal oxide photoanodes, primarily Fe_2O_3 and $BiVO_4$, with the highest reported STH efficiency being just over 5 % [53].

It is interesting to compare these STH efficiency values to those achieved with PV-biased electrosynthetic and PV+electrolyzer approaches. Table 2 compares demonstrations with >10 % reported STH efficiency. In general, the STH efficiency

Table 2 Summary of demonstrations of overall, unassisted solar water splitting using PV-biased electrosynthetic and PV+electrolyzer approaches with reported efficiency >10 %

Publication institute(s)	Device structure (cathode//anode) and catalysts electrolyte pH	Reported STH efficiency
Licht et al. (2000) [50] Technion, Israel	$Al_{0.15}Ga_{0.85}As$(pn)//Si(pn) monolithic PV integrated Pt HER catalyst integrated RuO_2 OER catalyst PV is not in contact with electrolyte pH 0	18.3 %
Khaselev et al. (2001) [46] NREL	$GaInP_2$(pn)//GaAs(pn) monolithic PV wired Pt HER catalyst integrated Pt OER catalyst pH 14	16.5 %
Fujii et al. (2013) [61] U. Tokyo	GaInP(pn)/InGaAs(pn)/Ge(pn) 3 J wired Pt HER catalyst wired Pt OER catalyst Nafion MEA pH 7	15.3 % at 10x
Peharz et al. (2007) [62] ISE Fraunhofer, Freiburg, Germany	$Ga_{0.83}In_{0.17}As$(pn)//$Ga_{0.35}In_{0.65}P$(pn) monolithic PV wired Pt HER catalyst wired IrO_2 OER catalyst Nafion MEA PV +electrolyzer pH 7	15 % 500x optical concentration
Luo et al. (2014) [63] EPFL	Halide perovskite ($CH_3NH_3PbI_3$) cells wired side-by-side wired NiFe HER catalyst wired NiFe OER catalyst PV+electrolyzer pH 13.6	12.3 %
Jacobsson et al. (2013) [64] Uppsala University	$CuIn_xGa_{1-x}Se_2$(pn) 3 cells wired side-by-side wired Pt HER catalyst wired Pt OER catalyst pH 0	10 %
Cox et al. (2014) [65] Harvard and MIT	Si minimodules 4 modules side by side wired NiMoZn HER catalyst wired NiB OER catalyst pH 9.2	9.7 %

MEA membrane electrode assembly
Illumination was simulated AM1.5G unless otherwise stated. Values are listed in decreasing order of reported STH efficiency

values are higher for this type of approach compared to photoelectrochemical water splitting using solid-liquid junctions, as opposed to buried junctions, for charge separation.

4 Summary and Outlook

The operating principles of photoelectrochemical approaches for water splitting are discussed. In contrast to PV-biased electrosynthetic approaches, these types of devices employ charge-separating junctions created by the electrolyte-absorber interface to provide photovoltage. While a number of materials have the requisite band alignment with the H^+/H_2 and/or the O_2/H_2O redox potentials that make them potential candidate absorbers, stability considerations greatly constrain the choices. While there have been some demonstrations with use III–V semiconductors in a PEC geometry, most demonstrations employ metal oxides such as Fe_2O_3 or $BiVO_4$. Solar to hydrogen efficiencies of over 5 % have been achieved with devices which use these oxide materials in combination with other PV elements. Use of III–V materials can produce STH efficiencies of over 10 %.

Acknowledgments This work was supported by the Joint Center for Artificial Photosynthesis, a DOE Energy Innovation Hub, supported through the Office of Science of the U.S. Department of Energy under Award Number DE-SC0004993.

References

1. Ciamician G (1912) The photochemistry of the future. Science 36:385–394. doi:10.1126/science.36.926.385
2. Graves C, Ebbesen SD, Mogensen M, Lackner KS (2011) Sustainable hydrocarbon fuels by recycling CO_2 and H_2O with renewable or nuclear energy. Renew Sustain Energy Rev 15:1–23. doi:10.1016/j.rser.2010.07.014
3. Chu S, Majumdar A (2012) Opportunities and challenges for a sustainable energy future. Nature 488:294–303
4. Turner JA (2004) Sustainable hydrogen production. Science (80-) 305:972–974. doi:10.1126/science.1103197
5. Barbir F (2005) PEM fuel cells: theory and practice. Theory Pract. doi:10.1016/B978-012078142-3/50013-6
6. Wang Y, Chen KS, Mishler J et al (2011) A review of polymer electrolyte membrane fuel cells: technology, applications, and needs on fundamental research. Appl Energy 88:981–1007. doi:10.1016/j.apenergy.2010.09.030
7. Edwards PP, Kuznetsov VL, David WIF, Brandon NP (2008) Hydrogen and fuel cells: towards a sustainable energy future. Energy Policy 36:4356–4362. doi:10.1016/j.enpol.2008.09.036
8. Evans A, Strezov V, Evans TJ (2012) Assessment of utility energy storage options for increased renewable energy penetration. Renew Sustain Energy Rev 16:4141–4147. doi:10.1016/j.rser.2012.03.048

9. Lewis NS (2001) Frontiers of research in photoelectrochemical solar energy conversion. J Electroanal Chem 508:1–10. doi:10.1016/S0022-0728(01)00399-0
10. Walter MG, Warren EL, McKone JR et al (2010) Solar water splitting cells. Chem Rev 110:6446
11. Nielander AC, Shaner MR, Papadantonakis KM et al (2015) A taxonomy for solar fuels generators. Energy Environ Sci 8:16–25. doi:10.1039/C4EE02251C
12. Bak T, Nowotny J, Rekas M, Sorrell C (2002) Photo-electrochemical hydrogen generation from water using solar energy. Materials-related aspects. Int J Hydrogen Energy 27:991–1022. doi:10.1016/S0360-3199(02)00022-8
13. Maeda K, Domen K (2010) Photocatalytic water splitting: recent progress and future challenges. J Phys Chem Lett 1:2655–2661. doi:10.1021/jz1007966
14. Kudo A (2007) Recent progress in the development of visible light-driven powdered photocatalysts for water splitting. Int J Hydrog Energy 32:2673–2678. doi:10.1016/j.ijhydene.2006.09.010
15. Kudo A, Miseki Y (2009) Heterogeneous photocatalyst materials for water splitting. Chem Soc Rev 38:253–278. doi:10.1039/b800489g
16. Chen X, Shen S, Guo L, Mao SS (2010) Semiconductor-based photocatalytic hydrogen generation. Chem Rev 110:6503–6570. doi:10.1021/cr1001645
17. Osterloh FE (2008) Inorganic materials as catalysts for photochemical splitting of water. Chem Mater 20:35–54. doi:10.1021/cm7024203
18. Chen Z, Jaramillo TF, Deutsch TG et al (2010) Review: accelerating materials development for photoelectrochemical hydrogen production: standards for methods, definitions, and reporting protocols. J Mater Res 25:3–16. doi:10.1557/JMR.2010.0020
19. Dotan H, Mathews N, Hisatomi T et al (2014) On the solar to hydrogen conversion efficiency of photoelectrodes for water splitting. J Phys Chem Lett 5:3330–3334. doi:10.1021/jz501716g
20. Fujishima A, Honda K (1972) Electrochemical photolysis of water at a semiconductor electrode. Nature 238:37. doi:10.1038/238037a0
21. Wrighton MS, Ellis AB, Wolczanski PT et al (1976) Strontium titanate photoelectrodes. Efficient photoassisted electrolysis of water at zero applied potential. J Am Chem Soc 98:2774–2779. doi:10.1021/ja00426a017
22. Nozik AJ (1976) p-n photoelectrolysis cells. Appl Phys Lett 29:150–153. doi:10.1063/1.89004
23. Kainthla RC (1987) Significant efficiency increase in self-driven photoelectrochemical cell for water photoelectrolysis. J Electrochem Soc 134:841. doi:10.1149/1.2100583
24. Khaselev O, Turner JA (1998) A monolithic photovoltaic-photoelectrochemical device for hydrogen production via water splitting. Science (80-) 280:425–427. doi:10.1126/science.280.5362.425
25. Golbeck JH (2006) Photosystem I: the light-driven plastocyanin: ferredoxin oxidoreductase. Springer, Dordrecht
26. Wydrzynski T, Satoh K (2006) Photosystem II: the light-driven water: plastoquinone oxidoreductase. Photosynth Res. doi:10.1007/s11120-006-9035-2
27. Shockley W, Queisser HJ (1961) Detailed balance limit of efficiency of p-n junction solar cells. J Appl Phys 32:510–519. doi:10.1063/1.1736034
28. Bolton JR, Strickler SJ, Connolly JS (1985) Limiting and realizable efficiencies of solar photolysis of water. Nature 316:495–500. doi:10.1038/316495a0
29. Weber M, Dignam M (1986) Splitting water with semiconducting photoelectrodes—Efficiency considerations. Int J Hydrogen Energy 11:225–232. doi:10.1016/0360-3199(86)90183-7
30. Licht S (2001) Multiple band gap semiconductor/electrolyte solar energy conversion. J Phys Chem B 105:6281. doi:10.1021/jp010552j
31. Döscher H, Geisz J, Deutsch T, Turner J (2014) Sunlight absorption in water−efficiency and design implications for photoelectrochemical devices. Energy Environ Sci 7:2951–2956. doi:10.1039/c4ee01753f
32. Hu S, Xiang C, Haussener S et al (2013) An analysis of the optimal band gaps of light absorbers in integrated tandem photoelectrochemical water-splitting systems. Energy Environ Sci 6:2984–2993

33. Grätzel M (2001) Photoelectrochemical cells. Nature 414:338–344. doi:10.1038/35104607
34. Chen S, Wang L-W (2012) Thermodynamic oxidation and reduction potentials of photocatalytic semiconductors in aqueous solution. Chem Mater 24:3659–3666. doi:10.1021/cm302533s
35. Walukiewicz W (2001) Intrinsic limitations to the doping of wide-gap semiconductors. Phys B Condens Matter 302–303:123–134. doi:10.1016/S0921-4526(01)00417-3
36. Van de Walle CG, Neugebauer J (2003) Universal alignment of hydrogen levels in semiconductors, insulators and solutions. Nature 423:626–628. doi:10.1038/nature01665
37. Fujishima A (1975) Hydrogen production under sunlight with an electrochemical photocell. J Electrochem Soc 122:1487. doi:10.1149/1.2134048
38. Yoneyama H, Sakamoto H, Tamura H (1975) A Photo-electochemical cell with production of hydrogen and oxygen by a cell reaction. Electrochim Acta 20:341–345. doi:10.1016/0013-4686(75)90016-X
39. Ohashi K, McCann J, Bockris JO (1977) Stable photoelectrochemical cells for the splitting of water. Nature 266:610. doi:10.1038/266610a0
40. Kohl PA, Frank SN, Bard AJ (1977) Semiconductor electrodes. J Electrochem Soc 124:225. doi:10.1149/1.2133270
41. Gerischer H (1977) On the stability of semiconductor electrodes against photodecomposition. J Electroanal Chem Interfacial Electrochem 82:133–143. doi:10.1016/S0022-0728(77)80253-2
42. Delahoy AE, Gau SC, Murphy OJ et al (1985) A one-unit photovoltaic electrolysis system based on a triple stack of amorphous silicon (pin) cells. Int J Hydrogen Energy 10:113–116. doi:10.1016/0360-3199(85)90043-6
43. Lin GH, Kapur M, Kainthla RC, Bockris JOM (1989) One step method to produce hydrogen by a triple stack amorphous silicon solar cell. Appl Phys Lett 55:386–387. doi:10.1063/1.101879
44. Gramaccioni C, Selvaggi A, Galluzzi F (1993) Thin film multi-junction solar cell for water photoelectrolysis. Electrochim Acta 38:111–113. doi:10.1016/0013-4686(93)80016-S
45. Rocheleau RE, Miller EL, Misra A (1998) High-efficiency photoelectrochemical hydrogen production using multijunction amorphous silicon photoelectrodes. Energy Fuels 12:3–10. doi:10.1021/ef9701347
46. Khaselev O, Bansal A, Turner JA (2001) High-efficiency integrated multijunction photovoltaic/electrolysis systems for hydrogen production. Int J Hydrogen Energy 26:127–132. doi:10.1016/S0360-3199(00)00039-2
47. Reece SY, Hamel JA, Sung K et al (2011) Wireless solar water splitting using silicon-based semiconductors and earth-abundant catalysts. Science (80-) 334:645–648. doi:10.1126/science.1209816
48. Appleby AJ, Delahoy AE, Gau SC et al (1985) An amorphous silicon-based one-unit photovoltaic electrolyzer. Energy 10:871–876. doi:10.1016/0360-5442(85)90120-3
49. Sakai Y, Sugahara S, Matsumura M et al (1988) Photoelectrochemical water splitting by tandem type and heterojunction amorphous silicon electrodes. Can J Chem 66:1853–1856. doi:10.1139/v88-299
50. Licht S, Wang B, Mukerji S et al (2000) Efficient solar water splitting, exemplified by RuO 2—catalyzed AlGaAs/Si photoelectrolysis. J Phys Chem B 104:8920–8924. doi:10.1021/jp002083b
51. Brillet J, Yum J-H, Cornuz M et al (2012) Highly efficient water splitting by a dual-absorber tandem cell. Nat Phot 6:824–828
52. Abdi FF, Han L, Smets AHM et al (2013) Efficient solar water splitting by enhanced charge separation in a bismuth vanadate-silicon tandem photoelectrode. Nat Commun 4:2195. doi:10.1038/ncomms3195
53. Han L, Abdi FF, van de Krol R et al (2014) Efficient water-splitting device based on a bismuth vanadate photoanode and thin-film silicon solar Cells. ChemSusChem 7:2832–2838. doi:10.1002/cssc.201402456

54. Sabba D, Mulmudi HK et al (2015) Perovskite–hematite tandem cells for efficient overall solar driven water splitting. Nano Lett 150330053659007. doi:10.1021/acs.nanolett.5b00616
55. Bornoz P, Abdi FF, Tilley SD et al (2014) A bismuth vanadate-cuprous oxide tandem cell for overall solar water splitting. J Phys Chem C 118:16959–16966. doi:10.1021/jp500441h
56. Liu C, Tang J, Chen HM et al (2013) A fully integrated nanosystem of semiconductor nanowires for direct solar water Splitting. Nano Lett 13:2989–2992. doi:10.1021/nl401615t
57. Shaner MR, Fountaine KT, Ardo S et al (2014) Photoelectrochemistry of core–shell tandem junction n–p^+ -Si/n-WO_3 microwire array photoelectrodes. Energy Environ Sci 7:779. doi:10.1039/c3ee43048k
58. Rongé J, Bosserez T, Martel D et al (2014) Monolithic cells for solar fuels. Chem Soc Rev 43:7963–7981. doi:10.1039/c3cs60424a
59. Ager JW III, Shaner M, Walczak K et al (2015) Experimental demonstrations of spontaneous, solar-driven photoelectrochemical water splitting. Energy Environ Sci. doi:10.1039/C5EE00457H
60. Gaillard N, Chang Y, Kaneshiro J et al (2010) Status of research on tungsten oxide-based photoelectrochemical devices at the University of Hawai'i. Proc SPIE 7770:77700 V–77700 V–14. doi:10.1117/12.860970
61. Fujii K, Nakamura S, Sugiyama M et al (2013) Characteristics of hydrogen generation from water splitting by polymer electrolyte electrochemical cell directly connected with concentrated photovoltaic cell. Int J Hydrogen Energy 38:14424–14432. doi:10.1016/j.ijhydene.2013.07.010
62. Peharz G, Dimroth F, Wittstadt U (2007) Solar hydrogen production by water splitting with a conversion efficiency of 18 %. Int J Hydrogen Energy 32:3248–3252. doi:10.1016/j.ijhydene.2007.04.036
63. Luo J, Im J-H, Mayer MT et al (2014) Water photolysis at 12.3 % efficiency via perovskite photovoltaics and earth-abundant catalysts. Science (80-) 345:1593–1596. doi:10.1126/science.1258307
64. Jacobsson TJ, Fjallstrom V, Sahlberg M et al (2013) A monolithic device for solar water splitting based on series interconnected thin film absorbers reaching over 10 % solar-to-hydrogen efficiency. Energy Environ Sci 6:3676–3683. doi:10.1039/C3EE42519C
65. Cox CR, Lee JZ, Nocera DG, Buonassisi T (2014) Ten-percent solar-to-fuel conversion with nonprecious materials. Proc Natl Acad Sci 111:14057–14061. doi:10.1073/pnas.1414290111

Photoelectrochemical Water Splitting Using Photovoltaic Materials

Nicolas Gaillard and Alexander Deangelis

Abstract In this chapter, we will focus on high performance PV (photovoltaic)-grade thin film materials for photoelectrochemical (PEC) water splitting applications. Using techno-economic analysis tools, we will show that achieving a high solar-to-hydrogen (STH) efficiency is by far the most important device attribute for affordable PEC hydrogen production. We will then introduce the concept of multi-junction PEC devices and establish their theoretical STH upper limit. Finally, we will present the PEC performances of one specific PV material with high potential for PEC hydrogen production: the copper chalcopyrite class.

1 Photovoltaic-Grade Photoelectrochemical Devices: Economic and Opto-Electronic Considerations

1.1 Introduction

With uncertain financial systems and chronic energy crises, the ability to produce energy from reliable, affordable and sustainable sources is critical to address the economic and environmental challenges that the world is currently facing. Among all possible thermodynamic pathways, direct solar-to-electron conversion is considered as one of the most efficient routes to supply our energy needs. However, location and time of photovoltaics (PV) operation require storage and transport of produced electricity, which both incur energetic losses. Converting solar power directly into chemical energy is an elegant way to address both energy storage and transport issues. So-called "solar fuels" can take various forms (e.g. H_2, CH_4, C_2H_4) and be converted on demand into mechanical or electrical power. Several photoelectrochemical (PEC) technologies (i.e. complete systems capable of splitting water without external bias) have been developed for hydrogen production over the past four decades since

N. Gaillard (✉) · A. Deangelis
Hawaii Natural Energy Institute, 1680 East-West Road, Honolulu, HI 96822, USA
e-mail: ngaillar@hawaii.edu

M. Sugiyama et al. (eds.), *Solar to Chemical Energy Conversion*,
Lecture Notes in Energy 32, DOI 10.1007/978-3-319-25400-5_16

Fujishima and Honda's first report of solar-assisted water splitting on titania [1]. Amongst them, two noticeable PEC technologies have emerged as the most promising approaches for economical hydrogen production: fixed-bed photocatalytic reactors [2, 3] (i.e. nanoparticles in suspension) and hybrid photoelectrodes (HPE) [4, 5] (i.e. multi-junction devices made of thin film materials), two robust designs whose effectiveness has been largely demonstrated for environmental remediation [6, 7] and photovoltaics [8] applications, respectively. Although extremely attractive in terms of hydrogen production cost (<$2 per Kg H_2) [9], fixed-bed photocatalytic reactors present numerous physical challenges that could limit future large-scale implementation with current designs, including hazardous combustible gas mixture in single-bed with co-generation of oxygen and hydrogen and mass transport limitation over large distances in dual-bed systems with separate O_2 and H_2 evolution. In the present chapter, we will focus exclusively on the second approach, thin film-based PEC systems, with which solar-to-hydrogen (STH) efficiency exceeding 10 % has been demonstrated [5].

1.2 Techno-Economic Considerations

The *Holy Grail* of PEC research on thin film photoelectrodes is to develop materials that are corrosion-resistant, inexpensive to manufacture and have suitable opto-electronic properties for efficient un-biased water splitting. To date, no known semiconductor material fulfills all aforementioned criteria simultaneously and debates are still ongoing in the PEC community between those in favor of durable but not so efficient PEC systems and others opting for highly efficient but not stable semiconductors. A key metric to consider in the selection of material classes is the cost required to produce a specific amount of hydrogen by a PEC material system. In fact this parameter includes not only the impact of the durability and the efficiency of the PEC system but also their manufacturing and maintenance costs. Such a parameter would be analogous to the dollar per watt ($/Watt) of installed photovoltaic systems (cost of module plus balance of systems). In the context of PEC, the dollar per kg of H_2 ($/kg H_2) is usually the metric of choice. Note that 1 kg of H_2 contains as much energy as 1 gal of gasoline, making the cost comparison between renewable hydrogen production technologies and fossil fuels straightforward. Several techno-economic analyses of PEC systems have been reported in the past 5 years on the cost of H_2 produced for various thin film materials and PEC reators [9–11]. Each scenario presented in these studies provides insights regarding the impact of material manufacturing costs, efficiency and durability on the cost of hydrogen produced via PEC water splitting. It is important to keep in mind that these techno-economic studies are not cost prediction tools, but rather ways to compare existing materials and device integrations against each other.

The results of a techno-economic analysis done for a hypothetical centralized PEC plant with a production capacity of 50 ton per day (TPD) are presented in Fig. 1. Two PEC reactors were considered here. In the first case, the PEC reactor is made of one photocathode and one photoanode stacked on top of each other,

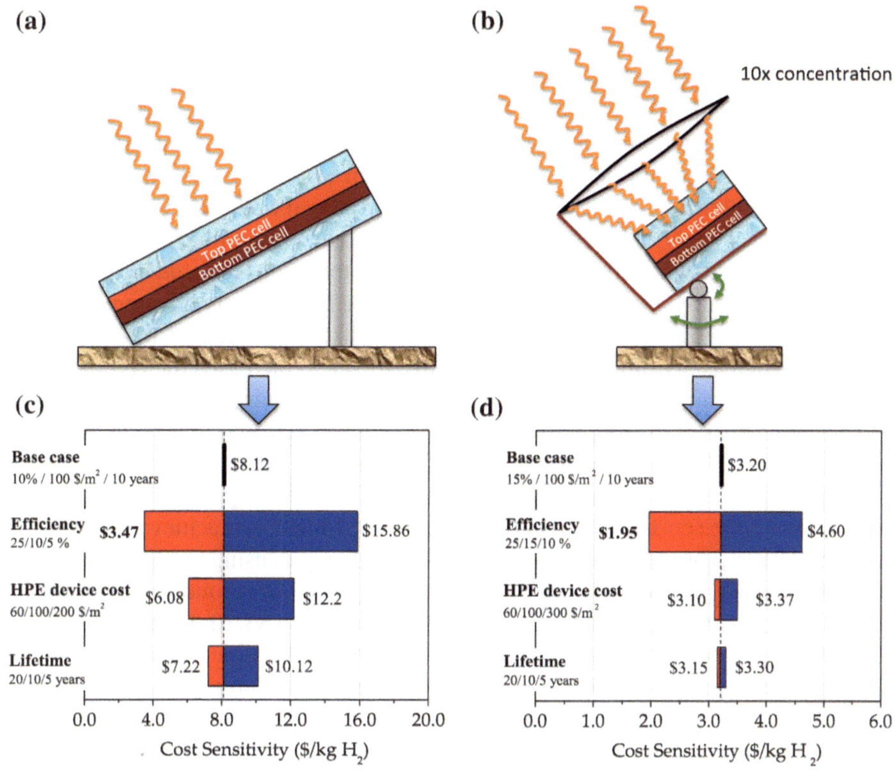

Fig. 1 Schematics of thin film-based PEC reactors. **a** fixed planar structure and **b** 2-axis tracking system with concentrating optics. Effect of efficiency, panel cost, and component lifetime on the cost of H_2 for PEC reactors either **c** fixed or **d** equipped with 10x concentrator and solar tracking system. Each calculation represents the variation of a single parameter from the base case scenario

encapsulated into a reservoir made of transparent material such as polymethyl methacrylate and filled with an electrolyte (Fig. 1a). This planar reactor is fixed on the ground and tilted at the appropriate angle (approx. 35°) to maximize solar energy capture throughout the year. In the second approach, a similar PEC reactor is used except that this time it is mounted on a 2-axis solar tracking system and equipped with concentrating optics. Unlike in concentrated photovoltaics where concentration factors of several hundreds can be used, the one used in this calculation is only 10x to prevent the electrolyte from boiling. The number of PEC reactors required to generate 50 TPD of hydrogen was calculated based on a theoretical STH efficiency and an average solar radiation of 6.55 kWh/m^2. The cost required to build all PEC reactors was based on values reported for various thin film technologies [9, 10]. PEC material lifetime was considered in order to schedule reactor replacement. The PEC reactor manufacturing and replacement costs were finally plugged into the U.S. Department of Energy "H2A model" in order to estimate the cost of kg of H_2 for each scenario [12].

Figure 1c shows the results obtained for a fixed planar PEC reactor, presented in a "tornado plot" fashion, where each calculation represents the variation of a single parameter from the base case scenario. The base case chosen here was a PEC reactor capable of splitting water with a STH efficiency of 10 %, a manufacturing cost of \$100/m^2 and a panel lifetime of 10 years. The cost per kg of H_2 for the base case was \$8.12. The impact of the STH efficiency was then evaluated. Decreasing the STH efficiency to 5 % increased the cost to \$15.86. Increasing the efficiency to 25 % led to a cost per kg H_2 of \$3.47. The device cost was then altered: with a value of \$200/m^2, a cost of \$12.20 is found, while lowering the material cost to \$60/m^2 reduced the production cost to \$6.08. Finally, the impact of the PEC material lifetime was assessed. Doubling the lifetime of the PEC material (20 years) compared to the base case lowered the H_2 cost by \$1.10, whereas dividing it by a factor of two (5 years) increased the H_2 cost by \$2.

The same exercise was repeated for the PEC reactor equipped with solar trackers and concentrated optics (Fig. 1d). The base case chosen here was a system with a STH efficiency of 15 %, a manufacturing cost of \$100/m^2 and a panel lifetime of 10 years. These parameters led to a cost per kg of H_2 of \$3.20. Decreasing the STH efficiency to 10 % increased the cost to \$4.60. Ultimately, a cost of \$1.95 was achieved with a 25 % efficient device. The manufacturing cost was then changed: using projected values for dual absorber PV technologies (\$300/m^2) [9] a cost of \$3.70 is found, while lowering the material cost to \$60/m^2 reduced the production cost to \$3.10. Finally, the impact of the PEC material lifetime was assessed. Doubling the lifetime of the PEC material (20 years) compared to the base case lowered the cost by 5 cents, whereas dividing it by a factor of two (5 years) increased the cost by 10 cents. The rather low impact of the PEC material lifetime on the hydrogen production costs when compared to the planar fixed reactor can be explained by a reduction of the capital cost for the PEC electrode by a factor of 10 with 10x concentrating optics. As a consequence, the concentrator design is advantageous as it lessens the burden of material durability on the cost of hydrogen when compared to the fixed planar design.

An important conclusion one can draw from this techno-economic analysis is that, regardless of the PEC reactor design, the STH efficiency is *the* variable with the most leverage on the cost of hydrogen production, followed by the manufacturing costs. Surprisingly, the PEC material durability has the lowest impact. However, one should note that the durability values used in this analysis (on the order of years) exceed significantly those achieved by any PEC material system. Nonetheless, these calculations show that to be considered a potential candidate for low-cost hydrogen generation, a PEC material system should have first and foremost a high STH efficiency.

1.3 Semiconductor Down-Selection Based on Optoelectronic Properties

We have shown in the previous paragraph that the STH efficiency is the parameter with the greatest leverage on the cost of H_2 production with PEC technologies. It is important at this point to define precisely the STH efficiency in order to understand which parameter drives it. The STH conversion efficiency of a PEC water-splitting system is defined as the ratio of the useable chemical energy in the generated hydrogen gas over the total solar energy delivered to the system [13]. For steady-state operations, this is equivalent to the ratio of the power output to the power input of the system:

$$\frac{P_{out}}{P_{in}} = \frac{(hydrogen\,production\,rate) \times (hydrogen\,energy\,density)}{solar\,flux\,integrated\,over\,illuminated\,area} \quad (1)$$

In the context of thin film-based PEC cells, the hydrogen production rate is tied directly to the steady state short-circuit photocurrent flowing between the (photo-) anode and (photo-) cathode, generating oxygen and hydrogen, respectively. At the (photo-) cathode surface, 2 electrons will be required to reduce 2 protons in order to form one molecule of hydrogen:

$$2e^- + 2H^+ \rightarrow H_2 \quad (2)$$

Thus, the hydrogen production rate can be described as:

$$R(H_2) = \frac{I_{photo} \times \eta_F}{2e} = \frac{(J_{photo} \times A)\eta_F}{2e}, \quad (3)$$

where $R(H_2)$ is the hydrogen production rate (mol/s), I_{photo} is the short-circuit photo-current (A), e is the charge of an electron (C), A is the area of the illuminated photoelectrode (m^2), J_{photo} is the short-circuit photo-current density (A/m^2) and η_F the Faradaic efficiency. Using the hydrogen production rate from Eq. (3) and the molar Gibbs energy of formation of water as the useful molar energy of the hydrogen, the STH efficiency for a thin film-based PEC system can be defined as:

$$STH(\%) = \frac{\Delta G \times R(H_2)}{P_{solar} \times A} = \frac{\Delta G\left(\frac{J_{photo} \times A \times \eta_F}{2e}\right)}{P_{solar} \times A}, \quad (4)$$

where P_{solar} is the solar flux of energy in W/m^2. Under $AM1.5_G$ illumination (1,000 W/m^2), Eq. (4) can be simplified to:

$$STH\,(\%) = J_{photo}(mA/cm^2) \times 1.23\,(V) \times \eta_F \quad (5)$$

It is important to note that Eq. (5) is valid only if the following conditions are satisfied:

- Light (natural or simulated) is the only source of energy input. Therefore, neither an electrical (power supply) nor a chemical (pH gradient) bias can be used during the test,
- The photocurrent density value used to determine the STH efficiency is measured in short-circuit condition. This implies that such measurement is carried

out under 2-electrode configuration at zero bias. No 3-electrode test may be used to defined the STH efficiency, as a bias is always applied under such conditions, even at 0 V versus a reference electrode,

- Sacrificial electron donors or acceptors cannot be used when measuring STH efficiency. Molecules such as CH_3OH or HBr in solution will be oxidized to provide protons. However, these reactions are by definition not water splitting and require much less voltage in comparison. One should note also that these reactants are not sustainable.

Equation (5) reveals that the STH efficiency depends essentially on two characteristics. First, the Faradaic efficiency η_F, which defines the fraction of photo-generated electrical charges participating in the oxygen evolution (OER) or hydrogen evolution (HER) reaction. For an ideal photoelectrode with no interfacial losses, η_F would be unity. However, catalysts are often deposited at PEC electrode surfaces to improve charge transfer and facilitate the water splitting process [5, 14]. The second key parameter in Eq. (5) is the photocurrent density obtained under short-circuit conditions by the PEC device. This metric, which describes the semiconductor's ability to effectively absorb photons, separate electron-hole pairs and transport them though its bulk, is an excellent parameter to identify potential PEC candidates. Figure 2 presents the maximum photocurrent density and STH efficiency as function of semiconductor bandgaps assuming the best-case scenario (quantum efficiency = 100 % and $\eta_F = 1$).

Amongst all possible semiconductor candidates, titanium dioxide and tungsten trioxide have been considered as the most promising water splitting systems since the early ages of PEC research, mainly due to their low-cost, good electrical

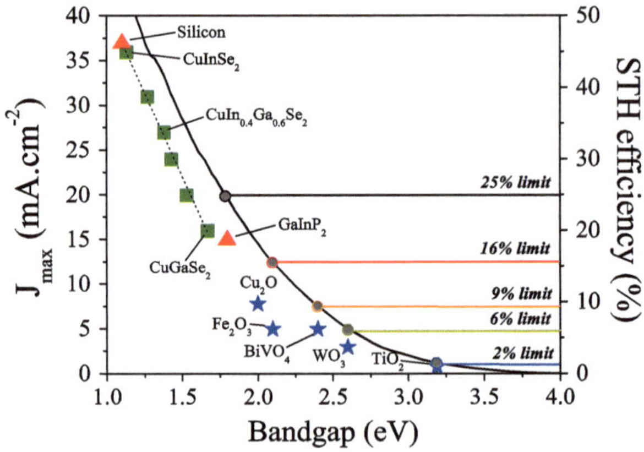

Fig. 2 Maximum photocurrent density and STH efficiency as function of PEC material bandgap. The photocurrent density was calculated using the optical limit (100 % quantum efficiency for photons with energy greater or equal to the bandgap). Experimental photocurrent densities measured on various materials, including metal oxides (*stars*), chalcopyrites (*squares*) and mono-crystalline semiconductors (*triangles*), are plotted for comparison. The STH efficiency limits were calculated using Eq. (5), assuming 100 % Faradaic efficiency

conductivity and resistance to corrosion. However, their relatively large optical band-gap energies ($E_{G,TiO2}$ = 3.2 eV [1], $E_{G,WO3}$ = 2.6 eV [15]) severely limit their absorption of the solar spectrum. As a consequence, the maximum photocurrent densities reported for TiO_2 [16] and WO_3 [15] -based PEC cells are rather small, 1 mA/cm^2 and 3.6 mA/cm^2, respectively. The reduction of transition metal oxide band-gaps has been the *Holy Grail* of PEC research in the past two decades and many attempts have been reported, primarily via incorporation of foreign elements [17–19]. Unfortunately, metal oxide band-gap reduction is usually accompanied by a decrease in PEC performances [20], as the required foreign element concentration (in the percent range) leads in most cases to major microstructural defects [20, 21]. With a bandgap of 2.4 eV [22], bismuth vanadate has emerged recently as a new PEC material candidate, showing rather impressive photo-conversion efficiency with maximum photocurrent densities of 4 mA/cm^2 [23]. Hematite (α-Fe_2O_3) is a material class that inherently has the ideal absorption characteristics for solar-powered water splitting. Indeed, its band-gap (2.0–2.2 eV) [24] is considered ideal since it allows an optimal balance between photocurrent density and generated photo-voltage [25]. However, hematite falls short on electrical performances, with hole-diffusion length in the order of 20 nm [26], though significant progress in Fe_2O_3 nano-structuring has been made to resolve this issue [27, 28]. Cuprous oxide (Cu_2O) is another 2.0 eV metal oxide material actively studied for PEC H_2 production. Recent work has shown that cuprous oxide can generate up to 7 mA/cm^2 at saturation, making this material the best binary oxide ever synthesized for solar conversion applications [29]. It should be noted that the maximum photocurrent densities reported for the vast majority of metal oxides represent at best 50 % of their theoretical maximum value (optical limit), limiting STH efficiencies to the 3–5 % range [23, 30, 31].

In contrast to metal oxides, mono-crystalline semiconductors are exceptional photo-convertors and can generate photocurrent densities close to their optical limits. Crystalline silicon is an emblematic material whose conversion efficiency has been largely demonstrated in PV applications, with short-circuit photocurrent densities exceeding 40 mA/cm^2 [32]. This value, if achieved by a silicon-based PEC device short-circuited with an anode, would set the STH efficiency limit to 50 %. However, with a bandgap of 1.1 eV, silicon cells generate too little photo-voltage, requiring at least 3 cells connected in series to split water. Another drawback of crystalline silicon comes from its fixed optical properties (not bandgap tunable). As a consequence, crystalline silicon-based PEC multi-junction devices can be realized only with a coplanar approach where each of the cells is sharing a fraction of the device real estate, resulting in a lower short-circuit photocurrent density. Using such coplanar design, Reece and coworkers have reported an STH efficiency of 4.7 % with mono-crystalline solar cells [33]. The National Renewable Energy Laboratory currently holds the world record STH efficiency with "III-V" semiconductors. This material class, which includes mono-crystalline GaAs, InP and $GaInP_2$, has the advantage of being bandgap tunable. This feature has been largely exploited to create multi-junction solar cells with conversion efficiencies exceeding 40 % [34]. Using $GaInP_2$ as a top PEC material, the NREL team has achieved an STH efficiency of 12.4 % [5].

Finally, the copper chalcopyrite material class (henceforth chalcopyrite), typically identified by its most popular PV-grade alloy $CuInGaSe_2$, is another material system with great candidates for PEC water splitting. A key asset of this band-gap tunable, direct absorber, thin film semiconductor material is its outstanding photon-to-electron conversion efficiency, as demonstrated with the short-circuit photocurrent densities achieved with low bandgap $CuInGaSe_2$-based PV cells [35]. As shown in Fig. 2, the short-circuit photocurrent density measured on chalcopyrite solar cells [36] is roughly 80 % of the theoretical value regardless of the alloy composition and bandgap. In the context of PEC devices, most of the research work has been done on 1.65 eV $CuGaSe_2$ photocathodes [37–39]. The maximum photocurrent densities achieved with $CuGaSe_2$ photocaothdes are in line with the values measured with PV devices, i.e. 14–16 $mA.cm^{-2}$, setting the STH limit to 20 %. A more detailed discussion on the PEC characteristics of the chalcopyrite material class will be presented later in this chapter.

Based on the techno-economic analysis presented in paragraph 2 of this chapter, it appears that bandgap tunable PV-grade semiconductors, such as the III-V and the chalcopyrite material classes, offer the best candidates for affordable PEC hydrogen production, as these material systems are the only ones capable of reaching the photocurrent values required for high STH efficiency. Moreover, their tunable bandgap allows for efficient device integration, such as the multi-junction stacked approach, in which each cells with various optical characteristics convert a specific portion of the solar spectrum.

2 Multi-junction PEC Devices

2.1 *Multi-junction PEC Reactor Designs*

Several PEC standalone devices made of multiple PEC cells, or a combination of PEC cells and PV devices, have been explored [40–43]. Typically, one can classify these designs into three main categories:

- Integrated PV-electrolyzers, in which PV drivers are connected to an anode and a cathode [33, 44, 45],
- PEC-PEC tandem device, where two photoactive semiconducting materials are coupled [46, 47],
- PV-PEC multi junction structure, in which one or two PV drivers are connected to one PEC photoactive material and a passive electrode [5, 23, 30, 37].

The vast majority of multi-junction PEC devices are either coplanar (Fig. 3a) or stacked (Fig. 3b) multi-junction devices. With the coplanar scheme, several solar absorbers are connected in series and share the real estate of the PEC device,

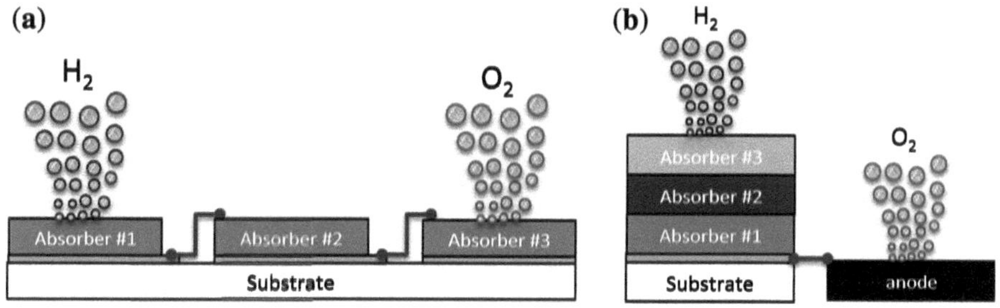

Fig. 3 **a** Coplanar and **b** stacked multi-junction PEC devices

yielding to lower overall photocurrent density. As an example, the maximum photocurrent density generated by a coplanar multi-junction device made of 3 silicon cells is a third of the maximum photocurrent density of each cell. The coplanar scheme is often used as proof of concept, where the aim of the research is to demonstrate the performance of new passive electro-catalysts, rather than the conversion efficiency of the solar absorber [33, 48].

The biggest challenge in PEC research is to create multi-junction devices using solar absorbers that are instead stacked onto each other. This design, whose effectiveness has already been demonstrated in PV applications, allows for more compact devices yielding higher photocurrent density output. However, this integration scheme implies that each of the absorber's bandgaps and thicknesses must be optimized in order to maximize the PEC device efficiency.

2.2 *Principals of Spectral Splitting in Multi-junction Devices*

A multi-junction structure is a simple and straightforward approach to overcome the fundamental conversion limits of a single junction device [49]. This approach relies on a concept often referred to as "spectrum splitting", where each of the junctions absorb a specific range of the solar spectrum. This concept is depicted in Fig. 4 with a multi-junction made of two absorbers: a 1.8 eV semiconductor sitting on top of a 1.4 eV. Photons with energy equal or greater than 1.8 eV (300–690 nm range) are absorbed by the top junction. The top junction generates a photocurrent density J_{top} and a photovoltage V_{top}. The photons with energy lower than 1.8 eV are transmitted to the underlying second junction. Here, photons with energy lower than 1.8 eV but greater than 1.4 eV (690–885 nm range) are absorbed. The bottom junction generates a photocurrent density J_{bottom} and a photovoltage V_{bottom}. Photons with energy lower than 1.4 eV are transmitted through in the present case.

A simplified equivalent circuit of a multi-junction device is shown in Fig. 4c. Since the two junctions are electrically in series, the overall photovoltage of the multi-junction device is equal to the sum of the sub-cell voltages, here

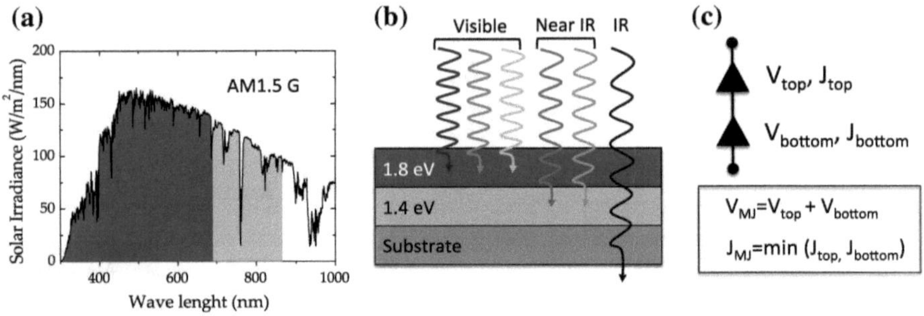

Fig. 4 **a** AM1.5_G spectrum, **b** schematic of a multi-junction device made of 1.8 eV *top* and 1.4 eV *bottom* solar absorbers and **c** simplified equivalent circuit of a multi-junction device

$V_{top} + V_{bottom}$. In virtue of Kirchhoff's circuit law, the photocurrent generated by the bottom junction must be equal to one produced by the top junction. As a consequence, the multi-junction device's net photocurrent density corresponds to the smaller of the two photocurrents generated by each of the two sub-cells (J = min [J_{top}, J_{bottom}]). An important aspect of multi-junction research is not only to combine thin film materials with complementary bandgaps but also optimize each film thickness to ensure that all sub-cells are generating an equivalent photocurrent density, preventing the so-called "bottleneck" effect.

2.3 *Modeling Achievable STH Efficiency with Multi-junction PEC Devices*

Modeling the maximum achievable STH efficiency of multi-junction PEC device candidates can give valuable information on their potential for hydrogen production and also provide guidance for improvement. Such modeling is basically performed using various sets of initial conditions on each of the semiconductor's optical absorption characteristics (bandgap), photo-conversion characteristics (quantum efficiency), usable potential (photovoltage, known value or expressed as a fraction of the bandgap), HER and OER overpotentials, catalysts' Tafel slope and exchange current density as well as the electrochemical load, just to name a few.

Figure 5a presents an example of such calculations done by Rocheleau and Miller in 1997 for a dual absorber PEC device made with high-quality III–V materials, assuming the optical limit (100 % conversion above bandgap), a Faradaic efficiency equal to unity and sets of modeled solid-state (Voc = 70 %*Eg) and electrochemical loses (approx. 400 mV total overpotentials) [50]. As expected, a large range of STH efficiencies are obtained based on the optical characteristics of both the bottom and top absorbers. A high photocurrent density is expected for a device made with low bandgap absorbers (e.g. 1.7 eV/0.7 eV). Nevertheless, the open circuit potential generated by such structures (70 % of $E_{g,top} + E_{g,bottom}$ = 1.7 V) is too low to ensure efficient water

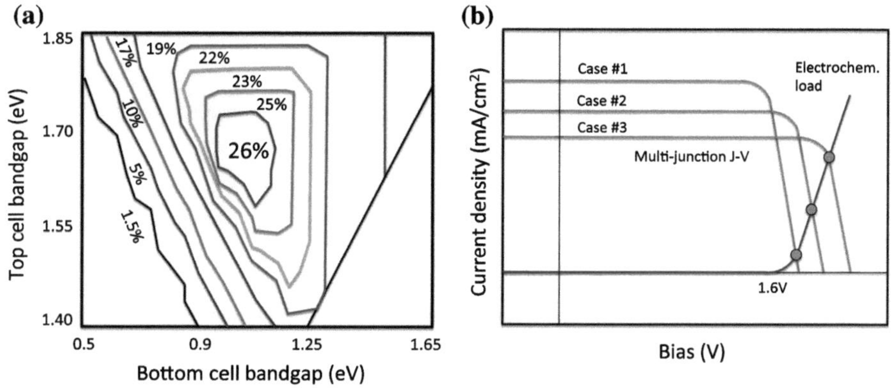

Fig. 5 **a** Maximum achievable STH efficiencies in a dual absorber stack as a function of *top* and *bottom* cell bandgaps. Adapted from [50]. **b** "Load line" analysis using J-V characteristics of multi-junction cells and an electrochemical load. The green circles represent the operating points for each scenario

splitting, as illustrated by the "load line" analysis in Fig. 5b for case #1. Thus, the bandgap of each sub-cell must be increased to provide a larger voltage required for efficient water splitting (1.8–2.0 V). In the present simulation, the optimum bandgap for the top and bottom III-V sub-cells are 1.65 and 1.1 eV, respectively. The maximum STH efficiency such high-quality device could reach is 26 %.

Seitz, Jaramillo and co-workers from Stanford University recently published a study in which the reported solid-state and electrochemical properties of physical absorbers and catalysts were used to establish the "practical" performance limits of multi-junction PEC devices [51]. Three fundamental losses were included in their study: absorbers' free energy losses (FEL) corresponding to the difference between the bandgap and the useable photovoltage (Fig. 6a), kinetic losses (over potentials) of known precious and non-precious catalysts (Fig. 6b) as well as shunt losses through the PEC materials. The individual current-voltage characteristics of the stacked photoanode and photocathode were modeled using each semiconductors' reported bandgap and FEL as well as the kinetic losses of the catalysts present at the surface of each photoelectrodes. The operating point of the standalone PEC device was defined by the intersection of the photocathode and photoanode J-V characteristics. The results of these simulations are presented in Fig. 6c for semiconductors having free energy losses of 0.49 eV (such as high-quality crystalline silicon and GaAs) and decorated with Pt and RuO_2 as HER and OER catalysts, respectively. The maximum practical STH efficiency achieved with such high-efficiency semiconductors and precious catalysts is 22.8 %. To achieve this efficiency, the bandgap of the top and bottom absorbers should be 1.84 and 1.23 eV, respectively. Also, indicated in this Fig. 6c is the 10 % STH efficiency limit, represented by the white triangle. All sets of materials with bandgaps within these limits are potentially viable candidates for low-cost hydrogen production, as defined by the techno-economic analysis presented earlier.

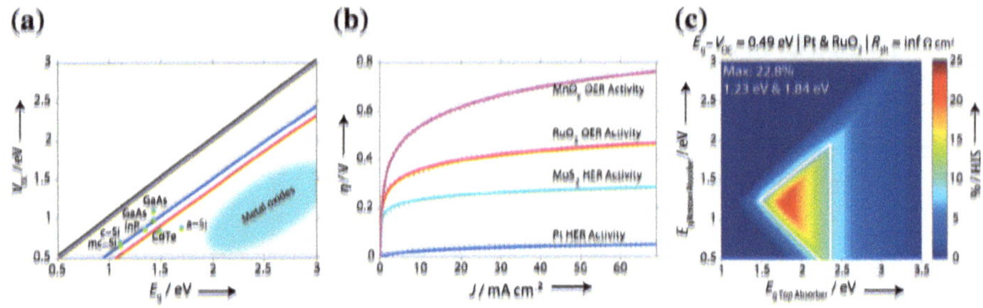

Fig. 6 **a** Open circuit potential versus bandgap reported for various semiconductors. **b** Catalyst activity represented as overpotential versus current density. Experimental data are shown in dashes and Butler-Volmer fits in *solid lines*. **c** STH efficiency contour plots for a dual stacked absorber configuration as function of *bottom* and *top* bandgap absorbers. See reference [51] for more details

These two modeling studies demonstrate that STH efficiencies greater than 10 % can be achieved with existing absorbers and catalysts, a prerequisite for affordable PEC hydrogen production. However, absorbers with compatible optical characteristics and excellent solid-state properties are necessary to achieve this goal. We have seen earlier that the III-V material class offers exceptional candidates for dual-absorber PEC devices, as demonstrated by NREL's world record STH efficiency of 12.4 % with $GaInP_2$/GaAs tandem device [5]. We will show in the next section that another PV-grade class, the copper chalcopyrite, could potentially achieve comparable efficiencies with the promise of lower cost and higher durability.

3 Engineering Future PV-Grade PEC Devices for Affordable Hydrogen Production: The Case of the Chalcopyrite Material Class

3.1 Basic Characteristics of Chalcopyrites

The chalcopyrite material class provides exceptionally good candidates to meet the requirements identified for efficient, cheap and sustainable hydrogen production. Below are four key points that make chalcopyrites an ideal class for PEC applications:

- *High photo-conversion efficiency*: as described earlier in this chapter, chalcopyrites can generate tremendous amounts of photocurrent, as observed in high-performance Cu(In,Ga)Se_2 solar cells (Fig. 7a),
- *Tolerance to defects*: chalcopyrites are highly tolerant to microstructural and electronic defects, allowing the use of rapid non-vacuum-based deposition techniques [52],

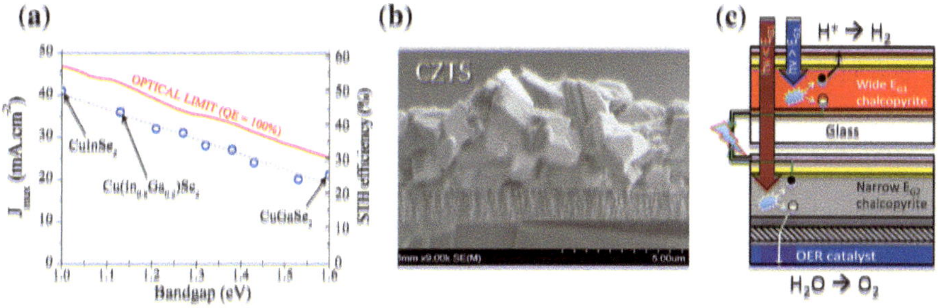

Fig. 7 **a** Reported photocurrent density of chalcopyrite-based PV cells as function of the absorber bandgap, adapted from [36], **b** SEM cross section of a Cu_2ZnSnS_4 thin film obtained from printable nano-crystalline inks (courtesy of Dr. Marina Chong from the University of Hawai'i) and **c** schematic of an all-chalcopyrite based PEC tandem device

- *Potentially low-cost system*: the chalcopyrites structure allows for element substitution with cheaper elements. Replacement of indium and gallium with zinc or tin in $CuInGaSe_2$ to form Cu_2ZnSnS_4 [53] is a perfect example (Fig. 7b).
- *Bandgap-tunable*: it is possible to tune the optical bandgap of chalcopyrites by changing their composition. Thus, the same material class can be used to make each sub-cell of a multi-junction device (Fig. 7c), simplifying the integration process.

We will now introduce the basic PEC characteristics of the copper chalcopyrite class, starting with 1.1 eV $CuInGaSe_2$. Figure 8 presents the PV and PEC performances of two $CuInGaSe_2$ samples fabricated from the same batch. These films were deposited using the so-called "three stage" co-evaporation technique [37]. For the PV cell, a 70 nm thick CdS buffer layer was deposited on top of the absorber, followed by the 80 nm ZnO resistive film and 150 nm ITO conductive layer. Finally, metallic grids were evaporated to form the front contacts. In the case of the

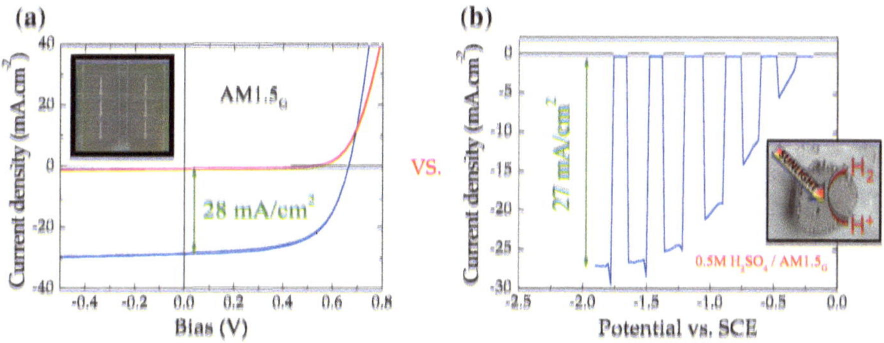

Fig. 8 Current density versus voltage measured on a $CuInGaSe_2$ thin film material integrated either as **a** a PV cell or **b** a PEC photocathode

PEC sample, no buffer, TCO nor catalysts were deposited on top of $CuInGaSe_2$. A metal wire was simply soldered on the sample back contact (Mo) and all other exposed conductive parts were coated with insulating epoxy. Both tests were done under simulated $AM1.5_G$ illumination. PEC tests were done in 0.5 M H_2SO_4 electrolyte in a three-electrode setup, using a saturated calomel electrode as reference and a platinum counter electrode.

Although PEC and PV tests are quite different by nature, this comparison provides insights on the potential of chalcopyrites for PEC applications. In fact, one can see that the saturation photocurrent density obtained with both PV and PEC devices are nearly identical, i.e. 27–28 mA/cm2. This rather simple test clearly confirms that chalcopyrites not only generate ample photocurrent densities, they can also effectively transfer photo-generated electrons at the semiconductor/electrolyte interface to reduce protons into hydrogen. It is important to recall that no catalysts were deposited at the surface of the $CuInGaSe_2$-based photocathode. Nonetheless, as discussed above, the optical bandgap of PV-grade $CuInGaSe_2$ (1.1 eV) is too low for the multi-junction approach and other alloys must be considered.

With a bandgap of 1.65 eV, $CuGaSe_2$ is an ideal candidate to form the top junction of a dual-absorber stacked PEC device. In fact, modeling studies presented in the previous paragraph show that STH efficiency limits in the 20 % [51] to 25 % [50] range can be achieved with such optical characteristics. Figure 9a presents the incident photon-to-current efficiency (IPCE) measured on a 500 nm thick co-evaporated [37, 39] $CuGaSe_2$ electrode in 0.5 M H_2SO_4 (pH 0.5) at a potential sufficient to reach photocurrent saturation ($-0.6\ V_{RHE}$), showing that $CuGaSe_2$ effectively absorbs and converts photons in the visible portion of the solar spectrum with a quantum efficiency that exceeds 60 %. Integrating the IPCE curve over $AM1.5_G$ gives a photocurrent density of 14 mA/cm^2.

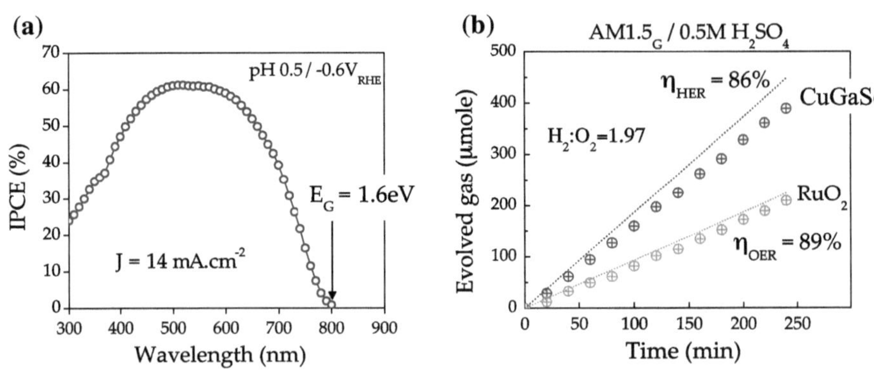

Fig. 9 **a** Incident photon-to-current efficiency analysis measured on a $CuGaSe_2$-based photocathode at $-0.6\ V_{RHE}$. **b** Moles of hydrogen and oxygen evolved as function of time by a $CuGaSe_2$-based photocathode and RuO_2 anode, respectively. The circles represent the experimental data and the dotted lines the expected value (100 % Faradaic efficiency)

Figure 9b presents the number of moles of hydrogen and oxygen evolved as function of time at the surface of a $CuGaSe_2$ photocathode and a RuO_2 anode, respectively. This analysis was done in 0.5 M H_2SO_4 under $AM1.5_G$ illumination. The number of moles of H_2 and O_2 gas evolved was calculated by measuring the volume of water displaced in a Hoffman apparatus (circles in the figure). These data were compared to moles of gas expected from the number of electrical charges passed through the system during the experiment (dotted lines in the figure). The Faradaic efficiency was calculated by dividing the number of moles measured by the number of moles expected. Faradaic efficiencies of 86 and 89 % were obtained for the bare $CuGaSe_2$ photocathode and the RuO_2 anode, respectively. Finally, a H_2 to O_2 volume ratio of nearly 2 (1.97) was measured, validating the methodology used in this experiment.

We have just shown that bare $CuGaSe_2$ thin films are capable of generating large photocurrent densities in a PEC configuration. In addition, evolved gas analysis indicates that a large majority (86 %) of the photo-generated electrons participate to the hydrogen evolution reaction. Although integral for efficient solar-powered hydrogen production, these characteristics do not guaranty that standalone water splitting will be easily achieved with a $CuGaSe_2$-based dual-absorber stacked device. In fact, the band diagram presented in Fig. 10a, obtained from impedance spectroscopy analysis [39], shows that $CuGaSe_2$ surface energetics are far from ideal for water splitting, as the valence band maximum (VBM) sits only 100 mV "below" the RHE. Since, in first approximation, water splitting will not occur until both the conduction band and valence band straddle the water redox potentials, an additional bias of 1.13 V (1.23 V–100 mV) will be required just to initiate the water splitting process. In addition to these undesired surface energetics, linear sweep voltammetry (LSV) reveals that an additional 500 mV is required to turn on the hydrogen evolution half reaction (Fig. 10b). Strategies to improve the surface energetics of $CuGaSe_2$ have been proposed, including the creation of a p-n junction

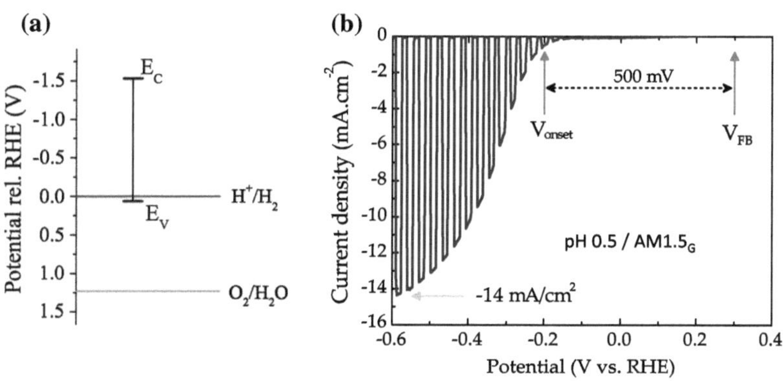

Fig. 10 **a** Position of the valence band and conduction band of $CuGaSe_2$ photocathode with respect to the water oxidation and reduction potentials. **b** Current density versus potential measured on a $CuGaSe_2$-based photocathode

using CdS as n-type buffer [38]. However, with the current state-of-the-art of PEC research on chalcopyrites (as of 2014), no spontaneous water splitting has been achieved with chalcopyrite-based dual-absorber stacked devices.

3.2 *Chalcopyrite-Based Standalone PEC Devices*

We report in this section an alternative approach to incorporate chalcopyrite photocathodes into standalone PEC devices. Specifically, we present a structure made with one 1.65 eV $CuGaSe_2$ photocathode and low-cost amorphous silicon solar cells. It should be noted that the dual-absorber stacked approach is not compatible with these materials as they share similar bandgaps. One way to circumvent this issue is to take advantage of the large photocurrent density generated by $CuGaSe_2$ and sacrifice part of the device estate to form a coplanar device. The LSV characteristic of $CuGaSe_2$ measured in a 2-electrode fashion against RuO_2 and the J-V curve of an a-Si-based triple junction cell are presented in Fig. 11a. Here, the relative area occupied by each device (PEC or PV) was altered. As an example, 50 % PEC/50 % PV relative area would mean that half of the total device area is occupied by the PEC electrode, whereas the other half is used by the 3 a-Si solar cells. Both LSV and PV characteristics where then plotted with different relative areas. The optimum area ratio was determined by the maximum photocurrent density achievable by each coplanar system, as defined by the intersection of both current-voltage characteristics. This analysis revealed that best water splitting performances would be achieved with a 30/70 PEC/PV coplanar structure, operating at 1.6 V (2-electrode vs. RuO_2) and generating a photocurrent density of 3.7 mA/cm^2. Such a device was built and tested under AM1.5$_G$ outdoor conditions. As predicted by the load-line analysis, a photocurrent density of 3.53 mA/cm^2 was achieved under true short circuit conditions. The STH efficiency could be then calculated using Eq. (5) and the measured Faradaic efficiency (86 %):

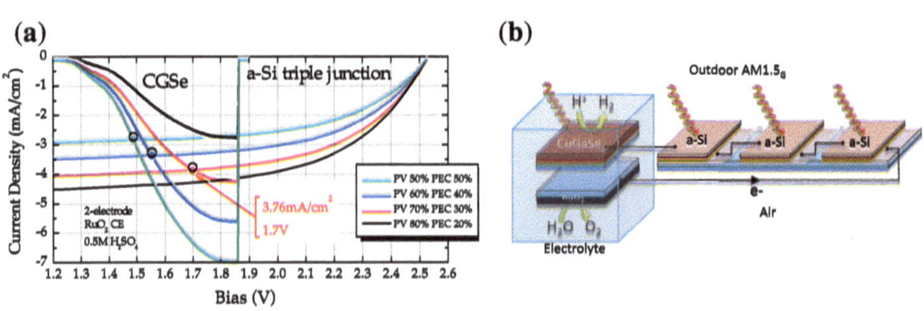

Fig. 11 **a** Load line analysis of a $CuGaSe_2$ photocathode with a triple junction a-Si solar cell. The relative area of each device (PEC or PV) was altered to define the optimum area ratio. **b** Schematic of the proposed planar structure

$$STH(\%) = J_{photo}(mA/cm^2) \times 1.23(V) \times \eta_F$$
$$= 3.53 \times 1.23 \times 0.86$$
$$= 3.7\%$$

Although 3.7 % STH efficient corresponds to an important milestone for chalcogenide-based PEC devices, one could argue that $CuGaSe_2$ is used here as a non-precious electrocatalyst rather than a photocatalyst. However, additional load line analyses revealed that $CuGaSe_2$ *outperforms* platinum at bias lower than 1.6 V [37]. However, at higher bias, one would be better off using a Pt foil to evolve H_2. Efforts to engineer chalcopyrites with appropriate band-edges for PEC water splitting will be discussed next.

3.3 Future Development for Efficient Water Splitting with Chalcopyrites

We demonstrated in the previous section that the properties of "conventional" chalcopyrites are not optimum for efficient standalone water splitting due to a valence band maximum that is too close to RHE. It has been reported that the position of CBM is set by the indium to gallium orbitals, whereas the position of VBM is defined Cu and Se orbitals. [54] Thus, substituting Cu and/or Se with other elements could theoretically change the position of VBM.

Copper substitution: using liquid phase reaction method, Beck and co-workers successfully synthesized single-phase $Cu_{1-x}Ag_xGaSe_2$ material [55]. The authors confirmed that Ag was replacing Cu in the chalcopyrite lattice while creating selenium vacancies (donor), reducing the net hole concentration. Using a co-evaporation method, Kaneshiro performed a partial substitution of Cu with Ag in copper chalcopyrites to form $AgCuInGaSe_2$ thin films [56]. Current-voltage characteristics measured on PV devices have shown a net increase in conversion efficiency with $CuAgInGaSe_2$ when compared to $CuGaSe_2$. However, no photocurrent onset shift was observed between the two materials when tested in PEC configuration. Recently, Zhang reported on $AgCuGaSe_2$ photocathodes synthesized using a process similar to the one reported by Kaneshiro (co-evaporation) [57]. A 140 mV onset potential anodic shift was measured with $AgCuGaSe_2$ films containing 5.9 % silver when compared to $CuGaSe_2$.

Selenium substitution: partial or total substitution of selenium with sulfur has been reported exclusively for PV applications so far. Initially, sulfur was incorporated only at the surface of $CuInGaSe_2$ to form a $CuInGa(S,Se)_2$ thin layer in order to increase solar cells' open circuit potential [58]. Recently, several groups have been actively participating in the development of pure sulfide solar cells [59, 60]. Solar cells with efficiencies in the 10–11 % range have been achieved routinely with

1.5 eV $CuInS_2$ thin film absorbers. Adding a small amount of gallium (about 30 %) boosted the efficiency to 12.6 %. It should be noted that sulfide-based chalcopyrites ($CuInGaS_2$) provides material systems with a wide range of bandgaps, ranging from 1.5 eV ($CuInS_2$) to 2.4 eV ($CuGaS_2$), making this material class ideal for PEC applications.

References

1. Fujishima A, Honda K (1972) Nature 238:37–38
2. Braham RJ, Harris AT (2009) Ind Eng Chem Res 48:8890–8905
3. Zou Z, Ye J, Sayama K, Arakawa H (2001) Nature 414:625–627
4. Miller E, Marsen B, Paluselli D, Rocheleau R (2005) Electrochem Solid-State Lett 8: A247–A249
5. Khaselev OK, Turner JA (1998) Science 280:425–427
6. McCullagh C, Skillen N, Adams M, Robertson P (2011) J Chem Technol Biotechnol 86:1002–1017
7. Chong MN, Jin B, Chow CW, Saint C (2010) Water Res 44:2997–3027
8. King RR, Fetzer CM, Colter PC, Edmondson KM, Ermer JH, Cotal HL, Hojun Y, Stavrides AP, Kinsey G, Krut DD, Karam, NH (2002) PVSC 776–781
9. James BD, Baum GN, Perez J, Baum KN (2009) Technoeconomic analysis of PEC hydrogen production. Directed Technologies, Inc. https://www1.eere.energy.gov/hydrogenandfuelcells/pdfs/pec_technoeconomic_analysis.pdf
10. Pinaud BA, Benck JD, Seitz LC, Forman AJ, Jaramillo TF (2013) Energy Environ Sci 6:1983
11. Sathre R, Scown CD, Morrow WR, Stevens JC, Sharp ID, Ager JW, Walczak K, Houle FA, Greenblatt JB (2014) Energy Environ Sci 7:3264–3278
12. http://www.hydrogen.energy.gov/h2a_prod_studies.html
13. Dohrmann JK, Schaaf NS (1992) Energy conversion by photoelectrolysis of water: determination of efficiency by in situ photocalorimetry. J Phys Chem 96:4558–4563
14. Abdi FF, Van de Krol R (2012) J Phys Chem C 116(17):9398–9404
15. Gaillard N, Cole B, Marsen B, Kaneshiro J, Miller EL, Weinhardt L, Bär M, Heske C, Ahn K-S, Yan Y, Al-Jassim MM (2010) J Mater Res 25:45
16. Cho IS, Chen Z, Forman AJ, Kim DR, Rao PM, Jaramillo TF, Zheng X (2011) Nano Lett 11 (11):4978–4984
17. Asahi R, Morikawa T, Ohwaki T, Aoki K, Taga Y (2001) Science 293:269–271
18. Morikawa T, Asahi R, Ohwaki T, Aoki K, Taga Y (2001) J J Appl Phys 40:561–563
19. Huda MN, Yan Y, Moon C, Wei S, Jassim MM (2008) Phys Rev B 77:195102–195114
20. Cole B, Marsen B, Miller EL, Yan Y, To B, Jones K, Ai-Jassim M (2008) J Phys Chem C 112:5213–5220
21. D'Arienzo M, Siedl N, Sternig A, Scotti R, Morazzoni F, Bernardi J, Diwald OJ (2010) Phys Chem C 114:18067–18072
22. Tokunaga S, Kato H, Kudo A (2001) Chem Mater 13:4624
23. Abdi FF, Han L, Smets HM, Zeman M, Dand B, van de Krol R (2013) Nature Commun 4:2195
24. Hardee KL, Bard AJ (1977) J Electrochem Soc 124:215
25. Chen Z, Jaramillo TF, Deutsch TG, Shwarsctein AK, Forman AJ, Gaillard N, Garland R, Takanabe K, Heske C, Sunkara M, McFarland EW, Domen K, Miller EL, Turner JA, Dinh HN (2010) J Mater Res 25:3–16
26. Dare-edwards MP, Goodenough JB, Hamnett A, Trevellick PR (1983) J Chem Soc, Faraday Trans 79:2027–2041
27. Sivula K, Formal FL, Gratzel M (2011) ChemSusChem 18:432–449

28. Vayssieres L, Sathe C, Butorin S, Shuh D, Nordgren J, Guo J (2005) J Adv Mater 17: 2320–2323
29. Paracchino A, Laporte V, Sivula K, Grätzel M, Thimsen E (2011) Nat Mater 10:456–461
30. Gaillard N, Chang Y, Kaneshiro J, Deangelis A (2010) Miller EL Solar Hydrogen and Nanotechnology V. Proc SPIE 7770:77700 V–14
31. Lee WJ, Shinde PS, Go GH, Ramasamy E (2011) Int J Hydrogen 36:5262–5270
32. Green MA (2009) Prog Photovoltaics 17:183
33. Reece SY, Hamel JA, Thomas KS, Jarvi D, Esswein AJ, Nocera DG (2011) Science 334:645
34. Cotal H, Fetzer C, Boisvert J, Kinsey G, King R, Hebert P, Yoonnad H, Karam N (2009) Energy Environ Sci 2:174–192
35. Jackson P, Hariskos D, Pwalla M (2011) Prog Photovoltaics Res Appl 19:894
36. Contreras M, Mansfield L, Egaas B, Li B, Romero M, Noufi R (2011) In: Proceeding of the 37th IEEE PVSC
37. Gaillard N, Prasher D, Kaneshiro J, Mallory S, Chong M (2013) MRS Online Proceedings Library 1558
38. Moriya M, Minegishi T, Kumagai H, Katayama M, Kubota J, Domen K (2013) J Am Chem Soc 135:3733–3755
39. Marsen B, Cole B, Miller EL (2008) Solar Mat Solar Cells 92:1054–1058
40. Miller EL, Gaillard N, Kaneshiro J, DeAngelis A, Garland R (2010) Progress in new semiconductor materials classes for solar photoelectrolysis. Int J Energy Res 34:1215–1222
41. Kelly NA, Gibson TL (2008) Solar energy concentrating reactors for hydrogen production by photoelectrochemical water splitting. Int J Hydrogen Energy 33:6420–6643
42. Walter MG, Warren EL, McKone JR, Boettcher SW, Mi QX, Santori EA, Lewis NS (2010) Solar Water Splitting Cells. Chem Rev 110:6446–6473
43. Minggu LJ, Daud WRW, Kassim MB (2010) Int J Hydrogen Energy 35:5233
44. Rocheleau R, Miller EL, Misra A (1998) Energy Fuels 12(1):3–10
45. Jacobsson TJ, Fjällström V, Sahlberg M, Edvinsson T (2013) Energy Environ Sci 6:3676–3683
46. Wang H, Deutsch T, Turner JA (2008) J Electrochem Soc 155:F91
47. Ingler WB, Khan SUM (2006) Electrochem. Solid State Lett 9:G144–G146
48. Luo J, Im JH, Mayer MT, Mayer M, Schreier M, Nazeeruddin MK, Park NG, Tilley SD, Han HJ, Gratzel M (2014) Science 345(6204)
49. Shockley W, Queisser H (1961) J Appl Phys 32:510–519
50. Rocheleau RE, Miller EL (1997) Int J Hydrogen Energy 22:771–782
51. Seitz LC, Chen Z, Forman AJ, Pinaud BA, Benck JD, Jaramillo TF (2014) ChemSusChem 7:1372
52. Akhavan VA, Goodfellow BW, Panthani MG et al (2012) J Solid State Chem 189:2–12
53. Todorov TK, Reuter KB, Mitzi DB (2010) Adv Mater 22:E156–E159
54. Bär M, Bohne W, Rohrich J, Strub E, Lindner S, Lux-Steiner MC, Fischer ChH, Niesen TP, Karg F (2004) Appl Phys 96:3857
55. Beck ME, Weiss T, Fischer D, Fiechter S, Jäger-Waldau A, Lux-Steiner M Ch (2000) Thin Solid Films 130–134
56. Kaneshiro J, Deangelis A, Gaillard N, Chang YC, Kowalczyk J, Miller E (2010) Proceedings of the 35th IEEE Photovoltaic Specialists Conference, 002448–002451
57. Zhang L, Minegishi T, Kubota J, Domen K (2014) Phys Chem Chem Phys 16(13):6167–6174
58. Nakada T, Ohbo H, Watanabe T, Nakazawa H, Matsui M, Kunioka A (1997) Solar Mat Solar Cells 49:285
59. Merdes S, Mainz R, Klaer J, Meeder A, Rodriguez-Alvarez H, Schock HW, Lux-Steiner MCh, Klenk R (2011) Solar Mat Solar Cells 95:864
60. Mainz R, Klenk R, Lux-Steiner MCh (2007) Thin Solid Films 515:5934

CO_2 Reduction by Photoelectrochemistry

Takeshi Morikawa

Abstract CO_2 reduction using sunlight energy is one of the ultimate methods for addressing issues related to global warming and a fossil fuel shortage, and constructing a carbon-neutral society in the future. Therefore, the amount of research in this field has been increasing in the 2010s. Photoelectrochemical CO_2 reduction is generally conducted using a semiconductor or a combination of a semiconductor and cocatalyst. Material design and systematic utilization of semiconductors are thus important for effective use of solar photons. In addition, the CO_2 molecule is highly stable; therefore, catalyst design is much more important than in the case for hydrogen generation by water splitting. In this chapter, the methodology for the catalysis of CO_2 reduction by photoelectrochemical means is explained.

1 Introduction

The motivation for photoelectrochemical CO_2 reduction is the recycling of CO_2 molecules that are mainly generated from the combustion of fossil fuels. If CO_2 can be reduced by the utilization of water as an electron donor and a proton source at normal pressure and temperature, then such a reaction would mimic photosynthesis in plants and be considered an industrially useful process to produce renewable energy. The species produced from CO_2 reduction can be a valuable feedstock for other chemical processes. Considering future energy sources, solar hydrogen generation by water splitting is one of the most important approaches. However, one of the main concerns associated with hydrogen is the difficulty with storage. Therefore, solar fuel generation from CO_2 and H_2O, as with photosynthesis in plants, is a very attractive approach. However, it is considerably more difficult to photoreduce CO_2 than to photoreduce water to generate hydrogen, as discussed later.

T. Morikawa (✉)
Toyota Central R&D Labs., Inc, 41-1 Yokomichi, Nagakute, Aichi, Japan
e-mail: morikawa@mosk.tytlabs.co.jp

M. Sugiyama et al. (eds.), *Solar to Chemical Energy Conversion*,
Lecture Notes in Energy 32, DOI 10.1007/978-3-319-25400-5_17

Table 1 CO_2 reduction potentials (reported at pH 7)

	E^0 /V versus NHE
$CO_2 + e^- \rightarrow CO_2^-$	≥1.9
$CO_2 + 2e^- + 2H^+ \rightarrow HCOOH$	−0.61
$CO_2 + 2e^- + 2H^+ \rightarrow CO + H_2O$	−0.53
$CO_2 + 4e^- + 4H^+ \rightarrow HCHO + H_2O$	−0.48
$CO_2 + 6e^- + 6H^+ \rightarrow CH_3OH + H_2O$	−0.38
$CO_2 + 8e^- + 8H^+ \rightarrow CH_4 + 2H_2O$	−0.24

The potentials for CO_2 reduction to various products are given in Table 1. Single-electron reduction of CO_2 to a CO_2 radical anion (CO_2^-) is unfavorable, because it requires a highly negative potential of −1.9 V versus NHE due to the large reorganization energy between the linear CO_2 molecule and the bent CO_2 radical anion [1]. Therefore, it is difficult to reduce CO_2 molecules using only electrical energy. A more favorable pathway is the reduction of CO_2 molecules though a proton-assisted multiple-electron reaction, as shown in Table 1.

Electrocatalysts can facilitate the proton-coupled multi-electron reactions that occur at lower potentials than that for the single-electron reaction. The potentials for the reduction of CO_2 to various organic substances decrease with an increase in the numbers of electrons and protons involved in the reactions. However, many electrocatalysts necessitate higher electrical energy (overpotential) than the theoretical values shown in Table 1; therefore, the development of electrocatalysts that can reduce CO_2 at lower overpotentials is also required. Transition metals and transition metal compounds are at the forefront of catalyst research. Metal-complex catalysts are also feasible because their reduction potentials can be controlled through ligand modification to match the potential required for CO_2 reduction and provide selectivity toward specific target products from CO_2 molecules.

2 Photoelectrochemistry for CO_2 Reduction Using Semiconductors

The first important mechanism that should be taken into consideration with photoelectrochemical CO_2 reduction is the electronic state at the semiconductor/liquid interface. This characteristic for semiconductor/liquid junctions is similar to that for a semiconductor/metal junction. When a semiconductor is placed in contact with a liquid containing a redox species, interfacial charge transfer occurs until equilibrium is reached. The direction and magnitude of this charge flow are determined by the energy band structure of the semiconductor and the redox species in the liquid. Figure 1 shows that the key factors for the semiconductor are the energy of the conduction band minimum E_{CBM}, the energy of the valence band maximum E_{VBM}, the Fermi level E_F, and the band gap energy Eg (= $E_{VBM} - E_{CBM}$), while that for the redox species is its redox potential. At semiconductor/liquid junctions,

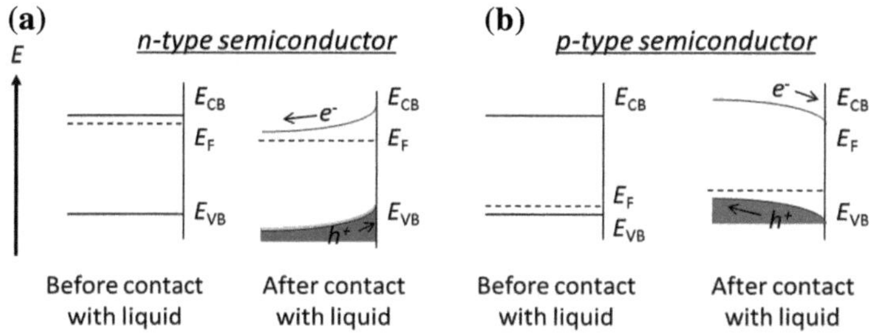

Fig. 1 Band structures at semiconductor/liquid interfaces, **a** *n*-type semiconductor, **b** *p*-type semiconductor

interfacial charge transfer generally induces downward band bending toward the surface of an *n*-type semiconductor (Fig. 1a) and upward band bending toward the surface of a *p*-type semiconductor (Fig. 1b).

Figure 1b shows a particular case for a *p*-type semiconductor/liquid junction, where photogenerated electrons in the conduction band migrate to the surface, due to the downward band bending, and are available for reductive chemical reaction at the semiconductor/liquid interface under illumination. Another important factor with regard to photoelectrochemical CO_2 reduction at the semiconductor is the band positions of the semiconductor. Figure 2 shows band-edge positions for selected semiconductors together with redox potentials for CO_2 and protons. It should also be recognized that the uncertainty in the band-edge positions is a few tenths of an electron volt for most of the semiconductors shown here. To reduce CO_2 molecules efficiently using photoexcited conduction band electrons, E_{CBM} must be located at a

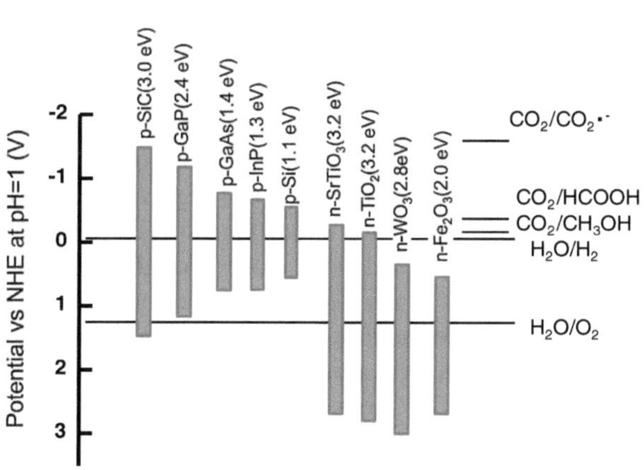

Fig. 2 Band-edge positions for selected semiconductors at pH 1, together with some important redox potentials. It should be realized that the uncertainty in the band-edge positions can amount to a few tenths of an electron volt for most semiconductors

more negative position than the CO_2 reduction potentials to facilitate proton-coupled electron transfer to the CO_2 molecules. This figure shows that some of the proton-coupled multi-electron CO_2 reduction reactions are located between E_{CBM} and E_{VBM} of the semiconductors. When E_{CBM} is located at a more negative position than the CO_2 reduction potential, electron transfer from the conduction band to the CO_2 molecule is thermodynamically favorable, depending on the quality of the semiconductor as a catalyst for CO_2 reduction. Therefore, photo-electrochemical reduction of CO_2 on *p*-type semiconductors is feasible. For example, *p*-type GaP, and *p*-type SiC possess highly negative E_{CBM} positions, which is beneficial for CO_2 reduction, and photoelectrochemical CO_2 reduction over these semiconductors has been reported [2, 3].

To realize the CO_2 reduction reaction by utilizing water molecules as an electron donor and proton source as with photosynthesis in plants, E_g must be generally greater than the energy difference between the CO_2 reduction and oxidation potentials. In addition, E_{CBM} and E_{VBM} must straddle the CO_2 reduction and oxidation potentials, respectively. When the overpotentials available for both the reduction and oxidation reactions are insufficient, a combination of two semiconductors must be adopted to realize the so-called Z-scheme, which is a two-step photoexcitation. In this case, one semiconductor acts as a regular photoanode or photocathode, while the other semiconductor functions as a complementary photoelectrode. Metal electrodes are sometimes employed as complementary electrodes, for which an additional external electrical bias is necessary.

3 Photoelectrochemical CO_2 Reduction at Metal Cathode Connected to Semiconductor Photoanode

Figure 3 shows examples of possible semiconductor configurations for photoelectrochemical CO_2 reduction. In Fig. 3a, b, where a single semiconductor is used, a metal electrode is installed as a catalyst for the other half reaction. The configuration in Fig. 3a has an *n*-type semiconductor as a photoanode for water oxidation, while a metal electrode is adopted as the cathode for CO_2 reduction. The configuration in Fig. 3b has a *p*-type semiconductor as a photocathode for CO_2 reduction, while a metal electrode is adopted as a water oxidation anode. For the configuration shown in Fig. 3c, the energy of electrons in the two semiconductors is increased in two steps where E_{CBM} of the photocathode and E_{VBM} of the photoanode should straddle the CO_2 reduction potential and the water oxidation potential, respectively. Semiconductor p-n junctions that efficiently separate the photogenerated electrons and holes in the semiconductor are also applicable to photocathodes and photoanodes for configurations (a)–(c). In photoelectrochemical systems, an external electrical bias voltage can be applied between the two electrodes to assist the chemical reaction. An external chemical bias is also applicable by producing a pH difference between two compartments in a reactor with the two electrodes immersed in each compartment.

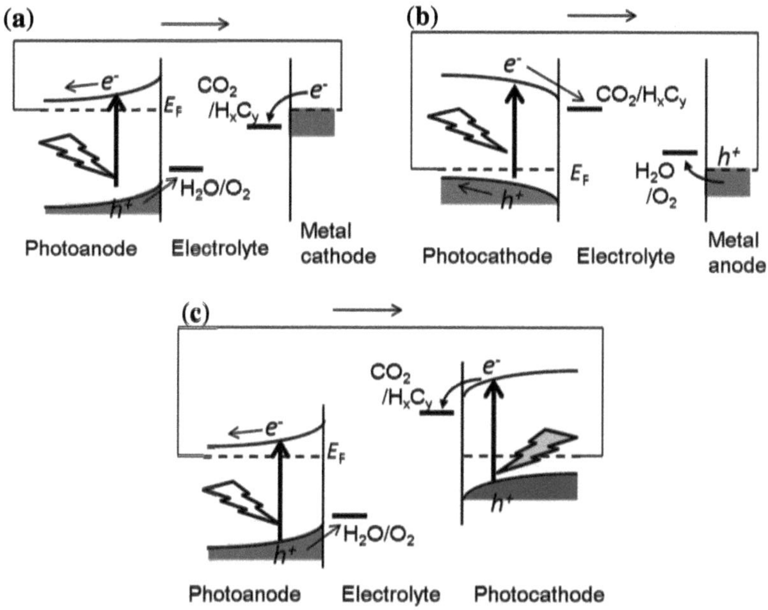

Fig. 3 Examples of possible configurations for photoelectrochemical CO_2 reduction using H_2O as an electron donor. **a** Combination of a semiconductor photoanode for water oxidation and a metal catalyst cathode for CO_2 reduction. **b** Combination of a semiconductor photocathode for CO_2 reduction and a metal catalyst for water oxidation. **c** Combination of a semiconductor photoanode for water oxidation and a photocathode for CO_2 reduction

Photoelectrochemical CO_2 reduction as shown in Fig. 3a utilizes metal catalyst cathodes such as Ag, Au, Cd, In, Pb, Sn, and Cu because they exhibit selective CO_2 reduction reaction and suppress proton reduction to produce hydrogen, even in aqueous solutions. Hori et al. extensively studied electrochemical CO_2 reduction over these metal electrodes at very high potentials between −1.39 and −1.66 V versus SCE, and reported that Cu is the most suitable metal catalyst for multi-electron reduction to produce useful products such as methane [4, 5]. To realize the CO_2 reduction reaction, a semiconductor photoanode that extracts electrons and protons from water molecules is necessary. TiO_2, WO_3 and Fe_2O_3 are well-known photoanodes for photoelectrochemical hydrogen production by water splitting, and they are also applicable to the CO_2 reduction system. However, very high additional external electrical and/or chemical biases that are much larger than those presented in Table 1 are necessary to facilitate CO_2 reduction reaction because the metal catalysts require a sufficiently high overpotential to reduce CO_2 molecules. Recently, it was reported that modification of metals is effective for decreasing the overpotential for CO_2 reduction [6]. The electronic structure of metals such as Cu and Au is influenced by grain boundaries formed in these polycrystalline electrodes, which alters the degree of adsorption and desorption of the CO_2 molecules and reaction intermediate species, so that electron transfer to CO_2 and reaction intermediate species is a possible reason for the reduced overpotential.

4 Photoelectrochemical CO_2 Reduction at Photocathodes

Systems for photoelectrochemical CO_2 reduction that utilize photoresponsive cathodes such as those shown in Fig. 3b, c can be categorized into 3 configurations with respect to the photoelectrode materials used, which are inorganic, organic, and inorganic-organic hybrid materials, as shown in Fig. 4a–d. Inorganic photoelectrodes made from the semiconductors shown in Fig. 4a are the most well known and have been extensively studied to date [2]. Photoelectrodes that exploit the technological advantages of both photoactivity in semiconductors and CO_2 reduction activity in inorganic metals or metal compound cocatalysts, as shown in Fig. 4b, have been developed [7]. Inorganic-organic hybrid systems with the advantages of both photoactivity from a semiconductor and very high selectivity for CO_2 reduction at a metal complex, as shown in Fig. 4c, have been recently proposed [8–11]. A photoelectrochemical system that uses a photoactive metal-complex catalyst, as shown in Fig. 4d, was proposed with the recent development of a CO_2 reduction photocatalyst that is active under visible light irradiation and also in the presence of water [12].

The advantage of the photoelectrochemical CO_2 reaction over electrochemical reactions in darkness is the lower overpotential required. Figure 5 shows an energy

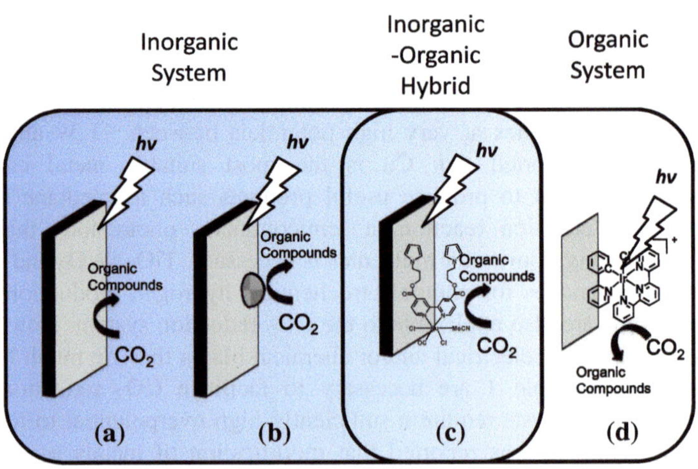

Fig. 4 Photoelectrochemical CO_2 reduction over various photocathodes that function in aqueous solutions. Photoinduced CO_2 reduction at a **a** semiconductor, **b** inorganic cocatalyst supported on a semiconductor, **c** metal-complex cocatalyst coated on a semiconductor, and **d** metal-complex photocatalyst supported on a conducting electrode

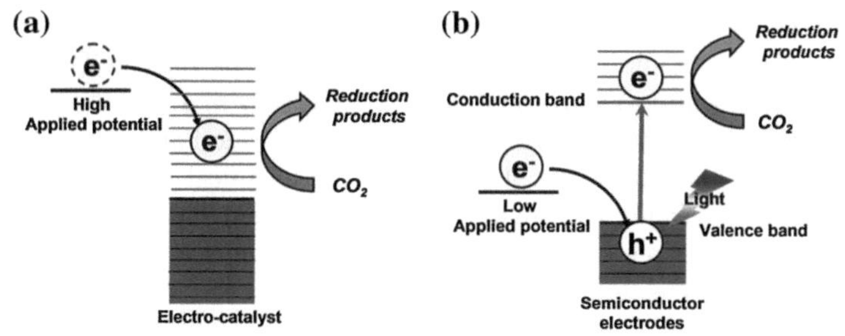

Fig. 5 Energy diagrams for electrochemical CO_2 reduction over, **a** metal cathodes as electrocatalysts, and **b** photocathodes as photocathodes that function in aqueous solution. Photocathodes such as semiconductor electrodes are advantageous because electron energy is increased at a lower applied potential with the assistance of photon energy than that with only an electrocatalyst

diagram for electrochemical CO_2 reduction over metal cathodes as electrocatalysts (a) and photocathodes as photocathodes (b) that operate in aqueous solution. Semiconductor photocathodes are advantageous because the electron energy is increased to the level of E_{CBM} at a low applied potential with the assistance of photon energy, compared with that for electrocatalysts. Figure 6 shows typical current-voltage characteristics for a metal cathode and a semiconductor photocathode measured during electrochemical and photoelectrochemical reduction of CO_2 in aqueous solution. A *p*-type InP electrode modified with Ru-complex catalyst (RCP/*p*-InP-Zn) is measured in the dark and under visible-light irradiation, in purified water saturated with CO_2 bubbling. Ag/AgCl and glassy carbon are used as reference and counter electrode, respectively. A Pyrex glass cell is used as a reactor, and a Xenon light source equipped with an optical filter ($\lambda > 400$ nm) and a cold mirror is used to irradiate visible light [11]. A cathodic reaction current is observed in darkness at potentials more negative than −0.8 V versus Ag/AgCl for the metal

Fig. 6 Technological advantage of CO_2 reduction by photoelectrochemistry over electrochemistry. External electrical bias voltage (∇) can be reduced to a potential near the valence band maximum E_{VBM} (▼) by photoirradiation

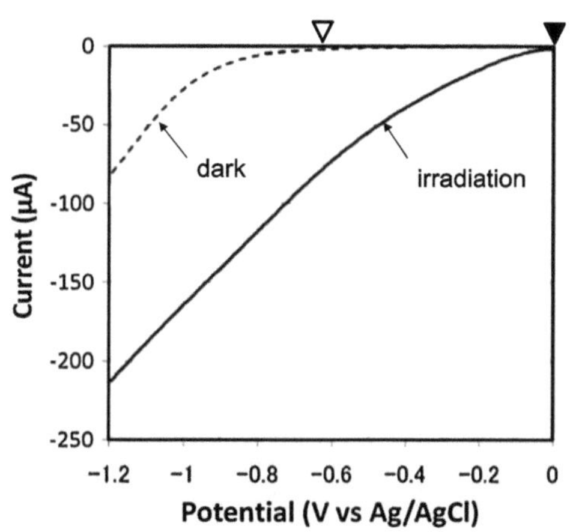

cathode in aqueous solution containing CO_2, while a cathodic photocurrent is observed at potentials more negative than −0.0 V versus Ag/AgCl. The threshold voltage of −0.0 V almost corresponds to the position of E_{VBM} for the semiconductor photocathode. The positive shift of the applied external electrical bias, which lowers the dissipation of electrical energy, is the technical advantage of the photoelectrochemical system. The amount of external bias decrease and the applicable solar energy spectral region are determined by the energy potential E_{CBM}, and E_{VBM} and by E_g of the semiconductor, respectively.

4.1 Photoelectrochemical CO_2 Reduction at Semiconductor Surfaces

Photoelectrochemical CO_2 reduction using a semiconductor material such as that shown in Fig. 3a was first reported by Halmann in 1978. Photoassisted electrolytic reduction of aqueous CO_2 was achieved using *p*-type GaP as a photocathode with part or all of the energy being supplied by light, as shown in Fig. 4a [2]. Honda and colleagues reported the photoelectrocatalytic reduction of CO_2 to form organic compounds in the presence of photosensitive semiconductors (TiO_2, ZnO, CdS, GaP, SiC and WO_3) as photocatalysts in 1979 [3]. Since then, the concept has been applied to many semiconductors; however, the reaction rates are generally low because of the poor surface properties of the CO_2 reduction catalysts.

4.2 Photoelectrochemical CO_2 Reduction Over Semiconductors Loaded with Inorganic Cocatalysts

As explained in Sect. 4.1, the photoelectrochemical CO_2 reduction over semiconductor photocathodes is feasible. However, these reactions are extremely slow on given semiconductor surfaces, which leads to a significant overpotential for CO_2 reduction. Apart from the high overpotential, these systems have a few advantages including sustainability (nothing is consumed in this system apart from light energy), direct conversion of solar energy to chemical energy, utilization of renewable energy resources for an energy intensive process, and the stability of the process (semiconductors are very stable under illumination). In cases where activity at a semiconductor surface is poor for a specific reaction, the corresponding photoelectrochemical reaction rate is very slow, irrespective of the level of applied potential and light intensity. Therefore, control of the semiconductor surface chemistry appropriate to each specific reaction is crucial in order to achieve high efficiency photoelectrochemical reaction. Loading of cocatalysts on the surfaces of photoelectrodes has been conducted, as shown in Fig. 4b, to enhance the catalytic activity. Metallic species such as Ag, Au, Cd, In, Sn, and Cu, and their complexes are potential candidate cocatalysts. The size or thickness of such cocatalysts is on the nanoscale, so that their resistivity are no more detrimental than when they are used non-photoactive cathodes, even in the case of resistive oxide cocatalysts.

4.3 *Photoelectrochemical CO_2 Reduction Over Semiconductors Loaded with Metal-Complex or Organic Cocatalysts*

Transition metal complexes are also at the forefront of research on potential catalysts that can have multiple and accessible redox states, which promote multi-electron transfer reactivity. Furthermore, the formal reduction potentials can be systematically tuned through ligand modification to match the potential required for CO_2 reduction. Most of the metal-complex catalysts for CO_2 reduction have poor activity under irradiation due to the short lifetime of the photoexcited state; however, they exhibit excellent selectivity for CO_2 molecules in electrochemical systems. It is essential that photoexcited electrons be transferred from the conduction band of the semiconductor to the metal-complex to promote selective CO_2 reduction on the complex. It is considered that E_{CBM} should be located at a more negative position than the CO_2 reduction potential (or lowest unoccupied molecular orbital (LUMO)) of the metal-complex catalyst to facilitate sufficient electron transfer to promote the reaction. In 2010, a breakthrough concept in the combination of metal complex electrocatalysts and semiconductor photosensitizers was demonstrated to realize high product selectivity. A Ru(II)-complex electrocatalyst ($[Ru(dcbpy)_2(CO)_2]^{2+}$; dcbpy = 2,2′-bipyridine-4,4′-dicarboxylic acid) linked with a *p*-type semiconductor (N-doped Ta_2O_5) exhibited selective CO_2 reduction to produce formate under visible light irradiation in a mixture of acetonitrile and triethanolamine containing CO_2 molecules [10]. Isotope tracer analysis confirmed that CO_2 was the carbon source for the formate and that triethanolamine was an electron donor and proton source for formate production. The concept of the organic/inorganic hybrid was applied to a photoelectrochemical system composed of InP and a Ru(II)-polymer catalyst ($[Ru(L\text{–}L)(CO)_2]_n$, in which L–L is a diimine ligand). In 2010, an organic/inorganic hybrid photocathode for CO_2 reduction in water, which consisted of InP connected with a Ru-complex polymer catalyst, was developed [11]. This system produces formic acid from CO_2 and H_2O under visible light irradiation with high selectivity under an electrical bias of −0.6 V (vs. Ag/AgCl). Isotope tracer analyses with $^{13}CO_2$ and D_2O verified that the carbon and proton sources for the formate were CO_2 and H_2O, respectively.

Another system employs photoelectrochemical CO_2 reduction to synthesize methanol and consists of a combination of a semiconductor electrode and a molecular catalyst that does not contain a metallic species [8]. In 2008, Bocarsly and colleagues developed a CO_2 reduction system consisting of a GaP photocathode and pyridinium ions dissolved in aqueous solution [8]. It is speculated that

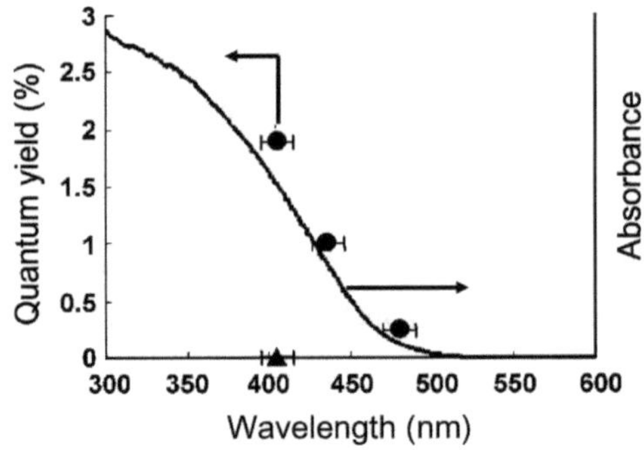

Fig. 7 Quantum efficiency of HCOOH generation and optical absorption of N-Ta_2O_5 (c) as a function of wavelength of incident light. Aliquots of acetonitrile/triethanolamine (5:1 v/v) containing [Ru-dcbpy]/N-Ta_2O_5 and [Ru-dcbpy] alone were irradiated under CO_2

redox shuttle with protons in pyridinium ions and conduction band electrons in visible light-excited GaP occur to realize the successive six-electron reduction of carbon dioxide to methanol in an acetate buffer under certain electrical bias. Reduction product selectivity is reported to range from 90 to 51 % under a low electric bias from −0.3 to −0.7 V (vs. SCE). Photoelectrochemical CO_2 reduction was achieved at wavelengths less than 480 nm, which almost corresponds to the bandgap of GaP.

In general, action spectra of the inorganic/organic systems, which are dependences of CO_2 reduction reaction on wavelength of light, correspond to photoresponses of semiconductors as shown in Fig. 7.

5 Photoelectrochemical CO_2 Reduction Over Photoactive Molecule Catalysts

A different approach for the photoreduction of CO_2 involves molecular photocatalysis, which operates in the absence of additional photosensitizers. Mononuclear iridium (III) terpyridine (tpy) 2-phenylpyridine (ppy) complex [Ir(tpy)(ppy)Cl]$^+$ ([Ir-ppy]), which selectively reduces CO_2 to CO under visible light irradiation without additional photosensitizers, was developed in 2013 [12]. The efficient visible light-induced photoreaction (quantum efficiency of 21 % at 480 nm) achieved with [Ir(tpy)(Meppy)Cl]$^+$, which possesses electron-donating methyl groups on ppy ligands, is due to strong spin-orbit coupling that coordinates with singlet and triplet excited states. The Ir-complex photocatalyst functions in solutions containing water, so that photoelectrochemical CO_2 reduction at the photocathode occurs when it is connected with a conductive electrode (Fig. 4d). Electrons that compensate holes in the photoexcited complex are supplied by the electrode, which enables consecutive photoreactions.

6 Z-Scheme Reaction for Photoelectrochemical CO_2 Reduction

The systems for photoelectrochemical CO_2 reduction introduced in the previous sections are designed for half-reactions. When a photoanode capable of CO_2 reduction is functionally coupled with a photocathode capable of H_2O oxidation under solar irradiation, then artificial photosynthesis, solar-driven CO_2 reduction that uses H_2O as both an electron donor and a proton source, by the Z-scheme mechanism (or two-step photoexcitation) shown in Fig. 8 is demonstrated. An example of a general setup for photoelectrochemical CO_2 reduction is schematically illustrated in Fig. 9a. A photocathode composed of InP coated with a Ru-complex-polymer for CO_2 reduction is connected to TiO_2 via a potentiostat, and immersed in a two-compartment reactor separated by a proton exchange membrane filled with an aqueous solution containing CO_2 and an electrolyte. A solar simulator equipped with an air mass 1.5 (AM 1.5) filter adjusted to a power of 1 Sun irradiates the system from the TiO_2 side, while the InP/Ru-polymer photocathode is irradiated with light transmitted through the TiO_2 photocatalyst electrode and proton exchange membrane. A photoreaction current flows between the photoanode and the photocathode, originating from electrons extracted from water molecules that react mainly with CO_2 molecules. To facilitate the charge transport from the photoanode to photocathode with no external electrical bias, the E_{VBM} of the photocatalyst for water oxidation must be more positive than the potential for water oxidation (theoretically 1.23 V versus NHE), at least thermodynamically, and the E_{CBM} of the photoanode must be more negative than the E_{VBM} of the photocathode. Figure 9b shows the photocurrent originating from CO_2 reduction reaction at applied potentials in the range from +0.5 to -1.0 V. Here, the most significant point is that the system operates at zero bias, which means that it is self-sustaining [13].

These results verified that CO_2 was reduced to formate by electrons extracted from H_2O during the oxidation process to O_2, and that protons also originated from H_2O, as shown by Eqs. 1 and 2:

$$H_2O + hv \rightarrow 1/2O_2 + 2H^+ + 2e^- \quad (1)$$

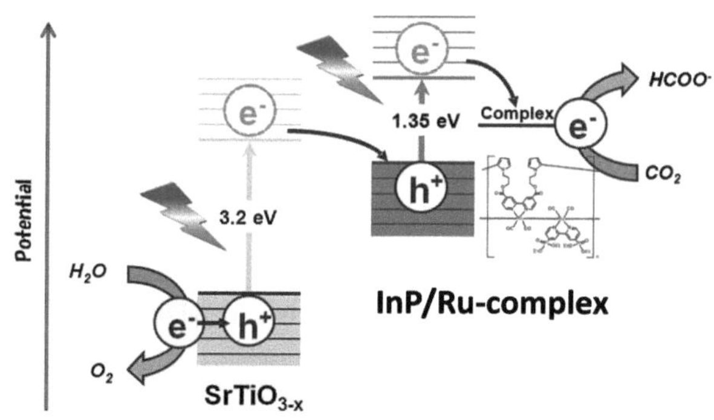

Fig. 8 Z-scheme reaction for solar-driven CO_2 reduction using H_2O as an electron donor and a proton source

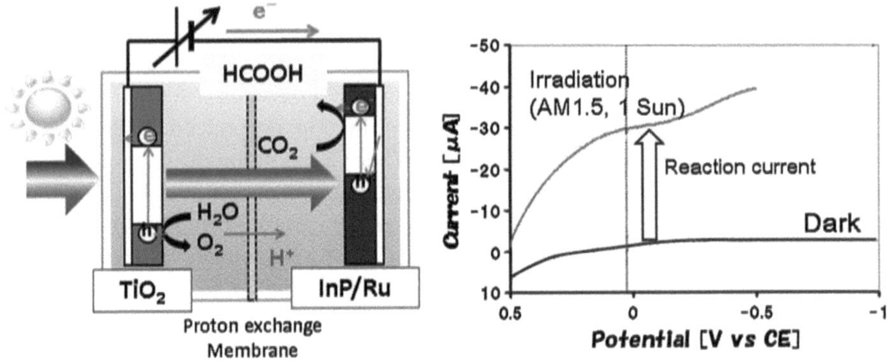

Fig. 9 Photoelectrochemical CO_2 reduction under simulated solar light irradiation using a tandem setup of photoelectrodes for CO_2 reduction and H_2O oxidation

$$CO_2 + 2H^+ + 2e^- + hv \rightarrow HCOOH \tag{2}$$

The band alignment of semiconductors is important for improvement of the reaction current. Photocurrent flowing between an InP/Ru-polymer photocathode and various oxide photoanodes in tandem setups are shown in Fig. 10. In the case of the WO_3 photoanode, no photocurrent was observed, because E_{CBM} of the WO_3 photoanode does not match E_{VBM} of InP, while for the TiO_2 photoanodes, photocurrent flows between the TiO_2 and InP. These results correspond to the energy difference between E_{CBM} of the photoanode and E_{VBM} of the InP photocathode.

A monolithic device is also feasible for solar-driven CO_2 reduction using water molecules due to its simplicity and lower impedance losses from electrical wiring. It is composed of a photoanode and a photocathode with a direct electrical connection. In this case, selective water oxidation and CO_2 reduction at each reaction site

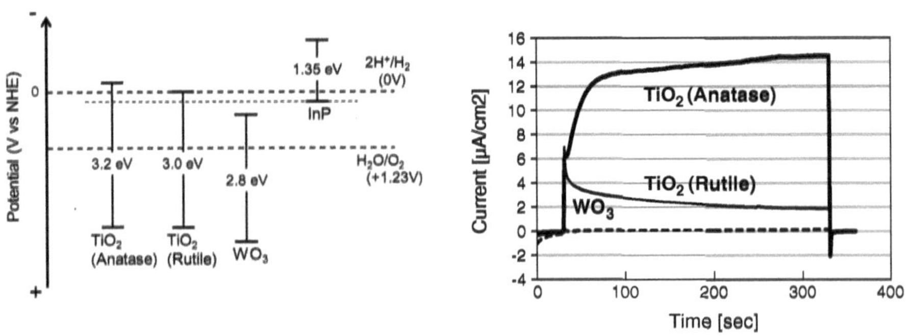

Fig. 10 Photocurrent flowing between InP/Ru-polymer photocathode and various oxide photoanodes in tandem setups

is necessary in order to realize long-term reaction in a one-compartment reactor with a separation membrane for reactants (water and CO_2) and products (oxygen and hydrocarbons) [14]. No separation is necessary at separate electrodes.

7 Selection of Cocatalysts for CO_2 Reduction

Carbon-based substances produced from photoelectrochemical CO_2 reduction are determined by the catalytic nature of the reaction sites at the semiconductors, cocatalysts, and photoactive complexes. Detailed analyses of electrochemical CO_2 reduction over metallic electrodes in aqueous solution have been conducted by Hori et al. [4, 5]. Electrochemical CO_2 reduction has also been reported over metal complexes [15–17]. The potentials necessary for photoelectrochemical CO_2 reduction over semiconductors and metal complexes [3, 11, 12, 15] are generally determined by the E_{CBM} and LUMO levels. In metallic catalysts, the main products are carbon monoxide or formic acid generated by the two-electron reduction of CO_2, while methane and C2 chemicals such as ethane can be produced over a Cu electrode by eight-electron reduction. To realize the multi-electron reactions to obtain more useful organic compounds in the photoactive systems, integration of these cocatalysts is necessary. In metal-complex catalysis systems available at the present, the products are carbon monoxide or formic acid. External electrical biases of over −1.0 V and −0.6 V versus NHE (pH = 7) are necessary for metallic catalysts and metal-complex catalysts, respectively. Modified crystalline electrodes of Au and Cu lower the CO_2 reduction potential. Thus, the modification of electronic structure at metallic ions by organic ligands and by the oxidative nature of grain boundaries is crucial in order to determine the CO_2 reduction potential and selectivity of products.

8 Isotope Tracer Analysis and Verification of Stoichiometric Reaction

There have been many reports on photocatalytic CO_2 reduction over TiO_2 loaded with metallic species, in which C2 and C3 chemicals were detected. However, because the amount of products was very small compared with the amount of catalyst in lots of papers, the results sometimes raise a controversy.

Mul and colleagues raised an alert that many research results reported to date may contain possible experimental errors, where surface contaminants were mistaken as products of the CO_2 reduction reaction [18]. Ishitani and coworkers also insisted that care should be taken with respect to surface contaminants when the amount of products is lower than that of the catalysts [19]. Therefore, long-term reactions help to avoid such misunderstandings. In this case, the turnover number calculated from the amount of carbon species produced divided by the number of active catalytic sites must be much larger than 1. To evaluate CO_2 reduction reactions, isotope tracer analyses (ion chromatography-time-of-flight mass spectrometry; IC-TOFMS) using $^{13}CO_2$ and D_2O are very useful. An example of such an analysis using $^{13}CO_2$ is shown in Fig. 11a. In this experiment, $^{12}CO_2$ was bubbled in an aqueous solution and formate with a mass number (m/z) of 45 ($H^{12}COO^-$) was detected. When $^{13}CO_2$ was bubbled in the solution, formate with a mass number (m/z) of 46 ($H^{13}COO^-$) was detected. These results indicate that the carbon source for formate generation is the CO_2 molecules dissolved in solution. To evaluate the CO_2 reduction reaction using H_2O as a proton source, isotope tracer analyses are conducted using $^{13}CO_2$ and D_2O. Furthermore, to verify that the electron source for CO_2 reduction is H_2O, isotope tracer analyses (gas chromatography-mass spectrometry; GC-MS) are conducted using $H_2^{18}O$ and measurement of the stoichiometric reaction from the ratios of products by oxidation (such as O_2) and reduction (such as formate or CO). An example of an analysis

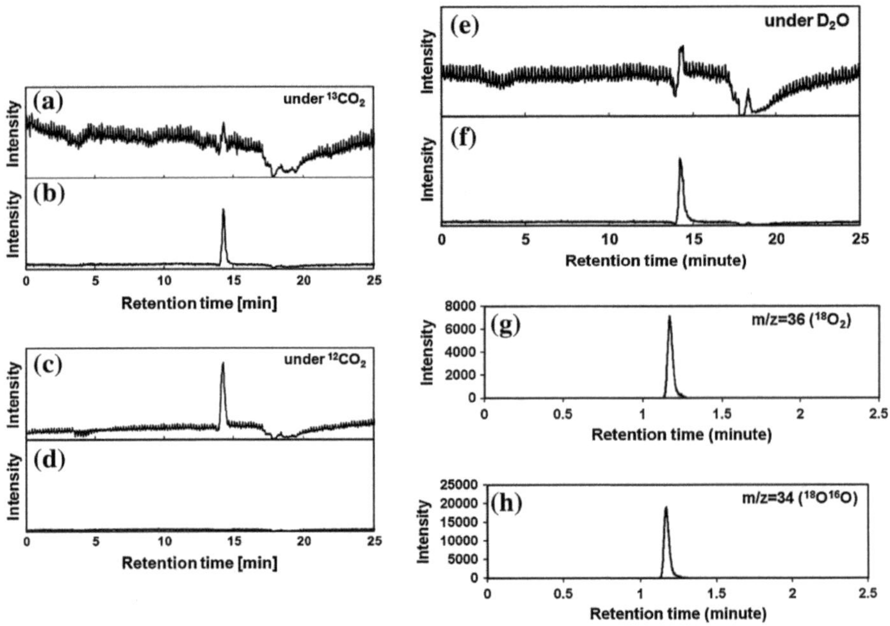

Fig. 11 IC-TOFMS spectra from a tracer analysis utilizing **a**, **b** $^{13}CO_2$ and **c**, **d** $^{12}CO_2$. The photoelectrochemical reaction utilizing the Z-scheme system was conducted in a closed system, purged with $^{13}CO_2$ (**a**) [12]. **e**, **f** IC-TOFMS spectra from a tracer analysis utilizing D_2O. The photoelectrochemical reaction utilizing the Z-scheme system was conducted using 10 mM $NaHCO_3$ solution containing D_2O as an electrolyte solution. GC-MS spectra from tracer analysis utilizing 0.1 M $NaHCO_3$ aqueous solution containing phosphoric acid and 25 % $H_2^{18}O$ (**g**, **h**). The spectra were calculated by subtracting the signal for air from the signal of the sample, corrected after the photocatalytic reaction in a $SrTiO_{3-x}$//InP/[Ru-complex polymer] Z-scheme system [13]

using $H_2^{18}O$ on a TiO_2 photoanode is shown in Fig. 11b, in which both $^{18}O_2$ and $^{16}O^{18}O$ was observed because lattice oxygen atoms are also involved in the water oxidation reaction at the primary stage of the overall reaction.

9 Future Prospects

One of the promising and useful goals of photoelectrochemical CO_2 reduction is realization of highly efficient direct hydrocarbon generation from CO_2 and H_2O similar to photosynthesis in plants. Recently, solar-driven CO_2 reduction using water molecules as both an electron donor and a proton source was realized. Solar conversion efficiency was demonstrated and the feasibility of the photoelectrochemical CO_2 reduction under sunlight irradiation at ambient pressure and temperature was verified. To further improve the system, efficient catalytic reaction, selectivity toward CO_2 conversion, efficient solar absorption, and efficient charge transfer are necessary. The basic construction of these systems is similar to that for photoelectrochemical hydrogen generation by water splitting. However, from a practical perspective, there are some differences with the CO_2 reduction system, such as the methods for product separation and CO_2 capture. Therefore, the feasibility of practical systems for solar fuel generation will have to be discussed further.

10 Conclusions

The impact of photoelectrochemical CO_2 reduction using water as an electron donor and a proton source is the ability to recycle CO_2 molecules that are generated from the combustion of fossil fuels. Although the photoelectrochemical CO_2 reduction is attractive due to possibility of direct solar-driven hydrocarbon generation using water and CO_2, similar to photosynthesis in plants, it has long been considered to be much more difficult than hydrogen generation by water splitting. The key technology is selective CO_2 reduction, which competes with proton reduction that generates hydrogen. Recently, some interesting concepts have been proposed: highly selective CO_2 reduction by a hybrid photoelectrode of semiconductor and a metal-complex catalyst, redox shuttle using a combination of semiconductor and molecular proton mediator, and low overpotential by a redox-treated metal-electrode. Therefore, rapid progress is expected in this field.

References

1. Schwarz HA, Dodson RW (1989) Reduction potentials of CO_2^{-} and the alcohol radicals. J Phys Chem 93:409–414
2. Halmann M (1978) Photoelectrochemical reduction of aqueous carbon dioxide on p-type gallium phosphide in liquid junction solar cells. Nature 275:115
3. Inoue T, Fujishima A, Konishi S, Honda K (1979) Photoelectrocatalytic reduction of carbon dioxide in aqueous suspensions of semiconductor powders. Nature 238:637–638
4. Hori Y, Kikuchi K, Murata A, Suzuki S (1986) Production of methane and ethylene in electrochemical reduction of carbon dioxide at copper electrode in aqueous hydrogencarbonate solution. Chem Lett 6:897–898
5. Hori Y, Murata A, Takahashi R, Suzuki S (1987) Electroreduction of carbon monoxide to methane and ethylene at a copper electrode in aqueous solutions at ambient temperature and pressure. J Am Chem Soc 109:5022–5023
6. Christina LW, Kanan MW (2012) CO_2 reduction at low overpotential on cu electrodes resulting from the reduction of thick Cu_2O films. J Am Chem Soc 134:7231–7234
7. Nakato Y, Tsubomura H (1992) Silicon photoelectrodes modified with ultrafine metal islands. Electrochim Acta 37:897–907
8. Kumar B, Smieja JM, Kubiak CP (2010) Photoreduction of CO_2 on p-type silicon using Re (bipy-Bu^t)$(CO)_3Cl$: photovoltages exceeding 600 mV for the selective reduction of CO_2 to CO. J Phys Chem C 114:14220–14223
9. Barton EE, Rampulla DM, Bocarsly AB (2008) Selective solar-driven reduction of CO_2 to methanol using a catalyzed p-GaP based photoelectrochemical cell. J Am Chem Soc 130:6342–6344
10. Sato S, Morikawa T, Saeki S, Kajino T, Motohiro T (2010) Visible-light-induced selective CO_2 reduction utilizing a ruthenium complex electrocatalyst linked to a p-type nitrogen-doped Ta_2O_5 semiconductor. Angew Chem Int Ed 49:5101–5105
11. Arai T, Sato S, Uemura K, Morikawa T, Kajino T, Motohiro T (2010) Photoelectrochemical reduction of CO_2 in water under visible-light irradiation by a p-type InP photocathode modified with an electropolymerized ruthenium complex. Chem Commun 46:6944–6946
12. Sato S, Morikawa T, Kajino T, Ishitani O (2013) A highly efficient mononuclear iridium complex photocatalyst for CO_2 reduction under visible light. Angew Chem Int Ed 125:1022–1026
13. Sato S, Arai T, Morikawa T, Uemura K, Suzuki TM, Tanaka H, Kajino T (2011) Selective CO_2 conversion to formate conjugated with H_2O oxidation utilizing semiconductor/complex hybrid photocatalysts. J Am Chem Soc 133:15240–15243
14. Arai T, Sato S, Kajino T, Morikawa T (2013) Solar CO_2 reduction using H_2O by a semiconductor/metal-complex hybrid photocatalyst: enhanced efficiency and demonstration of a wireless system using $SrTiO_3$ photoanodes. Energy Environ Sci 6:1274–1282
15. Lehn JM, Ziessel R (1982) Photochemical generation of carbon monoxide and hydrogen by reduction of carbon dioxide and water under visible light irradiation. Proc Natl Acad Sci 79:701–704
16. Ishida H, Tanaka K, Tanaka T (1987) Electrochemical CO_2 reduction catalyzed by ruthenium complexes $[Ru(bpy)_2(CO)_2]^{2+}$ and $[Ru(bpy)_2(CO)Cl]^{+}$: effect of pH on the formation of CO and $HCOO^{-}$. Organometallics 6:181–186
17. Chardon-Noblat S, Deronzier A, Ziessel R, Zsoldos D (1998) Electroreduction of CO_2 catalyzed by polymeric $[Ru(bpy)(CO)_2]$. films in aqueous media: parameters influencing the reaction selectivity. J Electroanal Chem 444:253–260
18. Yang CC, Yu YH, Linden B, Wu JCS, Mul G (2010) Artificial photosynthesis over crystalline TiO_2-based catalysts: fact or fiction? J Am Chem Soc 132:8398–8406
19. Yui T, Kan A, Saitoh C, Koike K, Ibusuki T, Ishitani O (2011) Photochemical reduction of CO_2 using TiO_2: effects of organic adsorbates on TiO_2 and deposition of Pd onto TiO_2. ACS Appl Mater Interfaces 3:2594–2600

Part VII
Chemical, Electrochemical and Photoelectrochemical Approach for Energy Conversion: Approach Using Photocatalysts

Semiconductor-Based Photocatalytic Water Splitting

Fuxiang Zhang and Can Li

Abstract Solar to chemical energy conversion from water by powdered photocatalyst is one of the most promising approaches. In this chapter, we will introduce some bases of photocatalytic water splitting, and key issues and challenges for solar water splitting. At the same time, the basic mechanism, processes, reaction systems as well as strategies for light absorption, charge separation and catalytic conversion will be summarized and discussed.

Keywords Powder photocatalyst · Water splitting · Hydrogen · Mechanism · Strategies

1 Introduction

Human beings have been exploiting a clean, renewable and a viable alternative to fossil fuels in order to address the depletion of fossil fuels and the serious environmental problems accompanying their combustion. Hydrogen is an ultimate clean energy, which can be used in fuel cells and chemical industries (i.e., industrial ammonia synthesis, hydrogenation of carbon dioxide to produce methanol). Currently, hydrogen is mainly produced from fossil fuels such as natural gas by steam reforming, but it suffers from consumption of fossil fuels and emission of CO_2 [1].

$$CH_4 + H_2O \rightarrow CO + 3H_2 \quad (1)$$

$$CO + H_2O \rightarrow CO_2 + H_2 \quad (2)$$

F. Zhang · C. Li (✉)
State Key Laboratory of Catalysis, Dalian Institute of Chemical Physics,
Chinese Academy of Sciences, Dalian National Laboratory for Clean Energy,
Dalian 116023, China
e-mail: canli@dicp.ac.cn

M. Sugiyama et al. (eds.), *Solar to Chemical Energy Conversion*,
Lecture Notes in Energy 32, DOI 10.1007/978-3-319-25400-5_18

Considering energy and environmental issues, achievement of hydrogen from water using natural energies such as sunlight is highly desirable. This can be generally called as solar hydrogen production, and whose typical ways include: (i) Electrolysis of water using a solar cell, a hydroelectric power generation, etc. (ii) Photocatalytic water splitting (artificial photosynthesis). (iii) Photoelectrochemical water splitting.

The advantage of water splitting using a powdered photocatalyst is the simplicity as shown in Fig. 1, where hydrogen will be readily obtained once photocatalyst powders dispersed in a pool with water are illuminated by sunlight. Because of the simplicity, the powdered photocatalytic system is expected for large-scale application of solar water splitting. The photocatalytic water splitting for hydrogen production is very similar to natural photosynthesis, as it can convert the photon energy to chemical energy but accompanied with a largely positive change in the Gibbs free energy, because these are uphill reactions. Thus, photocatalytic water splitting is regarded as an artificial photosynthesis and is an attractive and challenging theme (called as "holy grail") in the field of chemistry.

The research on photocatalytic water splitting was triggered by the Honda-Fujishima effect of water splitting using a TiO_2 electrode in the early 1970s [2]. Subsequently, extensive development of the necessary semiconductor photocatalysts has undergone considerable research. In the past 40 years, over a hundred semiconductors have been reported to be promising photocatalysts for solar hydrogen production, but up to date just a few of them have photocatalytic activities of splitting water into a stoichiometric mixture of H_2 and O_2 (2:1 by molar ratio). Moreover, most of them are only active in the ultraviolet (UV) light region, and the visible-light-responsive materials are still limited. In this section, we aim to give a fundamental introduction to photocatalytic water splitting and to summarize the research effort having been made so far, with a view of providing a good reference and inspiring new ideas for tackling this important challenge. Starting with a brief introduction to solar hydrogen production, we will introduce some bases of photocatalytic water splitting, and key issues and challenges for solar water splitting.

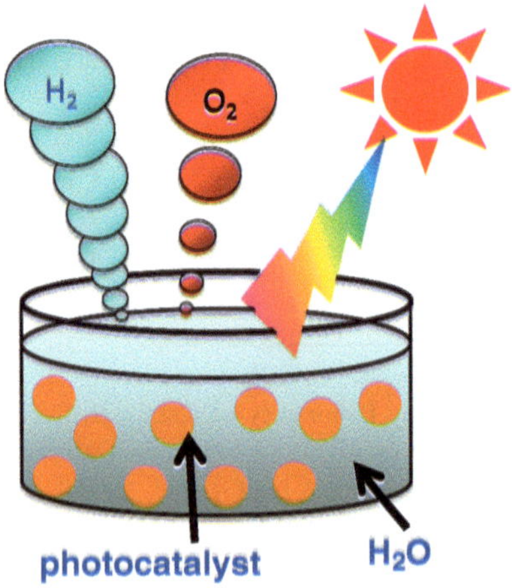

Fig. 1 Solar hydrogen production from water using a powdered photocatalyst

Meanwhile, the basic mechanism, processes, reaction systems as well as strategies for light absorption, charge separation and catalytic conversion will be summarized and discussed.

2 Bases of Photocatalytic Water Splitting

2.1 Fundamental Mechanism of Photocatalytic Hydrogen Generation

Thermodynamically, the overall water splitting reaction is an uphill reaction with a standard Gibbs free energy change ΔG^0 of 237 kJ/mol or 1.23 eV, as shown in Eq. 3.

$$H_2O \rightarrow 1/2O_2 + H_2; \Delta G^\theta = +237\ \text{kJ/mol} \quad (3)$$

A schematic illustration on the fundamental mechanism of overall water splitting on a powdered photocatalyst is given in Fig. 2. Typically, once a photocatalyst absorbs sunlight with an energy equivalent to or greater than its band gap, the electrons in the valence band of the photocatalyst will be excited to the conduction band accompanying with the holes left in the valence band, creating the negative-electron (e^-) and positive-hole (h^+) pairs. This stage is referred to the semiconductor's "photo-excited" state, and these photo-excited electrons and holes are subsequently transferred to surface of the photocatalyst for corresponding reduction and oxidation reactions, respectively. To achieve overall water splitting, the match of the band gap and the potentials of the conduction and valence bands are important. It is necessary that the bottoms of the conduction bands are more negative than the reduction potential of H^+ to H_2 (0 V versus NHE at pH 0), while the tops of the valence bands are more positive than the oxidation potential of H_2O to O_2 (1.23 V versus NHE). Theoretically, the minimum photon energy

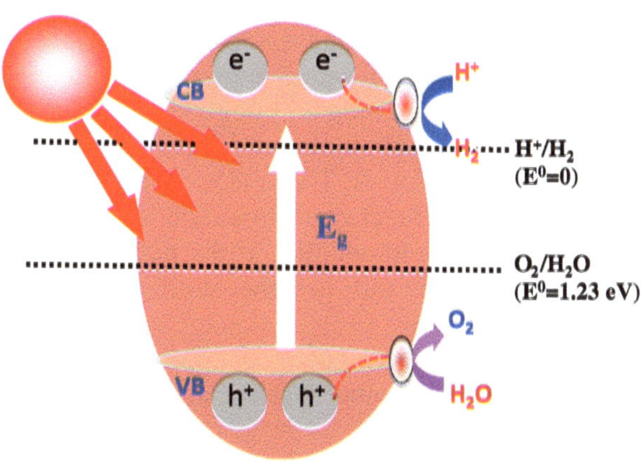

Fig. 2 A schematic illustration on the fundamental mechanism of overall water splitting on a powdered photocatalyst

thermodynamically required to drive the reaction is 1.23 eV, corresponding to a wavelength of ca. 1000 nm according to the following calculating formula.

$$\text{Band gap (eV)} = 1240/\lambda(\text{nm}) \quad (4)$$

It should be mentioned, however, that from the viewpoint of real reaction, a photon energy greater than the band gap of the photocatalyst is required to drive the overall water splitting reaction at reasonable reaction rates because of the factors such as overpotentials, charge separation, mobility, and activation barrier between photocatalysts and water molecules. The evaluated value of bandgap should be higher than 1.8 eV. In addition, the backward reaction, that is, water formation from H_2 and O_2, should be inhibited, and the photocatalysts themselves must be stable during the reaction.

2.2 *Main Processes of Photocatalytic Water Splitting*

According to the mechanism illustrated in Fig. 2, the photocatalytic processes are known to be mainly composed of three steps: (i) absorption of photons to form electron-hole pairs; (ii) charge separation and migration of photogenerated carriers; (iii) the surface chemical reactions. All of these processes integrally affect the final generation of hydrogen from water. If we define the above each step by conversion efficiency, the total conversion efficiency of solar energy will be calculated as follows.

$$\eta_{conv.} = \eta_{abs.} \times \eta_{sep.} \times \eta_{react.} \times 100\,\% \quad (5)$$

$\eta_{conv.}$: conversion efficiency of solar energy; $\eta_{abs.}$: efficiency of light absorption; $\eta_{sep.}$: efficiency of charge separation; $\eta_{react.}$: efficiency of surface reaction.

In order to improve the conversion efficiency of solar energy, the semiconductor photocatalyst should first have a relatively narrow band gap to absorb as much light as possible, and reflection or scattering of light by the photocatalyst should be minimized. Secondly, using the absorbed photons, the semiconductor photocatalyst should have a high efficiency in generating excited charges, instead of generating phonons or heat. The step is strongly affected by crystal structure, crystallinity and particle size of the photocatalyst. The higher the crystalline quality is, the smaller the amount of defects is. The defects commonly act as trapping and recombination centers of photogenerated electrons and holes, resulting in a decreased efficiency of charge separation. In addition, if the particle size becomes small, the distance of charge separation will become short, leading to a decreased recombination probability.

Thirdly, the important points for surface chemical reactions are surface character (active sites) and quantity (surface area). If the active sites for redox reactions do not exist on the surface, the photogenerated electrons and holes with thermodynamically sufficient potentials for water splitting will have to recombine with each other. In this case, loading of co-catalysts is usually necessary to introduce active sites. As for H_2 evolution, the conduction band levels of many photocatalysts are not efficient enough to reduce water to produce H_2 without assistance of cocatalyst. Comparatively, co-catalyst loading is not always necessary for water oxidation, even though it is considered as more challenging reaction involving 4-electron transfer. One of reasons may lie in the valence band of most photocatalysts is deep enough to oxidize water to form O_2. This is the characteristic point of heterogeneous photocatalysts being different from homogeneous photocatalysts.

2.3 Experimental Methods for Water Splitting

The efficiency of photocatalytic hydrogen generation from water splitting is normally evaluated by using a gas-closed circulation system, which is composed of a vacuum line, a reaction cell and a gas sampling port connected directly to a gas chromatograph as shown in Fig. 3. If a photocatalytic activity is too high to use a gas chromatograph, a volumetric method is employed for determination of evolved gases. The apparatus should be air-free because the detection of O_2 is very important for evaluation of photocatalytic water splitting. The reactors can be divided into inner irradiation reaction cells with a high-pressure mercury lamp as light source, and outer irradiation reaction cell normally with Xe-lamp as light source.

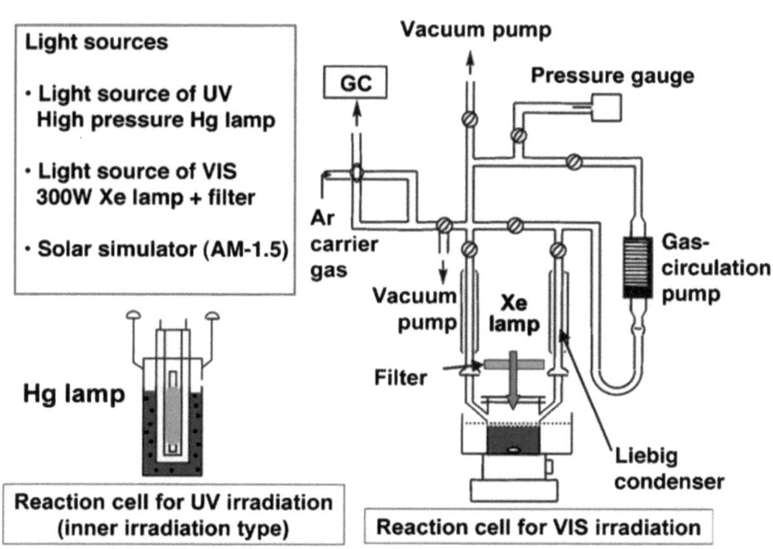

Fig. 3 An example of the experimental setup for photocatalytic water splitting. Reproduced from Ref. [1] with permission from the Royal Society of Chemistry

Depending on the requirement of tests, different cut-off filters and mirrors are usually employed. In the case of evaluation of solar hydrogen production, a solar simulator with an air-mass 1.5 filter (AM-1.5) and energy intensity of 100 mW/cm^2 should ideally be used.

The rate of gas evolution is usually indicated with a unit, for example μmol/h. Normalization of photocatalytic activity by weight or surface areas of used photocatalyst (for example, μmol.h^{-1}.g^{-1}, μmol.h^{-1}.m^{-2}) is not acceptable for water splitting because the activity of photocatalytic water splitting is not usually proportional to the weight or surface areas of photocatalyst. The amount of photocatalyst should be optimized for corresponding experimental setup. However, photocatalytic activity is dependent on the number of photons absorbed by the photocatalyst within a certain amount of light intensity. It also depends on experimental conditions such as a light source and a type of a reaction cell, so the photocatalytic performances should be given accompanying with the detailed experimental conditions containing amount of photocatalyst, volume of water, light source and intensity, and concentration of scavenger etc. Even so, the rate of gas evolution is too coarse to be compared with each other among different laboratories in consideration of the differences of reaction conditions. In this case, determination of a quantum yield is recommended for comparison among different laboratories.

The number of incident photons can be measured using a thermopile or Si photodiode. However, it is hard to determine the real amount of photons absorbed by a photocatalyst in a dispersed system because of scattering. Accordingly, the obtained quantum yield is an apparent quantum yield. The apparent quantum yield is estimated to be smaller than the real quantum yield because the number of absorbed photons is usually smaller than that of incident light. The overall quantum yield and apparent quantum yield are defined by Eqs. 6 and 7, respectively. In addition to the quantum yield, the conversion efficiency of solar energy that is usually used for evaluation of solar cells is also sometimes reported in the literature. It is defined as

$$\text{Overall quantum yield } (\%) = \text{Number of reacted electrons}/\text{Number of absorbed photons} \times 100\,\% \tag{6}$$

$$\begin{aligned}
&\text{Apparent Quantum yield } (\text{QY}, \%) \\
&\quad = \text{Number of reacted electrons}/\text{Number of incident photons} \times 100\,\% \\
&\quad = 2 \times \text{Number of evolved } H_2 \text{ molecules}/\text{Number of incident photons} \\
&\qquad\qquad \times 100\,\% (\text{for } H_2 \text{ evolution}) \\
&\quad = 4 \times \text{Number of evolved } O_2 \text{ molecules}/\text{Number of incident photons} \\
&\qquad\qquad \times 100\,\% (\text{for } O_2 \text{ evolution})
\end{aligned} \tag{7}$$

The conversion efficiency of solar energy on the current photocatalyst is mostly limited because of their insufficient activities, but it should finally be used to evaluate the photocatalytic water splitting if solar hydrogen production is considered.

2.4 Photocatalytic Systems

According to the basic principle of photocatalytic water splitting depicted in Fig. 2, the photogenerated electrons in the conduction band are used for water reduction to produce hydrogen, and the photogenerated holes can oxidize the reducing electron donors or H_2O. Strictly speaking, it is not really water splitting if the electron donors are not from water. However, it has been widely adopted for evaluation of photocatalysts and hydrogen generation by using hole scavengers. The typical electron donors are summarized in Table 1. According to the differences of electron donors, the photocatalytic hydrogen generation systems are correspondingly

Table 1 Typical sacrificial reagent contained photocatalytic hydrogen production system

Photocatalyst	Mass (g)	Sacrificial reagent	Light source	Hydrogen evolution rate (μmol. h^{-1})	Quantum yield (%)	Reference
Pt-PdS/CdS	0.3	SO_3^{2-}/S^{2-}	300 W Xe (>420 nm)	8770	93 @ 420 nm	[3]
Pt-PdS/CdS	0.025	H_2S	300 W Xe (>420 nm)	2350	30 @ 420 nm	[4]
MoS_2/CdS	0.1	Lactic acid	300 W Xe (>420 nm)	530		[5]
RuO_2/ $K_2La_2Ti_3O_{10}$:Zn	1	I^-	250 W Xe	55.5		[6]
Pt/TiO_2	0.1	Br^-	500 W Hg	~55		[7]
$KBi_3PbTi_5O_{16}$	1	Ce^{3+}	450 W Hg	35.21		[8]
Pt/$SrTiO_3$:Rh	0.1	Fe^{2+}	Xe 300 W (420 nm)	16.9		[9]
NiO/TiO_2	0.2	CN^-	450 W Hg	300	0.32	[10]
Pt/$H_2LaNb_2O_7$:In	1	Methanol	100 W Hg (290 nm)	5268	1.54 @ 290 nm	[11]
Pt-Ru/$Y_2Ta_2O_5N_2$	0.3	Ethanol	300 W Xe (420 nm)	250		[12]
NiO*x*/ CdS/$KNbO_3$	0.2	Isopropanol	500 W Hg–Xe (400 nm)	40.7	8.8 (400 nm)	[13]
Pt/TiO_2	0.08	1-propanol	500 W Xe	~12		[14]
Pt/TiO_2	0.02	Oxalic acid	250 W Hg	56		[15]

denoted as follows: (i) inorganic sacrificial reagent systems; (ii) organic sacrificial reagent system; (iii) pure water splitting system; (iv) biomimetic Z-scheme water splitting system.

The mostly studied inorganic sacrificial reagent systems contain commonly S^{2-}/$SO_3{}^{2-}$ sacrificial reagents or direct splitting of H_2S. Take the former as example, the reaction mechanism using S^{2-}/$SO_3{}^{2-}$ sacrificial reagents can be described by Eqs. 8–13.

$$\text{photocatalyst} + h\upsilon \rightarrow h^{+} + e^{-} \tag{8}$$

$$2e^{-} + 2H_2O \rightarrow H_2 + 2OH^{-} \tag{9}$$

$$SO_3^{2-} + 2OH^{-} + 2h^{+} \rightarrow SO_4^{2-} + 2H^{+} \tag{10}$$

$$2S^{2-} + 2h^{+} \rightarrow S_2^{2-} \tag{11}$$

$$S_2^{2-} + SO_3^{2-} \rightarrow S_2O_3^{2-} + S^{2-} \tag{12}$$

$$SO_3^{2-} + S^{2-} + 2h^{+} \rightarrow S_2O_3^{2-} \tag{13}$$

Organic compounds, such as alcohols (methanol, ethanol, isopropanol, etc.), organic acids (formic acid, acetic acid, etc.), and aldehydes (formaldehyde, acetaldehyde, etc.) have also been used as electron donors for photocatalytic hydrogen generation. Among them, methanol has been the most extensively studied, and the corresponding hydrogen generation processes can be described by Eqs. 14–18.

$$H^{+} + H_2O \rightarrow \cdot OH + H^{+} \tag{14}$$

$$CH_3OH + \cdot OH \rightarrow \cdot CH_2OH + H_2O \tag{15}$$

$$\cdot CH_2OH \rightarrow HCHO + H^{+} + e^{-} \tag{16}$$

$$2H_2O + 2e^{-} \rightarrow H_2 + 2OH^{-} \tag{17}$$

$$\text{Overall reaction}: CH_3OH \rightarrow HCHO + H_2 \tag{18}$$

As for the overall water splitting systems free of sacrificial reagents, both H_2 and O_2 should form with a stoichiometric amount of 2:1. Additional characterization or experiments should be carried out to confirm the reaction is really photocatalytic water splitting instead of a sacrificial reaction. For example, (i) the amount of gases evolved should be much more than the amount of photocatalyst to rule out the possibility of some stoichiometric reactions. (ii) the structure of photocatalyst before and after reactions should be characterized and compared. (iii) the rate of gases evolved should be checked by using cut-off filters with different transmission wavelength.

3 Key Issues of Photocatalytic Water Splitting

3.1 Light Absorption and Utilization

According to the photocatalytic mechanism, one of the key issues for photocatalytic water splitting is the light absorption and utilization. As shown in Fig. 4, the longer the absorption edge of the photocatalyst is, the higher the theory efficiency of solar energy conversion is. Accordingly, synthesis and development of heterogeneous photocatalyst materials have triggered extensive interest since the Honda-Fujishima effect finding [2], as evolves from UV only responsive to visible light or even near IR responsive. To date, about one hundred and fifty semiconductors have been reported for the water splitting, which are mostly metal oxides, (oxy)sulfides and (oxy)nitrides consisting of metal cations with d^0 and d^{10} configurations. Their conduction bands for the d^0 and d^{10} based photocatalysts are usually composed of d and sp orbitals, respectively. As shown in Fig. 5, many elements beyond d^0 and d^{10} are also employed to construct the heterogeneous photocatalyst materials, but their functions are not the same which can be classified into four groups: (i) to construct crystal and energy structure, (ii) to construct crystal but not energy structure, (iii) to form impurity levels as dopants, and (iv) to be used as cocatalyst.

Table 2 shows some typical oxide photocatalyst materials consisting of d^0 metal cations (Ti^{4+}, Zr^{4+}, Nb^{5+}, Ta^{5+} and W^{6+}) or d^{10} metal cations (Ga^{3+}, In^{3+}, Ge^{4+}, Sn^{4+} and Sb^{5+}) for water splitting with reasonable activities. Among them, NiO/La-$NaTaO_3$ and NiO/Zn-Ga_2O_3 are highly active. The photocatalytic activity of NiO/$NaTaO_3$ increased remarkably with doping of lanthanide ions, and the highest quantum yield of water splitting at the optimized conditions can reach as high as 56 %. The activity is stable for more than 400 h under irradiation of light from a

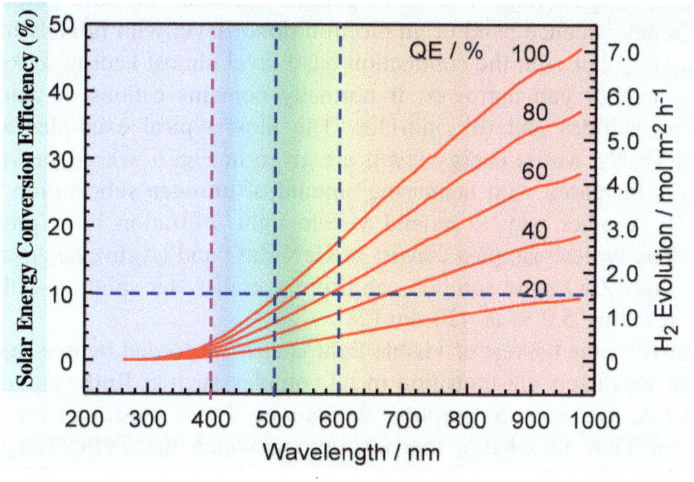

Fig. 4 Solar energy conversion efficiency as a function of absorption wavelength of materials

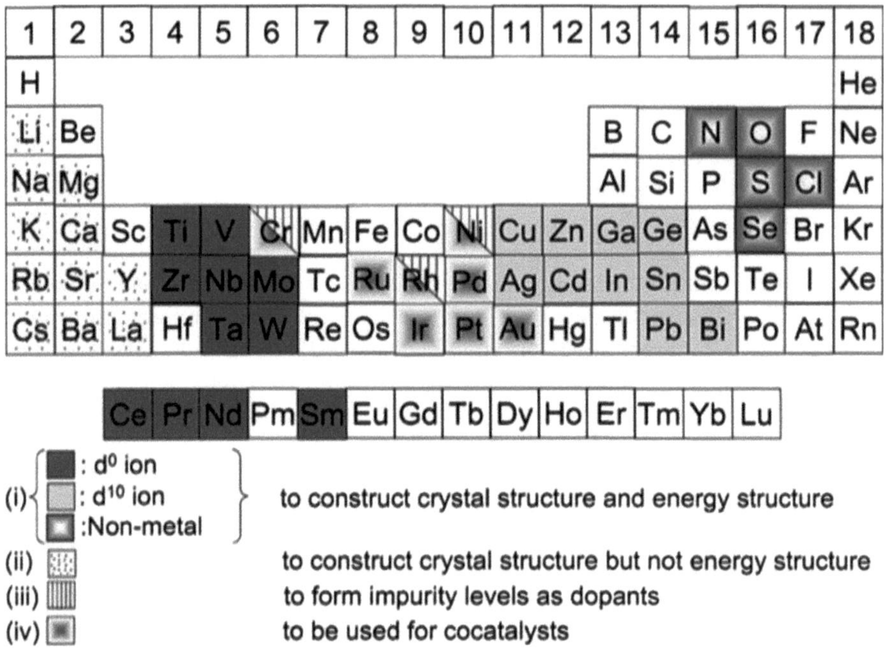

Fig. 5 Elements constructing heterogeneous photocatalysts. Reproduced from Ref. [1] with permission from the Royal Society of Chemistry

400 W high pressure mercury lamp. A much higher quantum yield (70 %) was achieved on the NiO/Zn-Ga_2O_3 photocatalyst with d^{10} electronic configuration.

Since the UV light just occupies less than 5 % of solar spectrum, the oxide photocatalysts have been extensively examined to extend the visible utilization by suitable band engineering. The most popular methods for this purpose are to introduce a new valence band or an electron donor level with orbitals of elements other than O 2p but with the conduction band level almost kept to make the band gap or the energy gap narrower. It normally contains cations or anions doped oxides, (oxy)sulfides and (oxy)nitrides. The most typical examples are Ta_2O_5, TaON and Ta_3N_5, whose energy levels are given in Fig. 6, where the visible light utilization is increased with increasing amount of nitrogen substitution to oxygen atoms [27]. Another way to extend visible light utilization is to prepare solid solutions. The typical samples consist of GaN-ZnO and $(AgIn)_xZn_{2(1-x)}S_2$, among which the GaN-ZnO solid solution can drive overall water splitting with quantum efficiency reaching 5.9 % at 420 nm [28].

Alternatively, the harvest of visible light is also succeeded by dye sensitization. The typical sensitizers are transition metal complex such as $Ru(bpy)_3^{2+}$ complexes, and metal free dyes such as porphine derivatives. These sensitizers are commonly adsorbed on TiO_2 for charge transfer, among which the TPPC/TiO_2/Pt system exhibited the highest H_2 evolution rate from water in the presence of EDTA as the electron donor (TPPC: tetrakis-(4-carboxyphenyl)porphine).

Table 2 Overall water splitting on some typical oxides with d^0 or d^{10} configuration

Photocatalyst	Crystal structure	BG/eV	Cocatalyst	Light source	Reactant solution	H_2 activity (μmol/h)	O_2 activity (μmol/h)	QY (%)	Reference
TiO_2	Anatase	3.2	Rh	Hg-Q	Water vapor	449		29	[16]
TiO_2	Anatase	3.2	Pt	Hg-Q	2.2 M Na_2CO_3	568	287		[17]
$SrTiO_3$	Perovskite	3.2	NiOx	Hg-P	5 M NaOH	40	19		[18]
$Rb_2La_2Ti_3O_{10}$	Layered Perovskite	3.4-3.5	NiOx	Hg-Q	0.1 M RbOH	869	430	5 @ 330 nm	[19]
$KLaZr_{0.3}Ti_{0.7}O_4$	Layered perovskite	3.91	NiOx	Hg-Q	Pure water	230	116	12.5	[20]
$Y_2Ti_2O_7$	Cubic pyrochlore	3.5	NiOx	Hg-Q	Pure water	850	420	6	[21]
$K_4Nb_6O_{17}$	Layered structuer	3.4	NiOx	Hg-Q	Pure water	1837	850	5 @ 330 nm	[22]
$Ba_5Nb_4O_{15}$	Layered perovskite	3.85	NiOx	Hg-Q	Pure water	2366	1139	7 @ 270 nm	[23]
$K_3Ta_3B_2O_{12}$	Tungsten bronze	4.0	None	Hg-Q	Pure water	2390	1210	6.5 @ 254 nm	[24]
$NaTaO_3$:La	Perovskite	4.1	NiO	Hg-Q	Pure water	19,800	9700	56 @ 270 nm	[25]
$Sr_2Ta_2O_7$	Layered perovskite	4.6	NiO	Hg-Q	Pure water	1000	480	12 @ 270 nm	[26]

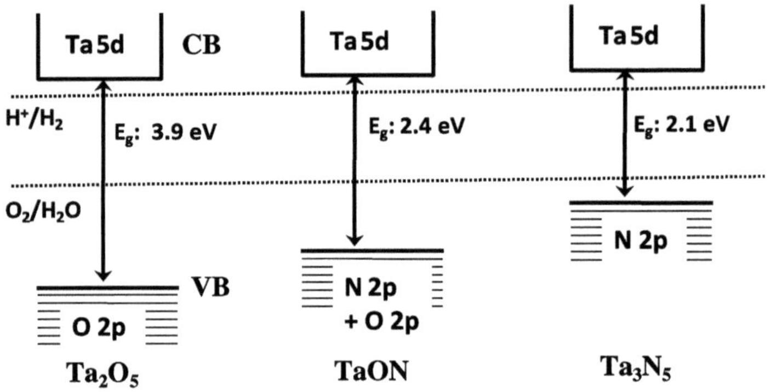

Fig. 6 Energy level structure of Ta_2O_5, TaON and Ta_3N_5 photocatalysts

Efforts have been also given to develop novel single-phase visible light responsive photocatalysts free of the modification technologies. Some heterometallic oxides have been successfully fabricated by using metal-mediated band structure engineering. According to elements used for band gap narrow, it can be divided into d-block, p-block and f-block metal oxides. The band gap narrowing of oxides by doping d-block transition metals result from formation of a separate donor level in the forbidden band, but the activities of the resulting doped photocatalysts are usually not high because the doping level can also serve as a recombination center for photogenerated electrons and holes. Different from the d-block transition metals, s-orbitals from the outer layer-orbital configuration of the p-block metal ions are commonly employed to narrow the band gap. The typical p-block metal ions contain Bi^{3+}, Pb^{2+}, Sn^{2+}, In^{3+} and Ag^{+} etc. The introduction of rare earth 4f orbitals can form a new energy level in the band structure, leading to narrowed bandgaps. For example, the pyrochlore-type compound Sm_2InMO_7 (M = Ta, Nb), with a 4f-d^{10}-d^{0} configuration, has become a new stable visible-light-responsive photocatalyst for H_2 evolution from water [29]. In addition, there also exist some other visible-light-responsive photocatalysts, such as phosphate Ag_3PO_4 [30], $TiSi_2$ [31], C_3N_4 [32] and a graphite oxide semiconductor photocatalyst etc. [33].

3.2 *Separation of Photogenerated Carriers*

The basic concept of promoting the separation of photogenerated carriers is to spatially separate the photogenerated electrons and holes in order to reduce their recombination possibility. According to the basic processes of photocatalysis, the photoenerated electrons and holes are first transferred from the bulk to the surface, and which will then be transferred into the reaction centers for corresponding reduction or oxidation reactions. Accordingly, the preferred efforts are to improve the charge separation efficiency in the bulk semiconductor. It is general to know that charge separation and transfer of photogenerated carriers are strongly influenced by the crystal, surface and electronic structural features of the materials. Such parameters as crystallinity, defects, crystal structure distortion, particle size, surface area, facets exposed and active reaction sites of the materials are commonly important. For example, the charge separation among the {001} and {110} facets of $BiVO_4$ has been experimentally proven [34]. Normally, enhanced crystallinity is in favor of charge separation, while defect sites act as recombination centers. In addition, decrease of particle size will reduce the transfer distance of carriers and will provide much more surface areas, but it reversely produces more surface defect sites. Thus, a balance consideration is required for fabrication of photocatalytic system with high efficiency. To date, a respectable research effort has been dedicated to both the modification of crystal structure and morphology of the photocatalysts with the view to improve more efficient charge separation. This has made great success in promoting photocatalytic activity for water splitting.

Besides the modification of crystal, surface and electronic structures of the bulk semiconductors, deposition of cocatalysts was mostly adopted to spatially separate the photogenerated carriers. The driving force of separating the carriers is mainly originated from their energy level differences. Taking noble metal as an example, since the Fermi energy level of noble metal is always lower than that of the semiconductor photocatalyst, the photogenerated electrons will transfer from the host photocatalyst to the noble metal cocatalyst. Meanwhile, the photogenerated holes stay at the host photocatalyst and migrate to its surface. This results in the efficient separation of the photogenerated electrons and holes. Subsequently, the separately localized electrons and holes become involved in their roles as the reducer and oxidizer, respectively, in the photocatalytic reactions. The cocatalysts developed so far will be introduced and summarized in the next part.

The semiconductor combination is another effective method to promote photogenerated charge separation with a formation of a heterojunction structure. The fabrication of heterojunction structure is commonly combined by using two n-type or p-type semiconductors, or using one n-type and one p-type semiconductor. The most typical examples are n-CdS/n-TiO_2 composite [35] and p-Cu_2O/n-WO_3 composite [36]. The basic processes for the charge separation are described in Fig. 7 by taking CdS/TiO_2 as an example, where electrons and holes are normally transferred into different semiconductors because of their distinct energy levels of conduction and valence bands.

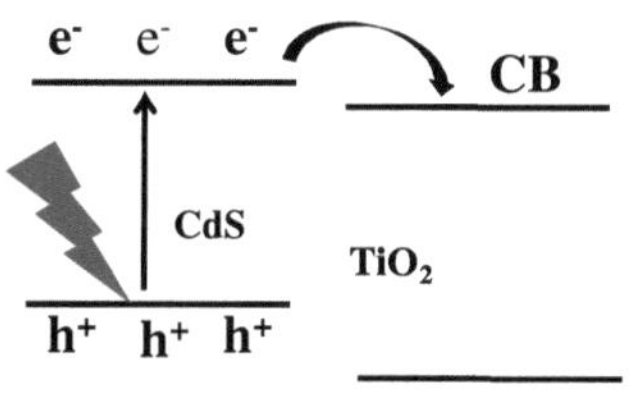

Fig. 7 Charge separation of CdS/TiO_2 semiconductor combination system under visible light irradiation

The photogenerated charge separation can be also promoted by using different phases of one semiconductor. It is indicated that a junction structure can be formed between the anatase and rutile phases of TiO_2 [37], and the migration of electrons across a phase junction is beneficial for charge separation, thereby enhancing the photocatalytic activity for hydrogen evolution. Similarly, the charge separation has been promoted by fabricating α/β-Ga_2O_3 phase junction, causing obviously enhanced overall water splitting performances [38].

3.3 Utilization of Photogenerated Carriers (Cocatalysts)

As the last step, the photogenerated electrons or holes reaching surface of photocatalyst will be finally utilized for water splitting. In most of cases, the oxidation reaction using the photogenerated holes can happen directly on the surface of photocatalyst, while the reduction reaction using the photogenerated electrons commonly requires the assistance of cocatalyst. As above mentioned, another function of cocatalyst is known to transfer electrons or holes for better charge separation. Thus, the cocatalyst loading becomes very important for the photocatalytic water splitting, and to date various cocatalysts have been extensively studied and developed.

First of all, noble metals have been widely used as the reduction cocatalyst in the photocatalytic water splitting over many kinds of photocatalysts: oxides, (oxy)-sulfides, and (oxy)nitrides. Up until now, the highest photocatalytic H_2 evolutions from water using visible light irradiation are from the Pt-loaded photocatalysts. Some other metals such as Au, Rh, Ag, Ni, Ru and Pd etc. have been also reported as efficient cocatalysts. In the case of noble metal coating, a schottky barrier will be formed at the interface of metal and semiconductor which could serve as an efficient electron trap preventing photogenerated electron-hole recombination. It was found that the smaller the Schottky barrier height at the metal/semiconductor junction is, the greater is the electron flow from semiconductor to metal, thus leading to higher photocatalytic activity for hydrogen production. On the other hand, the synergistic effect of two metal cocatalysts could contribute to much higher photocatalytic activity with respect to their corresponding single metal cocatalyst.

Some transition metal oxides such as RuO_2, IrO_2, CoO_x and NiO are also well-known as efficient cocatalysts in water splitting. It was demonstrated that both Ru(IV) species and bulk RuO_2 exhibited less activity for overall water splitting, whereas with RuO_2 nanoclusters as cocatalyst, $(Ga_{1-x}Zn_x)$–$(N_{1-x}O_x)$ displayed an improvement of H_2 and O_2 evolution. IrO_2 has been extensively used as oxidation cocatalyst for water oxidation. CoO or Co_3O_4 normally deposited is not active as IrO_2 for water oxidation, but on the oxynitrides photocatalysts, an ammonia-assisted thermal deposition of CoO_x exhibited much better promotion of water oxidation than IrO_2. NiO_x has been widely used for overall water splitting on oxides. However, activation pretreatment is generally necessary for NiO-loaded photocatalysts in order to obtain high activities. For example, a reduction-reoxidation treatment produces a double-layered NiO/Ni structure on the surface of the photocatalyst. In addition, Maeda and coworkers developed noble metal/Cr_2O_3 core/shell nanoparticles as a cocatalyst for overall photocatalytic water splitting [39]. The metal core can transfer photogenerated electrons to promote charge separation for hydrogen evolution, and the Cr_2O_3 shell prevents water formation from H_2 and O_2 (backward reaction) on the metal sites.

In addition to the noble metals and the transition-metal oxides, some other compounds, such as the transition-metal sulfides and nitrides, have also been developed as novel cocatalysts. Zong and co-workers developed MoS_2 as a novel cocatalyst for photocatalytic hydrogen evolution, even better than noble Pt [40]. Tabata et al. found that dispersion of transition metal sulfides such as NiS, FeS, Ru_2S_3, Ag_2S, CoS, and PdS, into the $CuGa_3S_5$ photocatalyst solution increased the photocatalytic activity [41]. Jang et al. proved that tungsten carbide (WC) had potential as an alternative cocatalyst for photocatalytic hydrogen production [42]. Under visible-light irradiation, the WC/CdS photocatalyst showed excellent photocatalytic activity for hydrogen production from water, comparable to that of conventional Pt/CdS.

In recent years, molecular photocatalytic systems based on biomimetic hydrogenase have been also studied for proton reduction to H_2 [43]. It has been found that hydrogenase, cobaloximes (Co(III) complexes) can be adsorbed into semiconductors for H_2 evolution in the presence of triethanolamine (TEOA) as sacrificial electron donor. Hydrogenase mimic $[(\mu\text{-SPh-4-NH}_2)_2Fe_2(CO)_6]$ (denoted as $[Fe_2S_2]$) was employed as the cocatalyst for H_2 evolution using semiconductor (ZnS) as the photoharvester and ascorbic acid (H_2A) as the electron donor. The photocatalytic H_2 production with more than 2600 turnover numbers (based on $[Fe_2S_2]$) and an initial TOF of 100 h^{-1} were achieved. Recently, Wang and coworker has achieved overall water splitting on the hybrid system [44].

The overall photocatalytic processes are composed of both oxidation and reduction reaction, and the slower one will be the rate determining step. In most of cases, single reduction or oxidation cocatalsyt was coated for the reduction or oxidation reaction. However, both reduction and oxidation centers exist in the

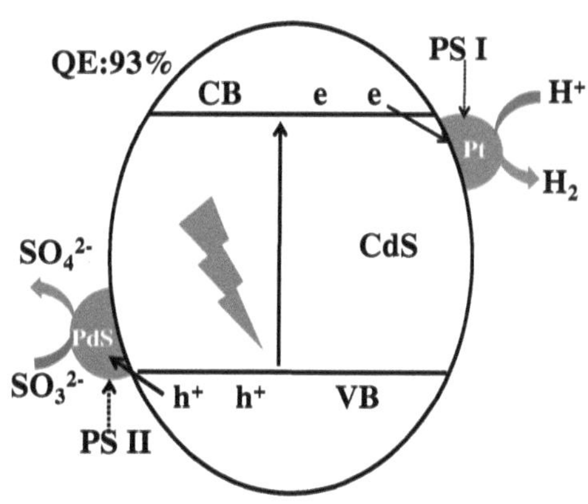

Fig. 8 Schematic illustration of dual cocatalysts modified photocatalyst Pt-PdS/CdS for highly efficient H_2 evolution under visible light irradiation

natural photosynthesis with high quantum efficiency. To mimic the natural processes, dual cocatalysts were then developed for the artificial photocatalytic water splitting. For example, when both Pt (reduction cocatalyst) and PdS (oxidation cocatalyst) were coloaded on the surface of CdS, the three component system Pt-PdS/CdS exhibits a high quantum efficiency of 93 % for H_2 production at 420 nm [45]. Figure 8 illustrates the tremendous effect of coloading oxidation and reduction cocatalysts on the exceptionally high QE.

4 Summary and Outlook

Significant advances have been made in photocatalytic water splitting, and the most efforts have been development of suitable semiconductor materials in the last century. To address the limited utilization of metal oxides to solar spectrum, various strategies have been explored to extend utilization of visible light in the past decade. This provides the basis for the design and assembly of highly efficient photocatalytic water splitting systems. It should be noted that although the theoretical requirement of the band gap for the photocatalytic one-step water splitting reaction is 1.23 eV, the actual band gap of the photocatalyst should be much larger than 1.8 eV because of the existence of band bending and surface overpotential. Figure 4 shows the calculated solar energy conversion efficiency as a function of wavelength for overall water splitting using photocatalysts with different quantum efficiencies. It was proposed that the solar energy conversion efficiency should reach at least 10 % for commercial application. Towards this target, one material with a 600 nm absorption edge and quantum efficiency of over 60 % is desirable. However, it still remains challenging to achieve overall water splitting with the 600 nm class materials, even though the overall water splitting on the materials with absorption edge of ca. 500 nm is feasible. Thus, more efforts should be given to understanding of the key issues related to charge separation and catalytic conversion. Novel cocatalysts as well as deposition methods are expected for promotion of conversion efficiency.

Photocatalytic water splitting for hydrogen production is a multidisciplinary topic, and it involves basic scientific principles from fields such as biological photosynthesis, photovoltaic cells and photoelectrocatalysis, etc. It is also of vital importance to understand the photocatalytic processes with the assistance of advanced techniques, such as theoretical calculation, ultrafast spectroscopy and in situ characterizations. Novel strategies and concepts similar to the cocatalyst and "junction" should be developed to aid the design and fabrication of highly efficient photocatalytic systems. In summary, research on visible light driven photocatalytic water splitting is a long standing challenge which needs scientists to persevere with patience and courage and to strive without relaxation.

References

1. Kudo A, Miseki Y (2009) Heterogeneous photocatalyst materials for water splitting. Chem Soc Rev 38:253–278
2. Fujishima A, Honda K (1972) Electrochemical photolysis of water at a semiconductor electrode. Nature 238:37
3. Yan H, Yang J, Ma G, Wu G, Zong X, Lei Z, Shi J, Li C (2009) Visible-light-driven hydrogen production with extremely high quantum efficiency on Pt-PdS/CdS photocatalyst. J Catal 266:165
4. Ma G, Yan H, Shi J, Zong X, Lei Z, Li C (2008) Direct splitting of H_2S into $H_2$and S on CdS-based photocatalyst under visible light irradiation. J Catal 260:134
5. Zong X, Yan H, Wu G, Ma G, Wen F, Wang L, Li C (2008) Enhancement of photocatalytic H_2 evolution on CdS by loading MoS_2 as cocatalyst under visible light irradiation. J Am Chem Soc 130:7176
6. Yang Y, Chen QY, Yin ZL, Li J (2009) Study on the photocatalytic activity of $K_2La_2Ti_3O_{10}$ doped with zinc(Zn). Appl Surf Sci 255:8419
7. Fujihara K, Ohno T, Matsumura M (1998) Splitting of water by electrochemical combination of two photocatalytic reactions on TiO_2 particles. J Chem Soc, Faraday Trans 94:3705
8. He CH, Yang OB (2003) Hydrogen evolution by photocatalytic decomposition of water under UV irradiation over $K[Bi_3PbTi_5O_{16}]$ perovskite: effect of cerium species. Ind Eng Chem Res 42:419
9. Kato H, Hori M, Konta R, Shimodaira Y, Kudo A (2004) Construction of Z-scheme type heterogeneous photocatalysis systems for water splitting into H_2 and O_2 under visible light irradiation. Chem Lett 33:1348
10. Lee SG, Lee S, Lee HI (2001) Photocatalytic production of hydrogen from aqueous solution containing CN^- as a hole scavenger. Appl Catal A 207:173
11. Wei Y, Li J, Huang Y, Huang M, Lin J, Wu J (2009) Photocatalytic water splitting with in-doped $H_2LaNb_2O_7$ composite oxide semiconductors. Sol Energy Mater Sol Cells 93:1176
12. Liu M, You W, Lei Z, Zhou G, Yang J, Wu G, Ma G, Luan G, Takata T, Hara M, Domen K, Li C (2004) Water reduction and oxidation on Pt–Ru/$Y_2Ta_2O_5N_2$ catalyst under visible light irradiation. Chem Commun 19:2192
13. Choi J, Ryu SY, Balcerski W, Lee TK, Hoffmann MR (2008) Photocatalytic production of hydrogen on Ni/NiO/$KNbO_3$/CdS nanocomposites using visible light. J Mater Chem 18:2371
14. Patsoura A, Kondarides DI, Verykios XE (2007) Photocatalytic degradation of organic pollutants with simultaneous production of hydrogen. Catal Today 124:94
15. Li YX, Lu GX, Li SB (2003) Photocatalytic production of hydrogen in single component and mixture systems of electron donors and monitoring adsorption of donors by in situ infrared spectroscopy. Chemosphere 52:843
16. Yamaguti K, Sato S (1985) Photolysis of water over metallized powdered titanium dioxide. J Chem Soc, Faraday Trans 81:1237
17. Sayama K, Arakawa H (1997) Effect of carbonate salt addition on the photocatalytic decomposition of liquid water over Pt–TiO_2 catalyst, J Chem Soc, Faraday Trans 93:1647
18. Domen K, Naito S, Soma S, Onishi M, Tamaru K (1980) Photocatalytic decomposition of water vapour on an NiO-$SrTiO_3$ Catalyst J Chem Soc, Chem Commun 543
19. Takata T, Furumi Y, Shinohara K, Tanaka A, Hara M, Kondo JN, Domen K (1997) Photocatalytic decomposition of water on spontaneously hydrated layered perovskites. Chem Mater 9:1063
20. Reddy VR, Hwang DW, Lee JS (2003) Effect of Zr substitution for Ti in $KLaTiO_4$ for photocatalytic water splitting. Catal Lett 90:39
21. Abe R, Higashi M, Sayama K, Abe Y, Sugihara H (2006) Photocatalytic activity of R_3MO_7 and $R_2Ti_2O_7$ (R = Y, Gd, La; M = Nb, Ta) for Water Splitting into H_2 and O_2. J Phys Chem B 110:2219

22. Domen K, Kudo A, Tanaka A, Onishi T (1990) Overall photodecomposition of water on a layered niobiate catalyst. Catal Today 8:77
23. Miseki Y, Kato H, Kudo A (2006) Water splitting into H_2 and O_2 over $Ba_5Nb_4O_{15}$ photocatalysts with layered perovskite structure prepared by polymerizable complex method. Chem Lett 35:1052
24. Kurihara T, Okutomi H, Miseki Y, Kato H, Kudo A (2006) Highly efficient water splitting over $K_3Ta_3B_2O_{12}$ photocatalyst without loading cocatalyst. Chem Lett 35:274–275
25. Kato H, Asakura K, Kudo A (2003) Highly efficient water splitting into H_2 and O_2 over lanthanum-doped $NaTaO_3$ Photocatalysts with high crystallinity and surface nanostructure. J Am Chem Soc 125:3082–3089
26. Yoshino M, Kakihana M, Cho WS, Kato H, Kudo A (2002) Polymerizable complex synthesis of pure $Sr_2Nb_xTa_{2-x}O_7$ solid solutions with high photocatalytic activities for water decomposition into H_2 and O_2. Chem Mater 14:3369–3376
27. Maeda K, Domen K (2007) New non-oxide photocatalysts designed for overall water splitting under visible light. J Phys Chem C 111:7851–7861
28. Maeda K, Teramura K, Domen K (2008) Effect of post-calcination on photocatalytic activity of $(Ga_{1-x}Zn_x)(N_{1-x}O_x)$ solid solution for overall water splitting under visible light. J Catal 254:198–204
29. Tang XD, Ye HQ, Liu H, Ma CX, Zhao Z (2009) A novel visible-light-driven photocatalyst Sm_2InNbO_7 for H_2 or O_2 evolution. Chem Phys Lett 484:48–53
30. Yi Z, Ye J, Kikugawa N, Kako T, Ouyang S, Stuart-Williams H, Yang H, Cao J, Luo W, Li Z, Liu Y, Withers RL (2010) An orthophosphate semiconductor with photooxidation properties under visible-light irradiation. Nat Mater 9:559–564
31. Ritterskamp P, Kuklya A, Wu"stkamp MA, Kerpen K, Weidenthaler C, Demuth A (2007) A titanium disilicide derived semiconducting catalyst for water splitting under solar radiation—reversible storage of oxygen and hydrogen. Angew Chem Int Ed 46:7770–7774
32. Wang X, Maeda K, Thomas A, Takanabe K, Xin G, Carlsson JM, Domen K, Antonietti M (2008) A metal-free polymeric photocatalyst for hydrogen production from water under visible light. Nat Mater 8:76–80
33. Yeh TF, Syu JM, Cheng C, Chang TH, Teng H (2010) Graphite oxide as a photocatalyst for hydrogen production from water. Adv Funct Mater 20:2255–2262
34. Li R, Zhang F, Wang D, Yang J, Li M, Zhu J, Zhou X, Han H, Li C (2013) Spatial separation of photogenerated electrons and holes among 010 and 110 crystal facets of $BiVO_4$. Nat Commun 4:1432
35. Zhang F, Yamakata A, Maeda K, Moriya Y, Takata T, Kubota J, Teshima K, Oishi S, Domen K (2012) Cobalt-modified porous single-crystalline $LaTiO_2N$ for highly efficient water oxidation under visible light. J Am Chem Soc 134:8348–8351
36. Baker DR, Kamat PV (2009) Photosensitization of TiO_2 nanostructures with CdS quantum dots: particulate versus tubular support architectures. Adv Funct Mater 19:805–811
37. Hu CC, Nian JN (2008) Electrodeposited p-type Cu_2O as photocatalyst for H_2 evolution from water reduction in the presence of WO_3. Sol Energy Mater Sol Cells 92:1071–1076
38. Zhang J, Xu Q (2008) Importance of the relationship between surface phases and photocatalytic activity of TiO_2. Angew Chem Int Ed 120:1766–1769
39. Wang X, Xu Q (2012) Photocatalytic overall water splitting promoted by an α–β phase junction on Ga_2O_3. Angew Chem Int Ed 51:13089–13902
40. Maeda K, Teramura K (2006) Noble-metal/Cr_2O_3 core/shell nanoparticles as a cocatalyst for photocatalytic overall water splitting. Angew Chem Int Ed 45:7806–7809
41. Zong X, Yan H (2008) Enhancement of photocatalytic H_2 evolution on CdS by loading MoS_2 as cocatalystunder visible light irradiation. J Am Chem Soc 130:7176–7177
42. Tabata M, Maeda K (2010) Photocatalytic hydrogen evolution from water using copper gallium sulfide under visible-light irradiation. J Phys Chem C 114:11215–11220
43. Jang JS, Ham DJ (2008) Role of platinum-like tungsten carbide as cocatalyst of CdS photocatalyst for hydrogen production under visible light irradiation. Appl Catal A 346: 149–154

44. Wen F, Li C (2013) Hybrid artificial photosynthetic systems comprising semiconductors as light harvesters and biomimetic complexes as molecular cocatalysts. Acc Chem Res 46: 2355–2364
45. Wang W, Chen J (2014) Achieving solar overall water splitting with hybrid photosystems of photosystem II and artificial photocatalysts. Nat Commun 5:4647
46. Yan H, Yang J (2009) Visible-light-driven hydrogen production with extremely high quantum efficiency on Pt-PdS/CdSphotocatalyst. J Catal 266:165–168

Photoelectrochemical Approach Using Photocatalysts

Jingying Shi and Can Li

Abstract Photoelectrode is the most crucial part of a photoelectrochemical (PEC) device for chemical conversion of solar energy. To use particulate photocatalysts for PEC approach, the photocatalysts should be deposited onto the conductive substrate to prepare photoelectrodes. In this chapter, physical and chemical routes for the fabrication of photoelectrodes from photocatalysts are introduced and the PEC performances for water splitting are discussed.

1 Introduction

1.1 Role of Photoelectrode

The photoelectrochemical (PEC) approach for 'artificial photosynthesis' is based on a cell which consists of two electrodes immersed in an aqueous electrolyte, the anode and the cathode and at least one of them is photoactive under irradiation [1–3].

In general, there are three options used for the arrangement of photo-electrodes (photo-anode or photo-cathode which is sensitive to the incident light) in the assembly of PEC cells [4–9]: photo-anode made of n-type semiconductor and cathode made of metal; photo-cathode made of p-type semiconductor and anode made of metal; photo-anode made of n-type semiconductor and photo-cathode made of p-type semiconductor. Setting the first option as example, water photo-electrolysis using a PEC cell involves several processes within photoelectrodes and at the photoelectrode/electrolyte interface, including: (1) light-induced intrinsic ionization of the semiconducting material (the photoanode), resulting in the formation of electronic charge carriers (quasi-free electrons and holes); (2) oxidation of water at the photoanode by holes to produce oxygen gas ($2H_2O + 4\ h^+ \rightarrow O_2 + 4H^+$);

J. Shi · C. Li (✉)
State Key Laboratory of Catalysis, Dalian National Laboratory for Clean Energy, Dalian Institute of Chemical Physics, Chinese Academy of Sciences, Dalian 116023, China
e-mail: canli@dicp.ac.cn

M. Sugiyama et al. (eds.), *Solar to Chemical Energy Conversion*,
Lecture Notes in Energy 32, DOI 10.1007/978-3-319-25400-5_19

(3) transport of H^+ ions from the photoanode to the cathode through the electrolyte and transport of electrons from photoanode to the cathode through the external circuit; (4) reduction of H^+ ions at the cathode by electrons to generate hydrogen gas ($2H^+ + 2e^- \rightarrow H_2$). Accordingly, the overall water splitting ($2H_2O \rightarrow 2H_2 + O_2$) continuously takes place under illumination of photoanode. From this working principle, it can be clearly seen that the photoelectrode is the most crucial part to a PEC cell, which directly decides the energy conversion efficiency of the whole PEC cell.

1.2 Electrical Resistance

Besides the intrinsic photo-physical and photo-chemical properties of semiconductor catalyst itself, electrical resistance is another importance factor, sometimes even a fateful factor to determine the photo-current output of a PEC cell, which is similar to that in electrochemistry. The major sources of energy losses derive from the ohmic resistances of the external and internal circuits of the PEC cells including electrodes, electrolytes, electrical leads (wires), electrical connections (contacts) as well as measuring and control equipment. In order to achieve the maximum conversion efficiency, the electrical resistance of all of these items must be minimized.

(1) Electrodes

The electrical resistance of the semiconducting photoanode is several orders of magnitude larger than that of the metallic cathode. The electrical conductivity of the photoanode, which is determined by the concentration of the charge carriers and their mobilities, is described in Eq. (4):

$$\sigma = en\mu_n + ep\mu_p + Z_i ei\mu_i \quad (4)$$

where n is the concentration of electrons, p the concentration of holes, i the concentration of ions, μ_n the mobility of electrons, μ_p the mobility of holes, μ_i the mobility of ion, Z_i charge number of ion.

At room temperature, the ionic component of the electrical conductivity may be ignored. The mobility terms do not change with concentration when interactions between the charge carriers are absent. However, at higher concentrations, these interactions result in a decrease in the mobilities. Therefore, the maximal σ is a compromise between the effect of increasing the concentrations while decreasing the mobilities. The optimal value of σ may be achieved through the imposition of a defect disorder that is optimal for conduction [10]. The defect disorder and electrical properties may be modified through the incorporation of aliovalent cations (forming donors and acceptors) and the imposition of controlled oxygen partial pressure during processing. Again, these required electrical properties may be achieved through in situ monitoring of the electrical conductivity, thermo-electric power, and work function during processing [10, 11].

(2) Electrical leads

Electrical leads usually are made of metal wires with resistances substantially lower than those of the photo-electrode and the electrolyte [3]. In this sense, selection of the wire is of secondary importance. Especially for the photoelectrode fabricated with single crystal semiconductors or powder photocatalysts, ohmic contact between the material and the connected conductive wire should be well build.

(3) Electrical connections

Connections, such as those between wires and those between wires and electrodes, may be sources of high resistance due to (i) contact potential difference, which develops between solids of different work function, and (ii) local corrosion resulting in the formation high-resistance scales. Therefore, it is desirable to minimize or, preferably, eliminate the number of inter wire connections. Also, the engineering of other types of connections, those between the leads and the other circuit elements, is of considerable importance.

1.3 Materials for Photoelectrode

Inorganic semiconductors are the most used materials for photoelectrodes and are selected as examples for discussion in this section.

1.3.1 Single Crystal Semiconductors

Single crystal semiconductors are used to fabricate photoelectrode in the early PEC studies. Rutile TiO_2 is the first and most investigated single crystal photoanode for PEC water splitting [1]. To increase the conductivity of crystal, the TiO_2 wafer was partly reduced at a high temperature as 700 °C in a hydrogen/argon mixture before use. In nonstoichiometric TiO_{2-x}, the higher x, the lower the resistance is [10, 11]. It would be expected that the resistance of the photo-anode during performance of a PEC in contact with oxygen would increase due to oxidation. Therefore, the electrical resistance of PEC's must be regenerated after oxidation by postreduction in a hydrogen/argon mixture. $SrTiO_3$ is another extensively investigated single crystal photoelectrode with wide band gap [12, 13]. Likewise, the single crystal wafer of $SrTiO_3$ was treated at a higher temperature (>1000 °C) with H_2 beforehand to improve its semiconductivity. Indium or gallium-indium eutectic is coated onto one side of surface to ensure ohmic contact and a copper leads wire usually is attached to this contact using conducting silver paste to complete conductive connection.

However, the application of single crystal photoelectrodes is very limited, mostly due to limited surface area and limited reaction sites as well as the complicated processing technic in the preparation of single crystal material and limited in variety. High resistance of bulk crystal greatly inhibited their PEC performance. Regardless of high cost of single crystal itself, elaborate formalities in electrical leads and connections make it inconvenient to fabricate a photoelctrode with an acceptable quality in electrical contact. Being lack of practical value, single crystal materials are seldom to be used as photoelectrode now.

1.3.2 Photovoltaic Materials

Since similar processes including light-induced charge carriers generating, separation and transportation are mainly involved in semiconductor photovoltaic cells for electricity generation, photovoltaic materials or even photovoltaic cells themselves can be used to fabricate photoelectrodes for highly efficient PEC water splitting owing to their excellent light harvesting and charge carriers transportation properties. A very efficient photoelectrolysis cell that utilizes a direct semiconductor/liquid electrolyte junction is the 12.4 % solar water splitting configuration based on a p-$GaInP_2$ photocathode connected in series to a p-n GaAs junction photovoltaic layer, reported by Khaselev and Turner [14]. It also produced to the artificial leaf, a triple junction amorphous silicon cell, capable of photoelectrolysing water with a solar energy efficiency of up to 4.7 % [15, 16]. But the performance cannot be sustained because the photoelectrode materials are not stable under operating conditions and undergo photocorrosion. Similar stability problems also exist for most other visible light absorbing II–VI, III–V, and group 14 element semiconductors. Moreover, complicated preparation technology and high cost are limiting factors for practical application. Detailed reviews on the photoelectrodes derived from photovoltaic materials have recently published [17].

1.3.3 Polycrystalline Ceramic Semiconductors (Photocatalysts)

Concerning the mechanisms of the reactions, the principle of photocatalytic water decomposition is similar to that of PEC water decomposition [18, 19]. The essential difference between the two consists of the location of the sites of oxidation and reduction half reactions. In the PEC process, these reactions take place at the photoanode and cathode, respectively. In the photocatalytic process, both oxidation and reduction occur on the surface of the photocatalyst, which exhibits the functions of both anode and cathode. It can be clearly seen that photocatalyst is able to play the role of photoelectrode so as to be good candidate materials for fabrication of photoelectrodes.

The photocatalytic effects of suspended semiconductor particles were firstly demonstrated by Bard in 1979 [20]. Since then, a large number of powder photocatalysts have been developed. Most of them show a polycrystalline phase and are more cost-effective than the as mentioned single crystal and photovoltaic materials. And facile synthesis methods make them more viable for application in large scale.

Accordingly, photoelectrodes fabricated from photocatalyst are more promising for commercialization. In this section, starting from photocatalysts directly or indirectly, the fabrication methods for typical semiconductor photoelectrodes with aggregated particles or porous morphology for water splitting are summarized. The representative visible-light-response photocatalysts of semiconductor including $BiVO_4$, WO_3, α-Fe_2O_3 and some nitrides/oxynitrides developed in recent years are demonstrated as the model materials for discussion.

2 Physical Routes for Fabrication of Photoelectrodes Using Photocatalysts

In physical routes, the powder photocatalyst with polycrystalline phase is directly used as the starting material for photoelctrode. In other words, the powder photocatalysts should be synthesized in advance before fabrication of phtoelectrodes.

2.1 Powder Spreading

With the powder spreading method, generally the slurry of photocatalyst particles in mixed solution containing organic solvent and water was spread or pasted onto the substrate following annealing at high temperatures to remove the residual solvent as well as improve adhesion [21]. This method is very simple and easy to be operated.

In 2008, Long et al. prepared $BiVO_4$ electrodes by this method, in which a suspension of the $BiVO_4$ powders in absolute ethanol was spread onto the ITO (indium tin oxide) glass substrate, and then was sintered at 400 °C in air [22]. However, the IPCE of the $BiVO_4$ electrode was very low and less than 2 % at 420 nm (even under a very high potential of 1.6 V_{RHE}). The low IPCE may be related to not only the poor contact between the particles, but also bad crystallinity of the $BiVO_4$ caused by a low calcining temperature. The powder spreading method was improved by using fine $BiVO_4$ powders in 2010 [23]. Unfortunately, the IPCE was still low, about 12 % at 440 nm and a very high potential of 2.1 V_{RHE}, possibly due to the poor contact and low carrier concentration.

The powder spreading method was also used to prepare TaON particulated photoelectrodes. TaON powder (1 g) was suspended into pure water (5 mL) by stirring. The resultant colloidal solution of TaON was then spin-coated at 2000 rpm onto flurione-doped tin-oxide (FTO) glass substrate [24]. To avoid oxidation at the TaON surface at elevated temperatures, the substrates were treated with an ethanol solution of 50 mM KOH in advance by a spin coater at 1000 rpm to get strong adhesion between TaON and FTO under low-temperature annealing. The TaON-coated glass was then dried in air and heated at 150 °C for 15 min with an electric furnace. The above TaON coating was repeated twice, and finally the TaON film was heated at 150 °C for 30 min. The values of IPCE for the as-fabricated photoelectrode in 0.1 M $HClO_4$ aqueous solution at 0.7 V versus Ag/AgCl are not more than 0.8 % even in the presence of I^- anions as electron donor.

Thus, although this fabrication method is very easy and cheap, the poor contacts between the particles and between the particles and substrate that makes it very limited in practical application.

2.2 Doctor Blade Technique

The doctor blading technique is the most commonly adopted method for the preparation of thin films fixed on conducting glass substrates [25]. The particle precursors are mixed with organic additives and a small amount of water to form a paste by grinding in mortar, followed by scraping the paste onto a substrate. A controllable thickness of film can be obtained; although the reproducibility is good, the precision of film thickness is in the micron range.

To fabricate WO_3 photoanodes for PEC cell system, Hong et al. firstly synthesized WO_3 nanocrystals using a hydrothermal process followed by a calcination [26]. The obtained WO_3 nanocrystals were then thoroughly mixed with polyethylene glycol (PEG, Mw 20,000) as a binder and cast on FTO coated glass substrates by the doctor blade method. The coated film was dried at room temperature and calcined in air to combust the binder before PEC tests. The photocurrent onset potential appeared at approximately 0.45 V (vs. NHE) with the maximum photocurrent of approximately 0.6 mA/cm^2 was achieved.

A porous $(Ga_{1-x}Zn_x)(N_{1-x}O_x)$ electrode was also prepared by pasting a viscous slurry onto conducting glass through doctor blading [27]. A mixture of 0.1 g of the as-prepared $(Ga_{1-x}Zn_x)(N_{1-x}O_x)$ powder, 10 μL of acetylacetone (Kanto Chemicals), 10 μL of TritonX (Aldrich, USA), and 200 μL of distilled water was ground in an agate mortar for preparation of the viscous slurry. Subsequently, the slurry was pasted on FTO glass slides following by calcinations in air at 623 K for 1 h. Regardless of the excellent photocatalytic performance for the particulate $(Ga_{1-x}Zn_x)(N_{1-x}O_x)$ [28], the photoelectrode prepared by this method show a very low photocurrent as about 10 μA/cm^2 at 0.6 V versus Ag/AgCl reference electrode in 0.1 M Na_2SO_4 aqueous solution (pH 4.5) [27]. $LaTiO_2N$ is another oxynitride photocatalyst which has been reported to show photocatalytic hydrogen and oxygen evolution activities in half reactions using sacrificial reagents [29, 30]. A slurry containing 0.1 g of $LaTiO_2N$ powder, distilled water, acetylacetone, and triton X-100 was pasted carefully by a glass rod onto a FTO transparent glass substrate, which was then heated in nitrogen or air at 573–773 K for an hour [31]. However, the obtained photocurrent density is less than 10 μA/cm^2 at 0.8 V versus Ag/AgCl reference electrode in 0.1 M Na_2SO_4 aqueous solution (pH 4.5). Even suffering $TiCl_4$ post-treatment, the obtained photocurrent is not more than 40 μA/cm^2.

TaON and Ta_3N_5 thin films have been achieved through this doctor blading technique [32, 33]. To achieve more contact between particles, Maeda et al.

prepared nanoparticular Ta_3N_5 for photoelectrode fabrication. Unfortunately, the obtained photocurrent is still very low, which only reaches 0.03 mA/cm^2 in Na_2SO_4 electrolyte at 0.8 V versus Ag/AgCl. The reason can be attributed to the serious aggregation of nano-particles and the limitations of the fabrication route itself. The biggest shortcoming of the doctor blading technique is the introduction of a large amount of high-boiling organic species, which can only be eliminated through high temperature calcinations. Unfortunately, high temperature treatment basically leads to disastrous damage to the final properties of the TaON and Ta_3N_5, stemming from their irreversible oxidation or decomposition of surface or bulk during the high temperature process, even under an inert atmosphere.

2.3 Electrophoresis Deposition

Electrophoretic deposition (EPD) is an alternative method for the deposition of nanoparticles films on conducting glass substrates, which may introduce fewer organic species [34]. The film thickness can be controlled in the submicron range, and the low boiling solvent can easily be removed by evaporation at low temperatures, thereby avoiding the high temperature calcinations. Meanwhile, EPD is also applicable to nonplanar and even complex multi-dimensional substrates, and enables facile deposition of particles on a variety of substrates [35].

TaON and Ta_3N_5 thin films spread on FTO conducting glass supports have been achieved by EPD with good reproducibility [36]. During the PEC water splitting process, Ta_3N_5 and TaON photoelectrodes have exhibited poor efficiencies (maximum IPCE 5.3 % at 450 nm, 0.5 V vs. Pt in 1 M KOH) [21] and terribly discouraging photochemical stabilities (persisting for only several minutes) until Abe and coworkers made a substantial contribution to the highly efficient Ta_3N_5 (ca. 31 % at 500 nm, 1.15 VRHE) and TaON (ca. 76 % at 400 nm, 1.15 VRHE) photoelectrodes by using the EPD method [34, 36]. They have highlighted the importance of efficient contacts among the particles, because efficient electron transport in the porous electrodes will be ensured by these bridge-like contacts [36, 37]. The rigid surfaces formed at high temperatures and the refractory characteristics of Ta_3N_5 or TaON materials need high energy input to promote the sintering and necking of Ta_3N_5 or TaON particles. However, owing to the temperature limit of conducting glass substrates and the bad antioxygenic properties of Ta_3N_5 or TaON, effective contacts among Ta_3N_5 or TaON particles cannot be successfully formed by simple heat treatment. The isolated particles greatly block the electron transport, which results in low PEC efficiencies. Necking treatment is therefore indispensable for forming interconnected particles film [37, 38]. $TiCl_4$ post-treatment for forming TiO_2 joints among the TaON and $LaTiO_2N$ particles is reported to improve evidently the photocurrent, while the post-calcination under an inert atmosphere is crucial to avoid the oxidation of oxynitrides and to transform $TiCl_4$ to TiO_2 [31, 32]. In this case, TiO_2 bridges among isolated particles behave as electron transfer media, which is favorable for the improvement of photocurrent, but the photocurrent benefits may be restricted by the hetero-junction interfaces between TiO_2 and these oxynitride particles. Ta_2O_5 bridges may also fail in the same way in obtaining desirable PEC properties on TaON or Ta_3N_5 photoelectrodes. Upon heat treatment under NH_3 atmosphere, TaON bridges formed between TaON or Ta_3N_5 particles by nitriding the amorphous Ta_2O_5 bridges from $TaCl_5$, producing a considerable increase in the photocurrent of TaON or Ta_3N_5 photoelectrodes. For TaON photoelectrodes, the transformation of bridges from Ta_2O_5 to TaON enables better electron transport pathways by forming more conductive TaON bridges and interconnected single phase TaON film [37]. Compared with TaON photoanodes, the inferior performance on Ta_3N_5 photoanodes may be associated with TaON bridges, in which relatively low electron transport may occur across the hetero-interfaces originating from the crystal structure discrepancy between TaON and Ta_3N_5. Transparent conducting oxide substrates which can withstand higher temperatures are desirable for further photocurrent improvement of Ta_3N_5 photoelectrodes by the formation of Ta_3N_5 bridges.

Lee and coworkers have prepared PO_4-doped $BiVO_4$ photoelectrodes on FTO glass substrate using the EPD technique [39]. However, the PEC performance of the $BiVO_4$ samples was also limited by the poor contact between $BiVO_4$ particles. Wang et al. compared the PEC activity of water oxidation over varied kind of electrocatalysts modified $BiVO_4$ particle photoanodes fabricated by EPD method [40]. All the photocurrents are less than 0.3 mA/cm^2 despite of a high bias (1.0 V vs. SCE) was applied.

2.4 Particle Transfer

Recently, the particle transfer method was developed as an alternative to necking treatment by Domen's group [41–44]. The particle transfer process is illustrated in Fig. 1 [41, 42]. First, a semiconducting powder is laid on a glass substrate. Then, a thin metal layer (approximately 100–300 nm) is deposited as a contact layer by the sputtering method. The choice of metal is considered to be crucial to establish a favourable electrical contact between the semiconductor particles and the metal layer. After that, continuous metal films several micrometers in thickness are deposited by the sputtering method for adequate conductivity and mechanical strength. The resulting metal film is transferred to another substrate by a resin, and the primary substrate is peeled off. The excess powder is removed by ultrasonication in water. This method ensures an adequate electrical contact between the semiconducting particles and metal layer. In addition, since the semiconductor particles attached loosely to the electrodes are removed, a mono-particle layer covers the metal conductor film. It should be noted that the process is applicable to a variety of powdered semiconductors in the fabrication of photoelectrodes.

The oxynitride photocatalyst of $LaTiO_2N$ was firstly used to fabricate photelctrode by particle transfer method [42]. Figure 2 shows the current–potential curves of $LaTiO_2N$ photoanodes in a 1 M aqueous Na_2SO_4 solution at pH = 13.5 [42]. This

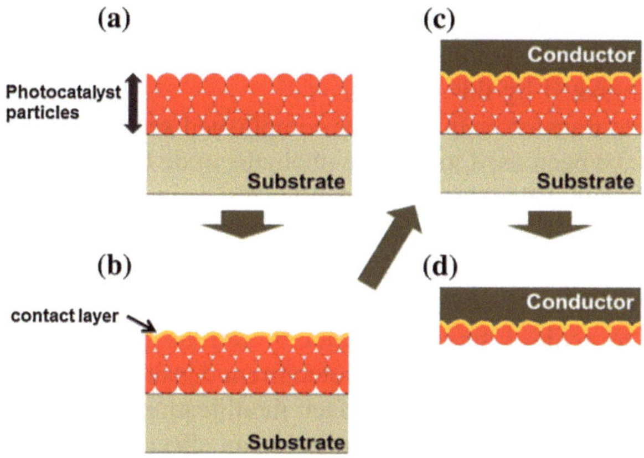

Fig. 1 Schematic diagram of the particle transfer (PT) method [42]. Reproduced from Ref. [42] with permission from the Royal Society of Chemistry. **a** Deposition of photocatalyst particles. **b** Formation of contact layer. **c** Formation of conductor layer. **d** Lift off

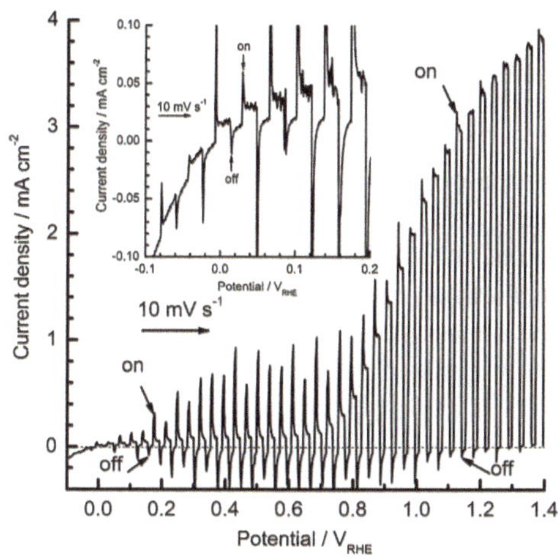

Fig. 2 Current–potential curve of IrO_2-loaded $LaTiO_2N$ electrodes with a Ta contact layer between the $LaTiO_2N$ particles and the Ti conductor layer: the electrolyte was a 1 M aqueous Na_2SO_4 solution with pH = 13.5, adjusted by the addition of NaOH. Simulated sunlight (AM1.5D) was turned on and off with a periodicity of 6 s. The electrode potential was swept in the positive direction at a rate of 10 mV s^{-1} [42]. Reproduced from Ref. [42] with permission from the Royal Society of Chemistry

photoelectrode was prepared by the particle transfer method using Ta and Ti layers as the contact layer and conductor layer, respectively. The surface of the photoanode was modified with IrO_2 as an oxygen evolution catalyst. The anodic photocurrent was observed below 0 V versus RHE and about a photocurrent density over 3 mA/cm^2 was found at 1.23 V versus RHE. This indicates that the potential of the conduction band minimum of $LaTiO_2N$ was more negative than the equilibrium potential of hydrogen evolution. It is suggests that an effective photocathode can be successfully fabricated by this method.

Using particle transfer method, other complicated semiconductor materials with multielement have been used to prepare photoelectrode by Domen's group [43, 44]. The copper gallium selenides (CGSe) powders prepared by solid-state-reaction were used to fabricate photocathodes by this method and the photocurrent reaches about 2 mA/cm^2 at −0.2 V versus RHE in an alkaline electrolyte (pH 9.5) [43]. It should be noted that a thin contact layer of Mo was employed for formation of an ohmic contact to achieve smooth transfer of carriers between CGSe and a thick conductor layer of Ti. The improved photocurrent density and positive-shifted onset potential were clearly observed by surface modification of platinum nanoparticles and (or) a CdS semiconductor layer, indicating an effective charge collection by the substrate for the as-fabricated photoelectrode. Another example is $La_5Ti_2CuS_5O_7$ photocathodes fabricated from the well-crystalline particles [44]. $La_5Ti_2CuS_5O_7$ is an oxysulphide p-type semiconductor that exhibits photocatalytic activity for both water reduction and oxidation under visible light irradiation in the presence of sacrificial reagents [45, 46]. A 2 μm thick Au layer was deposited on the sample as a back contact layer by vacuum evaporation. The photocurrent at 0 V versus RHE in an alkaline electrolyte was improved by about an order of magnitude compared with lower crystallinity prepared by pulsed laser deposition, which might suggest a lower series resistance. The photocurrent of a $La_5Ti_2CuS_5O_7$ photocathode was further boosted by p-type doping, indicating an effective electric contact between particles and substrate.

Based on these results, the particle transfer method can be regard as an effective way to prepare photoelectrode directly from powder photocatalysts. However, the metal materials for the construction of well ohmic contact need to be investigated for different photocatalyst materials and the fabrication process is cost and a bit tedious.

3 Chemical Routes for Fabrication of Photoelectrodes Using Photocatalysts

By chemical routes, semiconductor photoelectrodes are fabricated through growing the polycrystalline photocatalyst particles on conductive substrate in chemical reactions. Due to the in situ growth of semiconductor particles on substrate during chemical reaction, the connections between the particles and between the particles and substrate will be generally improved compared with those fabricated by physical routes. Thus, chemical methods are more often used to prepare photoelectrodes from photocatalysts.

3.1 Metal Organic Decomposition

Metal organic decomposition (MOD) is a easy method to synthesize $BiVO_4$ photoelectrodes, reported by various groups [47–52]. Up to date, the best performance of a $BiVO_4$ photoelectrode has been carried out by using metal–organic decomposition, which eases doping during the preparation. The samples prepared by this method possess good crystallinity and smaller grain size, though sintered under moderate high temperature.

Typically, organometallic species containing Bi^{3+} and V^{3+} or V^{5+} ions are dissolved into a solution and the resulting solution is deposited on a conducting substrate by spin coating or spray pyrolysis [47–53]. A commonly used Bi precursor is bismuth-2-ethylhexanote while V precursors include vanadium acetylacetone, vanadium (oxy)acetylacetonate or vanadium-tri-isopropoxy oxide. In some cases, bismuth (III) nitrate or vanadium (III) chloride was used with vanadium organometallic or bismuth organometallic species, respectively [49–52]. After these precursor molecules are coated on a conducting substrate, a heat treatment at moderately high temperatures (350–500 °C) is followed to decompose the organic precursors forming crystalline $BiVO_4$.

Among various deposition methods employed for MOD, spin coating is the most widely used for the preparation of $BiVO_4$ electrodes, which was first reported by Sayama et al. in 2003 [47]. The resultant $BiVO_4$ photoanode shows a considerably high photocurrent close to 3 mA/cm^2 in Na_2SO_4 (pH 6.6) aqueous solution at 1.0 V versus Ag/AgCl with an excellent efficiency (IPCE = 29 % at 420 nm) for the decomposition of water under visible light. Since then several studies using similar spin coating methods have been reported where the compositions of $BiVO_4$ were modified to improve its photoelectrochemical properties for water oxidation. One of the advantages of the MOD methods is the easy composition tuning, which can be achieved by simply adding dopant ions to the precursor solution. For example, Luo et al. reported doping $BiVO_4$ with Mo^{6+} by adding molybdenyl acetylacetonate to the precursor solution and the porous morphology of the as-prepared photoelectrodes were shown in Fig. 3 [49]. Similarly, Zhong et al. reported doping $BiVO_4$ with W^{6+} by incorporating tungstic acid into the precursor solution [50]. Also, surfactants have been added to the precursor solution to form micellar-inorganic units to produce porous $BiVO_4$ electrodes [51]. After the spin coating or spray pyrolysis deposition, the electrodes were heat treated in air resulting in crystalline $BiVO_4$ electrodes. These doped $BiVO_4$ photoanodes all show higher activity for PEC water splitting.

3.2 Sol–Gel Route

The sol–gel route is the most common method for preparing WO_3 films [25]. The W precursors were primarily prepared in the following two ways: (1) H_2WO_4

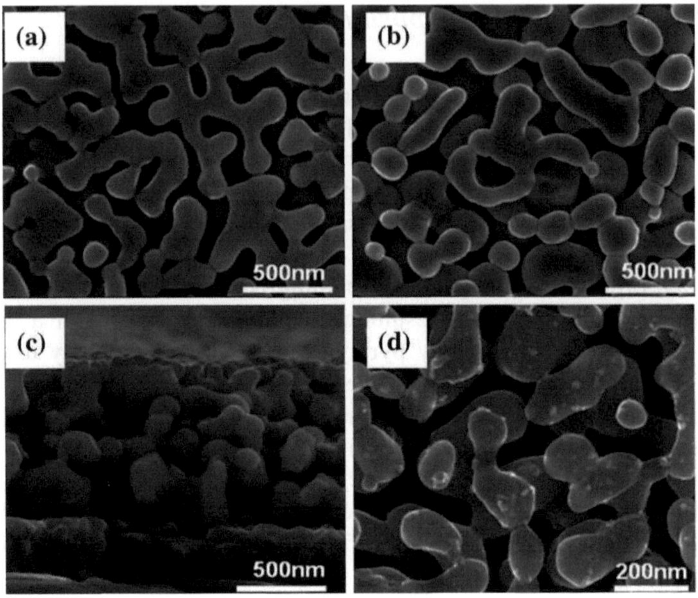

Fig. 3 SEM images of typical samples: **a** the surface of pure $BiVO_4$ film. **b** The surface of Mo-doped $BiVO_4$ film. **c** A cross-section of Mo-doped $BiVO_4$ on a FTO (F-doped tin oxide) substrate. **d** RhO_2 particles on the surface of a Mo-doped $BiVO_4$ film [49]. Reproduced from Ref. [49] with permission from the Royal Society of Chemistry

was obtained by the acidification of Na_2WO_4 solutions using a proton exchange resin. (2) Peroxopolytungstic acids (PPTA) was synthesized via the reaction of tungsten metal with hydrogen peroxide.

Wang et al. prepared a nanostructured WO_3 film using the sol–gel process (1) with the addition of carbowax and triton as the surfactants [53]. The photocurrent onset potential is approximately 0.1 V (vs. Ag/AgCl). A saturated photocurrent was reached at approximately 0.8 V (vs. Ag/AgCl). The photocurrent densities increased with increasing light intensity and temperature, and exhibited a linear relationship with the light intensity in the saturation regime.

Santato, Ulmann and Augustynski also prepared preferentially orientated WO_3 films using the sol–gel process with poly(ethylene glycol) 300 (PEG 300) as a stabilizing regent [54–56]. The size of the nanoparticles and the porosity of the films can be controlled by altering the annealing conditions and the ratio of tungsten acid to PEG 300. The sample without PEG 300 exhibited a morphology of irregular plate-like particles with substantial porosity of 200–400 nm after annealing at 500 °C. The sample with PEG 300 had a morphology of plate-like particles of 10–30 nm at the same annealing temperature. The maximum IPCE for a 2 mm-thick WO_3 film photoanode was approximately 75 % in a 1 M $HClO_4$ solution without CH_3OH at 1 V (vs. RHE) under illumination (100 mW/cm^2). The saturation photocurrent density reached

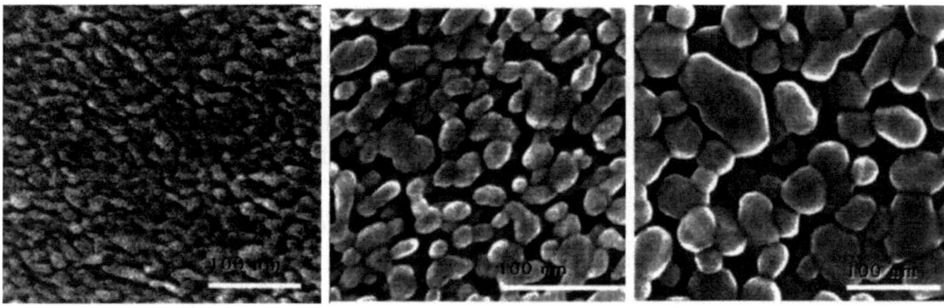

Fig. 4 SEM of WO_3 films prepared by the deposition of a tungsten acid/PEG 300 colloidal solution with annealing at 450 °C (*left*), at 500 °C (*middle*) and at 550 °C (*right*) for 30 min [56]. Reproduced from Ref. [56] with permission from the American Chemical Society

2.4 mA/cm^2 at 1.23 V (vs. RHE). Both the size and crystallinity of the WO_3 nanocrystals increased with increasing annealing temperature (Fig. 4 [56]).

Yang et al. prepared mesoporous WO_3 films using the sol–gel process [57–59] and created a comparison between the photocatalytic performance of the semi-transparent films and that of the transparent films. SEM images of the semi-transparent film indicated a network structure consisting of interconnected droplet-like nanoparticles with diameters of 100–300 nm. The average pore size and crystalline diameter were 12.5 and 17 nm, respectively. The transparent film exhibited smooth surfaces with occasional cracks, and its average pore diameter and crystallite size were 7.3 and 30 nm, respectively. The maximum photocurrent density of a 2.3 mm-thick semi-transparent WO_3 film was approximately 2.2 mA/cm^2 in a 1 M H_2SO_4 solution under light irradiation (100 mW/cm^2). 111 The IPCE of the semi-transparent WO_3 film at 410 nm was approximately 40 % in a 1.0 M H_2SO_4 solution at 1.0 V (vs. SCE).

3.3 Electrodeposition

Electrochemical synthesis is another simple solution-based method recently explored for the synthesis of crystalline semiconductor particles film [60]. In typical electrochemical synthesis methods, a working electrode and a counter electrode are immersed into a plating solution that contains ions and molecules that will participate in the electrochemical reactions. (A reference electrode is also added for a three electrode system.) Upon application of an electrochemical potential, a reduction (cathodic deposition) or oxidation reaction (anodic deposition) occurs on the working electrode. The species produced by the oxidation and reduction reactions trigger the deposition of desired materials on the working electrode surface. Sometimes, the counter electrode or the working electrode is not inert and also participates in the electrochemical reaction, for example, by oxidizing itself to provide ions necessary for the formation of the desired product.

$BiVO_4$ photoelectrode is often fabricated by this method and the first electrochemical synthesis of $BiVO_4$ was achieved by Myung et al. [61, 62]. In their synthesis, anodization of a bismuth metal coated Pt electrode resulted in the generation of Bi^{3+} ions, which reacted with V^{5+}-containing oxyanions in solution and precipitated crystalline $BiVO_4$ on the working electrode. In a more recent synthesis by Seabold et al.,46 a plating solution containing Bi^{3+} and V^{4+} ions (i.e., $VOSO_4$) was used. When V^{4+} ions were oxidized to V^{5+} ions upon application of an appropriate bias, they instantaneously reacted with Bi^{3+} ions and formed a film of an amorphous Bi–V–O phase having a Bi:V ratio of 2:3. After deposition, the films were thermally converted to crystalline $BiVO_4$ and V_2O_5. The V_2O_5 phase was easily removed with 1 M KOH leaving only a pure $BiVO_4$ film. The documented photocurrents for the as-prepared $BiVO_4$ photoanodes present at a lower level, generally being less than 0.5 mA/cm^2 in neutral or alkaline electrolyte which may be attributed to larger particle size ($\sim$1 μm) and compact morphology with limited surface area [61–64].

Recently, Choi et al. reported an improved electrochemical deposition to fabricate nanoporous $BiVO_4$ photoelectrode [65]. BiOI electrodes are firstly prepared by electrodeposition and then a dimethylsulfoxide (DMSO) solution of vanadyl acetylacetonate [VO(acac)$_2$] is applied onto their surface following by heating in air at 450 °C for 2 h. The specific advantage of using BiOI is that its two-dimensional crystal structure enables electrodeposition of extremely thin plates (*ca.* 20 nm) with sufficient voids between them. These voids inhibit grain growth of $BiVO_4$ during the conversion precess, resulting in nanoporous $BiVO_4$ electrodes. The top-view and side-view SEM images show the formation of much smaller $BiVO_4$ nanoparticles (mean particle size = 76 ± 5 nm) creating a three-dimensional nanoporous network. Photocurrent from the nanoporous $BiVO_4$ for water oxidation in phosphate buffer (pH 7) under AM 1.5 G illumination exceeds 1 mA/cm^2 when the applied bias ≥1.1 V versus RHE.

3.4 Anodization

Anodization technique is well known to prepare semiconductor materials with 3-dimensional morphology such as TiO_2 nanotube, nanowire, etc. In fact, this method can also be used to prepare some photoelectrodes with porous structure as demonstrated below [25].

A WO_3 porous nanostructure was grown on tungsten coated Ti foil using an anodization technique with a mixture of 22.5 ml H_2O, 1.8 g NH_4F^+ and 427.5 ml ethylene glycol as the electrolytes at 35 V for 1 or 2 h, followed by annealing for 30 min at 350 and 500 °C, respectively [66]. SEM images indicated that nanoporous sponge-like morphologies were formed with a diameter of 20–50 or 60 nm for 1 and 2 h anodization, respectively. The thickness of the porous nanostructures of the WO_3 layers approached 187 and 400 nm. The W layers, which were located between Ti substrate and WO_3 layers, were approximately 200 and 1.08 mm, respectively, The surface corrosion of the porous nanostructures of WO_3 in alkaline electrolytes was drastically reduced by coating the WO_3 surface with an ITO film (100 nm) using a DC magnetron sputtering technique at 300 °C, and the photocurrent density of the ITO-coated samples increased to 2.08 mA/cm^2 at 0.7 V (vs. Ag/AgCl) under 30 mW/cm^2. However, under the same conditions, the photocurrent density was approximately 0.61 mA/cm^2 for samples without the ITO-coating.

To study the effect of anodization conditions on the photoelectrochemical response of WO_3 films, Tacconi et al. prepared WO_3 films by anodization of tungsten foils using various electrolyte media in a two-electrode electrochemical cell [67]. The anodization voltages were controlled in the range of 10–60 V. SEM images indicated that a WO_3 film with a nanobowl morphology, with pore diameters of 44–85 nm and a photocurrent density of approximately 2.9 mA/cm^2, was formed by the anodization of tungsten foils in 0.3 M oxalic acid at constant voltage for 1 h. The WO_3 film grown from glycerol-H_2O (25:75) in 0.3 M oxalic acid at 35 V for 1 h had the highest photocurrent density of 2.8 mA/cm^2. The photocurrent density of the best WO_3 film anodized in 0.15 M NaF at 60 V for 3 h reached 3.52 mA/cm^2, and the maximum IPCE of the nanoporous film was up to 180 % in a 0.5 M Na_2SO_4 + 0.1 M HCOONa solution at 2.0 V (vs. Ag/AgCl).

Berger et al. prepared a WO_3 film with a porous structure by the anodization of tungsten foils at 40 V (vs. Ag/AgCl) in 1 M H_2SO_4 + 0.5 wt% NaF for 1 h [68]. SEM images indicated that a uniform porous layer was formed with a pore diameter of approximately 70 nm, pore spacing of 80 nm and a pore wall thickness of approximately 10–15 nm. After annealing at 500 °C, the as-anodized amorphous WO_3 was transformed into crystalline monoclinic WO_3. For comparison with the amorphous WO_3, approximately 90 nm compact WO_3 layers were prepared on W in a 1 M H_2SO_4 fluoride-free solution at 40 V (vs. Ag/AgCl). The photoelectrochemical studies indicated that films of both amorphous and compact layer structures resulted in a significantly enhanced photoresponse after the annealing process. The photocurrent efficiency of the amorphous WO_3 films was much greater than that of the compact layer films within comparable wavelengths.

Li et al. prepared nanoporous WO_3 films by anodization using tungsten foils in 1 M Na_2SO_4 with 0.5 wt% NaF for 30 min followed annealing at 450 °C for 3 h [69]. The compact WO_3 films for use as a reference were obtained by anodization in 1 M Na_2SO_4 without F^- anions. SEM images indicated that nanoporous WO_3 films with a pore diameter of 70–90 nm were formed at 50 V in 1 M Na_2SO_4 with 0.5 wt% NaF. The photocurrent densities of the nanoporous and compact WO_3 films were approximately 3.45 mA/cm^2 and 0.73 mA/cm^2, respectively, at 1.6 V (vs. Ag/AgCl) in a 0.5 M H_2SO_4 solution (pH = 0) under illumination (100 mW/cm^2). The maximum conversions of light energy to chemical energy for the annealed nanoporous and compact WO_3 films were approximately 0.91 and 0.20 %, respectively. The maximum IPCE values for the annealed nanoporous and compact WO_3 films were approximately 92 and 19 %, respectively, obtained at a wavelength of 340 nm at 1.2 V (vs. Ag/AgCl) in a 0.5 M H_2SO_4 solution. The annealed nanoporous WO_3 film demonstrated the smallest arc radius in the EIS Nyquist plot, indicating that it possessed the highest efficiency of charge separation and conductivities.

3.5 Sputtering

Sputtering technique has been widely used in industry for preparing large area and uniform films well connected to the substrates. The structure of the deposited film can be easily controlled by adjusting the process parameters, and it is also quite easy to repeat the producing process. Hence, this technique shows huge potential for practical large-scale manufacturing of thin film materials [70]. Some metal oxide or oxynitride photoelectrodes [70–75] had been prepared by this method for PEC water splitting, which showed good stability and anti-corrosion properties in electrolyte.

Marsen et al. reported the preparation of the WO_3 films with thicknesses of 1.2–1.6 mm by reactive RF magnetron sputtering from a metallic W target, in which the sputtering was carried out in an argon/oxygen (or nitrogen for N-doped films) atmosphere with the desired percentage of oxygen or nitrogen at low processing temperatures (100–250 °C) [72]. The film deposition rate, thickness, grain size, crystallinity and nanostructure morphology can be precisely controlled by adjusting the parameters (applied sputtering power, temperature, gas percentage, time and pressure). For a sample with a Scherrer grain size of 27 and 49 nm in the (002) and (200) planes, the highest photocurrent density for the WO_3 film was approximately 2.7 mA/cm^2 at 1.6 V (vs. SCE) in 0.33 M H_3PO_4 aqueous solution under simulated AM 1.5 global illumination. Using the same sputter target, they prepared another sample with a Scherrer grain size of 13 and 35 nm in the (002) and (200) planes, and its photocurrent density was approximately 1.5 mA/cm^2 under the same conditions. The maximum photocurrent density of a nitrogen-doped sample at low dopant concentration was approximately 2.3 mA/cm^2 at 1.6 V (vs. SCE).

Owing to the rigorous reaction conditions, the synthesis of Ta_3N_5 and TaON photoelectrodes by nitridation of Ta_2O_5 are restricted to several limited techniques. A simple procedure involves the surface oxidation of Ta foil followed by nitridation of the as-formed Ta_2O_5 layer on Ta foil under NH_3 flow [73]. The drawback of this method is the poor control of the film thickness as well as the structural discontinuities (cracks at the surface and gaps between the product layer and the substrate), which develop from expansion and contraction of the Ta sheet during the oxidation and nitridation. In an attempt to eliminate these drawbacks, radio-frequency magnetron sputtering technology is adopted to control precisely the composition and thickness of a film [74]. Packed Ta_3N_5 film can be directly grown on the substrate without undesirable surface cracks or gaps between the product layer and the substrate. However, the TaON byproduct must be removed, since the structure imperfections may behave as recombination centers for photogenerated electrons and holes, thus lowering the PEC performance. In addition, there are reports on atomic layer deposition for the direct generation of Ta_3N_5 or TaON films, and pulsed laser deposition for the growth of Ta_2O_5 film followed by thermal nitridation under ammonia to produce Ta_3N_5 and TaON films [75, 76].

However, the films prepared by sputtering technique generally present mirror-like surface as well as compact structure that result in high reflectivity and

Fig. 5 SEM images of W–Cu–O film etched in **a** 0.5 M HCl, **b** 1 M HCl, **c** 2 M HCl **d** 4 M HCl at room temperature; scale bars: 1 μm [70]. Reproduced from Ref. [70] with permission from the Royal Society of Chemistry

small contact surface area with electrolyte, which make their PEC water splitting performances greatly limited. To overcome this shortcoming, Wang et al. reported a two-step method to obtain porous WO_3 photoanodes by magnetron co-sputtering followed chemical etching to remove the impurity phase [70]. Some extra metals (M = Cu, Zn or Al) were sputtered together with tungsten to produce the mixed W-M-O compounds thin films on FTO glass substrates. After soaking in acid solution of HCl, H_2SO_4 or HNO_3 for etching at room temperature, these extra metal oxides disappeared due to dissolution in acidic media while a porous plate-like WO_3 phase (Fig. 5), distinct from those conventional compact WO_3 thin films prepared by a magnetron sputtering, were left behind with twice photocurrent enhancement for PEC water oxidation (about 1.0 mA/cm^2 at 1.2 V vs. SCE in aqueous Na_2SO_4 solution with pH 3). This approach can be general and useful for the fabrication of some other semiconductor photoelectrodes.

3.6 Hydrothermal Synthesis

Hydrothermal synthesis is well known as an effective method for preparing crystalline ceramic powders at moderate temperatures [77, 78]. In addition to the moderate synthesis conditions, the hydrothermal synthesis provides easy control of geometric properties such as surface morphology, surface area, and particle size, through the appropriate selection of starting materials and simple adjustment of synthesis conditions, i.e., temperature, time, and mixing ratio. Some porous films typical as WO_3 can also in situ grow under hydrothermal environment on the desired substrates to fabricate photoelectrodes [79–81].

Starting from tungsten sheets, Amano et al. prepared the porous plate-like WO_3 films on metal tungsten substrate in a dilute nitric acid solution at 100–180 °C with subsequent calcination at 450 °C via hydrothermal reaction [79]. The optimized thickness of the film was approximately 1.6 mm with a compact layer under the plate-like structures, which was obtained for 1 h at 180 °C. The photocurrent onset potential was approximately 0.2 V (vs. Ag/AgCl). The IPCE was up to 66.2 %. The photocurrent density reached approximately 4 mA/cm^2 in a 0.1 M H_2SO_4 solution at 1.2 V (vs. Ag/AgCl) under photoirradiation. However, the photocurrent density will be significantly diminished if the reaction times are extended beyond the optimum time; prolonged reaction times at a higher temperature resulted in a rapid increase in the thickness of the compact layer, which impeded the electron transport.

The porous WO_3 films can also be deposited on a common substrate as FTO by one-step hydrothermal method. Jiao et al. selectively fabricated the plate-like, wedge-like, and sheet-like nanostructured WO_3 films on FTO glasses by a crystal-seed-assisted hydrothermal method using $Na_2WO_4{\cdot}2H_2O$ as a tungsten source and Na_2SO_4, $(NH_4)_2SO_4$, and CH_3COONH_4 as capping agents followed by calcination in air at 500 °C for 30 min, as schematically shown in Fig. 6 [80]. The morphology could be controlled by altering the crystal seeds layers. With CH_3COONH_4 as a capping agent, the highest photocurrent density was approximately 0.5 mA/cm^2 at 1.45 V (vs. Ag/AgCl) in a 1 M H_2SO_4 solution under illumination (100 mW/cm^2). The highest efficiency for the photoconversion of light energy to chemical energy for water splitting was approximately 0.3 % at 0.65 V (vs. Ag/AgCl).

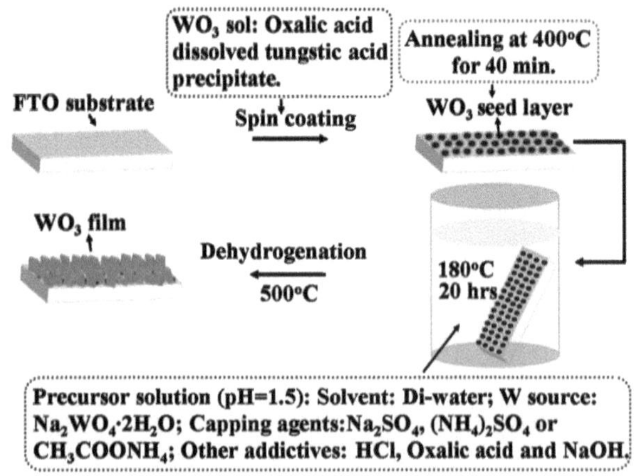

Fig. 6 Schematic illustration of the film fabrication process [80]. Reproduced from Ref. [80] with permission from the American Chemical Society

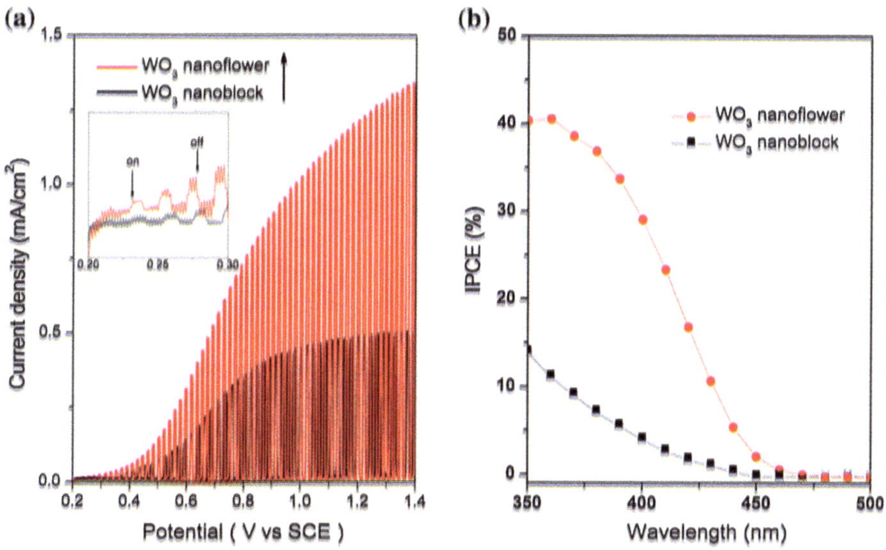

Fig. 7 **a** Linear sweep voltammetric scans of WO_3 photoanodes in 1 M H_2SO_4 under chopped light illumination; light source: AM 1.5G; scanning rate: 20 mV s^{-1}. **b** IPCE for WO_3 photoelectrodes at 0.8 V applied potentials (vs. SCE) in 1 M H_2SO_4 [81]. Reproduced from Ref. [81] with permission from the Royal Society of Chemistry

By applying microwave irradiation, Wang et al. synthesized the hexagonal nanoflower WO_3 particle on FTO glass substrate by using $RCOO^-$ as the structure directing agent in the hydrothermal synthesis process [81]. It is found that the perpendicular nano-flower arrays (~1 μm in thickness) well grow on pre-seeding FTO, while random growth is observed on naked substrates. Without additive of capping agent in synthesis, the WO_3 film with nanoblock morphology was formed. The PEC measurements indicated that the photocurrent density of nanoflower WO_3 is almost three times that of the nanoblock one at 1.4 V versus SCE in 1 M H_2SO_4 aqueous solution (Fig. 7a). The incident photon-to-current efficiencies (IPCEs) of flower-like and block-like electrodes are 29 and 4 % at the wavelength of 400 nm, respectively, indicating that the nanoflower electrode is more effective for PEC water oxidation than the nanoblock one (Fig. 7b).

3.7 *Spray Pyrolysis*

Spray pyrolysis is a convenient, fast and economical technique to prepare powders and films. Using this method, films with desired stoichiometric ratio can be easily deposited and film thickness can be convenient controlled [82]. Moreover, it is easy to dope films with various elements by simply adding them into the spray precursor solution. Furthermore, this method is easy to scale up for large area film preparation.

The spray pyrolysis method has been introduced to prepare pure and W-doped $BiVO_4$ photoelectrodes by Guo and his coworkers [82]. The $BiVO_4$ electrode is less than 150 nm in thickness and highly transparent to wavelengths longer than 500 nm. The IPCE of $BiVO_4$ photoelectrodes at 420 nm and 0.9 V versus RHE is only about 1 % in 0.5 M Na_2SO_4 solution, not only because the $BiVO_4$ photoelectrode is too thin and cannot absorb more light, but also because the doping concentration is very low. Some ion-doped Fe_2O_3 photoanodes prepared by spray pyrolysis [83, 84] also seldom achieve high photocurrent in mA scale.

$BiVO_4$ electrodes have also been deposited using a spray pyrolysis MOD method by Adbi et al. where Bi and V containing precursor solutions are sprayed onto heated substrates to instantaneously evaporate the solutions and decompose the precursor molecules [52]. After the spin coating or spray pyrolysis deposition, the electrodes were heat treated in air resulting in crystalline $BiVO_4$ electrodes. The photocurrent density demonstrated at 1.0 and 1.5 V vs RHE in K_2SO_4 electrolyte (pH 5.6) are about 0.25 and 1.20 mA/cm^2, respectively.

3.8 Chemical Bath Deposition (CBD)

A CBD method is cheap and easy to scale up for industrial production. $BiVO_4$ photoelectrodes have been prepared onto the FTO substrates, using the CBD method by the present authors [85]. $Bi(NO_3)_3 \cdot 5H_2O$ and NH_4VO_3 as Bi and V sources were dissolved in different aqueous solution, and ethylenediaminetetraacetic acid (EDTA) was added as a stabilizer. However, the IPCE of as-prepared $BiVO_4$ photoelectrodes is still low (ca. 5 % at 400 nm and high potential 1.6 V vs. RHE), owing to large particle and poor contact between the particles. A $BiVO_4$ seed layer has been introduced in the CBD method to prepared $BiVO_4$ photoelectrodes on the FTO glass, thus obtaining the smaller particle size of $BiVO_4$, because the $BiVO_4$ seed layer favors its nucleation [86]. However, the PEC performance is still low, possibly due to its bad crystallinity.

In 2012, Kudo et al. reported a facile method to fabricate $BiVO_4$ thin film electrode with greatly improved PEC water oxidation activity based on CBD route [87]. In their synthesis, an aqueous nitric acid solution was used as solvent to dissolve $Bi(NO_3)_3$ and NH_4VO_3 sources and the resultant yellow precursor solution was applied to FTO glass substrate for film. After dryness at room temperature, an orange film was obtained and subsequently calcination of this film at 573–723 K in air to give a yellow $BiVO_4$ thin film with 300–400 nm of the thickness. Anodic photocurrent of the $BiVO_4$ thin film electrode depended on the concentration of a precursor solution and calcination temperature. The increasing anodic photocurrent with the elevating calcination temperature (I–V curves in Fig. 8) was due to the

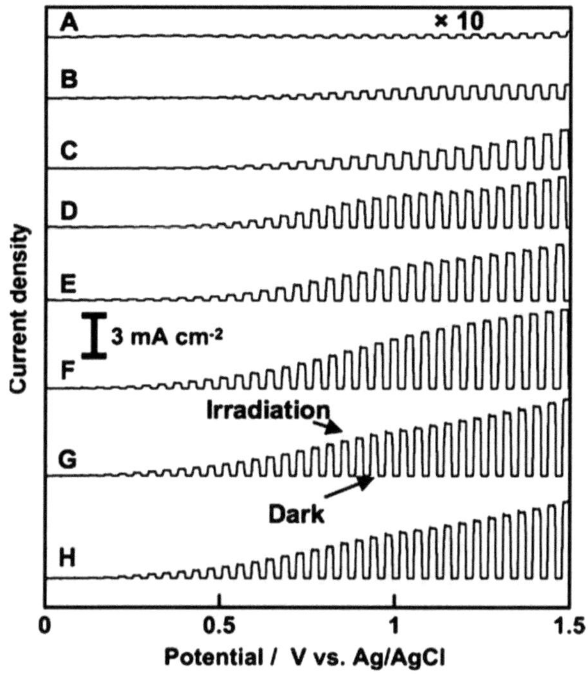

Fig. 8 Current versus potential curves of $BiVO_4$ thin film electrodes under visible light irradiation. Concentration of precursor solution and calcinations temperature, *A* 200 mmol/l, 573 K, *B* 200 mmol/l, 623 K, *C* 20 mmol/l, 673 K, *D* 50 mmol/l, 673 K, *E* 100 mmol/l, 673 K, *F* 200 mmol/l, 673 K, *G* 300 mmol/l, 673 K, and *H* 200 mmol/l, 723; calcination time, 2 h. Electrolyte, 0.1 mmol/l of an aqueous K_2SO_4 solution; sweep rate, 20 mV/s; light source, 300 W Xe-lamp with a cut-off filter ($\lambda > 420$ nm) [87]. Reproduced from Ref. [87] with permission from the National Academy of Sciences of the United States of America

improvement of the denseness of the thin film, and the purity and crystallinity of scheelite-monoclinic phase. On the other hand, as the thickness of the $BiVO_4$ film was increased, the anodic photocurrent became large. The $BiVO_4$ thin film prepared using 200–300 mmol/l of precursor solutions gave an excellent photoelectrochemical property. Thicknesses of these thin films were sufficient for light absorption. Consequently, the $BiVO_4$ thin film electrode prepared at 673–723 K using 200–300 mmol/l of the precursor solution gave an excellent anodic photocurrent with 73 % of an IPCE at 420 nm at 1 V versus Ag/AgCl. Photoelectrochemical water splitting using the $BiVO_4$ thin film proceeded with applying an external bias smaller than 1.23 V, which is a theoretical voltage required for electrolysis of water. The amount of H_2 evolved was similar to the half of the amount of electron passing through the outer circuit, indicating that the photocurrent was due to water splitting.

3.9 Other Methods

Some other preparation methods such as molecular beam epitaxy (MBE), pulsed laser deposition (PLD), atomic layer deposition (ALD) can also used to fabricate thin film photoelectrodes although their applications in the photoelectrodes are very limited now. More interestingly, the fabrication of nitride semiconductor materials such as Ta_3N_5 often has to carry out by multi-step methods because of the difficulty in one-step nitridization. For example, the tantalum foil was firstly anodized to give Ta_2O_5 film and further reactive with NaOH in hydrothermal environment to produce perovskite $NaTaO_3$ before it finally converted into porous cubic Ta_3N_5 thin film on Ta metal substrate through nitridization [88].

4 Conclusion

For using photocataysts in PEC approach, many synthesis routes including physical and chemical processes have been developed to fabricate photoelectrodes from powdered photocatalysts directly and indirectly, respectively. However, most methods are limited to prepare photoelectrodes from one or few kinds of materials. Moreover, the energy conversion efficiencies for PEC water splitting based on the as-fabricated photoelectrodes are much lower than their theoretical maximums (generally less than one tenth). Thus, considerable efforts are still required for the development of new methods and improvement of known methods.

References

1. Fujishima A, Honda K (1972) Electrochemical photolysis of water at a semiconductor electrode. Nature 238:37–38
2. Walter MG, Warren EL, McKone JR, Boettcher SW, Mi Q, Santori EA, Lewis NS (2010) Solar water splitting cells. Chem Rev 110:6446–6473
3. Bak T, Nowotny J, Rekas M, Sorrell CC (2002) Photo-electrochemical hydrogen generation from water using solar energy. Materials-related aspects. Int J Hydrogen Energy 27:991–1022
4. Honda K (1979) In: Ohta T (ed) Solar-hydrogen energy systems. Pergamon Press, Oxford, pp 137–169
5. Seraphin BO (ed) (1979) Solar energy conversion. Springer, Berlin, pp 5–56
6. Chandra S (1985) Photoelectrochemical solar cells. Gordon and Breach, NewYork
7. Morrison SR (1980) Electrochemistry at semiconductor and oxidized metal electrodes. Plenum Press, New York, pp 1–401
8. Nozik AJ (1977) In: Heller A (ed) Semiconductor liquid-junction solar cells: proceedings of the conference on the electrochemistry and physics of semiconductor liquid interfaces under illumination. Airlie, Virginia, pp 272–292
9. Minggu LJ, Daud WRW, Kassim MB (2010) An overview of photocells and photoreactors for photoelectrochemical water splitting. Int J Hydrogen Energy 35:5233–5244

10. Bak T, Nowotny J, Sorrell CC (1997) In Nowotny J, Sorrell CC (eds) Electrical properties of oxide materials. Trans Tech Publ., Enfield, p 1
11. Nowotny J, Radecka M, Rekas M (1997) Semiconducting properties of undoped TiO_2. J Phys Chem Solids 58:927
12. Wrighton MS, Ellis AB, Wolczanski PT, Morse DL, Abrahamson HB, Ginley DS (1976) Strontium- titanate photoelectrodes—efficient photoassisted electrolysis of water at zero applied potential. J Am Chem Soc 98:2774–2779
13. Mavroides JG, Kafalas JA, Kolesar DF (1976) Photoelectrolysis of water in cells with $SrTiO_3$ anodes. Appl Phys Lett 28:241–243
14. Khaselev O, Turner JA (1998) A monolithic photovoltaic-photoelectrochemical device for hydrogen production via water splitting. Science 280:425–427
15. Reece SY, Hamel JA, Sung K, Jarvi TD, Esswein AJ, Pijpers JJH, Nocera DG (2011) Wireless solar water splitting using silicon-based semiconductors and earth-abundant catalysts. Science 334:645–648
16. Nocera DG (2012) The artificial leaf. Acc Chem Res 45:767–776
17. Frank E (2013) Osterloh, Inorganic nanostructures for photoelectrochemical and photocatalytic water splitting. Chem Soc Rev 42:2294–2320
18. Hoffmann MR, Martin ST, Choi W, Bahnemann DW (1995) Environmental applications of semiconductor photocatalysis. Chem Rev 95:69–96
19. Hagfeldt A, Grätzel M (1995) Light-induced redox reactions in nanocrystalline systems. Chem Rev 95:49–68
20. Bard AJ (1979) Photoelectrochemistry and heterogeneous photocatalysis at semiconductors. J. Photochem. 10:59–75
21. Li Z, Luo W, Zhang M, Feng J, Zou Z (2013) Photoelectrochemical cells for solar hydrogen production: current state of promising photoelectrodes, methods to improve their properties, and outlook. Energy Environ Sci 6:347–370
22. Long M, Cai W, Kisch H (2008) Visible light induced photoelectrochemical properties of n-$BiVO_4$ and n-$BiVO_4$/p-CO_3O_4. J Phys Chem C 112:548–554
23. Iwase A, Kudo A (2010) Photoelectrochemical water splitting using visible-light-responsive $BiVO_4$ fine particles prepared in an aqueous acetic acid solution. J Mater Chem 20:7536–7542
24. Matsumura M, Saho Y, Tsubomura H (1983) Photocatalytic hydrogen production from solutions of sulfite using platinized cadmium sulfide powder. J Phys Chem 87:3807–3808
25. Liu X, Wang F, Wang Q (2012) Nanostructure-based WO_3 photoanodes for photoelectrochemical water splitting. Phys Chem Chem Phys 14:7894–7911
26. Hong SJ, Jun H, Borse PH, Lee JS (2009) Size effects of WO_3 nanocrystals for photooxidation of water in particulate suspension and photoelectrochemical film systems. Int J Hydrogen Energy 34:3234–3242
27. Maeda K, Hashiguchi H, Masuda H, Abe R, Domen K (2008) Photocatalytic activity of $(Ga_{1-x}Zn_x)(N_{1-x}O_x)$ for visible-light-driven H_2 and O_2 evolution in the presence of sacrificial reagents. J Phys Chem C 112:3447–3452
28. Maeda K, Teramura K, Lu D, Takata T, Saito N, Inoue Y, Domen K (2006) Photocatalyst releasing hydrogen from water—Enhancing catalytic performance holds promise for hydrogen production by water splitting in sunlight. Nature 440:295–295
29. Maeda K (2013) Z-scheme water splitting using two different semiconductor photocatalysts. ACS Catal. 3:1486–1503
30. Takanabe K, Domen K (2012) Preparation of inorganic photocatalytic materials for overall water splitting. ChemCatChem 4:1485–1497
31. Nishimura N, Raphael B, Maeda K, Gendre LL, Abe R, Kubota J, Domen K (2010) Effect of $TiCl_4$ treatment on the photoelectrochemical properties of $LaTiO_2N$ electrodes for water splitting under visible light. Thin Solid Films 518:5855–5859
32. Abe R, Takata T, Sugihara H, Domen K (2005) The use of $TiCl_4$ treatment to enhance the photocurrent in a TaON photoelectrode under visible light irradiation. Chem Lett 34:1162–1163

33. Maeda K, Nishimura N, Domen K (2009) A precursor route to prepare tantalum (V) nitride nanoparticles with enhanced photocatalytic activity for hydrogen evolution under visible light. Appl Catal A 370:88–92
34. Abe R, Higashi M, Domen K (2010) Facile fabrication of an efficient oxynitride TaON photoanode for overall water splitting into H_2 and O_2 under visible light irradiation. J Am Chem Soc 132:11828–11829
35. Sarkar P, Nicholson PS (1996) Electrophoretic deposition (EPD): mechanisms, kinetics, and application to ceramics. J Am Ceram Soc 79:1987–2002
36. Higashi M, Abe R, Domen K (2011) Fabrication of efficient TaON and Ta_3N_5 photoanodes for water splitting under visible light irradiation. Energy Environ Sci 4:4138–4147
37. Abe R (2010) Recent progress on photocatalytic and photoelectrochemical water splitting under visible light irradiation. J Photochem Photobiol C 11:179–209
38. Moon S-J, So W-W, Chang H-Y (2001) Effect of $TiCl_4$ treatment on photoelectrochemical properties of nanocrystalline CdS particulate films. J Electrochem Soc 148:E378–E381
39. Jo WJ, Jang J-W, Kong K-J, Kang HJ, Kim JY, Jun H, Parmar KPS, Lee JS (2012) Phosphate doping into monoclinic $BiVO_4$ for enhanced photoelectrochemical water oxidation activity. Angew Chem Int Ed 51:3147–3151
40. Wang D, Li R, Zhu J, Shi J, Han J, Zong X, Li C (2012) Photocatalytic water oxidation on $BiVO_4$ with the electrocatalyst as an oxidation cocatalyst: essential relations between electrocatalyst and photocatalyst. J Phys Chem C 116:5082–5089
41. Hisatomi T, Kubota J, Domen K (2014) Recent advances in semiconductors for photocatalytic and photoelectrochemical water splitting. Chem Soc Rev 43:7520–7535
42. Minegishi T, Nishimura N, Kubota J, Domen K (2013) Photoelectrochemical properties of $LaTiO_2N$ electrodes prepared by particle transfer for sunlight-driven water splitting. Chem Sci 4:1120–1124
43. Kumagai H, Minegishi T, Moriya Y, Kubota J, Domen K (2014) Photoelectrochemical hydrogen evolution from water using copper gallium selenide electrodes prepared by a particle transfer method. J Phys Chem C 118:16386–16392
44. Liu J, Hisatomi T, Ma G, Iwanaga A, Minegishi T, Moriya Y, Katayama M, Kubota J, Domen K (2014) Improving the photoelectrochemical activity of $La_5Ti_2CuS_5O_7$ for hydrogen evolution by particle transfer and doping. Energy Environ Sci 7:2239–2242
45. Meignen V, Cario L, Lafond A, Moëlo Y, Guillot-Deudon C, Meerschaut A (2004) Crystal structures of two new oxysulfides $La_5Ti_2MS_5O_7$ (M = Cu, Ag): evidence of anionic segregation. J Solid State Chem 177:2810–2817
46. Katayama M, Yokoyama D, Maeda Y, Ozaki Y, Tabata M, Matsumoto Y, Ishikawa A, Kubota J, Domen K (2010) Fabrication and photoelectrochemical properties of $La_5Ti_2MS_5O_7$ (M = Ag, Cu) electrodes. Mater Sci Eng, B 173:275–278
47. Sayama K, Nomura A, Zou Z, Abe R, Abe Y, Arakawa H (2003) Photoelectrochemical decomposition of water on nanocrystalline $BiVO_4$ film electrodes under visible light. Chem Commun 2908–2909
48. Sayama K, Nomura A, Arai T, Sugita T, Abe R, Yanagida M, Oi T, Iwasaki Y, Abe Y, Sugihara H (2006) Photoelectrochemical decomposition of water into H_2 and O_2 on porous $BiVO_4$ thin-film electrodes under visible light and significant effect of Ag ion treatment. J Phys Chem B 110:11352–11360
49. Luo WJ, Yang ZS, Li ZS, Zhang JY, Liu JG, Zhao ZY, Wang ZQ, Yan SC, Yu T, Zou ZG (2011) Solar hydrogen generation from seawater with a modified $BiVO_4$ photoanode. Energy Environ Sci 4:4046–4051
50. Zhong DK, Choi S, Gamelin DR (2011) Near-complete suppression of surface recombination in solar photoelectrolysis by "Co-Pi" catalyst-modified W:$BiVO_4$. J Am Chem Soc 133:18370–18377
51. Pilli SK, Furtak TE, Brown LD, Deutsch TG, Turner JA, Herring AM (2011) Cobalt-phosphate (Co-Pi) catalyst modified Mo-doped $BiVO_4$ photoelectrodes for solar water oxidation. Energy Environ Sci 4:5028–5034

52. Abdi FF, van de Krol R (2012) Nature and light dependence of bulk recombination in Co-Pi-Catalyzed $BiVO_4$ photoanodes. J Phys Chem C 116:9398–9404
53. Wang HL, Lindgren T, He JJ, Hagfeldt A, Lindquist SE (2000) Photolelectrochemistry of nanostructured WO_3 thin film electrodes for water oxidation: mechanism of electron transport. J Phys Chem B 104:5686–5696
54. Santato C, Ulmann M, Augustynski J (2001) Enhanced visible light conversion efficiency using nanocrystalline WO_3 films. Adv Mater 13:511–514
55. Santato C, Ulmann M, Augustynski J (2001) Photoelectrochemical properties of nanostructured tungsten trioxide films. J Phys Chem B 105:936–940
56. Santato C, Odziemkowski M, Ulmann M, Augustynski J (2001) Crystallographically oriented Mesoporous WO_3 films: synthesis, characterization, and applications. J Am Chem Soc 123:10639–10649
57. Yang B, Zhang YJ, Drabarek E, Barnes PRF, Luca V (2007) Enhanced photoelectrochemical activity of sol-gel tungsten trioxide films through textural control. Chem Mater 19:5664–5672
58. Yang B, Barnes PRF, Zhang YJ, Luca V (2007) Tungsten trioxide films with controlled morphology and strong photocatalytic activity via a simple sol-gel route. Catal Lett 118:280–284
59. Yang B, Barnes PRF, Bertrama W, Luca V (2007) Strong photoresponse of nanostructured tungsten trioxide films prepared via a sol-gel route. J Mater Chem 17:2722–2729
60. Park Y, McDonald KJ, Choi K-S (2013) Progress in bismuth vanadate photoanodes for use in solar water oxidation. Chem Soc Rev 42:2321–2337
61. Myung N, Ham S, Choi S, Chae Y, Kim W-G, Jeon YJ, Paeng K-J, Chanmanee W, de Tacconi NR, Rajeshwar K (2011) Tailoring interfaces for electrochemical synthesis of semiconductor films: $BiVO_4$, Bi_2O_3, or composites. J Phys Chem C 115:7793–7800
62. Dall'Antonia LH, de Tacconi NR, Chanmanee W, Timmaji H, Myung N, Rajeshwar K (2010) Electrosynthesis of Bismuth vanadate photoelectrodes. Electrochem Solid-State Lett 13:D29–D32
63. Seabold JA, Choi K-S (2012) Efficient and stable photo-oxidation of water by a bismuth vanadate photoanode coupled with an iron oxyhydroxide oxygen evolution catalyst. J Am Chem Soc 134:2186–2192
64. Ding C, Shi J, Wang D, Wang Z, Wang N, Liu G, Xiong F, Li C (2013) Visible light driven overall water splitting using cocatalyst/$BiVO_4$ photoanode with minimized bias. Phys Chem Chem Phys 15:4589–4595
65. Kim TW, Choi K-S (2014) Nanoporous $BiVO_4$ photoanodes with dual-layer oxygen evolution catalysts for solar water splitting. Science 343:990–994
66. Ishihara H, Kannarpady GK, Khedir KR, Woo J, Trigwell S, Biris AS (2011) A novel tungsten trioxide (WO_3)/ITO porous nanocomposite for enhanced photo-catalytic water splitting. Phys Chem Chem Phys 13:19553–19560
67. de Tacconi NR, Chenthamarakshan CR, Yogeeswaran G, Watcharenwong A, de Zoysa RS, Basit NA, Rajeshwar K (2006) Nanoporous TiO_2 and WO_3 films by anodization of titanium and tungsten substrates: Influence of process variables on morphology and photoelectrochemical response. J. Phys. Chem. B 110:25347–25355
68. Berger S, Tsuchiya H, Ghicov A, Schmuki P (2006) High photocurrent conversion efficiency in self-organized porous WO_3. Appl Phys Lett 88:203119
69. Li WZ, Li J, Wang X, Luo S, Xiao J, Chen QY (2010) Visible light photoelectrochemical responsiveness of self-organized nanoporous WO_3 films. Electrochim Acta 56:620–625
70. Wang N, Zhu J, Zheng X, Xiong F, Huang B, Shi J, Li C (2014) A facile two-step method for fabrication of plate-like WO_3 photoanode under mild conditions. Faraday Discuss 176:185–197
71. Chen L, Alarcón-Lladó E, Hettick M, Sharp ID, Lin Y, Javey A, Ager JW (2013) Reactive sputtering of bismuth vanadate photoanodes for solar water splitting. J Phys Chem C 117:21635–21642
72. Marsen B, Miller EL, Paluselli D, Rocheleau RE (2007) Progress in sputtered tungsten trioxide for photoelectrode applications. Int J Hydrogen Energy 32:3110–3115

73. Ishikawa A, Takata T, Kondo JN, Hara M, Domen K (2004) Electrochemical behavior of thin Ta_3N_5 semiconductor film. J Phys Chem B 108:11049–11053
74. Yokoyama D, Hashiguchi H, Maeda K, Minegishi T, Takata T, Abe R, Kubota J, Domen K (2011) Ta_3N_5 photoanodes for water splitting prepared by sputtering. Thin Solid Films 519:2087–2092
75. Ritala M, Kalsi P, Riihelä D, Kukli K, Leskelä M, Jokinen J (1999) Controlled growth of TaN, Ta_3N_5, and TaO_xN_y thin films by atomic layer deposition. Chem Mater 11:1712–1718
76. Kerlau M, Merdrignac-Conanec O, Guilloux-Viry M, Perrin A (2004) Synthesis of crystallized TaON and Ta_3N_5 by nitridation of Ta_2O_5 thin films grown by pulsed laser deposition. Solid State Sci 6:101–107
77. Yokoi T, Sakuma J, Maeda K, Domen K, Tatsumia T, Kondo JN (2011) Preparation of a colloidal array of $NaTaO_3$ nanoparticles via a confined space synthesis route and its photocatalytic application. Phys Chem Chem Phys 3:2563–2570
78. Shi J, Liu G, Wang N, Li C (2012) Microwave-assisted hydrothermal synthesis of perovskite $NaTaO_3$ nanocrystals and its photocatalytic properties. J Mater Chem 22:18808–18813
79. Amano F, Tian M, Wu GS, Ohtani B, Chen AC (2011) Facile preparation of platelike tungsten oxide thin film electrodes with high photoelectrode activity. ACS Appl Mater Interfaces 3:4047–4052
80. Jiao ZH, Wang JM, Ke L, Sun XW, Demir HV (2011) Morphology-tailored synthesis of tungsten trioxide (hydrate) thin films and their photocatalytic properties. ACS Appl Mater Interfaces 3:229–236
81. Wang N, Wang D, Li M, Shi J, Li C (2014) Photoelectrochemical water oxidation on photoanodes fabricated with hexagonal nanoflower and nanoblock WO_3. Nanoscale 6:2061–2066
82. Li MT, Zhao L, Guo LJ (2010) Preparation and photoelectrochemical study of $BiVO_4$ thin films deposited by ultrasonic spray pyrolysis. Int J Hydrogen Energy 35:7127–7133
83. Sartoretti CJ, Alexander BD, Solarska R, Rutkowska IA, Augustynski J (2005) Photoelectrochemical oxidation of water at transparent ferric oxide film electrodes. J Phys Chem B 109:13685–13692
84. Satsangi V, Kumari S, Singh A, Shrivastav R, Dass S (2008) Nanostructured hematite for photoelectrochemical generation of hydrogen. Int J Hydrogen Energy 33:312–318
85. Luo WJ, Wang ZQ, Wan LJ, Li ZS, Yu T, Zou ZG (2010) Synthesis, growth mechanism and photoelectrochemical properties of $BiVO_4$ microcrystal electrodes. J Phys D Appl Phys 43:405402
86. Su J, Guo L, Yoriya S, Grimes CA (2010) Aqueous growth of pyramidal-shaped $BiVO_4$ nanowire arrays and structural characterization: application to photoelectrochemical water splitting. Cryst Growth Des 10:856–861
87. Jia Q, Iwashina K, Kudo A (2012) Facile fabrication of an efficient $BiVO_4$ thin film electrode for water splitting under visible light irradiation. Proc Natl Acad Sci USA 109:11564–11569
88. Liu G, Shi J, Zhang F, Chen Z, Han J, Ding C, Chen S, Wang Z, Han H, Li C (2014) A tantalum nitride photoanode modified with a hole-storage layer for highly stable solar water splitting. Angew Chem Int Ed 53:7295–7299

Solar Hydrogen Production on Photocatalysis-Electrolysis Hybrid System Using Redox Mediator and Porous Oxide Photoelectrodes

Kazuhiro Sayama

Abstract Solar hydrogen production from water using semiconductor photocatalysts and photoelectrodes is one of the important artificial photosynthesis technologies to achieve a sustainable hydrogen society based on solar energy conversion. It is necessary to develop new systems with practical efficiency and significantly simpler and lower cost technologies than those that combine "photovoltaic generation and electrolysis". A photocatalysis-electrolysis hybrid system is a breakthrough system that solves almost all the disadvantages of conventional photocatalysis reactions by replacing the photocatalysis reaction on the hydrogen-production side of the Z-scheme reaction with electrolysis. Solar splitting of water using porous oxide photoanodes prepared by simple wet process is also a promising system to produce low cost hydrogen. It should be noted that maintaining the external bias at 1.23 V or less means that the apparent electrolytic efficiency can be 100 % or higher in the photoanode system that uses light energy as well as in the photocatalysis-electrolysis hybrid system.

1 Introduction

The term "solar hydrogen production," which refers to the use of photocatalysts and photoelectrodes, is becoming popular in the area of artificial photosynthesis technology. Solar hydrogen production focuses on the production of hydrogen from water using solar energy to achieve a clean and sustainable hydrogen society. While the ultimate goal of research on artificial photosynthesis and solar hydrogen production technology is to solve energy problems on a global scale, it is important to note that solar hydrogen production technology is strongly goal-oriented. Figure 1 shows a technological map in which the vertical and horizontal axes represent the

K. Sayama (✉)
Energy Technology Research Institute, AIST Tsukuba Central 5, 1-1-1 Higashi, Tsukuba 305-8565, Japan
e-mail: k.sayama@aist.go.jp

M. Sugiyama et al. (eds.), *Solar to Chemical Energy Conversion*, Lecture Notes in Energy 32, DOI 10.1007/978-3-319-25400-5_20

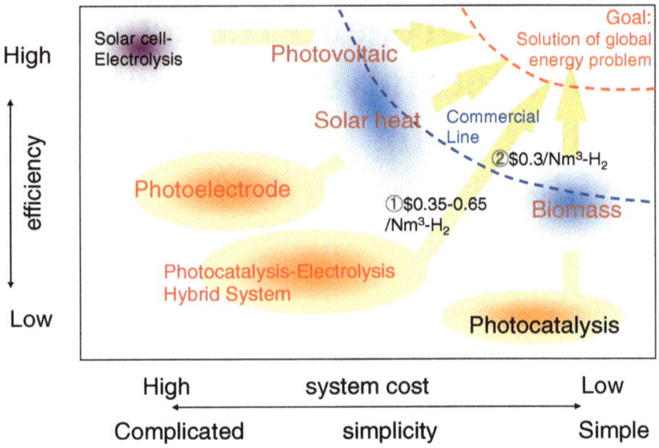

Fig. 1 Technology map for various solar energy conversion systems

solar energy conversion efficiency and cost or system complexity, respectively, of various solar energy conversion technologies. Photovoltaic generation, the use of solar heat, and fuel production using biomass are currently suitable for practical application. While current solar hydrogen production technologies that use photocatalysts and photoelectrodes are still far from being practically applied, it is necessary to investigate their future potentials for practical efficiency and significantly simpler and lower cost systems than those that combine "photovoltaic generation and electrolysis". The application and spread of solar hydrogen production technology as well as the future realization of a renewable energy society that does not depend on fossil resources are challenging and may require time. Therefore, it is desirable to strategically conduct research with an eye to future goals, i.e., it is necessary to determine which technology can be applied practically in order to achieve the ultimate goal of solving the global energy problems.

For long-term research, it is extremely important to set the research scope and design a roadmap for practical use, interim goals, and long-term prospects in addition to clarifying the ultimate significance of the research.

We set the following two clear interim goals, as shown in Fig. 1: Reduce the cost of generating hydrogen to lower than that obtained ① by simply combining photovoltaic generation and electrolysis, and ② from reforming fossil-fuel resources. The two interim goals that we present are greatly dependent on the costs of photovoltaic generation and fossil resources. The cost of hydrogen production has been calculated to be less than $\$0.35$–$0.65/\mathrm{Nm}^3$ for interim goal ① and $\$0.30/\mathrm{Nm}^3$ for interim goal ②. In the future, to achieve these cost goals and a clean hydrogen society using renewable energy, it is necessary to promote the early development of innovative technology. Even for ideal technologies, it is necessary to consider the time that is needed for realization, as there may be a number of very difficult issues to overcome.

2 Comparison of Various Solar Hydrogen Production Technologies Using Semiconductors

Figure 2 shows various water decomposition technologies involving photocatalysts and electrolysis and their potential diagram. During water decomposition using photocatalysts, the potential of the conduction band must be more negative than the oxidation-reduction potential of hydrogen ($E°(H^+/H_2) = 0$ V), and the valence band potential must be more positive than the potential at which oxygen is produced from water ($E°(O_2/H_2O) = +1.23$ V). Also, photoelectrodes use external bias (external power), which provides the following advantages: There are no limitations to the levels of semiconductors used in the photoelectrodes, charge separation is promoted, and hydrogen and oxygen can be produced separately. With the n-type semiconductor, the potential of the necessary external bias is the difference between the theoretical minimum potential of the conducting band and H^+/H_2 potential, and the voltage can be lower than that for the ordinary electrolysis of water. On the other hand, a short charge transfer distance and simplification also provide advantages because the reaction is completed at each semiconductor particle in the photocatalyst.

Currently, photocatalyst and photoelectrode systems are diverse and progressing in a variety of forms. Photocatalyst systems can be roughly categorized as conventional photocatalysts (single-step photoexcitation-type), double-step photoexcited photocatalysts using a redox mediator (Z-scheme-type), and photocatalysis-electrolysis hybrid systems [1–4]. A redox mediator is a substance that transfers the electron while undergoing oxidation and reduction. The double-step photoexcitation reaction is similar to photosynthesis in plants and is called the Z-scheme reaction, which describes the processes in which the electrons are photoexcited twice and zigzag between the redox mediators. In the photocatalysis-electrolysis hybrid system, external bias is used,

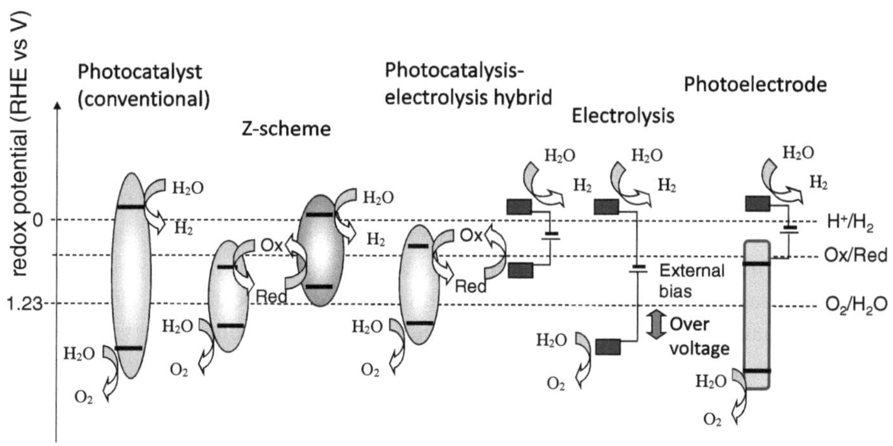

Fig. 2 Various hydrogen production technologies from water and their potential

as shown in Fig. 2; however, because the external bias is the theoretical difference between the oxidation-reduction potential of the redox mediator and hydrogen, the voltage can be lower than that required for the ordinary electrolysis of water.

Photoelectrode systems are categorized as n-type semiconductors, p-type semiconductors, p+n-type semiconductors, and pn bonding films. The combination of n- and p-type semiconductors can result in water decomposition without external bias. However, when hydrogen is generated by the p-type semiconductor, a cocatalyst with a low overvoltage, such as Pt, is required for hydrogen production over a large area, and the development of non-precious-metal cocatalysts becomes extremely important. Also, a hydrogen-accumulating hood is necessary for systems with large surface areas. Currently, p-type semiconductor photoelectrodes have a higher efficiency than n-type semiconductor photoelectrodes; however, they must be produced under film-forming conditions using the same materials and methods as solar cells, which significantly increases the cost and makes it difficult to achieve hydrogen production costs of \$0.4/Nm^3 or less [5]. For the n-type semiconductor electrode, while a clear cost estimate has not been published, it can be assumed that it would be less expensive than the p-type semiconductor photoelectrode because the method and hydrogen accumulation are simple and precious metals are only required at the counter-electrode.

3 Photocatalytic Water Decomposition Using a Z-Scheme Reaction and Redox Mediator

The complete decomposition of water using ultraviolet light (i.e., the steady production of H_2 and O_2 in stoichiometric proportions) has been realized using many photocatalysts; however, it is difficult to achieve similar results using visible light. We shifted our perspective and attempted the complete decomposition of water using two types of photocatalysts and redox mediators by simulating the Z-scheme reaction, i.e., double-step photoexcitation, which occurs during photosynthesis. As a result, we succeeded in complete decomposition using the Z-scheme reaction by combining Fe^{3+}/Fe^{2+} redox, photocatalyst, and ion photoreactions using ultraviolet light in 1997 [6]. Moreover, in 2001, we reported the first successful example of photocatalytic water decomposition using only visible light [7]. This was achieved using a system that comprised a Pt–$SrTiO_3$ (Cr-doped) photocatalyst for hydrogen production, Pt–WO_3 photocatalyst for oxygen production, and IO_3^-/I^- as the redox mediator. This system was interesting academically as an artificial photosynthesis model; subsequently, other groups reported improved photocatalysts. However, the apparent quantum efficiency (QE) was about 6 % and the solar energy conversion efficiency (η_{sun}) remained at ~0.1 % [8]. Particularly, as shown in Fig. 2, suitable semiconductor materials were extremely limited because of the problem of potential limitations of the conducting band, and it was difficult to increase the efficiency of the photocatalyst on the hydrogen production side.

4 Photocatalysis-Electrolysis Hybrid System [1–4]

4.1 Principles of the Photocatalysis-Electrolysis Hybrid System

In water decomposition using a photocatalyst, there are several major hurdles that currently prevent practical use besides the low efficiency; these include the fact that hydrogen and oxygen are produced as detonating gas, it requires a transparent and large hydrogen-accumulating cover without gas leaks, and a large amount of precious metal cocatalyst is necessary to enhance the hydrogen production performance, etc. Unless these issues are resolved, practical use will be difficult even if a high performing photocatalyst is discovered. Therefore, the authors considered a future system that would be highly practical, and found an article regarding a pilot plant for low-voltage electrolytic hydrogen production using Fe^{2+} ions that focused on recovering energy from H_2S as hydrogen rather than incinerating the H_2S gas emitted from a petrochemical plant [9]. Bubbling H_2S through a pool containing Fe^{3+} ions resulted in the production of sulfur and Fe^{2+}; the sulfur is removed via filtration. Since the redox potential (E°) of Fe^{3+}/Fe^{2+} is +0.77 V, the electrolytic voltage can be held at 1 V or less when hydrogen is produced by oxidizing Fe^{2+} to Fe^{3+} (Fig. 2). During ordinary water electrolysis, a total of ~1.6–2 V is necessary because the overvoltage of oxygen production is large and in addition to the theoretical electrolytic voltage (1.23 V). The majority of the cost of hydrogen production via general electrolysis is electricity; therefore, if a large amount of Fe^{2+} is present, the hydrogen production cost can be significantly reduced via the reduction in the electrolytic voltage.

The authors applied some of the ideas in this article to devise a "photocatalysis-electrolysis hybrid system" that produces Fe^{2+} from Fe^{3+} while photocatalytically producing oxygen via water oxidation, as shown in Fig. 3, and then combining it with hydrogen-producing low-voltage electrolysis [1, 4]. During electrolysis, Fe^{2+} is regenerated to Fe^{3+} to produce hydrogen. The overall reaction equations are

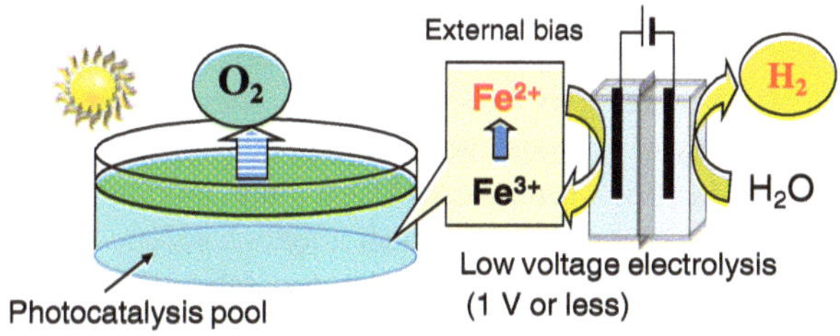

Photocatalysis :	$2H_2O + 4Fe^{3+} \rightarrow O_2 + 4Fe^{2+} + 4H^+$
Electrolysis :	$4Fe^{2+} + 4H^+ \rightarrow 4Fe^{3+} + 2H_2$
Total :	$2H_2O \rightarrow O_2 + 2H_2$

Fig. 3 Photocatalysis-electrolysis hybrid system using redox mediator, Courtesy of AIST [4]

shown at the bottom of Fig. 3. In conventional photocatalysis, there are limitations to the conduction- and valence-band potentials, as described earlier; however, in this hybrid system, the potential of the semiconductor is less restricted and several visible-light responsive materials can be used. Moreover, since hydrogen is not produced at the surface of the photocatalyst, there is no need for a precious metal cocatalyst and hydrogen trapping is very simple. Various ion pairs can be used as the redox mediator. If the redox mediator has a redox potential close to 0 V (RHE), the electrolytic voltage becomes close to 0 V and the theoretical marginal efficiency of the photocatalyst of this redox reaction becomes equivalent to that of a conventional, single-step photoexcitation photocatalyst. The reaction of the redox mediator itself has the same function as a storage battery. As is evident, the photocatalysis-electrolysis hybrid system is a breakthrough system that solves almost all the issues of conventional photocatalysis reactions by replacing the photocatalysis reaction on the hydrogen-production side of the Z-scheme reaction with electrolysis.

Although the concept of external bias used in the photocatalysis-electrolysis hybrid system may be difficult to understand, we would like to emphasize that it is not mere energy loss: Most of the energy supplied by the external bias is converted into hydrogen. When ordinary water is electrolyzed at 1.23 V, the conversion of energy from electric power to hydrogen is 100 % efficient, i.e., the energy loss is only of the overvoltage. Since the overvoltage of Fe^{3+}/Fe^{2+} is smaller than that of O_2/H_2O, the energy loss decreases. It is important to note that maintaining the external bias at 1.23 V or lower results in a possible apparent electrolytic efficiency of 100 % or higher using light energy.

4.2 Photocatalysis Using Redox Mediators

To achieve a photocatalysis-electrolysis hybrid system, it is particularly important and difficult to develop a photocatalyst that undergoes high-performance redox reactions. The present status of the development of this type of photocatalyst will be explained.

During the iron redox reaction, the Fe^{3+} ion must preferentially adhere to the photocatalyst surface and receive the electron in order for Fe^{3+} reduction to progress with high efficiency; therefore, the reaction activity is significantly dependent on the condition of the Fe^{3+} ion. Accordingly, the authors studied the oxygen production reaction using a variety of iron salt aqueous solutions with TiO_2 powder, which is a representative oxygen-producing photocatalyst [10]. We determined that the oxygen production activity of the perchlorate salt was more than 10 times higher than those of conventionally used sulfate salts. The apparent QE of the TiO_2 photocatalyst at optimal conditions was 55 % (365 nm); this is the highest QE value obtained at the ultraviolet wavelength in sunlight for the reversible redox reaction. To achieve such a high QE using a simple photocatalyst is highly significant considering the future potential of the reaction.

Next, visible light-responsive WO_3 photocatalysts were investigated to determine the effect of the counter-anion of the iron salt. The results showed that the oxygen production activity was highest when the perchlorate salt was used. In an iron perchlorate solution, the iron ion is preferentially coordinated by water rather than the perchlorate ion; in contrast, the sulfate ion strongly coordinates to the iron ion. This difference in the coordination behavior of water and the anion to the iron ion is thought to affect oxygen production activity. Also, surface treatment of the WO_3 powder with aqueous solutions containing various metal salts, revealed that the WO_3 photocatalyst that was surface-treated with a cesium salt solution showed extremely high oxygen production activity [2, 3]; the QE at 420 nm reached 31 %, which was the highest value obtained in the visible light range. Thin layers of cesium tungstate with different crystal configurations were produced all over the Cs–WO_3 surface. It is interesting that H_2O oxidation to O_2 of other redox reactions, such as IO_3^-/I^- and VO_2^+/VO^{2+}, was also improved by Cs treatment. The degree of improvement for the $Fe^{3+/2+}$ reaction was much higher than those obtained in other redox reactions. We investigated the mechanism of Cs^+ treatment using photoelectrochemical measurements on a WO_3 photoelectrode. The anodic photocurrent increased after Cs^+ treatment and the cathodic current increased after Cs^+ and Fe^{2+} treatment, suggesting that both the anodic and cathodic reactions were improved by Cs^+ treatment. The mechanism is thought to involve ion exchange sites that form on the WO_3 surface by treatment with cesium, which facilitate the adherence and reactions of Fe^{3+} and H_3O^+, as shown in Fig. 4. Calculation of the solar energy conversion efficiency (η_{sun}) revealed that the conversion of sunlight energy into chemical energy of the Fe^{2+} ion reached 0.38 %. This value surpasses the value for switchgrass (0.2 %), a plant that serves as a potential source of biofuel. $BiVO_4$ is another semiconductor that can absorb longer wavelengths than WO_3 thereby enabling the use of light up to 520 nm.

Research into redox mediators other than iron ions is extremely important, and several redox mediators have been developed recently including IO_3^-/I^- ($E° = +1.086$ V), VO_2^+/VO^{2+} ($E° = +1.00$ V), [11] and I_3^-/I^- ($E° = +0.545$ V) [12]. There is an advantage that the voltage necessary for electrolysis decreases as E° approaches zero. Currently, however, iron is considered to be an optimal redox mediator from the perspective of photocatalytic activity, cost, stability, and lack of toxicity, etc.

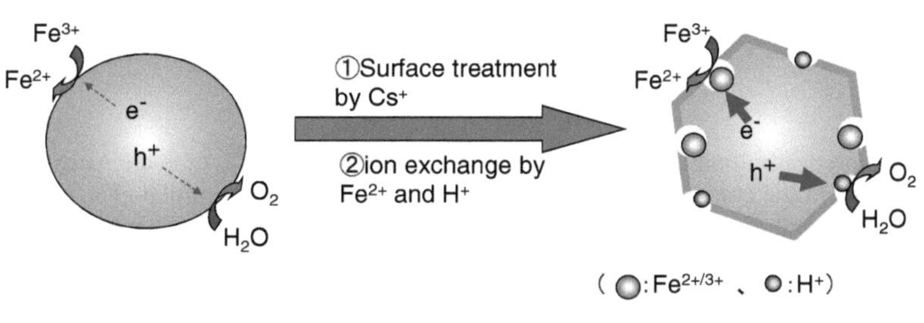

Fig. 4 Speculated reaction mechanism on Cs^+ ion treatment

4.3 Evaluation of the Theoretical Limit of Solar Energy Conversion Efficiency [4]

To elucidate the potential practical use of a photocatalysis-electrolysis hybrid system, we estimated the theoretical limit of the solar energy conversion efficiency (η^{m}_{sun}). Figure 5 shows the photoabsorption wavelength threshold (L_{max}, nm) of the semiconductor and η^{m}_{sun} when the QE is assumed to be 100 % based on the solar spectrum data for AM-1.5 total solar irradiation. For the photoabsorption rate, light with a shorter wavelength than L_{max} is assumed to have 100 % absorption (i.e., undergoes no loss by light reflection or permeation). As examples, the theoretical η^{m}_{sun} limits of photocatalysis reactions when iron and an ideal redox mediator are used at an oxidation-reduction level of 0 V are shown. Theoretically, long wavelengths of light up to 2700 and 1000 nm, respectively, can be used; however, it is not realistic to set the loss (U_{loss}) from the reaction overvoltage to zero. To estimate U_{loss}, an electrolytic voltage of 1.6 V or less (which corresponds to a U_{loss} of 0.37 V or less) has been achieved using the hydrolysis method so far. During photosynthesis, it is notable that the potential difference of the individual redox mediators in various electron transfer processes is about 0.2 V. If iron and ideal redox mediators are used with a minimum U_{loss} value (i.e., 0.4 V as for two electron—transfer

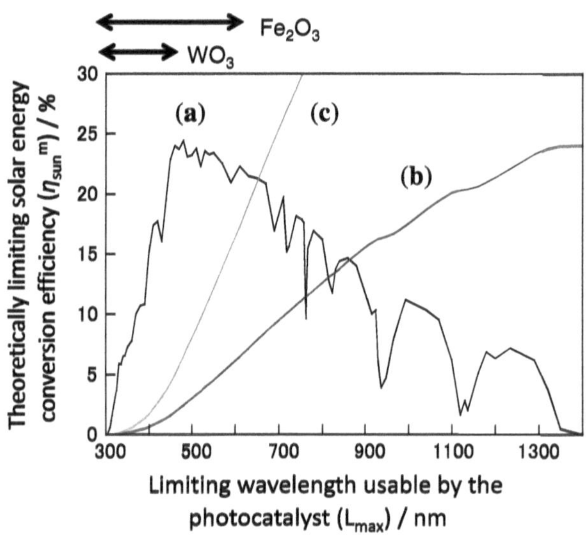

Fig. 5 Theoretically limiting solar energy conversion efficiency (η^{m}_{sun}) for the redox reaction of photocatalyst. In the case where the quantum efficiency and photoabsorption efficiency up to the limiting wavelength that can be used by photocatalyst is 100 %; **a** Sunlight spectrum, **b** iron ion redox reaction, and **c** E° = 0 redox reaction, Courtesy of AIST [4]

processes), the photocatalytic reactions can use light up to 1440 and 760 nm, respectively, and their η_{sun}^{m} values reach maximums of 24 and 30 %, respectively. As such, η_{sun}^{m} is much larger in theory than the actual η_{sun} values that have been achieved; the experimental values are likely to increase in the future.

For current semiconductor materials, η_{sun}^{m} is about 2.4 % when the iron redox mediator is reduced at a QE value of 100 % because all the wavelengths of light up to 480 nm are entirely absorbed, as in the WO_3 photocatalyst. If the light from 520 to 600 nm can also be harnessed for this reaction by using $BiVO_4$ or Fe_2O_3, it can be seen that η_{sun}^{m} values from 3.6 to 6.2 % could be achieved. The relationship of L_{max} and η_{sun}^{m} when QE is the variable using the iron redox mediator is shown in Fig. 6; it is evident that an η_{sun}^{m} of 3 % can be achieved if the QE is set at about 80 % using light up to 520 nm. The major characteristic of photocatalysis is that the reaction is complete with one particle, which enables easy mixing and layering of several photocatalysts. When different semiconductors are layered in the photo-electrode, it is necessary to match the conduction and valence bands, but this does not have to be considered in photocatalysts. This means that the performance of the entire system can be increased by concurrently conducting research to improve the QE of each photocatalyst for multiple semiconductor photocatalysts and then combining them.

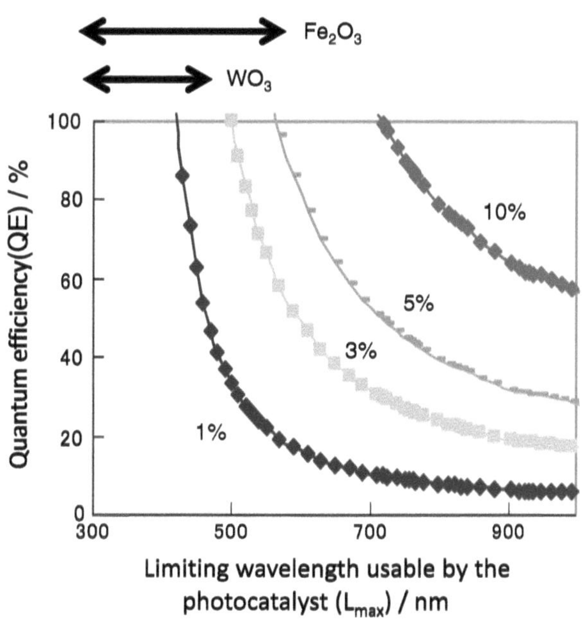

Fig. 6 Quantum efficiency (QE) of the iron ion redox reaction of photocatalyst (E^o = +0.77 V) and the theoretically limiting solar energy conversion efficiency (η_{sun}^{m}). In the case where the photoabsorption efficiency up to the limiting wavelength that can be used by the photocatalyst is 100 %, Courtesy of AIST [4]

4.4 Cost Estimation of the Photocatalysis-Electrolysis Hybrid System

Ultimately, when assessing practical use, it is necessary to estimate the cost of the whole system and compare it to the cost of hydrogen production. For the photocatalysis-electrolysis hybrid system, we must determine whether hydrogen production is possible at a lower cost than the system that combines photovoltaic generation and water electrolysis, as in interim goal ①, and whether the cost will be $0.3/Nm^3 or less, which is the interim goal ②. For the solid polymer film electrolysis used in the cost estimation, the report of the Central Research Institute of Electric Power Industry was used as a reference [13], and the DOE report was used to determine the photocatalysis pool and land price [5].

A large-scale solid polymer hydrolysis device (32,000 Nm^3/h) was chosen as the water electrolysis device for comparison. For the electricity cost, the least expensive power during the time period was selected and $0.08/kWh (which corresponds to the night-time electricity rate) at 40 % operation was assumed. For the photocatalysis-electrolysis hybrid system, iron-ion redox is used with a solar energy conversion efficiency (η_{sun}) of the photocatalyst of 3 %. The other assumptions include the following: The cost of the photocatalyst is twice that of WO_3, the photocatalyst pool is made of polyethylene, the system depreciates over 10 years, and Fe^{2+} is produced during the day and electrolysis is done during the night (10 h). The results and an example of the assumptions are shown in Fig. 7. The area of the photocatalyst pool required for the above production of hydrogen is 3 km^2. The photocatalysts pool costs $2.7/m^2, which is the same level shown in the DOE report. The additional costs related to the photocatalysts pool (i.e., facility, pump, cost of labor, land cost, management fee, interest, etc.) are about $0.03/Nm^3. If the electrolysis voltage of the electrolyte solution containing Fe^{2+} is set at 0.8 V, the cost of electric power can be reduced by half, which significantly reduces the cost of this system versus that of

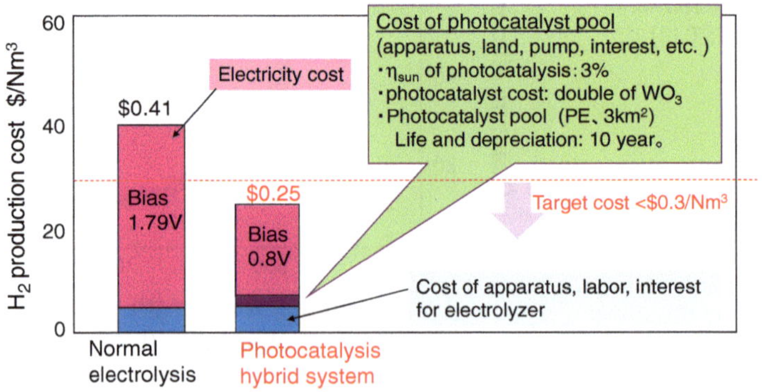

Fig. 7 Detail cost evaluation on photocatalysis-electrolysis hybrid system, Large solid polymer electrolyzer (32,000 Nm^3/h), Electricity cost: $0.08/kWh (operation rates: 40 %), Courtesy of AIST [4]

traditional electrolysis. From the above assumption, the hydrogen production cost of the photocatalysis-electrolysis hybrid system was estimated to be about $0.25/Nm^3. Under the same conditions, hydrogen production cost using usual large-scale hydrolysis with power at $0.08/kWh would be about $0.41/Nm^3. The above calculation revealed that interim goal ① could be reached, i.e., the photocatalysis-electrolysis hybrid system is capable of producing hydrogen at a lower cost than a system that simply combines photovoltaic generation and water electrolysis. Moreover, it is possible to lower the cost of hydrogen production of the photocatalysis-electrolysis hybrid system by using alternate electric power including photovoltaic generation. Also, the required cost of $0.30/Nm^3 or less to reach interim goal ② can be achieved depending on the conditions. This is the basic data for the cost estimation. If the electrolyte power can be brought close to zero using an ideal redox mediator, the hydrogen production cost will be nearly $0.14/Nm^3. Considering both the cost estimation results and the low barrier to realization (i.e., the high possibility of realization with time), it can be concluded that it is worth concentrating on this system.

5 Solar Hydrogen Production Using Porous Oxide Photoelectrodes

Solar splitting of water into H_2 and O_2 using a photoelectrochemical cell composed of an n-type semiconductor photoanode and H^+-reducing cathode has been widely investigated for solar energy conversion and storage [14–16]. Some nanocrystalline oxide semiconductor photoelectrodes with narrow band-gap energy (E_g) values [15, 17, 18], such as Fe_2O_3 ($E_g = 2.1$ eV), WO_3 ($E_g = 2.7$ eV), and $BiVO_4$ ($E_g = 2.4$ eV) on a conductive glass substrate, are easily prepared using a wet coating process followed by calcination under an air atmosphere. These photoelectrodes have significant advantages for the practical production of solar hydrogen, including simple preparation, H_2 gas accumulation, and large-area production. These nanocrystalline and porous photoelectrodes offer higher photocurrent because the small semiconductor particles are covered with the electrolyte solution and the hole diffusion length is very short, as shown in Fig. 8. Figure 9 shows the significance of n-type photoelectrode water splitting for low-cost hydrogen production. It should be noted that maintaining the external bias at 1.23 V or less means that the apparent electrolytic efficiency can be 100 % or higher in photoelectrode systems that use light energy as well as the photocatalysis-electrolysis hybrid system.

Augstynski et al. reported a photocurrent of ~2.7 mA cm^{-2} (at 1.23 V_{RHE}) using a nanocrystalline WO_3 photoelectrode [19]. Grätzel et al. reported that, in the case of a modified Fe_2O_3 film photoelectrode consisting of a perpendicularly oriented dendritic nanostructure [20–23], the photocurrent reached ~3.2 mA/cm^2 (at 1.23 V_{RHE}) [22]. Moreover, Pt-doped Fe_2O_3/Pt nanorod arrays on a gold substrate photoelectrode were developed by Park et al., and the photocurrent at 1.23 V_{RHE} is

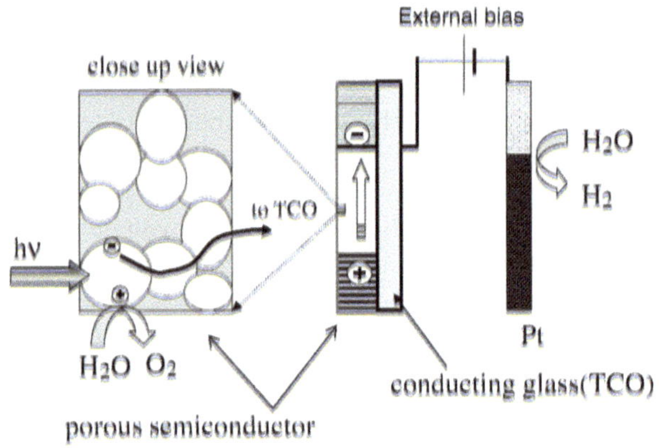

Fig. 8 Porous oxide semiconductor films on conducting glass photoelectrodes

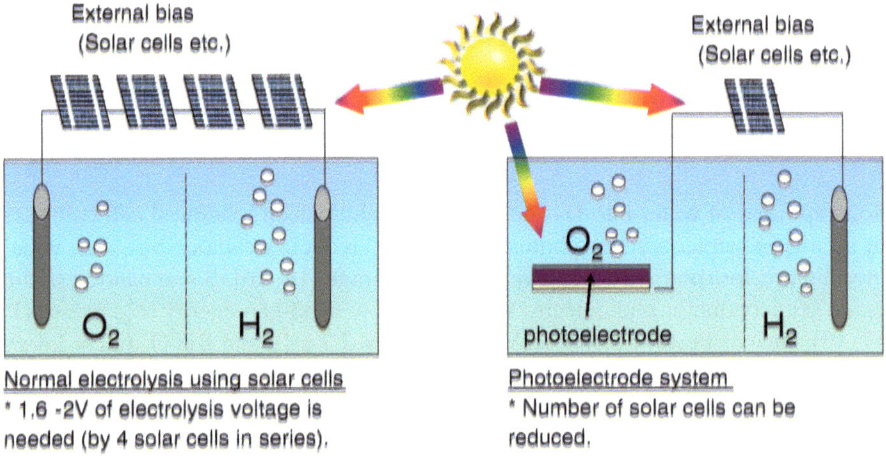

Fig. 9 Significance of n-type photoelectrode water splitting

very high (~7.0 mA/cm^2) [24]. On the other hand, the solar energy conversion efficiency (η_{sun}) value remains low; its improvement poses the greatest challenge. Several equations have been used to calculate the value of η_{sun} [8, 25, 26]: The η_{sun} value that results from the energy loss of the external bias in a two-electrode system, i.e., applied-bias photon-to-current efficiency (ABPE or η^{ex}_{sun}), is calculated using the following equation [18]:

$$\mathrm{ABPE\ or\ }\eta^{ex}_{sun}(\%) = [P_{max}/Int] \times 100 = [J_{opt} \times (1.23 - E_{opt})/Int] \times 100, \quad (1)$$

where P_{max} is the solar energy acquired by the photoelectrode (box in Fig. 10), J_{opt} (mA/cm^2) is the photocurrent density at E_{opt}, E_{opt} (V) is the applied voltage

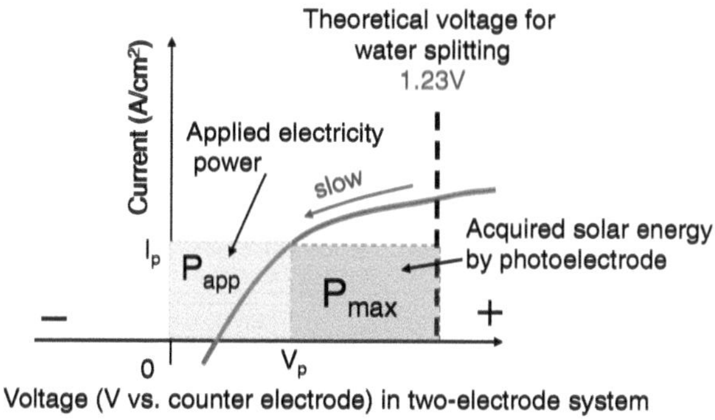

Fig. 10 How to evaluate the applied bias solar photon-to-current efficiency (ABPE) under 1 SUN irradiation

between the working and counter electrodes under optimal operating conditions; *Int* (mW/cm^2) is the intensity of incident solar light under A.M. 1.5, 1 sun conditions, and 1.23 V versus reversible hydrogen electrode (RHE) is the standard electrode potential of H_2O.

To improve the ABPE, it is very important to decrease the E_{opt} and onset potential and improve the *I–V* curve shape as well as increase the photocurrent. The applied bias and onset potential are influenced by the conduction-band potential (E_{CB}) values of semiconductors. In the case of Fe_2O_3 photoelectrodes, it has been reported that a large applied bias is necessary to generate high photocurrent because of the positive E_{CB}. Since the first report on the $BiVO_4$ photoelectrodes in 2003 [27], effective $BiVO_4$ photoelectrodes have been attracting considerable attention. Figure 11 shows the progress in the efficiencies of $BiVO_4$ photoelectrodes and publication number. Bi-Mixed oxide semiconductors, such as $BiVO_4$, have often been used for both photocatalysts and porous photoelectrodes because of their unique band structures. The valence band mainly consists of O2p in the case of conventional simple oxide semiconductors if the d-orbital is unoccupied, and the potential of the O2p orbital is very positive around 3 V (vs. RHE). The required external bias of the photoelectrode increases with increasing positive shift of the conduction band; therefore, to decrease the band gap, it is beneficial to negatively shift the valence band potential without positively shifting the conduction band. In the case of Bi-mixed oxide semiconductors, the valence band potential undergoes a negative shift when the hybrid orbitals of O2p and Bi6s are formed because of repulsion of two orbitals (Fig. 12), and the band gap becomes smaller than those of conventional oxide semiconductors. It is noteworthy that the E_{CB} of $BiVO_4$ (−0.4 V vs. NHE, pH = 7 [28]) is higher than that of WO_3 and similar to that of TiO_2. The photocurrent of $BiVO_4$ significantly increases upon the addition of an under-layer coating of WO_3 [28–30] or SnO_2 [31, 32] on the conducting glass. However, the photocurrent of theses $BiVO_4$ multi-composite photoelectrodes is still low. Lee et al. reported that the photocurrent at 1.23 V_{RHE} of heterojunction $BiVO_4/WO_3$ electrodes reached ~1.4 mA/cm^2 [28]. Moreover, various surface-treated and modified $BiVO_4$ photoelectrodes have been developed [33–39]: Choi et al. reported that the photocurrent at 1.23 V_{RHE} of $FeOOH/BiVO_4$ photoelectrodes reached ~2.3 mA/cm^2 [38].

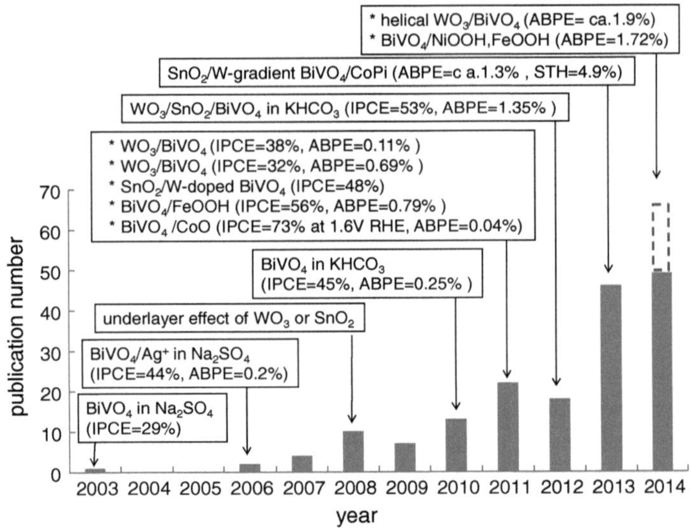

Fig. 11 The progress of efficiencies on $BiVO_4$ photoenelctrode and the publication number (until September in 2014)

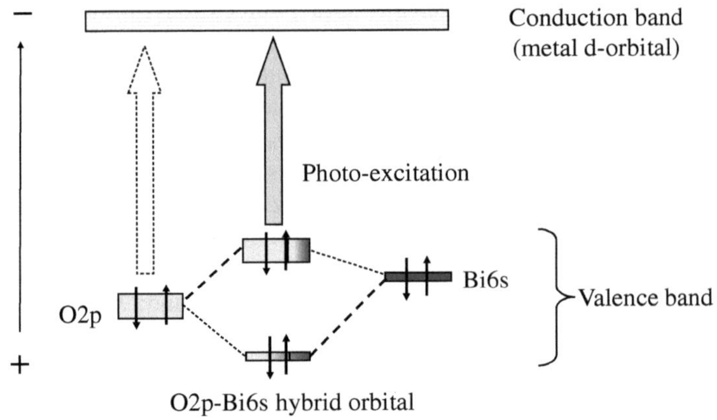

Fig. 12 The band structure of Bi-mixed oxides

We investigated the improvement of the photoelectrochemical properties of bare $BiVO_4$ photoelectrodes in a carbonate electrolyte [40] rather than the sulfate electrolyte that is generally used. Recently, we reported on $BiVO_4$ composite electrodes and the effects of introducing a SnO_2 intermediate layer between the $BiVO_4$ and WO_3 layers (i.e., a $BiVO_4/SnO_2/WO_3$ photoelectrode) in a carbonate electrolyte [41, 42]. The maximum ABPE values of the single- and double-stacked photoelectrode are 0.86 and 1.35 %, respectively. It was assumed that the excited electron transfer is positively affected by contact with the oxide semiconductors with different band structures shown in Fig. 13, and that the carbonate anion behaves as a catalyst for irreversible O_2 evolution on the $BiVO_4$ film surface. The ABPE of the single photoelectrode increases to 1.28 % using finely porous $BiVO_4$ prepared via an auto-combustion method [43].

Krol et al. reported [44] that the poor carrier-separation efficiency of the $BiVO_4$ photoanode on SnO_2/FTO could be overcome by implementing a multi-step gradient in the tungsten dopant concentration (Fig. 14), resulting in the formation of a series of homo-junctions in the bulk material. A ten-step gradient-doped Co-Pi-catalyzed W:$BiVO_4$ photoanode generates an AM 1.5 photocurrent of 3.6 mA/cm^2 (at 1.23 V RHE), representing a 60 % improvement over the same electrode with a homogeneous W-dopant concentration. Combining this electrode with a two-junction α-Si solar cell generates a water-splitting device with a 4.9 % STH efficiency. Choi et al. reported [45] that a nanoporous morphology with a high surface area effectively suppresses bulk carrier recombination without additional doping or an underlayer resulting in an electron-hole separation yield of 0.90 at 1.23 V RHE. BiOI film was electrodeposited and heated with VO(acac)2 to form $BiVO_4$. The resulting FTO/$BiVO_4$ photoanode with a FeOOH/NiOOH co-catalyst achieved a photocurrent density of 2.73 mA/cm^2 at 0.6 V RHE; the maximum ABPE obtained was 1.72 %. Recently, Shi et al. reported [46] the advantages of using helical WO_3 nanowire structures (5.5 μm length) with $BiVO_4$ particles for photoelectrochemical solar water splitting. A maximum photocurrent density of 5.35 mA/cm^2 was achieved at 1.23 V RHE, and the ABPE was ~1.9 %. The maximum photocurrent on $BiVO_4$ is >6 mA/cm^2, as calculated from the light harvesting efficiency of the thick film, and the maximum ABPE and STH values are >5 and 7 %, respectively. New methods should be developed to maximize the ability of $BiVO_4$ and other semiconductor materials.

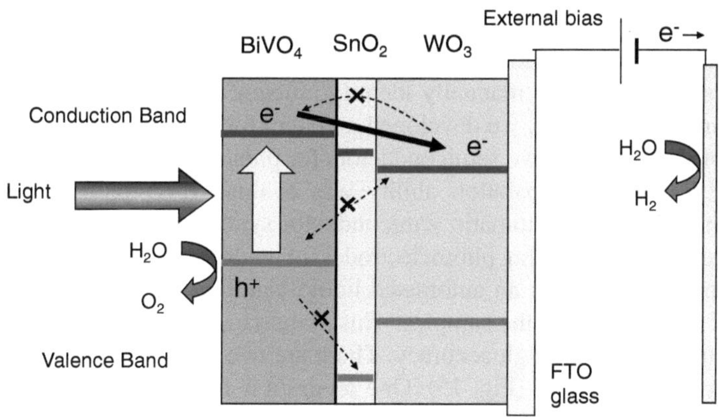

Fig. 13 Mechanism of multi-layer effect of oxide films on photoelectrode

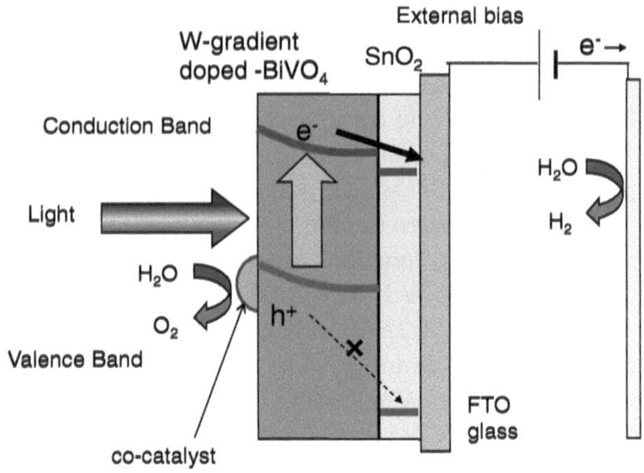

Fig. 14 Mechanism of W-gradient doped $BiVO_4$/SnO_2 photoanode with co-catalyst

6 Development of a System for High-Throughput Screening of Semiconductors for Photoelectrodes and Photocatalysts

In our study of porous photoelectrodes, we have mainly adapted semiconductors developed for photocatalysts into photoelectrodes, and we have recognized some relationships between the photoelectrodes and photocatalysts. For example, for O_2-evolution photocatalysts with sacrificial reagents, most semiconductors, such as $BiVO_4$ and WO_3, are adaptable into porous photoanodes for O_2 evolution. Therefore, we concluded that we should continue developing new semiconductor materials for use as both photoelectrodes and photocatalysts. Moreover, there are many semiconductor materials in addition to $BiVO_4$: Some mixed oxides of Pb, Sb, Sn, etc., in which the valence band potential is negatively shifted by the s-p hybrid orbitals are promising candidates.

It takes many years to manually identify efficient semiconductors composed of multiple metals; therefore, we developed a system for high-throughput screening of new visible light-responsive semiconductors for photoelectrodes and photocatalysts [47, 48]. The charge-separation ability was evaluated via photoelectrochemical measurements, and an automatic semiconductor synthesis system that can be used to prepare porous thin-film photoelectrodes of various materials was developed. The equipment comprises an automated liquid-handling platform, electric furnace, and robot arm to transfer the samples. This system can be used to prepare a variety of semiconductors with high accuracy. There are two types of evaluation programs and two kinds of libraries (Fig. 15): One program is for high-speed evaluation, i.e., numerous samples are printed on the same conductive glass substrate and their photocurrents are quickly evaluated by scanning light across the substrate. Another program enables detailed evaluation, i.e., the samples are printed on separate substrates and evaluated individually. The libraries are evaluated photoelectrochemically using focused light and an X-Y stage. As an example of the utilization of our system, an n-type semiconductor, i.e., Fe_2O_3, was selected as a target material and iron binary oxides, such as Fe–Ti, Fe–Nb, and Fe–V with various ratios were synthesized. The Fe–Ti and Fe–Nb binary oxide systems have been reported previously [49], and the results obtained from our system were consistent with the previously published results; this confirms the validity of our system. In the Fe–Sn–Ti system, the highest photocurrent was observed at an Fe/Sn/Ti ratio of 6:1:3 (Fig. 15); thus, this is a potential new visible light—responsive material. We also screened a p-type semiconductor material containing Bi binary oxides and found $CuBi_2O_4$ to be a new visible light-responsive p-type semiconductor [47, 50]. We developed a highly efficient and visible light-responsive photocatalyst by combining the p-type semiconductor $CuBi_2O_4$ and n-type semiconductor WO_3. $CuBi_2O_4/WO_3$ shows higher reactivity than typical TiO_2-based photocatalysts for the complete oxidation of acetaldehyde into CO_2 under both solar-simulating UV and visible light. From the photoelectrochemical measurements, it was determined that the photocatalytic reaction mechanism could be explained by the model of the p-n photochemical diode for reduction/oxidization improvement (i.e., the Z-scheme mechanism) [50].

More than 20,000 semiconductor samples can be evaluated over 3 years using our high-throughput screening system, and more than 100 special compositions of three metal-oxide semiconductors, which generate a higher photocurrent than two metal-oxide semiconductors, were found. Some strong patents based on real data were applied. We believe that new highly functional semiconductors can be found for photoelectrodes and photocatalysts using high-throughput screening systems in the future. Recently, we investigated the co-doping effect of α-Fe_2O_3 powder on photocatalytic water oxidation [51]. Pt-loaded Fe_2O_3 shows activity towards water

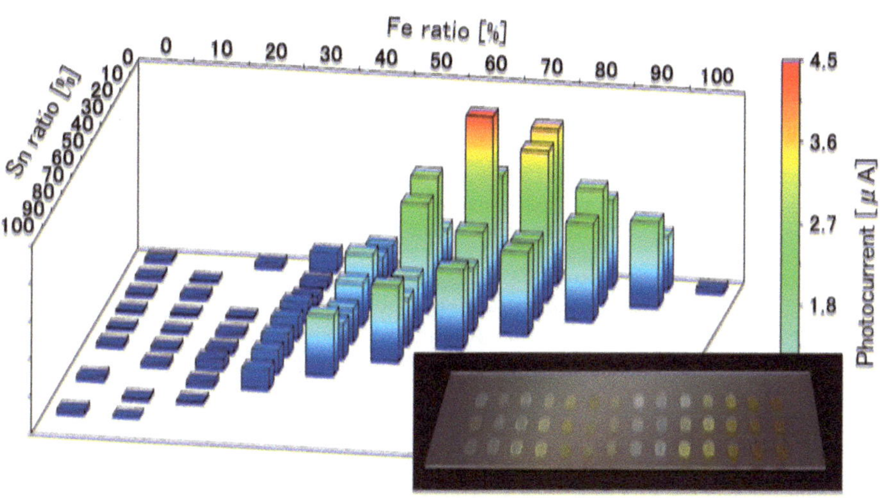

Fig. 15 The dependence of photocurrent in Fe–Sn–Ti complex oxide under visible light. *Lower right* libraries of photoelectrode films prepared by the high-throughput screening system

oxidation in an aqueous $NaIO_3$ solution. The activity was much improved by Sr and Ti co-doping. The optimized doping amount of Sr and Ti in the Fe_2O_3 photocatalyst is almost the same as that in a Fe_2O_3 photoelectrode. We developed a photocatalyst that mainly consisted of d^5 metal ions for water oxidation using a reversible redox ion for the first time.

7 Conclusion

It should be noted that maintaining an external bias of 1.23 V or less enables an apparent electrolytic efficiency of 100 % or higher in both the photocatalysis-electrolysis hybrid system and photoelectrode systems using light energy. The wide introduction of renewable energy results in the need for its storage. Hydrogen production by electrolysis is being studied as a technology to store the excess power gained from renewable energy such as wind and photovoltaic generation at lower costs than ordinary batteries. Depending on the location, this electrolysis device can be combined with photocatalysts with redox reactions and photoelectrodes. Also, combining daytime and nighttime electrolysis could increase the facility operation rate and further reduce the cost of hydrogen production. Large-scale smoothing of the power load over short and long cycles can be achieved. The hurdle of using the excess power was low for the initial introductory verification, and research for full verification of the hybrid system for such excess power electrolysis and photocatalysts and photoelectrodes will be the first step toward practical use.

References

1. Sayama K, Arakawa H, Okabe K, Kusama H (2001) Jpn Patent 3198298; U.S. Patent 09/028495, 1998
2. Miseki Y, Kusama H, Sugihara H, Sayama K (2010) Cs-modified WO_3 photocatalyst showing efficient solar energy conversion for O_2 production and Fe(III) ion reduction under visible light. J Phys Chem Lett 1:1196–1200
3. Miseki Y, Sayama K (2014) High-efficiency water oxidation and energy storage utilizing various reversible redox mediators under visible light over surface-modified WO_3. RSC Adv. 4:8308–8316
4. Sayama K, Miseki Y (2014) Research and development of solar hydrogen production, Toward the realization of ingenious photocatalysis-electrolysis hybrid system. Synthesiology 7:79–91
5. James BD, Baum GN, Perez J, Baum KN (2009) Technoeconomic analysis of photoelectrochemical (PEC) hydrogen production, DOE Report (2009), Contract No. GS-10F-009J
6. Sayama K, Yoshida R, Kusama H, Okabe K, Abe Y, Arakawa H (1997) Photocatalytic decomposition of water into H_2 and O_2 by a two-step photoexcitation reaction using a WO_3 suspension catalyst and an Fe^{3+}/Fe^{2+} redox system. Chem Phys Lett 277:387–391
7. Sayama K, Mukasa K, Abe R, Abe Y, Arakawa H (2001) Stoichiometric water splitting into H_2 and O_2 using a mixture of two different photocatalysts and an IO_3^-/I^- shuttle redox mediator under visible light irradiation. Chem Commun 2416–2417

8. Kudo A, Miseki Y (2009) Heterogeneous photocatalyst materials for water splitting. Chem Soc Rev 38:253–278
9. Mizuta S, Kondo W, Fujii K, Iida H, Isshiki S, Noguchi H, Kikuchi T, Sue H, Sakai K (1991) Hydrogen production from hydrogen sulfide by the Fe-Cl hybrid process. Ind Eng Chem Res 30:1601–1608
10. Miseki Y, Kusama H, Sugihara H, Sayama K (2010) Significant effects of anion in aqueous reactant solution on photocatalytic O_2 evolution and Fe(III) reduction. Chem Lett 39:846–847
11. Miseki Y, Kusama H, Sayama K (2012) Photocatalytic energy storage over surface-modified WO_3 using V^{5+}/V^{4+} redox mediator. Chem Lett 41:1489–1491
12. Miseki Y, Fujiyoshi S, Gunji T, Sayama K (2013) Photocatalytic water splitting under visible light utilizing I_3^-/I^- and IO_3^-/I^- redox mediators by Z-scheme system using surface treated PtO_x/WO_3 as O_2 evolution photocatalyst. Catal Sci Technol 3:1750–1756
13. Asaoka Y, Uotani M (2003) Feasibility study on hydrogen production with off-peak electricity —evaluation of the effects of availability and electric power transmission. CRIEPI Research Report T02039:1–16
14. Fujishima A, Honda K (1972) Electrochemical photolysis of water at a semiconductor electrode. Nature 238:37–38
15. Grätzel M (2001) Photoelectrochemical cells. Nature 414:338–344
16. Lin Y, Yuan G, Liu R, Zhou S, Sheehan SW, Wang D (2011) Semiconductor nanostructure-based photoelectrochemical water splitting: a brief review. Chem Phys Lett 507:209–215
17. Kudo A, Omori K, Kato H (1999) A novel aqueous process for preparation of crystal form-controlled and highly crystalline $BiVO_4$ powder from layered vanadates at room temperature and its photocatalytic and photophysical properties. J Am Chem Soc 121:11459–11467
18. Chen Z, Jaramillo TF, Deutsch TG, Kleiman-Shwarsctein A, Forman AJ, Gaillard N, Garland R, Takanabe K, Heske C, Sunkara M, McFarland EW, Domen K, Miller EL, Turner JA, Dinh HN (2010) Accelerating materials development for photoelectrochemical hydrogen production: standards for methods, definitions, and reporting protocols. J Mater Res 53:3–16
19. Alexander BD, Kulesza PJ, Rutkowska I, Solarska R, Augustynski J (2008) Metal oxide photoanodes for solar hydrogen production. J Mater Chem 18:2298–2303
20. Cesar I, Kay A, Martinez JAG, Grätzel M (2006) Translucent thin film Fe_2O_3 photoanodes for efficient water splitting by sunlight: nanostructure-directing effect of Si-doping. J Am Chem Soc 128:4582–4583
21. Kay A, Cesar I, Grätzel M (2006) New benchmark for water photooxidation by nanostructured α-Fe_2O_3 films. J Am Chem Soc 128:15714–15721
22. Tilley SD, Cornuz M, Sivula K, Grätzel M (2010) Light-induced water splitting with hematite: improved nanostructure and iridium oxide catalysis. Angew Chem Int Ed 49:6405–6408
23. Hisatomi T, Formal FL, Cornuz M, Brillet J, Tétreault N, Sivula K, Grätzel M (2011) Cathodic shift in onset potential of solar oxygen evolution on hematite by 13-group oxide overlayers. Energy Environ Sci 4:2512–2515
24. Mao A, Park NG, Han GY, Park JH (2011) Controlled growth of vertically oriented hematite/Pt composite nanorod arrays: use for photoelectrochemical water splitting. Nanotechnology 22:175703–175709
25. Brillet J, Cornuz M, Formal FL, Yum JH, Grätzel M, Sivula K (2010) Examining architectures of photoanode-photovoltaic tandem cells for solar water splitting. J Mater Res 25:17–24
26. Varghese OK, Grimes CA (2008) Appropriate strategies for determining the photoconversion efficiency of water photoelectrolysis cells: a review with examples using titania nanotube array photoanodes. Solar Energy Mater Solar Cells 92:374–384
27. Sayama K, Nomura A, Zou Z, Abe R, Abe Y, Arakawa H (2003) Photoelectrochemical decomposition of water on nanocrystalline $BiVO_4$ film electrodes under visible light. Chem Commun 2908–2909

28. Hong SJ, Lee S, Jang JS, Lee JS (2011) Heterojunction $BiVO_4/WO_3$ electrodes for enhanced photoactivity of water oxidation. Energy Environ Sci 4:1781–1787
29. Su J, Guo L, Bao N, Grimes CA (2011) Nanostructured $WO_3/BiVO_4$ heterojunction films for efficient photoelectrochemical water splitting. Nano Lett 11:1928–1933
30. Chatchai P, Murakami Y, Kishioka S, Nosaka AY, Nosaka Y (2009) Efficient photocatalytic activity of water oxidation over $WO_3/BiVO_4$ composite under visible light irradiation. Electrochim Acta 54:1147–1152
31. Liang Y, Tsubota T, Mooij LPA, van de Krol R (2011) Highly improved quantum efficiencies for thin film BiVO4 photoanodes. J Phys Chem C 115:17594–17598
32. Chatchai P, Murakami Y, Kishioka S, Nosaka AY, Nosaka Y (2008) $FTO/SnO_2/BiVO_4$ composite photoelectrode for water oxidation under visible light irradiation. Electrochem Solid State Lett 11:H160–H163
33. Sayama K, Nomura A, Arai T, Sugita T, Abe R, Yanagida M, Oi T, Iwasaki Y, Abe Y, Sugihara H (2006) Photoelectrochemical decomposition of water into H_2 and O_2 on porous $BiVO_4$ thin-film electrodes under visible light and effect of Ag ion treatment. J Phys Chem B 110:11352–11360
34. Zhong DK, Choi S, Gamelin DR (2011) Near-complete suppression of surface recombination in solar photoelectrolysis by "Co-Pi" catalyst-modified W:$BiVO_4$. J Am Chem Soc 133:18370–18377
35. Jeon TH, Choi W, Park H (2011) Cobalt-phosphate complexes catalyzed the photoelectrochemical water oxidation of $BiVO_4$ electrodes. Phys Chem Chem Phys 13:21392–21401
36. Pilli SK, Furtak TE, Brown LD, Deutsch TG, Turner JA, Herring AM (2011) Cobalt-phosphate (Co-Pi) catalyst modified Mo-doped $BiVO_4$ photoelectrodes for solar water oxidation. Energy Environ Sci 4:5028–5034
37. Luo W, Li Z, Yu T, Zou Z (2012) Effects of surface electrochemical pretreatment on the photoelectrochemical performance of Mo-doped $BiVO_4$. J Phys Chem C 116:5076–5081
38. Seabold JA, Choi KS (2012) Efficient and stable photo-oxidation of water by a bismuth vanadate photoanode coupled with an iron oxyhydroxide oxygen evolution catalyst. J Am Chem Soc 134:2186–2192
39. Jia Q, Iwashina K, Kudo A (2012) Facile fabrication of an efficient $BiVO_4$ thin film electrode for water splitting under visible light irradiation. PANS 109:11564–11569
40. Sayama K, Wang N, Miseki Y, Kusama H, Onozawa-Komatsuzaki N, Sugihara H (2011) Effect of carbonate ions on the photooxidation of water over porous $BiVO_4$ film photoelectrode under visible light. Chem Lett 39:17–19
41. Saito R, Miseki Y, Sayama K (2012) Highly efficient photoelectrochemical water splitting using a thin film photoanode of $BiVO_4/SnO_2/WO_3$ multi-composite in a carbonate electrolyte. Chem Commun 48:3833–3835
42. Saito R, Miseki Y, Sayama K (2013) Photoanode characteristics of multi-layer composite BiVO4 thin film in a concentrated carbonate electrolyte solution for water splitting. J Photochem Photobiol A: Chem 258:58–60
43. Fujimoto I, Wang N, Saito R, Miseki Y, Gunji T, Sayama K (2014) $WO_3/BiVO_4$ composite photoelectrode prepared by improved auto-combustion method for highly efficient water splitting. Int J Hydrogen Energy 39:2454–2461
44. Abdi F, Han L, Smets A, Zeman M, Dam B, Krol R (2013) Efficient solar water splitting by enhanced charge separation in a bismuth vanadate-silicon tandem photoelectrode. Nature Commun 4:2195–2201
45. Kim T, Choi K (2014) Nanoporous $BiVO_4$ photoanodes with dual-layer oxygen evolution catalysts for solar water splitting. Science 343:990–994
46. Shi X, Choi I, Zhang K, Kwon J, Kim D, Lee J, Oh S, Kim J, Park J (2014) Efficient photoelectrochemical hydrogen production from bismuth vanadate-decorated tungsten trioxide helix nanostructures. Nat Commun 5:5775–5782

47. Arai T, Konishi Y, Iwasaki Y, Sugihara H, Sayama K (2007) High-throughput screening using porous photoelectrode for the development of visible-light-responsive semiconductors. J Comb Chem 9:574–581
48. Kusama H, Wang N, Miseki Y, Sayama K (2010) Combinatorial search for iron/titanium-based ternary oxides with a visible-light response. J Comb Chem 12:356–362
49. Sartoretti CJ, Alexander BD, Solarska R, Rutkowska IA, Augustynski J, Cerny R (2005) Photoelectrochemical oxidation of water at transparent ferric oxide film electrodes. J Phys Chem B 109:13685–13692
50. Arai T, Yanagida M, Konishi T, Iwasaki Y, Sugihara H, Sayama K (2007) Efficient complete oxidation of acetaldehyde into CO_2 over $CuBi_2O_4/WO_3$ composite photocatalyst under visible and UV light irradiation. J Phys Chem C 111:7574–7577
51. Miseki Y, Majima Y, Gunji T, Sayama K (2014) Codoping effect of Sr and Ti for α-Fe_2O_3 photocatalyst on water oxidation utilizing IO_3^- as a reversible redox ion under visible light. Chem Lett 43:1560–1562

Part VIII
Energy Conversion Using Photosynthesis Mechanism: Learning from Nature

Fundamentals of Photosynthesis for Energy Storage

Z.-Y. Wang-Otomo

Abstract Photosynthesis is the most fundamentally important energy-converting process on Earth. It converts solar energy to chemical energy and provides all the food we eat, the fossil fuels we consume and the oxygen we breathe. The basic concepts underlying photosynthesis have been well established and a brief introduction is given in this chapter. The principles, especially those obtained from primitive photosynthetic organisms, are considered to serve as a guide for the development of artificial photosynthesis today.

1 Introduction

The sunlight reaching the earth's surface every year is estimated to bring about energy of some 2.5×10^{24} J. Only about 0.2 % is utilized by photosynthesis to produce organic matter [1]. This is partly because the sunlight has a broad spectrum and only the visible range of wavelength from 400 to 700 nm, called photosynthetically active radiation and comprising about 40 % of the solar irradiance, can be used by most photosynthesis. Despite the low efficiency, the amount of energy stored by photosynthesis each year in the biosphere is still roughly four times that of the annual consumption by humans [1].

The fossil fuels we use today are all made from ancient photosynthesis. Coal, petroleum, and natural gas are decomposition products of plants and animals. The energy stored in these organisms was harnessed from the solar radiation millions of years ago. Total resources of the fossil fuel stored under the earth's surface is equivalent to about 60 years of net photosynthesis [2].

Z.-Y. Wang-Otomo (✉)
Ibaraki University, Bunkyo 2-1-1, Mito, Japan
e-mail: otomo@mx.ibaraki.ac.jp

M. Sugiyama et al. (eds.), *Solar to Chemical Energy Conversion*,
Lecture Notes in Energy 32, DOI 10.1007/978-3-319-25400-5_21

2 The Energy Flow, Electron Sources and Carbon Circle

The solar energy captured through photosynthesis is stored in the form of chemical bonds, i.e., the formation of new C–C bonds. This process is also called carbon-fixation that converts atmospheric CO_2 into organic molecules (Fig. 1). The photosynthesizing reaction requires electrons provided from external sources. For primitive photosynthesis, as occurred in sulfur bacteria, the electron donor is the reduced sulfur compounds [3]:

$$nCO_2 + 2nH_2S \xrightarrow{\text{light}} (CH_2O)_n + 2nS + nH_2O \quad (1)$$

where $(CH_2O)_n$ represents a carbohydrate, typically the glucose ($C_6H_{12}O_6$). The S^{2-} in H_2S loses 2 electrons to become elemental S^0. This reaction has a midpoint redox potential $E_m^{'}$ of −0.23 V which is much lower than +0.64 V ($BChl^+ + e^- \rightleftharpoons BChl$, BChl stands for bacteriochlorophyll) of the special pair (see below) in the reaction center (RC) of purple bacteria (Fig. 2, left scheme). The green plants, algae and cyanobacteria, which developed much later, have evolved to acquire the ability to extract electrons from water molecules, but in essentially the same way as that of sulfur bacteria (Fig. 2, middle scheme):

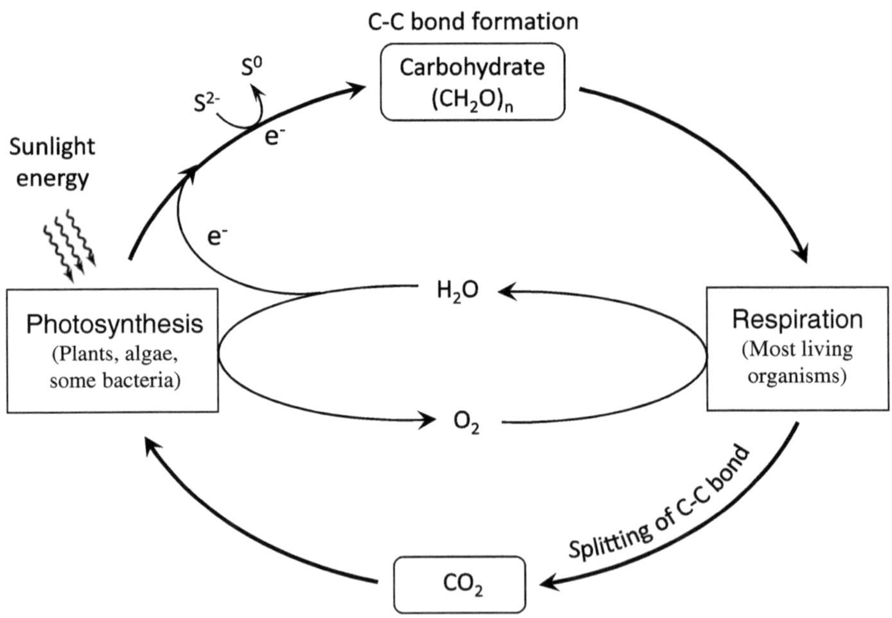

Fig. 1 A schematic diagram showing the relationship between photosynthesis and respiration. The energy storage by photosynthesis requires external electron sources. CO_2 and O_2 are circulated in the atmosphere and cells

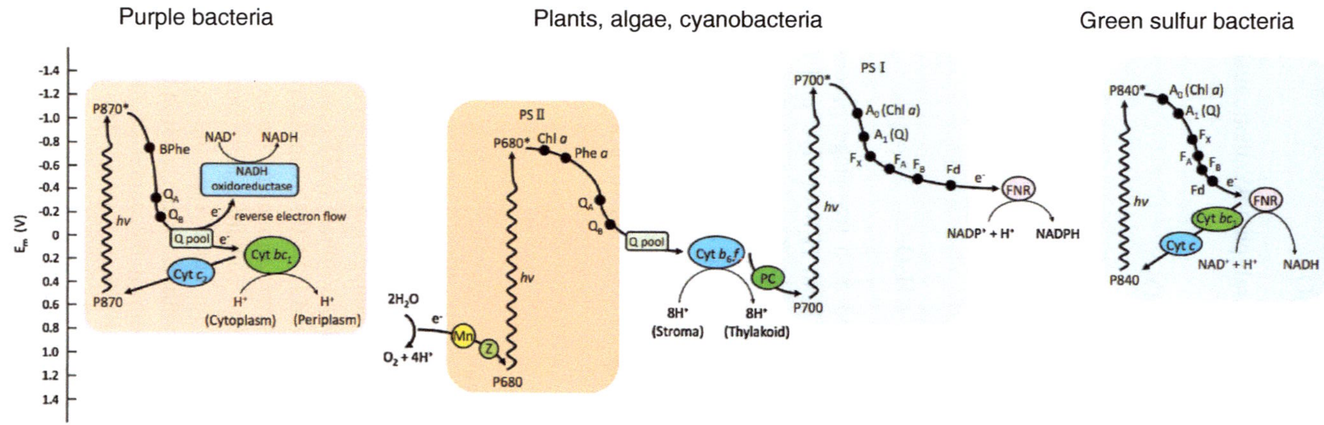

Fig. 2 Comparison of the electron transport chains between the photosynthetic bacteria (*left* and *right* schemes) and green plants (*middle* scheme). P870 and P870*: ground and excited states of the special pair in purple bacteria, respectively; *Q* quinone; *BPhe* bacteriopheophytin; *Cyt* cytochrome; *ATP* adenosine triphosphate; *NADH* nicotinamide adenine dinucleotide; *NADPH* nicotinamide adenine dinucleotide phosphate; P680 and P680*: ground and excited states of the special pair in photosystem II, respectively; *Chl* chlorophyll; *Phe* pheophytin; *PC* plastocyanin; P700 and P700*: ground and excited states of the special pair in photosystem I, respectively; *F* Fe-S cluster: *Fd* ferredoxin; *FNR* ferredoxin-NADP reductase; P840 and P840*: ground and excited states of the special pair in green sulfur bacteria, respectively; Q_A quinone-binding site A; Q_B quinone-binding site B

$$nCO_2 + nH_2O \xrightarrow{\text{light}} (CH_2O)_n + nO_2 \tag{2}$$

If we use glucose to represent the carbohydrate, the overall photosynthetic reaction can be written as follows:

$$6CO_2 + 6H_2O \rightarrow C_6H_{12}O_6 + 6O_2 \tag{3}$$

The reaction (2) is actually composed of two reactions, in which the water molecule is first split by light (photolyzed) and then the released electrons are used for carbohydrate synthesis:

$$2H_2O \rightarrow 4H^+ + O_2 + 4e^- \tag{4}$$

$$CO_2 + 4H^+ + 4e^- \rightarrow (CH_2O) + H_2O \tag{5}$$

Therefore, the photosynthesis can also be viewed as a two-stage process in which the light energy is utilized to gain electrons from H_2S or H_2O and these electrons are subsequently used to build C–C bonds with CO_2 as a substrate.

The photosynthesized organic compounds are consumed by animals through a process called "respiration" (Fig. 1). It is a reverse reaction of photosynthesis that use molecular oxygen, also released by photosynthesis, to burn (oxidize) the organic molecules into CO_2 and waters. This is also a decomposition process that splits the C–C chemical bonds. The photosynthesis and respiration are complementary processes.

3 Four Steps for the Energy Storage

The whole photosynthesis process can be divided into four steps as shown in Fig. 3: (1) light absorption and energy transfer; (2) energy conversion by charge separation; (3) electron transfer for producing adenosine triphosphate (ATP) and nicotinamide adenine dinucleotide phosphate (NADPH) and (4) CO_2 assimilation.

For the photosynthesis to occur, any photosynthetic organism needs first to collect sunlight energy. These organisms developed to contain a large amount of pigment molecules, typically the chlorophylls (Chl) for eukaryotes and cyanobacteria, and the bacteriochlorophylls (BChl) for photosynthetic bacteria. Both Chl and BChl are squarish planar molecules with a Mg atom at the center of the plane (Fig. 4). The central Mg atom is coordinated to four nitrogen atoms. The antenna pigments are in most cases associated with proteins to form a set of light-harvesting (LH) pigment-protein complexes that either are integrated into the membrane or attach to the membrane surface (see Chap. 22 for details on the case of purple bacteria). Since the Chls are highly conjugated molecules, they strongly absorb

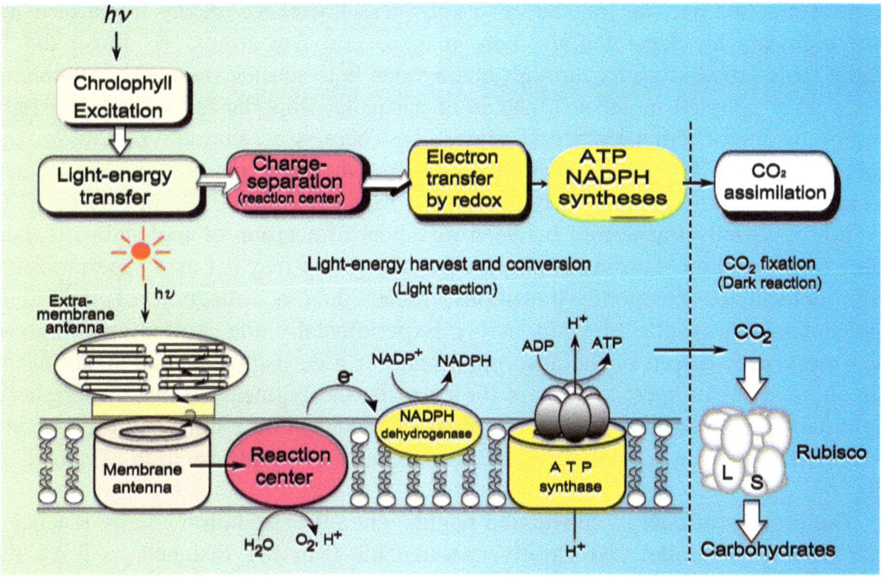

Fig. 3 A flow chart illustrating the whole process of photosynthesis. *ADP* adenosine diphosphate; *Rubisco* ribulose 1,5-bisphosphate carboxylase/oxygenase. Other symbols are the same as in Fig. 2

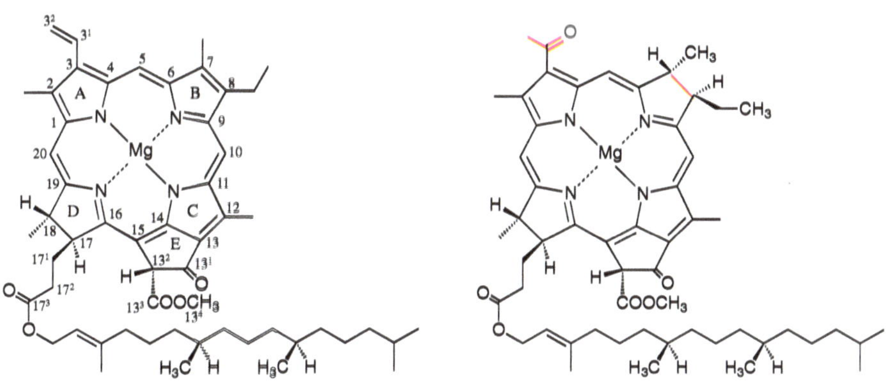

Chlorophyll *a* Bacteriochlorophyll *a*

Fig. 4 Chemical structures of chlorophyll *a* and bacteriochlorophyll *a* based on the numbering scheme of IUPAC nomenclature. Five rings are lettered *A* through *E*. The chemical differences between structures of Chl *a* and BChl *a* are marked by *red* color

visible light with the molecular extinction coefficients over 10^5 M^{-1} cm^{-1}, which are among the highest known for organic molecules [3].

When a Chl molecule is excited by a quantum of incident light (a photon), the absorbed energy is transferred directly to a neighboring unexcited Chl molecule by a process called exciton transfer, also known as resonance energy transfer, if the two pigments are separated by more than several Ångstroms. As illustrated in Fig. 5, this process occurs through interactions between the molecular orbitals of the weakly coupled molecules, and is of particular importance in funneling light energy to photochemical reaction center [4]. There is an alternative view for the energy transfer, known as exciton coupling, when the pigment molecules are located closely to each other, typically less than 10 Å for Chls. In this case, an exciton is serially transferred between members of a group of molecules. If their electronic coupling is strong enough, the entire group may act as a supermolecule with delocalized electronic transitions, rather than a collection of individual molecules with localized transitions [4]. Experimentally, the exciton coupling can be observed by a split of the absorption band or a derivative-shaped signal in the circular dichroism spectra. This is the case for the pigments B850 in peripheral light-harvesting complex (LH2), the pigments B880 in core light-harvesting complex (LH1) and the special pair in reaction center of purple bacteria as shown in Fig. 5.

The process of energy conversion begins when the excitation energy reaches a pair of nearly parallel and closely spaced Chls (Mg–Mg distance: ~7 Å), the so-called "special pair" (sp in Fig. 5), in reaction center where an electron is excited and delocalized over the Chl dimer to form an excited state. The special pair serves as an irreversible trap for excitation energy since its excited state is unstable and the excited electron is immediately transferred stepwise from the special pair to nearby pigments, resulting in a charge-separated state [5]. For purple photosynthetic

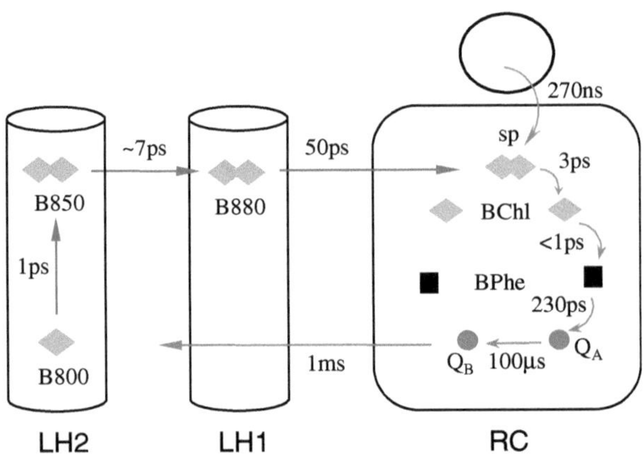

Fig. 5 Organization of a photosynthetic unit composed of light-harvesting complexes (*LH*) and reaction center (*RC*) in purple photosynthetic bacteria. The directions of excitation energy and electron transfers are indicated by *blue* and *purple arrows*, respectively. Typical time constants are shown for each step. *LH2* peripheral light-harvesting complex; *LH1* core light-harvesting complex; *B800* the monomeric pigments that absorb at 800 nm; *B850* the dimeric pigments that absorb at 850 nm; *B880* the dimeric pigments that absorb at 880 nm; *sp* special pair; Q_A quinone-binding site A; Q_B quinone-binding site B (Color figure online)

bacteria and evolutionarily related photosystem II in plants and cyanobacteria, the final electron acceptor in reaction center is an ubiquinone molecule (UQ). When the UQ receives two electrons induced by two photons to become anionic ubiquinol UQ^{2-}, it takes up two protons from solution to form UQH_2. Thus, UQ is a molecular transducer that converts two light-driven one-electron excitation to a two-electron chemical reduction [3]. The energy consumed in each photochemical reaction center is provided from several hundred Chl molecules, i.e. most Chls actually function as light-harvesting antennas. The energy conversion process in reaction center is finely tuned so that in most cases the quantum yield per photon absorbed is close to 100 % [4].

The consequence of UQH_2 formation and its transport to a membrane-bound quinone pool and other redox carriers is the formation of a proton gradient across the membrane. Synthesis of ATP, the cell's "energy currency", is driven by the dissipation of this pH gradient. However, ATP is not sufficient as the sole source of cellular free energy [4]. Reduction of CO_2 to form carbohydrates requires a second source, the reducing power that is the reduced form of NADPH (NADH in anoxygenic photosynthetic bacteria). Plants and cyanobacteria produce both ATP and NADPH directly by a two-step process called noncyclic photophosphorylation [5]. The two photosystems I and II in these organisms are used in a series to extract electrons from water and transfer it to NADPH. As the high-energy electrons pass through the coupled photosystems to generate NADPH, some of their energy is used for ATP synthesis. Since the UQH_2 is thermodynamically unable to directly reduce NAD^+ (+0.04 V vs. −0.32 V, Fig. 2), in purple photosynthetic bacteria a reverse electron flow takes place (Fig. 2, left scheme) [5, 6], in which reduced quinone is the electron donor and NAD^+ is the electron acceptor [4, 6]. The energy is supplied by a transmembrane chemiosmotic potential that is built up by the light-driven cyclic electron transport system [4, 7]. In green sulfur bacteria, the quinone molecule is a menaquinone instead of ubiquinone, the former has a much lower redox potential (−0.06 V, Fig. 2 right scheme). In this case, the reaction center can directly reduce ferredoxin and then NAD^+ without the need for energy-consuming reverse electron flow [4].

The final step of photosynthesis is the conversion of CO_2 to carbohydrates using the ATP and NADPH produced by photosynthetic electron transfer. The initial carbon-fixing reaction involves incorporation of one molecule of CO_2 from atmosphere into a five-carbon compound, ribulose 1,5-bisphosphate (RuBP), to yield two molecules of the three-carbon compound 3-phosphoglycerate (3PG) (Fig. 6). This carboxylation reaction is catalyzed by ribulose 1,5-bisphosphate carboxylase/oxygenase (Rubisco), the most abundant enzyme in the biosphere. Most Rubiscos are hexadecameric proteins, consisting of eight large catalytic subunits ($M_r \cong 53{,}000$) and eight small regulatory subunits ($M_r \cong 14{,}000$). In some purple photosynthetic bacteria, the Rubisco is comprised of only large subunits (either L_2, L_4, or L_8). Following the carboxylation reaction, the resulting 3PG is converted to 1,3-bisphosphoglycerate (BPG) using ATP and then to glyceraldehyde 3-phosphate (GAP) using NADPH. The GAP is an important intermediate for subsequent biosynthesis of various compounds, such as sugars, fatty acids and

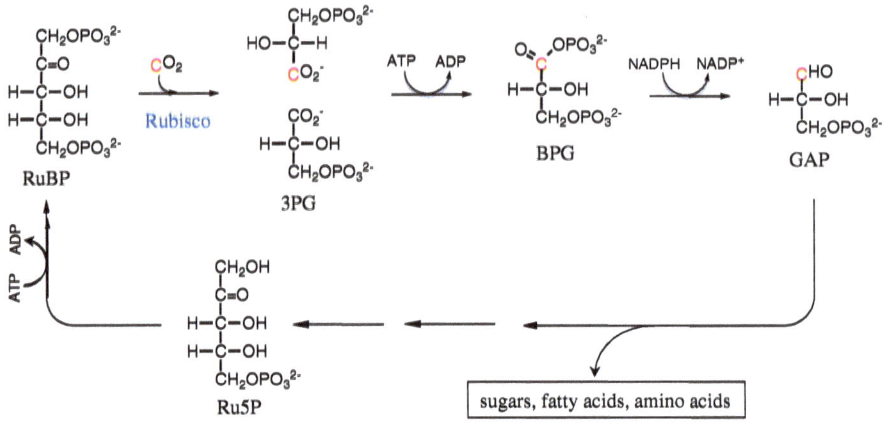

Fig. 6 The carbon-fixation cycle. A total of 3 molecules of ATP and 2 molecules of NADPH are consumed for each CO_2 molecule converted into carbohydrate

amino acids. A two-carbon keto unit is added to the GAP through reactions catalyzed by transketolase to form a five-carbon compound ribulose-5-phosphate (Ru5P) that is the precursor of RuBP.

4 Summary

All photosystems are composed of an antenna complex and a photochemical reaction center. Photosynthesis begins with absorption of sunlight energy by various pigment molecules. The absorbed energy is transferred to reaction center where the charge separation occurs. The electrons are utilized to generate reducing power (NADPH) and transmembrane electrochemical proton gradient that finally lead to generation of the high energy molecule ATP. These molecules serve as energy sources in carbon-fixation reaction. Photosynthetic bacteria contain only a single photosystem with a simple composition, whereas green plants, algae and cyanobacteria have two photosystems connected in series which are capable of extracting electrons from water molecules to generate molecular oxygen (O_2) as a by-product.

References

1. Sherman BD, Vaughn MD, Bergkamp JJ, Gust D, Moore AL, Moore TA (2014) Evolution of reaction center mimics to systems capable of generating solar energy. Photosynth Res 120:59–70
2. Hall DO, Rao KK (1994) Photosynthesis, Chapter 1, 5th edn. Cambridge University Press, Cambridge

3. For example: Voet D, Voet JG (1990) Biochemistry, Chapter 22. Wiley, New York
4. Blankenship RE (2002) Molecular mechanisms of photosynthesis. Blackwell Science, Oxford
5. For example: Alberts B, Johnson A, Lewis J, Raff M, Roberts K, Walter P (2002) Molecular biology of the cell, Chapter 14, 4th edn. Garland Science, New York
6. Herter SM, Kortlüke CM, Drews G (1998) Complex I of *Rhodobacter capsulatus* and its role in reverted electron transport. Arch Microbiol 169:98–105
7. Jackson JB, Obiozo UM (2009) Proton-translocating transhydrogenase in photosynthetic bacteria. In: Hunter CN, Daldal F, Beatty JT (eds) The purple phototrophic bacteria. Springer, The Netherlands, pp 495–508

Recent Understanding on the Photosystem of Purple Photosynthetic Bacteria

Z.-Y. Wang-Otomo

Abstract Bacterial photosynthesis provides a simplified model system ideally for studying the basic mechanism of light-energy harvest and conversion. The early events in this process are carried out by two distinct components, the light-harvesting (LH) complexes and the reaction center (RC). The LH complexes in purple photosynthetic bacteria are classified into two major types, the core LH1 complex that surrounds the RC and the peripheral LH2 complex that exists around the LH1. In addition to light-harvesting, the LH1 also plays a role in quinone (Q) transport between the RC and quinone pool in the cell membrane. While several high-resolution structures are known for the RC and LH2, the structures of LH1 remained at low resolutions. Here, the crystal structure of a LH1-RC complex from thermophilic purple sulfur bacterium *Thermochromatium tepidum* is described. This complex is characterized by an enhanced thermostability and an absorption maximum at 915 nm for the LH1. These properties have been shown to be regulated by Ca^{2+} ions. The structure reveals a closed arrangement of LH1 complex around the RC, and the LH1 BChl *a* molecules form a partially overlapping ring with a shorter Mg–Mg spacing compared with that of B850 in LH2. Structural evidence is for the first time provided for the possible ubiquinone pathway in the closed LH1 complex. The Ca^{2+}-binding sites are identified. Molecular mechanisms of quinone transport, Ca^{2+}-regulation and interaction between LH1 and RC are discussed.

1 Introduction

Our current understanding of the most fundamental mechanism on the photosynthesis has been largely relied on the information from photosynthetic bacteria. Bacterial photosynthesis has played an important role in simplifying the compli-

Z.-Y. Wang-Otomo (✉)
Ibaraki University, Bunkyo 2-1-1, Mito, Japan
e-mail: otomo@mx.ibaraki.ac.jp

M. Sugiyama et al. (eds.), *Solar to Chemical Energy Conversion*,
Lecture Notes in Energy 32, DOI 10.1007/978-3-319-25400-5_22

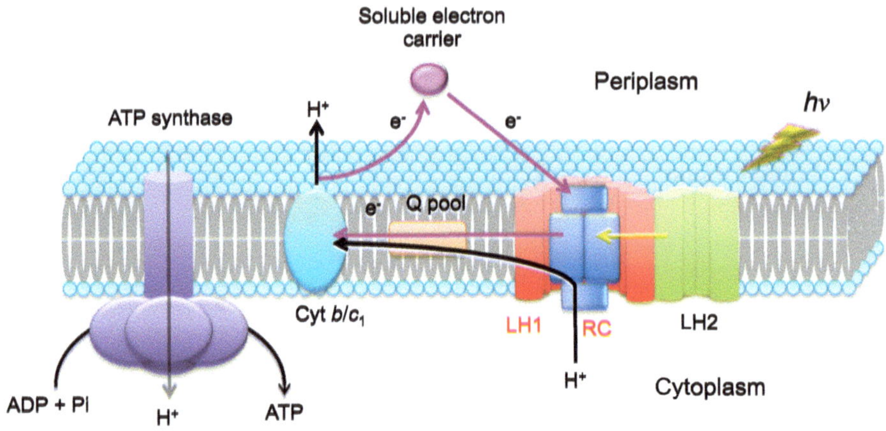

Fig. 1 Schematic representation showing the arrangement of various photosynthetic components in the cell membrane of purple photosynthetic bacteria

cated phenomena as observed in more evolved organisms such as green plants, algae and cyanobacteria. The principles of solar energy conversion obtained from primitive photosynthetic bacteria have been proved to be essentially the same in all photosynthetic organisms. The first step in photosynthesis is the capture of sunlight energy by a set of light-harvesting pigment membrane protein complexes. The light energy absorbed is transferred rapidly and efficiently to the reaction center where primary charge separation takes place and a cyclic electron transport chain is formed. The electron flow generates a proton gradient across the membrane, which is eventually utilized for producing adenosine triphosphate (ATP) (Fig. 1 and see Chap. 21). In purple photosynthetic bacteria, the antenna system mainly consist of two types of light-harvesting complexes, LH1 and LH2 (Fig. 1). The LH1 complex exists in all purple bacteria and surrounds the RC to form the so-called core complex (LH1-RC), whereas the LH2 complex is located in the periphery of the core complex and is absent in some species. Both LH1 and LH2 are large oligomers of a basic structural subunit composed of a heterodimer of two integral membrane polypeptides (α and β) associated with bacteriochlorophyll (BChl) and carotenoid molecules.

While a number of high-resolution structures have been available for the RC [1–3] and LH2 complexes [4–6], the structures of LH1 have remained at low resolutions. Projection maps of electron cryomicroscopy (cryoEM) at 8.5 Å on 2D crystals of LH1 and LH1-RC show a closed ring formed by 16 αβ pairs for the LH1 from *Rhodospirillum* (*R.*) *rubrum* [7, 8]. Similar structural feature was observed by atomic force microscopy (AFM) for the LH1-RC in the native photosynthetic membranes from *Blastochloris* (*Blc.*, formerly *Rhodopseudomonas*) *viridis* [9] and *R. photometricum* [10]. On the other hand, different structure has been reported for the LH1-RC from *Rhodobacter* (*Rb.*) species, where the LH1 αβ subunits show an S-shaped arrangement around the RC and an additional protein, PufX, appears to be involved in the quinone transport and assembly of the dimeric core complex [11–13]. Different from both of the LH1-RCs mentioned above, crystal structure of a third type of the core complex from *Rhodopseudomonas* (*Rps.*) *palustris* has been determined at 4.8 Å [14]. It shows that 15 pairs of the LH1 αβ-polypeptides form an incomplete ring around RC, and electron density near the gap of the ring is assigned to a PufX-like polypeptide.

Currently, at least two issues on the function of LH1 remain unsolved, and the answers to both of them require more detailed structural information. One is on the pigment arrangement in LH1 which determines the rate and efficiency of excitation energy transfer from LH2 to the special pair in RC and the absorption maximum (Q_y transition) in the near-infrared region. Although the structural subunits of LH1 and LH2 are considered to be highly similar, the LH1 complexes usually exhibit Q_y transitions at longer wavelengths (870–960 nm) than do the LH2 (800–850 nm) for the BChl *a*-containing species. Another is on the pathway of the final electron acceptor ubiquinol generated in RC to the ubiquinone (UQ) pool outside the LH1. Particular interest is how the ubiquinone/ubiquinol exchange occurs across a closed LH1 ring as observed for many species. Molecular dynamics simulation predicts that such ubiquinone shuttling is possible through a simple diffusion process based on the fluctuation theorem [15]. Here, the recently published crystal structure of the LH1-RC core complex from thermophilic purple sulfur bacterium *Thermochromatium* (*Tch.*) *tepidum* at 3.0 Å resolution is described [16]. This complex exhibits an enhanced thermal stability and a Q_y transition at 915 nm for the LH1 [17]. Both of these properties are shown to be regulated by Ca^{2+} ions [18, 19]. The Ca^{2+}-binding sites have been identified in the crystal structure and the issues mentioned above can be addressed on the basis of the structural information of this study.

2 Overall Structure of the LH1-RC

The LH1 subunits are uniformly distributed around the RC and form a closed and slightly elliptical cylinder composed of 16 pairs of helical αβ-polypeptides, 32 BChls *a*, 16 carotenoids (spirilloxanthin) and 16 Ca^{2+} ions (Fig. 2). The pigment and Ca^{2+} stoichiometry are consistent with those determined by biochemical analysis [17, 18]. Similar to LH2, the double-layered cylinder of LH1 has the α-polypeptides located inside and the β-polypeptides outside with the amino (N) termini on the cytoplasmic side. The long and short dimensions are 107 and 99 Å (distance between the centers of the opposite helices) for the outer ring, and 77 and 69 Å for the inner ring, respectively. Due to the reduced curvature of LH1 ring, the adjacent α-polypeptides in the inner ring are more separate and the average spacing (13.5 Å) is greater than that of the LH2 (10.9 Å for *Rps. acidophila* and 9.7 Å for *R. molischianum*), whereas the spacing between β-polypeptides is similar to that of LH2. Structures of the LH1 α- and β-polypeptides are similar to those of the corresponding LH2 polypeptides from *R. molischianum* [5] as expected from

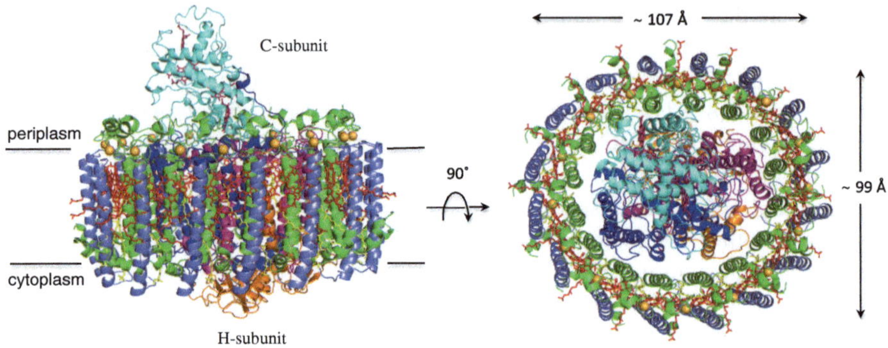

Fig. 2 Overall structure of the *Tch. tepidum* LH1-RC complex. *Left* side view parallel to the membrane plane with the periplasmic C-subunit (*cyan*) above and the cytoplasmic H-subunit (*orange*) below. *Right* top view from the periplasmic side of membrane [37]. LH1 α: *green*; LH1 β: *slate blue*; L-subunit: *magenta*; M-subunit: *blue*; Ca^{2+}: *orange ball*; BChl *a*: *red*; Spirilloxanthin: *yellow*

the high sequence similarity [20]. These polypeptides share sequence identities of 38 % (Tyr9-Trp46) for the α-polypeptides and 41 % (Ser5-Leu46) for the β-polypeptides. The α-helical regions span the membrane bilayer from α-Pro17 to α-Ser41 and from β-Asp11 to β-Tyr42.

The RC is accommodated in the LH1 ellipsoid and fits the shape of the inner ring with the L- and M-subunits aligned along the long axis. The overall structure of the RC in the LH1-RC complex is similar to that of the RC-only complex [3] with a root mean square deviation (rmsd) of 0.94 Å for the 1272 Cα atoms. Relatively large differences are found for the membrane-surface attached H-subunit and the membrane extruded C-subunit. All cofactors reported in the RC-only structure are identified, including the 15-*cis*-spirilloxanthin, menaquinone and phytol chains of BChls *a* and bacteriopheophytins (BPhe) *a*. The head group of newly identified ubiquinone is located at the same position (Q_B site) with the same orientation as those in *Blc. viridis* and *Rb. sphaeroides* RCs. Contrary to expectations, most of the transmembrane regions in RC seems not to specifically interact with the LH1 polypeptides. Only a few portions in the L- and M-subunits are in relatively close proximity (~3.7 Å) to the LH1 α-polypeptides, these include L36-51, L41-61, M54-65 and M115-123.

3 Pigment Organization

The BChl *a* molecules in LH1 form an elliptical and partially overlapping ring with the long and short dimensions of 93 and 83 Å (distance between the central Mg atoms of opposite BChls *a*), respectively (Fig. 3). The spirilloxanthins in LH1 span the transmembrane region in between the α- and β-polypeptides without apparent

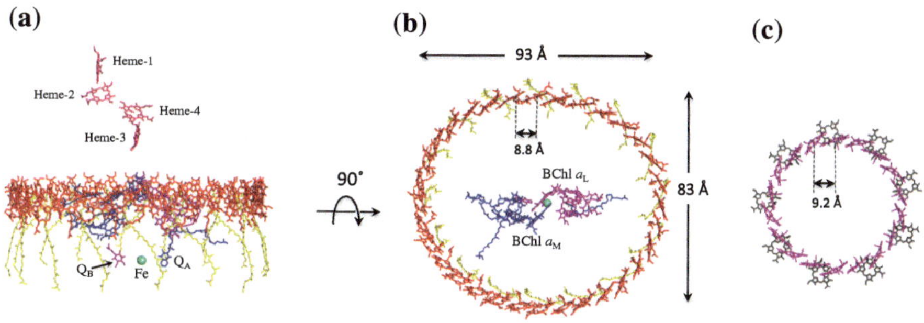

Fig. 3 Cofactor arrangement in LH1-RC and a comparison with that in LH2. **a** *Side view* along the membrane plane with the periplasm above and the cytoplasm below. **b** *Top view* from the periplasmic side of membrane (the hemes in C-subunit are omitted for clarity). BChls *a* and spirilloxanthins in LH1 are coloured in *red* and *yellow*, respectively. Cofactors in L- and M-subunits of RC are coloured in *magenta* and *blue*, respectively. **c** BChl *a* arrangement in LH2 of *Rps. acidophila* [37]

association with the polypeptides and BChl *a*. The average distance between the central Mg atoms of neighboring BChls *a* is 8.8 Å. This value is smaller than those of the LH2 (9.3 Å for *Rps. acidophila* and 9.2 Å for *R. molischianum*), indicating more extensive overlap and stronger coupling between the bacteriochlorin planes. The larger ring and shorter spacing between the BChls *a* may explain why the LH1 exhibits a more red-shifted Q_y transition than LH2 [21]. The central Mg atoms of LH1 BChls *a* are liganded by histidine residues with average distances of 2.6 and 2.3 Å for the Mg-Nε2 (α-His36) and the Mg-Nε2 (β-His36), respectively. The C3-acetyl oxygen atoms ($O3^2$) of LH1 BChls *a* are involved in hydrogen bonding with the indole nitrogen atoms (Nε1) of tryptophans (α46 and β45). This is in agreement with the resonance Raman result that shows a C3-acetyl C = O stretching band at 1637 cm^{-1} and indicates strong hydrogen bonding for the carbonyl oxygen [22]. It is noted that the Nε1 of α-Trp46 seems to form even stronger hydrogen bond, as the average distance between $O3^2$ (α-BChl *a*) and Nε1 (α-Trp46) is only 2.6 Å compared to the distance of 2.9 Å between $O3^2$ (β-BChl *a*) and Nε1 (β-Trp45). These tryptophan residues are conserved in almost all LH1 polypeptides of photosynthetic bacteria [23]. The C13-keto groups do not form hydrogen bonds.

The LH1 BChls *a* are aligned with the RC special pair (BChl a_L and BChl a_M) on the same plane that parallels to the membrane surface. The bacteriochlorin plane normals of special pair point toward approximately the middle of the long and short axes of LH1 BChl *a* ring. The distances between the central Mg atoms of RC special pair and those of the closest LH1 BChl *a* are 38–39 Å. This is slightly smaller than the value of 43 Å as estimated from computational modeling [24]. The spatial arrangement is considered as the optimized configuration for facilitating efficient migration of the excitation energy from LH1 to RC. The non-heme iron and the head groups of menaquinone and ubiquinone in RC are aligned along a line parallel to the long axis of the LH1 BChl *a* ring.

4 Ca^{2+}-Binding Sites

The LH1-RC of *Tch. tepidum* exhibits high thermostability and is characterized by an unusual LH1 Q_y transition at 915 nm, about 35 nm red-shift from those of most other mesophilic counterparts [25]. We previously demonstrated that the Ca^{2+} ions are responsible for these unique properties [18, 19]. Removal of the Ca^{2+} resulted in formation of a species with the LH1 Q_y transition at 880 nm and reduced thermostability, and addition of Ca^{2+} back to the 880-nm species restored its native 915-nm form with high stability. The putative Ca^{2+}-binding sites were subsequently identified by biochemical method to be located on the membrane surface region of the C-terminal domain of the LH1 [26]. In the crystal structure, the identity of Ca^{2+} along with their positions are unambiguously determined from the anomalous difference map. Sixteen Ca^{2+} ions are found in the LH1 distributing in the middle between the inner and outer rings (Fig. 4). The Ca^{2+}-binding sites are located in the C-terminal regions of the α- and β-polypeptides with both chains providing ligands for the coordination. Each Ca^{2+} is coordinated by the main chain oxygen atom of α-Trp46, side chain carbonyl groups of α-Asp49 and α-Asn50 together with the C-terminal carboxyl group of the β-Leu46 in the adjacent subunit. Water molecules, whose electron densities have not been clearly identified, may also be involved in the ligation to form a heptacoordinated structure as favored by the Ca^{2+} ion. Association of the α- and β-polypeptides through the Ca^{2+} ions stabilizes the LH1 structure and therefore accounts for its enhanced thermostability. Since the binding network is positioned close to the BChl *a* molecules, configuration of the coupled pigments could be modified through the nearby α-Trp46 and β-Trp45 whose side chains are hydrogen-bonded to the C3-acetyl oxygen atoms of BChls *a*. The Ca^{2+}-binding effect is considered to have additional contribution to the larger red-shift of the LH1 Q_y transition in *Tch. tepidum* compared with the Q_y transitions (~880 nm) in other species.

In addition to the Ca^{2+} ions in LH1, another Ca^{2+} is identified in the RC complex from the anomalous scattering measurement with strong and positive density in the difference Fourier map. It is located at the interface between the C- and M-subunits

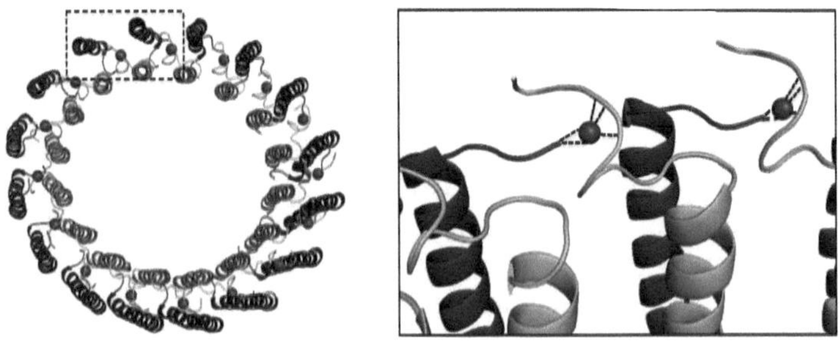

Fig. 4 Distribution of the Ca^{2+} ions (*red balls*) in LH1 (*left*) and an enlarged view of the region marked by *dashed rectangle* showing the Ca^{2+}-binding site (*right*) [37]

and this site is also exposed on the presumed membrane surface. The Ca^{2+} is coordinated by the carboxyl groups of two Glu residues (C230 and M96) and the side chain carbonyl oxygen of a Gln residue (C183). Although the function of this Ca^{2+} is unknown, it may play a role in stabilizing the structure of RC as implied from its unique position. It is noted that the coordinating Gln183 is in a loop region (C179-198) which is the only portion that largely differs from that in the RC-only structure. This suggests that the conformation of the loop region is maintained by the Ca^{2+}-binding.

5 Possible Ubiquinone Pathway

The crystal structure provides insights into the mechanism of ubiquinone exchange in the LH1-RC core complex with a closed LH1 ring. Inspection of the structure reveals that there exist channels formed between every adjacent pair of the LH1 αβ-polypeptides. An example is shown in Fig. 5. The channels in LH1 are located near to the N-terminal ends of transmembrane domain. The corresponding regions in LH2 are occupied by a set of monomeric BChl *a* (B800) that lie in between the β-polypeptides. It is noted that not only the size but also the shape of the channel well fit the head group of the ubiquinone. The channel opening has dimensions of ~9.7 Å in height and ~3.4 Å in width, implying that the head group must have correct orientation with respect to the channel opening to pass through. The path from Q_B site to the LH1 channel opening is almost parallel to the presumed membrane surface, and the closest distance between the benzoquinone head and the channel inner gate is about 30 Å. There is a large space between the Q_B site and LH1 inner wall. These structural features suggest a possibility that multiple channels in the LH1 could be available simultaneously for the ubiquinone transport and the head group should sufficiently often find the channel openings.

The putative ubiquinone channels seems to be mainly defined by the LH1 α-polypeptides as they form the narrowest portion of the openings, in which at least eight amino acids are involved. Most of these residues are highly hydrophobic. Residues Val20, Ser23, Ile24 and Phe27 are located on one side of the channel, and residues Leu21, Val22, Val25 and Ile29 are located on the opposite side. A similar number of residues in the same region of LH1 β-polypeptides form the outer gate of the channel. The interior of the channel is in hydrophobic environment. Existence of the channels and their role as potential pathways for the ubiquinone transport are supported by the molecular dynamics simulation [15] in which a mechanism of ubiquinone diffusion through the closed LH1 ring was proposed. The simulation estimated an upper limit of $\sim 8 \times 10^{-3}$ s for the passage time, which is slightly longer compared to the experimentally measured turnover time (3.5×10^{-3} s) for the monomeric LH1-RC complex with a closed LH1 ring from a *Rb. sphaeroides* $PufX^-$ strain [27]. The experiment by Comayras et al. strongly suggests that the closed LH1 ring is far from a tight barrier, but presents sufficient openings for the quinone diffusion [27].

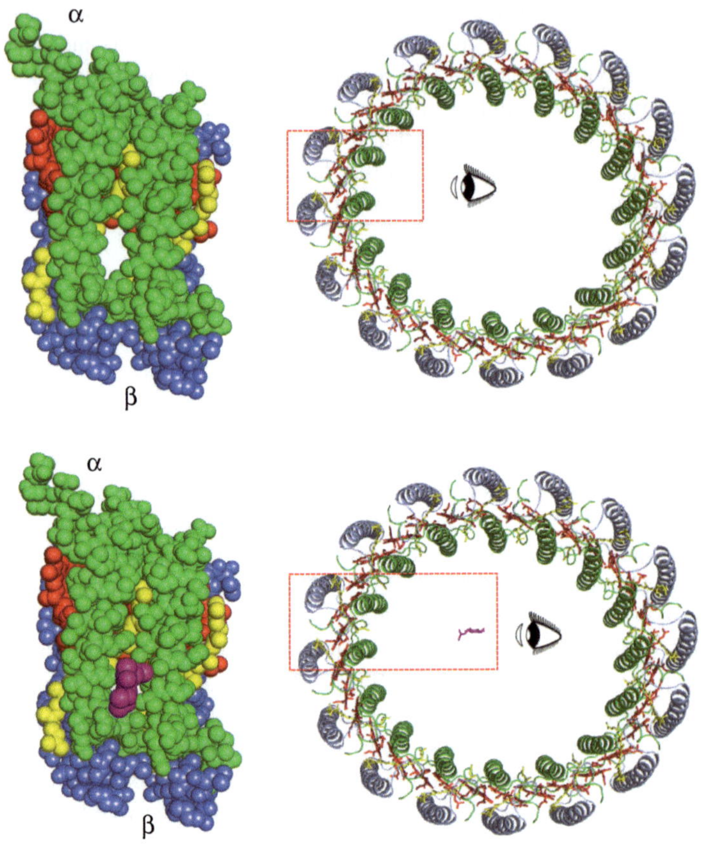

Fig. 5 A channel viewing from inside of the LH1 ring (*upper panel*). Its shape and size fit well the benzoquinone head group of an ubiquinone when the ubiquinone (*purple*) is positioned to its original Q_B binding site (*lower panel*)

6 Implications for Energy and Electron Transfers

Since the LH1 complex encircles RC, it serves as both the entrance of excitation energy and the exit of reduced ubiquinone from the RC to the quinone (Q) pool and cytochrome bc_1 complex (Fig. 6). As a result, the LH1 plays a crucial role in regulating the functions of RC. It has been known that energy transfer from LH1 to RC is the rate-limiting step of the overall energy-trapping process [28, 29]. The time constant of energy transfer from LH1 BChls *a* to the RC special pair in *Tch. tepidum* was measured to be ~50 ps in the photoactive state [30], which is comparable with the range of 35–80 ps for the core complexes in other photosynthetic bacteria [31, 32]. These values are significantly greater than the lifetime of about 3 ps for the excited state of special pair [33, 34] and the time constants of 3–5 ps for the inter-complex energy transfer from LH2(B850) to LH1 [29, 35]. The slow step

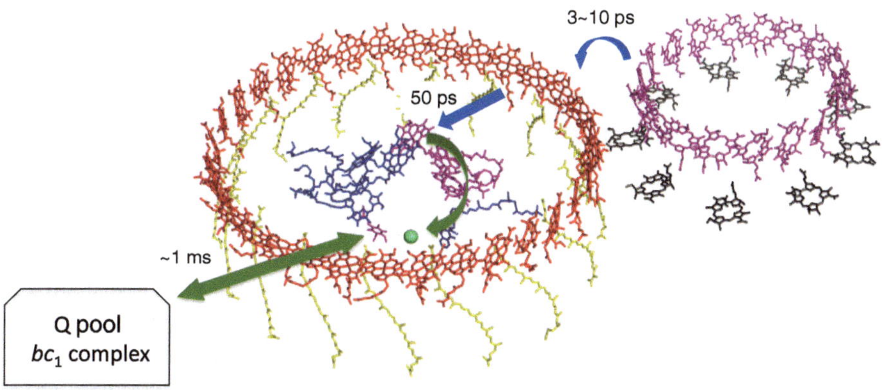

Fig. 6 Overview of the excitation energy and electron transfers in photosynthetic membrane of purple photosynthetic bacteria. Typical time constants are indicated for major steps

is mainly attributed to the long distance between the LH1 BChls *a* and special pair, which is determined to be 38 Å for the closest BChl *a* pair in this work and was estimated to be 42 Å by computational modeling [24]. Given the long distance, a circular organization of LH1 around the RC is considered as the only way to keep the trapping efficiency high [28, 29]. The energy transfer has also been demonstrated to be sensitive to the redox state of LH1. Oxidation of very few BChls *a* in LH1-only complex resulted in complete quench of the fluorescence and therefore lost its function as an antenna complex [36].

Despite the striking similarity in structures between LH1 and LH2, the most important difference between the two complexes is probably the absence of monomeric BChls *a* (B800) in LH1. These pigments in LH2 are located at the positions corresponding to the outer gates of the putative ubiquinone channels in LH1 and would otherwise block the gates if exist. Presence of the channels in LH1 may explain the fact that there is still a rapid passage ($\sim$1 ms) of the ubiquinone across the closed LH1 ring as observed from the PufX$^-$ mutant of *Rb. sphaeroides* [27], even though the diffusion is about twofold slower than that in the wild-type. On the other hand, the time constant of 1 ms is much greater compared to those of electron transfers in other steps in RC ranging from 1 ps to 200 µs. In this regard, the ubiquinone transport across LH1 can also be viewed as the rate-limiting step in the overall electron transfer process. A large number of studies using AFM and cryoEM have shown that LH1 complexes in both LH1-only and LH1-RC forms are not rigid assemblies but quite flexible with variations in the shape from circles to ellipses. The flexibility and the lack of apparent specific interaction between LH1 and RC as revealed in this study lead us to hypothesize that the RC might rotate within the LH1 ring in its working state without taking specific orientations with respect to the LH1 complex as the closed LH1 ring is composed of uniformly distributed subunits. This dynamic model might provide alternative interpretations for the way of optimization of the function of LH1-RC in adjusting the balance between excitation energy transfer and electron transport.

7 Summary

In the photosystem of purple bacterium *Tch. tepidum*, the LH1 forms a closed and slightly elliptical cylinder around the RC, and is composed of 16 pairs of αβ-polypeptides, 32 bacteriochlorophylls (BChl) *a*, 16 spirilloxanthins and 16 Ca^{2+} ions. The LH1 BChl *a* molecules form a partially overlapping ring with an average Mg–Mg distance of 8.77 Å, which is shorter than that of B850 in LH2. The crystal structure provides for first time the evidence for the possible quinone channels present in the closed LH1 complex. These channels exist between every adjacent pair of the LH1 αβ-polypeptides and are located at the N-terminal ends of trans-membrane regions. Each Ca^{2+} ion binds one LH1 α- and one β-polypeptide at their C-terminal domains. The Ca^{2+}-binding network is considered to modulate the Q_y transition and thermostability of the LH1 complex.

References

1. Deisenhofer J, Epp O, Miki K, Huber R, Michel H (1985) Structure of the protein subunits in the photosynthetic reaction centre of Rhodopseudomonas viridis at 3 Å resolution. Nature 318:618–624
2. Allen JP, Feher G, Yeates TO, Komiya H, Rees DC (1987) Structure of the reaction center from Rhodobacter sphaeroides R-26: the cofactors. Proc Natl Acad Sci USA 84:5730–5734
3. Nogi T, Fathir I, Kobayashi M, Nozawa T, Miki K (2000) Crystal structures of photosynthetic reaction center and high-potential iron-sulfur protein from Thermochromatium tepidum: Thermostability and electron transfer. Proc Natl Acad Sci USA 97:13561–13566
4. McDermott G, Prince DM, Freer AA, Hawthornthwaite-Lawless AM, Papiz MZ, Cogdell RJ, Isaac NW (1995) Crystal structure of an integral membrane light-harvesting complex from photosynthetic bacteria. Nature 374:517–521
5. Koepke J, Hu X, Muenke C, Schulten K, Michel H (1996) The crystal structure of the light-harvesting complex II (B800-B850) from Rhodospirillum molischianum. Structure 4:581–597
6. McLuskey K, Prince SM, Cogdell RJ, Isaac NW (2001) The crystallographic structure of the B800-820 LH3 light-harvesting complex from the purple bacteria Rhodopseudomonas acidophila strain 7050. Biochemistry 40:8783–8789
7. Karrasch S, Bullough PA, Ghosh R (1995) The 8.5 Å projection map of the light-harvesting complex I from Rhodospirillum rubrum reveals a ring composed of 16 subunits. EMBO J 14:631–638
8. Jamieson SJ, Wang P, Qian P, Kirkland JY, Conroy MJ, Hunter CN, Bullough PA (2002) Projection structure of the photosynthetic reaction centre-antenna complex of Rhodospirillum rubrum at 8.5 Å resolution. EMBO J 21:3927–3935
9. Scheuring S, Seguin J, Marco S, Levy D, Robert B, Rigaud J-L (2003) Nanodissection and high-resolution imaging of the Rhodopseudomonas viridis photosynthetic core complex in native membranes by AFM. Proc Natl Acad Sci USA 100:1690–1693
10. Scheuring S, Francia F, Busselez J, Melandris BA, Rigaud J-L, Levy D (2004) Structural role of PufX in the dimerization of the photosynthetic core complex of Rhodobacter sphaeroides. J Biol Chem 279:3620–3626
11. Jungas C, Ranck J-L, Rigaud J-L, Joliot P, Vermeglio A (1999) Supramolecular organization of the photosynthetic apparatus of Rhodobacter sphaeroides. EMBO J 18:534–542

12. Qian P, Hunter CN, Bullough PA (2005) The 8.5 Å projection structure of the core RC-LH1-PufX dimer of Rhodobacter sphaeroides. J Mol Biol 349:948–960
13. Scheuring S, Busselez J, Levy D (2005) Structure of the dimeric PufX-containing core complex of Rhodobacter blasticus by in situ atomic force microscopy. J Biol Chem 280:1426–1431
14. Roszak AW, Howard TD, Southall J, Gardiner AT, Law CJ, Isaac NW, Cogdell RJ (2003) Crystal structure of the RC-LH1 core complex from Rhodopseudomonas palustris. Science 302:1969–1972
15. Arid A, Wrachtrup J, Schulten K, Tietz C (2007) Possible pathway for ubiquinone shuttling in Rhodospirillum rubrum revealed by molecular dynamics simulation. Biophys J 92:23–33
16. Niwa S, Yu L-J, Takeda K, Hirano Y, Kawakami T, Wang-Otomo Z-Y, Miki K (2014) Structure of the LH1-RC complex from Thermochromatium tepidum at 3.0 Å. Nature 508:228–232
17. Suzuki H, Hirano Y, Kimura Y, Takaichi S, Kobayashi M, Miki K, Wang Z-Y (2007) Purification, characterization and crystallization of the core complex from thermophilic purple sulfur bacterium Thermochromatium tepidum. Biochim Biophys Acta 1767:1057–1063
18. Kimura Y, Hirano Y, Yu L-J, Suzuki H, Kobayashi M, Wang Z-Y (2008) Calcium ions are involved in the unusual red shift of the light-harvesting 1 Q_y transition of the core complex in thermophilic purple sulfur bacterium Thermochromatium tepidum. J Biol Chem 283:13867–13873
19. Kimura Y, Yu L-J, Hirano Y, Suzuki H, Wang Z-Y (2009) Calcium ions are required for the enhanced thermal stability of the light-harvesting-reaction center core complex from thermophilic purple sulfur bacterium Thermochromatium tepidum. J Biol Chem 284:93–99
20. Germeroth L, Lottspeich F, Robert B, Michel H (1993) Unexpected similarities of the B800-850 light-harvesting complex from Rhodospirillum molischianum to the B870 light-harvesting complexes from other purple photosynthetic bacteria. Biochemistry 32:5615–5621
21. Cogdell RJ, Howard TD, Isaac NW, McLuskey K, Gardiner AT (2002) Structural factors which control the position of the Q_y absorption band of bacteriochlorophyll a in purple bacterial antenna complexes. Photosynth Res 74:135–141
22. Kimura Y, Inada Y, Numata T, Arikawa T, Li Y, Zhang J-P, Wang Z-Y, Ohno T (1817) Metal cations modulate the bacteriochlorophyll-protein interaction in the light-harvesting 1 core complex from Thermochromatium tepidum. Biochim Biophys Acta 1022–1029:2012
23. Rücker O, Köhler A, Behammer B, Sichau K, Overmann J (2012) Puf operon sequences and inferred structures of light-harvesting complexes of three closely related Chromatiaceae exhibiting different absorption characteristics. Arch Microbiol 194:123–134
24. Hu X, Schulten K (1998) Model for the light-harvesting complex I (B875) of Rhodobacter sphaeroides. Biophys J 75:683–694
25. Wang Z-Y, Shimonaga M, Suzuki H, Kobayashi M, Nozawa T (2003) Purification and characterization of the polypeptides of core light-harvesting complexes from purple sulfur bacteria. Photosynth Res 78:133–141
26. Yu L-J, Kato S, Wang Z-Y (2010) Examination of the putative Ca^{2+}-binding site in the light-harvesting complex 1 of thermophilic purple sulfur bacterium Thermochromatium tepidum. Photosynth Res 106:215–220
27. Comayras F, Jungas C, Lavergne J (2005) Functional consequences of the organization of the photosynthetic apparatus in Rhodobacter sphaeroides. II. A study of $PufX^-$ membranes. J Biol Chem 280:11214–11223
28. Fleming GR, van Grondelle R (1997) Femtosecond spectroscopy of photosynthetic light-harvesting systems. Curr Opin Struct Biol 7:738–748
29. Robert B, Cogdell RJ, von Grondelle R (2003) The light-harvesting system of purple bacteria. In: Green BR, Parson WW (eds) Light-harvesting antennas in photosynthesis. Kluwer Academic Publishers, Dordrecht, pp 169–194

30. Ma F, Kimura Y, Zhao X-H, Wu Y-S, Wang P, Fu L-M, Wang Z-Y, Zhang J-P (2008) Excitation dynamics of two spectral forms of the core complexes from photosynthetic bacterium Thermochromatium tepidum. Biophys J 95:3349–3357
31. Beekman LMP, van Mourik F, Jones MR, Visser M, Hunter CN, van Grondelle R (1994) Trapping kinetics in mutants of the photosynthetic purple bacterium Rhodobacter sphaeroides: influence of the charge separation rate and consequences for the rate-limiting step in the light-harvesting process. Biochemistry 33:3143–3147
32. van Grondelle R, Novoderezhkin VI (2009) Spectroscopy and dynamics of excitation transfer and trapping in purple bacteria. In: Hunter CN, Daldal F, Beatty JT (eds) The purple phototrophic bacteria. Springer, Dordrecht, pp 231–252
33. Woodbury NW, Allen JF (1995) The pathway, kinetics and thermodynamics of electron transfer in wild type and mutant reaction centers of purple nonsulfur bacteria. In: Blankenship RE, Madigan MT, Bauer CD (eds) Anoxygenic photosynthetic bacteria. Kluwer Academic Publishers, Dordrecht, pp 527–557
34. Parson WW, Warshel A (2009) Mechanism of charge separation in purple bacterial reaction centers. In: Hunter CN, Daldal F, Beatty JT (eds) The purple phototrophic bacteria. Springer, Dordrecht, pp 355–377
35. Hess S, Chachisvilis M, Timpmann K, Jones MR, Fowler GJS, Hunter CN, Sundström V (1995) Temporally and spectrally resolved subpicosecond energy transfer within the peripheral antenna complex (LH2) and from LH2 to the core antenna complex in photosynthetic purple bacteria. Proc Natl Acad Sci USA 92:12333–12337
36. Law CJ, Cogdell RJ (1998) The effect of chemical oxidation on the fluorescence of the LH1 (B880) complex from the purple bacterium Rhodobium marinum. FEBS Lett 432:27–30
37. Wang-Otomo Z-Y (2014) The frontiers in photosynthesis research. NTS Inc., Tokyo (in Japanese)

Mn_4Ca Cluster in Photosynthetic Water Oxidation

Junko Yano

Abstract In natural photosynthesis, the water oxidation reaction takes place in the oxygen evolving complex (OEC) in Photosystem II (PSII), that cycles through a series of five intermediate S-states. Mechanistic studies of this catalytic reaction have been carried out with various spectroscopic methods, structural analysis, and theoretical studies. Among them, the structural information of the metal cluster has been obtained largely from the X-ray techniques that include X-ray crystallography and X-ray spectroscopy. This chapter describes the recent studies of the OEC in PSII with X-ray spectroscopy and crystallography.

1 Introduction

Oxygen, that supports all aerobic life, is produced by photosynthetic water oxidation in plants, algae, and cyanobacteria. The metal complex involved in oxygenic photosynthesis, the oxygen-evolving complex (OEC), consists of an oxo-bridged structure with four Mn and one Ca atoms. No variations have been observed so far among oxygenic photosynthetic organisms through higher plants and algae back to cyanobacteria, which represents the earliest oxygenic photosynthetic organisms. Oxygen itself is the by-product of the photosynthetic water oxidation reaction;

$$2H_2O \rightarrow O_2 + 4H^+ + 4e^-$$

However, it was this oxygen that enabled oxygenic life to evolve and that led to the current diverse and complex life on earth by dramatically increasing the metabolic energy that became available from aerobic respiration.

J. Yano (✉)
Molecular Biophysics and Integrated Bioimaging Division, Lawrence Berkeley National Laboratory, Berkeley, CA 94720, USA
e-mail: jyano@lbl.gov

M. Sugiyama et al. (eds.), *Solar to Chemical Energy Conversion*,
Lecture Notes in Energy 32, DOI 10.1007/978-3-319-25400-5_23

The OEC is embedded in Photosystem II (PS II), a membrane pigment-protein complex, where the primary charge separation by absorbed sunlight energy and successive electron transfer occurs through several pigment molecules (Fig. 1) [1]. The electrons and protons produced in the water oxidation reaction in PS II are ultimately used to store energy in the form of ATP and to reduce CO_2 to carbohydrates via the Calvin-Benson cycle, and the precursors for synthesis of the biological molecules needed by the organism. Nature has thus evolved a way to store sunlight energy in the form of chemical energy that all life depends on. One of the key questions, in relation to the development of artificial photosynthetic device for renewable energy resources, is how nature manages the uphill photo-induced water oxidation reaction by modulating the redox potential of each of the steps in the four electron redox reaction.

This chapter focuses on the recent development in the field of the OEC in PSII using X-ray based techniques. There are other important approaches taken by EPR/ENDOR studies, water exchange investigation, IR spectroscopy, and theoretical calculations. Those studies are not covered in this report, but the excellent reviews are available in literature (e.g. [2–5]).

2 Photosystem II

In PSII, the light-harvesting, charge separation, charge stabilization and electron transfer take place in the electron transfer chain shown in Fig. 1a. The reaction center, P_{680}, in the D1 and D2 subunits is the primary electron donor that traps the light energy delivered from the inner antenna subunits (CP43 and CP47 subunits) and the outer antenna complexes (LHC1 and LHC2) of PS II. Then the excited state, $P^{\cdot+}_{680}$, rapidly transfers the electron to Chl_{D1}, pheophytin ($Pheo_{D1}$), and to the acceptor, plastoquinone Q_A, and subsequently to the final electron acceptor plastoquinone Q_B. This fast electron transfer process stabilizes the charge separated state. Acceptor Q_B at the quinone pocket in PS II accepts two electrons by protonation as plastoquinol Q_BH_2, and it is released into the membrane matrix from PS II. Thus, the energy transfers to the cytb$_6$f complex that mediates electron between PS II and PS I. On the donor side of PS II, the cationic radical $P680^{\cdot+}$ is reduced by a tyrosine residue, Tyr_Z (D1Tyr161), to generate a neutral tyrosine radical $Tyr_Z^{\bullet}$ which acts as an oxidant for the water oxidation reaction at the OEC.

The OEC cycles through a series of five intermediate S-states (Si, i = 0 to 4), representing the number of oxidizing equivalents stored on the OEC driven by the energy of the four successive photons absorbed by the PS II reaction center (Fig. 1b) [6]. When PS II is dark-adapted, it relaxes to the S_1 state. Although S_0 is the most reduced state of the OEC, the S_0 state is oxidized by tyrosine D(Y_D^{+}) during the dark adaptation and therefore the S_1 state becomes the dark-stable state. Illumination of dark-adapted S_1 state with saturating flashes of visible light leads to a maximum O_2 yield after the 3rd flash, and then after every 4th flash. Each flash advances the oxidation state of the OEC by removal of one electron. Once four

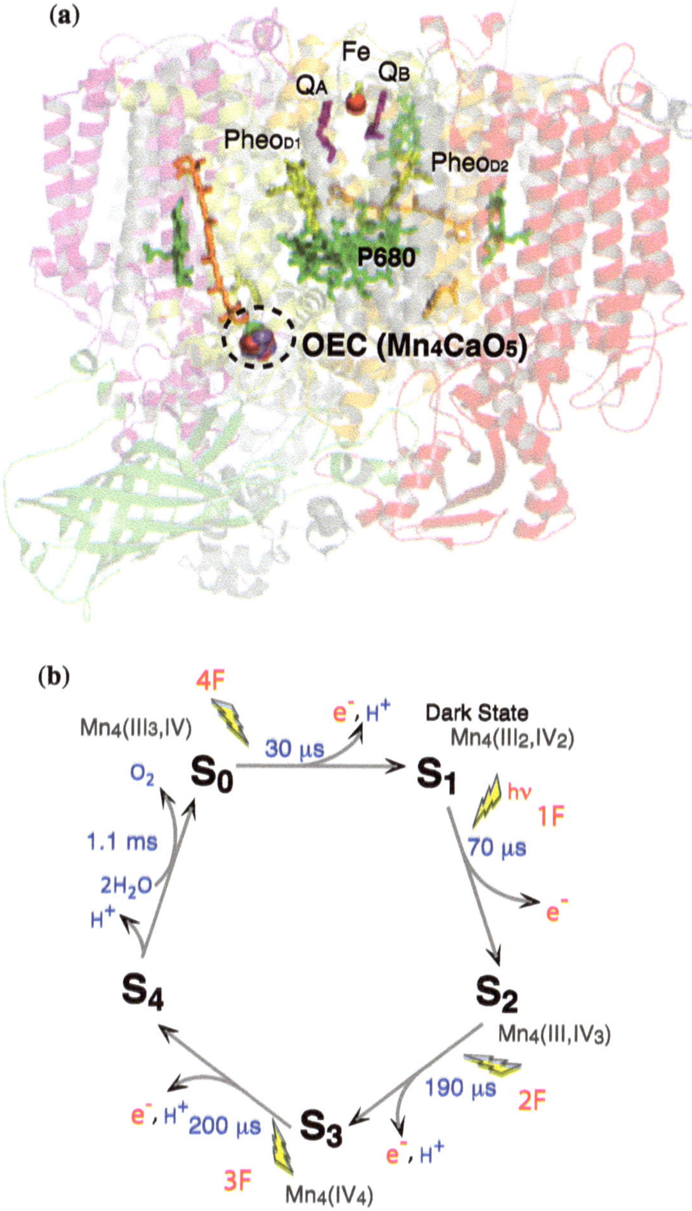

Fig. 1 **a** The Photosystem II structure from 1.9 Å data from X-ray crystallography showing the membrane spanning helices and the extrinsic polypeptides which are in lighter shade of color. The location of the cofactors involved in charge separation and the Mn_4Ca cluster in the membrane are shown highlighted against the polypeptide background. The Mn_4Ca cluster is on the lumenal side of the membrane with the acceptor quinones on the stromal cytoplasmic side of the membrane. **b** The S-state Kok scheme for oxygen evolution along with 'formal' oxidation states of Mn in the intermediate S-states. Figure is adapted from Ref. [59]

oxidizing equivalents are accumulated in the OEC (S_4-state), a spontaneous reaction occurs that results in the oxidation of water, release of O_2 and the formation of the S_0-state. During the reaction, the Mn cluster provides a high degree of redox and chemical flexibility, while the protein residues are critical for mediating the reaction by modulating the redox potentials and providing pathways for electrons, protons, substrate H_2O, and product O_2 [7–9].

3 Oxygen Evolving Complex and the Mn_4Ca Cluster

The geometry of the Mn_4CaO_5 cluster in the OEC in the dark S_1 state has been studied by X-ray crystallography [10–12], and finally the radiation-damage-free structure has been confirmed in the 1.95 Å crystal structure taken at SACLA (Japan) (Fig. 2) [13]. Prior to these recent X-ray diffraction (XRD) studies, detailed extended X-ray absorption fine structure (EXAFS) studies [14–16] using solutions [17], oriented membranes [18–22], single crystals [23], and range-extended methods [24, 25]. The presence of Ca–Mn interactions is supported also by compelling evidence from EXAFS measurements obtained at the Ca- and Sr edges of Sr-substituted PS II [14, 26, 27]. The Mn EXAFS studies showed that there are two Mn–Mn distances at ~2.7 Å, one Mn–Mn distance at ~2.8 Å, and one Mn–Mn distance at ~3.3 Å [17, 23], corresponding to the di-μ-oxo and mono-μ-oxo bridges. While the synchrotron XRD studies had evidences of a partial Mn reduction during the data collection, and as a consequential elongation of the metal-metal and metal-ligand distances, the data collected at SACLA with femtosecond X-ray pulses at 100 K has finally confirmed the EXAFS distances [13]. There are remaining differences in the metal ligand distances obtained from the

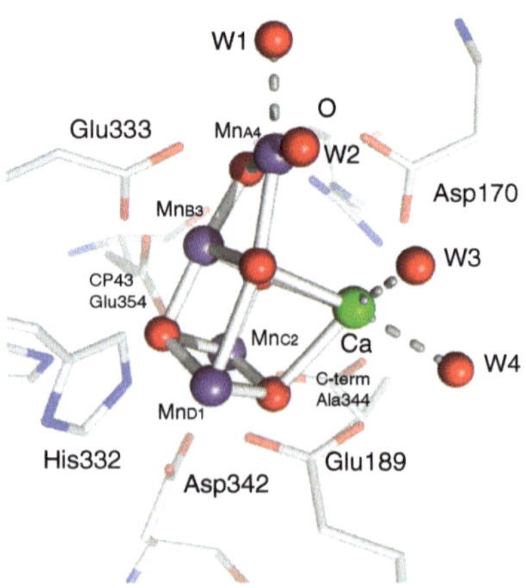

Fig. 2 The OEC S_1 state structure of the OEC in PSII (PDB: 4UB8) [13]

EXAFS and the XRD analysis, that has not been clearly explained, but it may be partly due to the errors in each method.

There are 7 ligands directly ligated to the Mn_4CaO_5 cluster in the S_1 structure; 6 carboxyl ligands (aspartate and glutamate) and one imidazole ligand (histidine). These ligands are from side chains from two domains of the D1 subunit; the interhelical CD luminal loop (residues Asp170 and Glu189) and the C-terminal region (between residues His332 and Ala344), and one domain of CP43—the large luminal EF loop. Most of the ligands are arranged in a bidentate fashion bridging 2 metals. D1-Asp170 and the C-terminal carboxylate group of D1-Ala344 bridge Ca with Mn_{A4} and Mn_{C2}, respectively, and D1-Glu189 provides monodentate ligation to Mn_{D1}. The only His providing direct Mn ligation is D1-His332, binding Mn_{D1}. The second His in close proximity to the Mn_4CaO_5 cluster is D1-His337, which is too far for direct ligation of a Mn, but is in hydrogen bonding distance to the bridging oxygen. In a similar manner CP43-Arg357 ligates one of the oxygen bridges connecting Mn_{B3}-Mn_{A4}. In addition to the μ-oxo bridges and protein ligands, Umena et al. [12] and Suga et al. [13] located four water molecules bound to Mn/Ca in the electron density . Two of them, W1 and W2 are bound to Mn_{A4}, the other two, W3 and W4, are bound to Ca. With the proposed ligation scheme, all Mn are fully coordinated, having six ligands each, and the Ca has seven ligands.

4 Structural Changes of the Mn_4CaO_5 Cluster

Structural changes of the Mn_4CaO_5 cluster have been studied primarily by EXAFS methods, using native PS II as well as Sr EXAFS using Sr-substituted PS II instead of Ca. The EXAFS spectra of the PS II S states show that the structure of the Mn_4CaO_5 cluster changes during the catalytic cycle. In particular, the short Mn–Mn interactions undergo distance changes in the range of 2.7–2.8 Å [17, 28–31]. Such distance changes can reflect several chemical parameters; Mn oxidation state changes, protonation state changes of bridging oxygens, ligation modes (e.g. bidentate/monodentate), as well as fundamental changes in geometry (i.e. dimeric, trimeric, or cubane-like structure). Possible structural changes of the Mn_4CaO_5 cluster during the S state transitions are derived from the Mn EXAFS and Sr EXAFS results shown in Fig. 3.

Among all the S states, the S_2 state is the most studied state as it is characterized by a rich EPR signal at g = 2. A distance change in the dark stable S_1 to S_2 state transition is observed, with a shortening of one Mn–Mn interaction around 2.7 Å, that is likely due to the oxidation state change of one Mn (formally Mn(III) to Mn (IV)). The study supports the open-cubane like structure that was also suggested by the theoretical studies of Siegbahn for S_1 and S_2 [32] and Neese's group for the S_2 state [33] characterized by the g = 2 multiline EPR signal (MLS).

The S_2 state can also be prepared in a different spin state characterized by an EPR signal at g = 4.1 in spinach preparations or at g ~ 6–10 for preparations from cyanobacteria. It has been suggested by Pantazis et al. that the g = 4.1 S_2 state of the

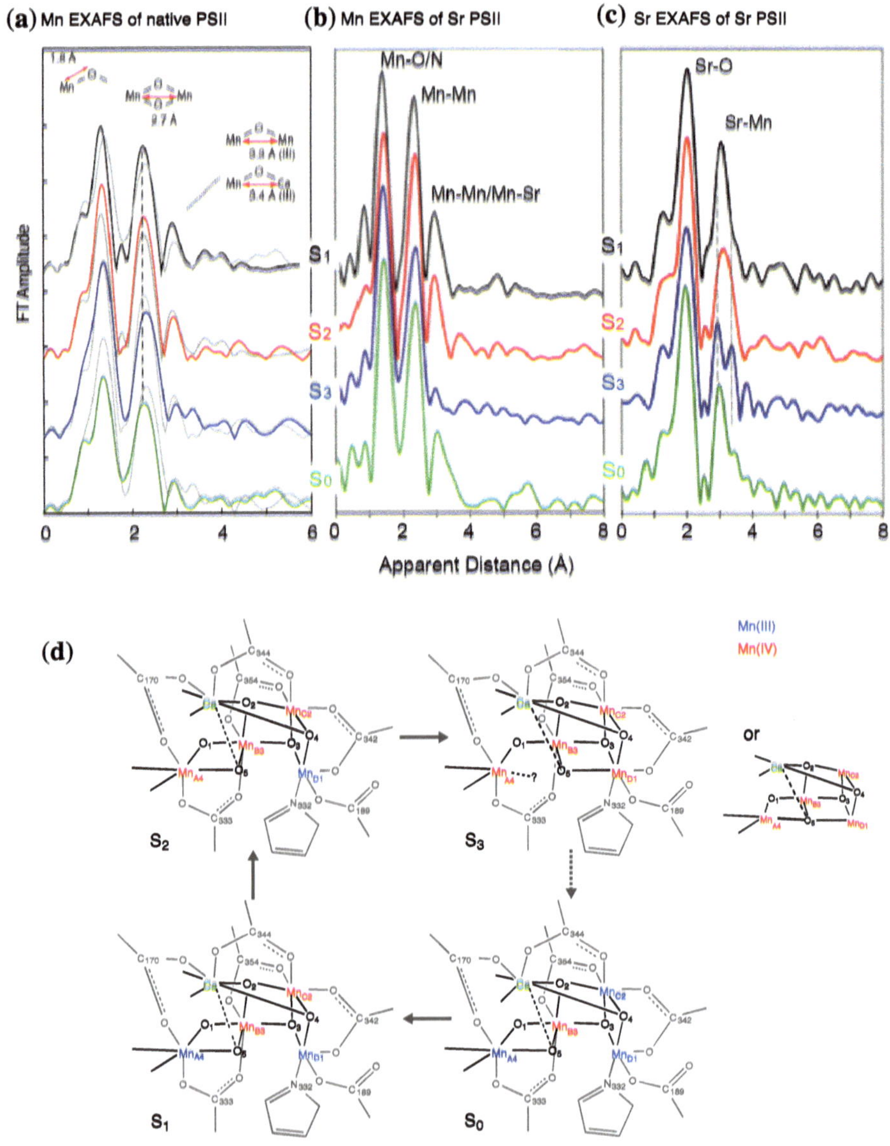

(a) Mn EXAFS of native PSII
(b) Mn EXAFS of Sr PSII
(c) Sr EXAFS of Sr PSII
FT Amplitude
Mn-O/N
Mn-Mn
Mn-Mn/Mn-Sr
Sr-O
Sr-Mn
S1
S2
S3
S0
0
2
4
6
8
Apparent Distance (Å)
(d)
Mn(III)
Mn(IV)
or
S2
S3
S1
S0

◀ **Fig. 3** **a** Fourier-transformed spectra of native PS II solutions in the S_0 (*green*), S_1 (*black*), S_2 (*red*), and S_3 (*blue*) states are shown. For comparison, the spectrum of the S_{n-1} state is overlaid in the S_1, S_2, and S_3 spectra (*gray*). Prominent changes between the S_2 and the S_3 state and the S_3 and the S_0 state in the peak II of the FT spectra are indicated by a dashed line. All spectra are shown in the same scale but with a vertical offset. **b** The FTs from Mn EXAFS of the S-states from Sr-substituted PSII. **c** The FTs from Sr-EXAFS of Sr-substituted PSII shows the first FT peak from Sr-O and the second FT peak from Sr-Mn. The FT peak corresponding to Sr-Mn changes during the S-state cycle and most significantly for the S_2 to S_3 transition. **d** Possible structural changes of the Mn_4CaO_5 cluster during the S-state transitions. We note that some updates are made based on the recent works, that includes the assignment of the Mn(III) position in the S_1 state from the new crystal structure [13]. (**a**)–(**c**) are from Yano and Yachandra [59]. (**d**) is adapted from Ref. [59], by changing the numbering of Mn for an easy comparison to other references

OEC arises from the total spin 5/2 configuration, and the g = 2 (low spin form, the total spin of S_{total} = 1/2) and g = 4.1 (high spin form, the total spin of S_{total} = 5/2) forms are energetically close and interexchangeable [34, 35]. The theoretical studies suggest that the S_2 g = 4 state may be actually a precursor of the S_3 state [36]. More detailed studies are required to understand the transition phenomena of these structural isomers, and how it plays a role during the catalytic cycle. Earlier EXAFS studies showed that there is a structural change between the S_2-g2 and the S_2-g4 states [37].

In the S_2 to S_3 transition, elongations of the Mn–Mn interactions are observed, unlike the S_0 to S_1 or S_1 to S_2 transitions [17, 28–31]. This suggests that the S_2 to S_3 step is not a simple one-oxidation state change of Mn, but is accompanied by fundamental changes of the Mn_4CaO_5 geometry. Elongation of Mn–Mn due to the oxo-bridge protonation is unlikely at the S_2 to S_3 transition, unless protons from terminal water molecules are transferred to the neighboring bridging oxygens. EXAFS studies suggest such structural change could occur by the shift of oxygen (O-5) position from Mn $_{A4}$ side to Mn $_{D1}$ side. Such O-5 shuffling possibility has been suggested by Pantazis et al. [34] but as a reason for the S_2 low spin (S = 1/2) to S_2 high spin (S = 5/2) state changes, and by Isobe et al. [38] in the S_2 to S_3 state transition. The repositioning of O-5 could be accompanied by the ligand symmetry changes of Mn_{D1}, that becomes 6 coordinate from a 5 coordinate geometry. An insertion of a water molecule in the cubane site has been suggested by Cox and Siegbahn, that also fills up the coordination site of Mn4 [32, 36]. From the EPR studies, the heterogeneity of the S_3 state is reported recently [39]. Further study is required to consider a consequence of such heterogeneity, if exists, in the spectroscopic data.

Upon S_3 to S_0 transition via the S_4 state, the Mn-ligand and the ~2.7 Å Mn–Mn distances are shortened [17]. This is counterintuitive if the Mn oxidation state changes from the most oxidized form (S_3) to the most reduced state (S_0). However, such changes could be explained if the Mn_4CaO_5 geometry in the S_0 state is changed back to the one similar to the S_1 and S_2 states, where the Mn_3Ca moiety takes the open-cubane like structure.

4.1 Structural Role of Calcium

Calcium is an essential element for the function of the OEC [37, 40]. It has been speculated that Ca controls substrate water binding to the catalytic Mn site [41] and proposed mechanisms have suggested the involvement of Ca [42–46]. Ca^{2+} can be depleted from the OEC in PS II membrane preparations, using two different methods, (i) high ionic strength during a salt wash, using 1–2 M NaCl, or (ii) a pH drop to 3 using citric acid, which also functions as a Ca^{2+} chelator simultaneously. Reactivation of oxygen evolution in inhibited preparations can be achieved by addition of Ca^{2+}, or by addition of Sr^{2+} but with slower turnover in the S-state cycle that produces a lower overall rate of oxygen evolution at saturating light intensities [47, 48]. A recent study has reported that with the exception of the S_0 to S_1 transition, Ca is required for all the other S-state transitions [40].

The structural consequences of calcium depletion of PS II has been determined by ENDOR and Mn XAS on PS II solutions [49]. Mn ENDOR results have shown that removal of Ca does not perturb the magnetic properties of the Mn cluster, suggesting that the main bridging structure of Mn atoms is not changed by Ca removal [50]. Polarized EXAFS data of Ca^{2+}-depleted PS II showed only minor changes in distances and orientations of the Mn–Mn vectors as in the Ca^{2+}-containing OEC [51]. Thus, while removal of the Ca^{2+} ion does not lead to fundamental distortion or rearrangement of the tetranuclear Mn cluster, it results in some loss of rigidity of its structure, which indicates that the Ca^{2+} ion in the OEC is not critical for structural maintenance of the cluster at least in the S_1 and S_2 states, but fulfills a crucial function in the catalytic mechanism of the water oxidation reaction. The fact that Ca can be removed more easily in the S_3 state (or that Ca can be more easily exchanged in the higher S state) [40, 52, 53] compared to the S_1 and the S_2 state, together with the above observations, implies that the Mn–Ca binding modes are changed upon S_2 to S_3 transition.

The Ca^{2+} in the OEC is not essential for structural integrity of the cluster at least up to the S_2 state, but it is necessary for the formation of the S_3 state. A recent study by Rappaport et al. showed that Ca and a tyrosine residue of the D1 polypeptide (Y_Z) are involved in the common hydrogen-bond network; Ca^{2+} facilitating the correct configuration of the hydrogen-bond network for proton transfer and therefore important for the S_2 to S_3 transition that is accompanied by the proton transfer [54]. On the other hand, there is no proton release upon S_1 to S_2 transition.

4.2 Ligands of Mn and Ca and Site-Specific Mutants

In addition to the Mn_4CaO_5 core structural changes discussed above, we expect that terminal ligands that are derived from carboxylates, histidine, and water/hydroxo ligands also likely change during the catalytic process. As shown in Fig. 2, there are

6 carboxyl and one histidine residues directly ligated to the Mn_4CaO_5 cluster. Except for one glutamate residue (CP43-Glu354), other ligands to the Mn_4CaO_5 cluster are provided from the D1 subunit. There has been a series of mutagenesis studies to identify those amino acid residues that control the assembly and functioning of the Mn_4CaO_5 cluster and to gain insight on which Mn are involved in the different steps during the water oxidation reaction [8, 55]. In the OEC, it has been shown that the hydrogen bonding network is extended around the Mn_4CaO_5 cluster, that makes up the water access and the proton release pathway. Mutations of amino acid residues that disturb such network have a strong effect in the function of the OEC [2, 5].

4.3 *Protein and Chemical Dynamics*

X-ray-based techniques (crystallography and spectroscopy) have been the major tools for studying the architecture of PS II and its water oxidation reaction in the OEC. In many cases, studies at the synchrotron facilities have been done with a cryo-trapped method. However, accessing the kinetically unstable S_4 state is difficult using the traditional cryo-trapping methodology, and characterizing the nature of such transient state requires time-resolved detection at room temperature. Within the total time scale of $\sim$1.3 ms of the S_3-S_4-S_0 transition after initiated by the 3rd flash, sequential events occur, that include (a) release of one electron, (b) release of two protons, (c) release of molecular oxygen, and (d) two water substrates binding along with four electron reduction of Mn. Following such sequential chemistry during the S_3-S_4-S_0 transitions is therefore at the heart of understanding the water splitting mechanism. In addition, the understanding the changes of the protein environment that provides the environment for stabilizing the wide range of Mn redox-states and a pathway for protons, water, and oxygen, are the key questions for understanding the mechanism of this metalloenzyme.

Recent development of serial femtosecond XRD at X-ray free electron laser (XFEL) [56] opens a new way of studying the structural dynamics of enzyme systems under functional conditions at ambient conditions. The very short, intense fs pulses make it possible to outrun damage and collect data.

Our group has developed a simultaneous data collection of XRD and X-ray spectroscopy, that has several important advantages; the method probes the overall protein structure from XRD and the electronic structure of the Mn_4CaO_5 cluster from the spectroscopic data (Fig. 4). The use of XFEL pulses is critical for this approach. Among the various spectroscopic methods, XES using an energy-dispersive X-ray spectrometer is well suited for such combined shot-by-shot studies, as excitation energies above the 1 s core hole of first row transition metals are also ideal for XRD, and therefore neither incident nor emitted photon energy have to be scanned [57]. $K\beta_{1,3}$ and $K\beta$' lines of XES are a probe of the number of unpaired 3d electrons, hence providing information about the oxidation and/or spin state. The structural and the electronic structural changes of the OEC and PSII

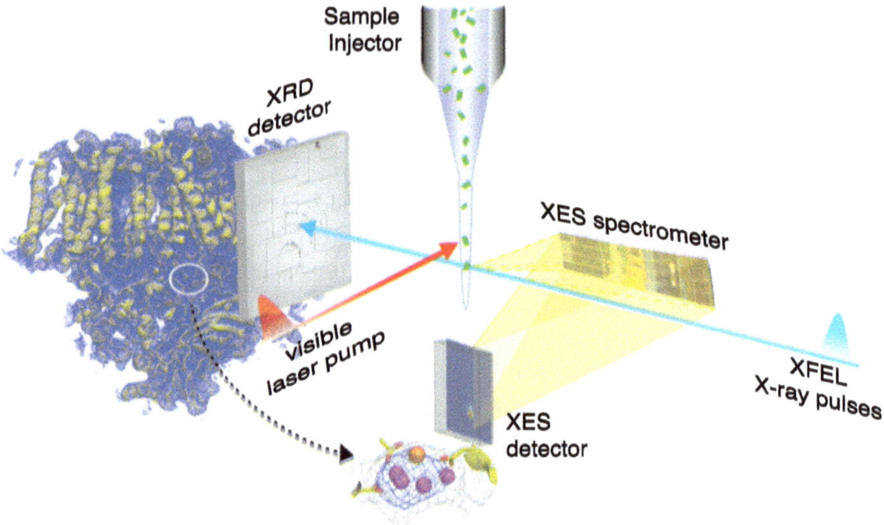

Fig. 4 Schematic of the simultaneous detection of X-ray diffraction and X-ray emission spectra of photosystem II crystals using the femtosecond pulses from a X-ray free electron laser (XFEL) at room temperature. The ultra-short, intense pulses from the XFEL allows us to collect data at room temperature without radiation damage, thus opening up possibilities for conducting time-resolved studies. The crystal suspension is injected using a microjet that intersects the X-ray pulses. XRD data from a single crystal are collected downstream and XES data from the same crystal are collected at ~90° to the beam using an XES spectrometer and a position sensitive detector. A visible laser (527 nm) is used to illuminate the crystals to advance the PS II crystals through the S-states [58, 60]

during the catalytic cycle can be studied after the multiple photo-excitation, and such effort is on-going [58, 60].

Acknowledgments The research presented here was supported by the Director, Office of Science, Office of Basic Energy Sciences (OBES), Division of Chemical Sciences, Geosciences, and Biosciences of the Department of Energy (DOE) under Contract DE-AC02-05CH11231, and by the NIH Grant GM 55302. X-ray facilities were provided by the Stanford Synchrotron Radiation Laboratory (SSRL) and the Linac Coherent Light Source (LCLS) at Stanford, the Advanced Light Source (ALS) at Berkeley, the Advanced Photon Source (APS) at Argonne, all funded by DOE OBES.

References

1. Wydrzynski T, Satoh S (eds) (2005) Photosystem II: the light-driven water: plastoquinone oxidoreductase. Springer, Dordrecht
2. Debus RJ (2015) Biochim Biophys Acta 1847:19
3. Cox N, Pantazis DA, Neese F, Lubitz W (2013) Acc Chem Res 46:1588
4. Pokhrel R, Brudvig GW (2014) Phys Chem Chem Phys 16:11812

5. Noguchi T (2015) Biochim Biophys Acta 1847:35
6. Kok B, Forbush B, McGloin M (1970) Photochem Photobiol 11:457
7. Chu H-A, Debus RJ, Babcock GT (2001) Biochemistry 40:2312
8. Debus RJ (2008) Coord Chem Rev 252:244
9. Ho FM (2008) Photosynth Res 98:503
10. Ferreira KN, Iverson TM, Maghlaoui K, Barber J, Iwata S (2004) Science 303:1831
11. Guskov A, Kern J, Gabdulkhakov A, Broser M, Zouni A, Saenger W (2009) Nat Struct Mol Biol 16:334
12. Umena Y, Kawakami K, Shen J-R, Kamiya N (2011) Nature 473:55
13. Suga M, Akita F, Hirata K, Ueno G, Murakami H, Nakajima Y, Shimizu T, Yamashita K, Yamamoto M, Ago H, Shen J-R (2015) Nature 517:99
14. Yano J, Yachandra VK (2008) Inorg Chem 47:1711
15. Dau H, Haumann M (2008) Coord Chem Rev 252:273
16. Penner-Hahn JE (1998) Struct Bond 90:1
17. Robblee JH, Messinger J, Cinco RM, McFarlane KL, Fernandez C, Pizarro SA, Sauer K, Yachandra VK (2002) J Am Chem Soc 124:7459
18. George GN, Prince RC, Frey TG, Cramer SP (1989) Phys B 158:81
19. Mukerji I, Andrews JC, DeRose VJ, Latimer MJ, Yachandra VK, Sauer K, Klein MP (1994) Biochemistry 33:9712
20. Cinco RM, Robblee JH, Messinger J, Fernandez C, Holman KLM, Sauer K, Yachandra VK (2004) Biochemistry 43:13271
21. Dau H, Andrews JC, Roelofs TA, Latimer MJ, Liang W, Yachandra VK, Sauer K, Klein MP (1995) Biochemistry 34:5274
22. Dau H, Dittmer J, Iuzzolino L, Schiller H, Dörner W, Heinze I, Sole VA, Nolting H-FJ (1997) Phys IV France 7:C2–607
23. Yano J, Kern J, Sauer K, Latimer M, Pushkar Y, Biesiadka J, Loll B, Saenger W, Messinger J, Zouni A, Yachandra VK (2006) Science 314:821
24. Pushkar Y, Yano J, Glatzel P, Messinger J, Lewis A, Sauer K, Bergmann U, Yachandra VK (2007) J Biol Chem 282:7198
25. Yano J, Pushkar Y, Glatzel P, Lewis A, Sauer K, Messinger J, Bergmann U, Yachandra VK (2005) J Am Chem Soc 127:14974
26. Robblee JH, Cinco RM, Yachandra VK (2001) Biochim Biophys Acta 1503:7
27. Yachandra VK, Yano J (2011) Photochem Photobiol B-Biol 104:51
28. Liang W, Roelofs TA, Cinco RM, Rompel A, Latimer MJ, Yu WO, Sauer K, Klein MP, Yachandra VK (2000) J Am Chem Soc 122:3399
29. Pushkar YL, Yano J, Sauer K, Boussac A, Yachandra VK (2008) Proc Natl Acad Sci USA (105):1879
30. Glockner C, Kern J, Broser M, Zouni A, Yachandra V, Yano J (2013) J Biol Chem 288:22607
31. Haumann M, Muller C, Liebisch P, Iuzzolino L, Dittmer J, Grabolle M, Neisius T, Meyer-Klaucke W, Dau H (2005) Biochemistry 44:1894
32. Siegbahn PEM (2009) Acc Chem Res 42:1871
33. Ames W, Pantazis DA, Krewald V, Cox N, Messinger J, Lubitz W, Neese F (2011) J Am Chem Soc 133:19743
34. Pantazis DA, Ames W, Cox N, Lubitz W, Neese F (2012) Ang Chem Int Ed 51:9935
35. Bovi D, Narzi D, Guidoni L (2013) Ang Chem Int Ed 52:11744
36. Cox N, Retegan M, Neese F, Pantazis DA, Boussac A, Lubitz W (2014) Science 345:804
37. Liang MC, Latimer MJ, Dau H, Roelofs TA, Yachandra VK, Sauer K, Klein MP (1994) Biochemistry 33:4923
38. Isobe H, Shoji M, Yamanaka S, Umena Y, Kawakami K, Kamiya N, Shen JR, Yamaguchi K (2012) Dalton Trans 41:13727
39. Boussac A, Rutherford AW, Sugiura M (2015) Biochim Biophys Acta-Bioenerg 1847:576
40. Miqyass M, Marosvolgyi MA, Nagel Z, Yocum CF, van Gorkom HJ (2008) Biochemistry 47:7915
41. Chen C, Kazimir J, Cheniae GM (1995) Biochemistry 34:13511

42. Limburg J, Szalai VA, Brudvig GW (1999) J Chem Soc Dalton Trans 1353
43. Renger G (1997) Physiol Plant 100:828
44. Siegbahn PEM (2000) Inorg Chem 39:2923
45. Ananyev GM, Dismukes GC (1997) Biochemistry 36:11342
46. Cox N, Ames W, Epel B, Kulik LV, Rapatskiy L, Neese F, Messinger J, Wieghardt K, Lubitz W (2011) Inorg Chem 50:8238
47. Boussac A, Rutherford AW (1988) Biochemistry 27:3476
48. Ghanotakis DF, Babcock GT, Yocum CF (1984) FEBS Lett 167:127
49. Latimer MJ, DeRose VJ, Yachandra VK, Sauer K, Klein MP (1998) J Phys Chem B 102:8257
50. Lohmiller T, Cox N, Su JH, Messinger J, Lubitz W (2012) J Biol Chem 287:24721
51. Lohmiller T, Shelby ML, Long X, Yachandra VK, Yano J (2015) J Phys Chem B 119:13742
52. Vander Meulen KA, Hobson A, Yocum CF (2004) Biochim Biophys Acta-Bioenerg 1655:179
53. Miqyass M, van Gorkom HJ, Yocum CF (2007) Photosynth Res 92:275
54. Rappaport F, Ishida N, Sugiura M, Boussac A (2011) Eng Env Sci 4:2520
55. Hwang HJ, Dilbeck P, Debus RJ, Burnap RL (2007) Biochemistry 46:11987
56. Chapman HN, Fromme P, Barty A, White TA, Kirian RA, Aquila A, Hunter MS, Schulz J, DePonte DP, Weierstall U, Doak RB, Maia FRNC, Martin AV, Schlichting I, Lomb L, Coppola N, Shoeman RL, Epp SW, Hartmann R, Rolles D, Rudenko A, Foucar L, Kimmel N, Weidenspointner G, Holl P, Liang MN, Barthelmess M, Caleman C, Boutet S, Bogan MJ, Krzywinski J, Bostedt C, Bajt S, Gumprecht L, Rudek B, Erk B, Schmidt C, Homke A, Reich C, Pietschner D, Struder L, Hauser G, Gorke H, Ullrich J, Herrmann S, Schaller G, Schopper F, Soltau H, Kuhnel KU, Messerschmidt M, Bozek JD, Hau-Riege SP, Frank M, Hampton CY, Sierra RG, Starodub D, Williams GJ, Hajdu J, Timneanu N, Seibert MM, Andreasson J, Rocker A, Jonsson O, Svenda M, Stern S, Nass K, Andritschke R, Schroter CD, Krasniqi F, Bott M, Schmidt KE, Wang XY, Grotjohann I, Holton JM, Barends TRM, Neutze R, Marchesini S, Fromme R, Schorb S, Rupp D, Adolph M, Gorkhover T, Andersson I, Hirsemann H, Potdevin G, Graafsma H, Nilsson B, Spence JCH (2011) Nature 470:73
57. Alonso-Mori R, Kern J, Gildea RJ, Sokaras D, Weng TC, Lassalle-Kaiser B, Tran R, Hattne J, Laksmono H, Hellmich J, Glockner C, Echols N, Sierra RG, Schafer DW, Sellberg J, Kenney C, Herbst R, Pines J, Hart P, Herrmann S, Grosse-Kunstleve RW, Latimer MJ, Fry AR, Messerschmidt MM, Miahnahri A, Seibert MM, Zwart PH, White WE, Adams PD, Bogan MJ, Boutet S, Williams GJ, Zouni A, Messinger J, Glatzel P, Sauter NK, Yachandra VK, Yano J, Bergmann U (2012) Proc Nat Acad Sci USA 109:19103
58. Kern J, Tran R, Alonso-Mori R, Koroidov S, Echols N, Hattne J, Ibrahim M, Gul S, Laksmono H, Sierra RG, Gildea RJ, Han G, Hellmich J, Lassalle-Kaiser B, Chatterjee R, Brewster AS, Stan CA, Glockner C, Lampe A, DiFiore D, Milathianaki D, Fry AR, Seibert MM, Koglin JE, Gallo E, Uhlig J, Sokaras D, Weng TC, Zwart PH, Skinner DE, Bogan MJ, Messerschmidt M, Glatzel P, Williams GJ, Boutet S, Adams PD, Zouni A, Messinger J, Sauter NK, Bergmann U, Yano J, Yachandra VK (2014) Nat Commun 5:4371
59. Yano J, Yachandra V (2014) Chem Rev 114:4175
60. Kern J, Alonso-Mori R, Tran R, Hattne J, Gildea RJ, Echols N, Glockner C, Hellmich J, Laksmono H, Sierra RG, Lassalle-Kaiser B, Koroidov S, Lampe A, Han G, Gul S, Difiore D, Milathianaki D, Fry AR, Miahnahri A, Schafer DW, Messerschmidt M, Seibert MM, Koglin JE, Sokaras D, Weng TC, Sellberg J, Latimer MJ, Grosse-Kunstleve RW, Zwart PH, White WE, Glatzel P, Adams PD, Bogan MJ, Williams GJ, Boutet S, Messinger J, Zouni A, Sauter NK, Yachandra VK, Bergmann U, Yano J (2013) Science 340:491

Recent Understanding on Photosystem I

Yuichiro Takahashi

Abstract In oxygenic photosynthesis, photosystem I (PSI) drives electron transfer from plastocyanin or cytochrome *c* to ferredoxin. PSI forms a core complex that consists of 11–14 subunits and associates more than one hundred cofactors such as chlorophyll *a*, carotenes, naphthoquinones, lipids, and iron-sulfur (Fe-S) clusters. PSI complex is embedded in the thylakoid membranes and harvests light using antenna pigments, transfers excitons from antenna to reaction center, where photochemical reaction occurs, and stabilizes the charge separation to allow for oxidation of plastocyanin or cytochrome *c* and reduction of ferredoxin. PSI core complex forms a trimer in cyanobacteria but associates light-harvesting complexes I (LHCI) to form a PSI-LHCI supercomplex in plants and algae. The crystal structures of PSI core trimer and PSI-LHCI supercomplex have revealed that the subunits and cofactors of PSI are organized to optimize the efficiency of light harvesting and photochemical reaction. However it remains elusive how these components of PSI are integrated into a functional structure in photosynthetic organisms.

1 Introduction

Oxygenic photosynthesis by plants, algae, and cyanobacteria produces carbohydrates from CO_2 and H_2O using light energy and generates O_2 as a by-product. Photosystem II (PSII) generates strong oxidants that split H_2O into O_2, H^+, and electrons, whereas photosystem I (PSI) produces strong reductants. Two photosystems operate in series to drive oxygenic photosynthetic electron transfer from H_2O to nicotinamide adenine dinucleotide phosphate ($NADP^+$) to produce NADPH. This linear electron transfer reactions concomitantly generate proton

Y. Takahashi (✉)
The Graduate School of Natural Science and Technology, Okayama University,
3-1-1 Tsusfiima-naka, Kita-ku, Okayama 708530, Japan
e-mail: taka@cc.okayama-u.ac.jp

M. Sugiyama et al. (eds.), *Solar to Chemical Energy Conversion*,
Lecture Notes in Energy 32, DOI 10.1007/978-3-319-25400-5_24

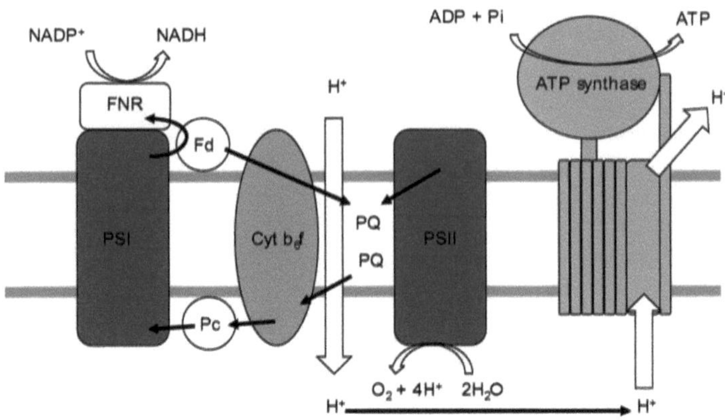

Fig. 1 Electron transfer system of oxygenic photosynthesis. Four multiprotein complexes (PSI, PSII, Cyt b_6f, and ATP synthase) are localized in the thylakoid membranes. PQ pool is present in the hydrophobic region of the thylakoid membranes and Pc, FNR, and Fd are located on the peripheral region of the thylakoid membranes. Electron transfer reaction is driven by PSI and PSII using light energy. Proton translocation across the thylakoid membranes is coupled to the electron transfer reactions. The resulting ΔpH across and membrane potential are used to produce ATP from ADP and Pi (inorganic phosphate) by ATP synthase

motive force, which is used to synthesize ATP from ADP and phosphate (Pi) by ATP synthase. NADPH and ATP are used to assimilate CO_2 (Fig. 1). Light energy is first absorbed by photosynthetic pigments such as chlorophylls and carotenoids, and the resulting excitons are transferred to reaction centers (RCs) where photochemical reaction causes charge separation. A series of electron donors and acceptors in RCs are engaged in the charge separation and its stabilization.

PSI accepts electrons from PSII through plastoquinone (PQ), cytochrome b_6f (Cyt b_6f) complex, and plastocyanin (Pc) or cytochrome *c* (Cyt *c*), and donates electrons to $NADP^+$ through ferredoxin (Fd) and ferredoxin-NADP oxidoreductase (FNR) (Fig. 1). PSI core complex consists of 11–14 subunits and a number of cofactors such as chlorophyll *a* (Chl*a*), carotenes, naphthoquinones, lipids, and iron-sulfur (Fe-S) clusters and is embedded in lipid bilayer of the thylakoid membranes as a large intrinsic complex. Thus PSI is a multi-protein complex that harvests light energy, causes photochemical reaction, and stabilizes the charge separation. In addition to the linear electron transfer, PSI is engaged in driving the cyclic electron transfer by which Fd reduced by PSI transfers an electron back to PSI through PQ, Cyt b_6f, and Pc (or Cyt *c*) to produce ATP without generating NADPH. This cyclic electron transfer plays important roles not only in producing ATP to optimize ATP/NADPH ratio but also in protecting PSI from excess light energy. The reductants generated by PSI are also used for assimilation of nitrogen and sulfur as well as for providing reducing power for redox regulation of enzymes in chloroplasts. Thus PSI plays multiple and pivotal functions in oxygenic photosynthesis.

2 PSI Forms a Chlorophyll-Protein Complex

The components of PSI are precisely organized in a chlorophyll-protein complex to carry out efficient light harvesting and photochemical reactions. PSI complex, which is embedded in the thylakoid membranes, can be solubilized using a mild detergent and can be biochemically purified without significantly affecting its structure and activity. Figure 2 shows the subunit composition of the PSI preparation purified from the thylakoid extracts of the green alga *Chlamydomonas reinhardtii.* SDS-polyacrylamide gel electrophoresis clearly revealed that the purified PSI complex contains a number of PSI subunits. The diffuse bands around 70 kDa are two homologous RC subunits encoded by *psaA* and *psaB* genes. In addition to these RC subunits, the PSI complex contains several small subunits that are encoded by *psaC*, *psaD*, *psaE*, *psaF*, *psaG*, *psaH*, *psaK*, and *psaL* genes. It was also confirmed by immunoblotting analyses using anti-PsaI and anti-PsaJ antibodies that the PSI complex also contains very small subunits (<5 kDa) encoded by *psaI* and *psaJ* genes. As summarized in Table 1, the chloroplast genes (*psaA*, *psaB*, *psaC*, *psaI* and *psaJ*) encode four subunits while the nuclear genes coded for the other subunits. Localization of PsaO in PSI complex has not yet been determined because this subunit is weakly bound to and is easily dissociated from PSI complex during solubilization and purification. However chemical cross-linking experiments revealed that PsaO interacts with PsaL [1]. In addition to the PSI subunits present in the PSI core complex, the PSI complex from *Chlamydomonas* contains nine subunits (Lhca1-9) of light-harvesting complex I (LHCI) [2]. Because their sizes are rather similar, some of them are not clearly separated on SDS-polyacrylamide gel. The PSI complex from *Chlamydomonas* is thus designated as a PSI-LHCI supercomplex. It is of note that plant PSI also forms PSI-LHCI supercomplex but the number of LHCI is four as will be described below. In summary, *Chlamydomonas* PSI-LHCI supercomplex has a larger structure of about 700 kDa with 40 transmembrane helices.

PSI core complex has a structure conserved among cyanobacteria, algae, and plants and is composed of PsaA, PsaB, PsaC, PsaD, PsaE, PsaF, PsaI, PsaJ, PsaK, and PsaL and associates cofactors necessary to harvest light energy, charge separation and its stabilization in common [3]. However, there are some minor differences in subunit composition between cyanobacteria and algae/plants; cyanobacterial PSI core complex additionally contains PsaM and PsaX (PsaX is found only in thermophilic cyanobacteria) while algae and plats contain PsaG, PsaH, PsaN, and PsaO (PsaP is found only in plants). It is of note that cyanobacterial PSI core complex forms a trimer [4], whereas in algae and plants PSI core complex is a monomeric and forms a PSI-LHCI supercomplex [5].

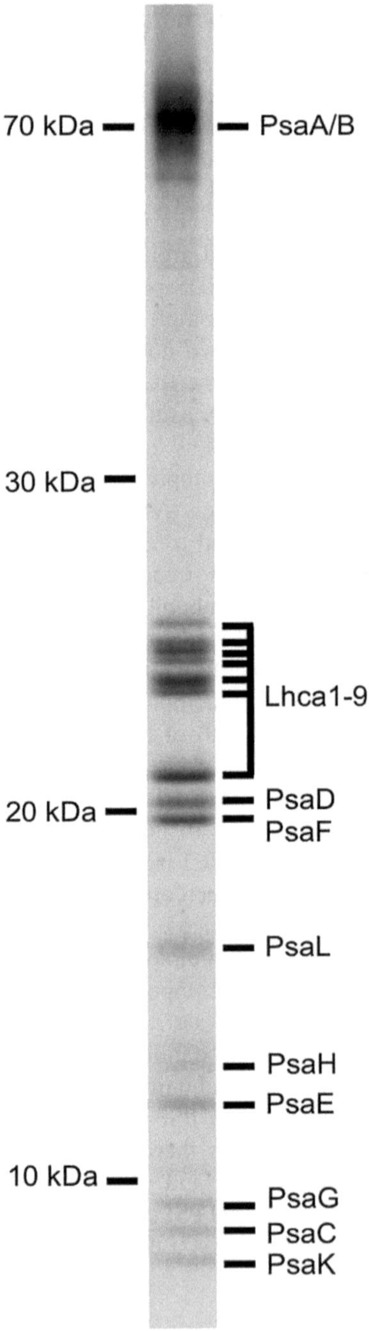

Fig. 2 Subunit composition of PSI-LHCI supercomplex purified from the green alga *Chlamydomonas reinhardtii*. The purified PSI-LHCI supercomplex from the thylakoid extracts was dissociated into polypeptides and the polypeptides were separated by SDS-polyacrylamide gel electrophoresis. The separated polypeptides were stained. Two large homologous polypeptides, PsaA and PsaB are separated around 70 kDa and several smaller polypeptides (PsaD, PsaF, PsaL, PsaH, PsaE, PsaG, PsaC, and PsaK) are resolved. Very small polypeptides, PsaI and PsaJ, are not observed under this condition. In addition to PSI polypeptides, nine LHCI polypeptides (Lhca1-9) are separated. Since some Lhca polypeptides are closely separated, some of them are not detected individually

Table 1 Subunits of PSI complex

Subunit	Gene location	No. of TM	Function and remarks
PsaA	cp	11	RC and core antenna
PsaB	cp	11	RC and core antenna
PsaC	cp	0	F_A and F_B, Fd docking site, located on stromal side
PsaD	nuc	0	Fd docking site, located on stromal side
PsaE	nuc	0	Fd docking site, located on stromal side
PsaF	nuc	1	Pc or Cyt *c* docking site (in algae and plants)
PsaG	nuc	2	LHCI stabilization (in algae and plants), not found in cyanobacteria
PsaH	nuc	1	State transitions (in algae and plants), not found in cyanobacteria
PsaI	cp	1	
PsaJ	cp	1	Encoded by nuclear gene in *Chlamydomonas*
PsaK	nuc	2	LHCI stabilization (in algae and plants)
PsaL	nuc	3	Stabilization of PSI core trimer (in cyanobacteria), state-transitions (in algae and plants),
PsaM	cyano	1	
PsaN	nuc	0	Located on the stromal surface of LHCIs, not found in cyanobacteria
PsaO	nuc	2	State-transitions, not found in cyanobacteria
PsaP	nuc	2	Phosphoprotein, found in plants
PsaX	cyano	1	Found only in the thermophilic cyanobacteria
Lhca1	nuc	3	Chl*a* and Chl*b*, found in plants and algae
Lhca2	nuc	3	Chl*a* and Chl*b*, found in plants and algae
Lhca3	nuc	3	Chl*a* and Chl*b*, found in plants and algae
Lhca4	nuc	3	Chl*a* and Chl*b*, found in plants and algae
Lhca5	nuc	3	Chl*a* and Chl*b*, found in *Chlamydomonas*
Lhca6	nuc	3	Chl*a* and Chl*b*, found in *Chlamydomonas*
Lhca7	nuc	3	Chl*a* and Chl*b*, found in *Chlamydomonas*
Lhca8	nuc	3	Chl*a* and Chl*b*, found in *Chlamydomonas*
Lhca9	nuc	3	Chl*a* and Chl*b*, found in *Chlamydomonas*

cp chloroplast-encoded, *nuc* nuclear encoded, *cyano* found only in cyanobacteria, *TM* transmembrane helix

3 Subunit Structure of Photosystem I Core Complex

The crystal structures of PSI core trimer from the thermophilic cyanobacteria *Thermosynechococcus elongatus* and PSI-LHCI supercomplex from *Pisus sativum* (pea) have been determined at 2.5 and 2.8 Å resolutions, respectively [4, 6, 7]. The subunit structure and cofactor organization are highly conserved between these two core complexes (Fig. 3a, b). The structure of the cyanobacterial core complex viewed from the stromal side shows that two homologous RC subunits, PsaA and

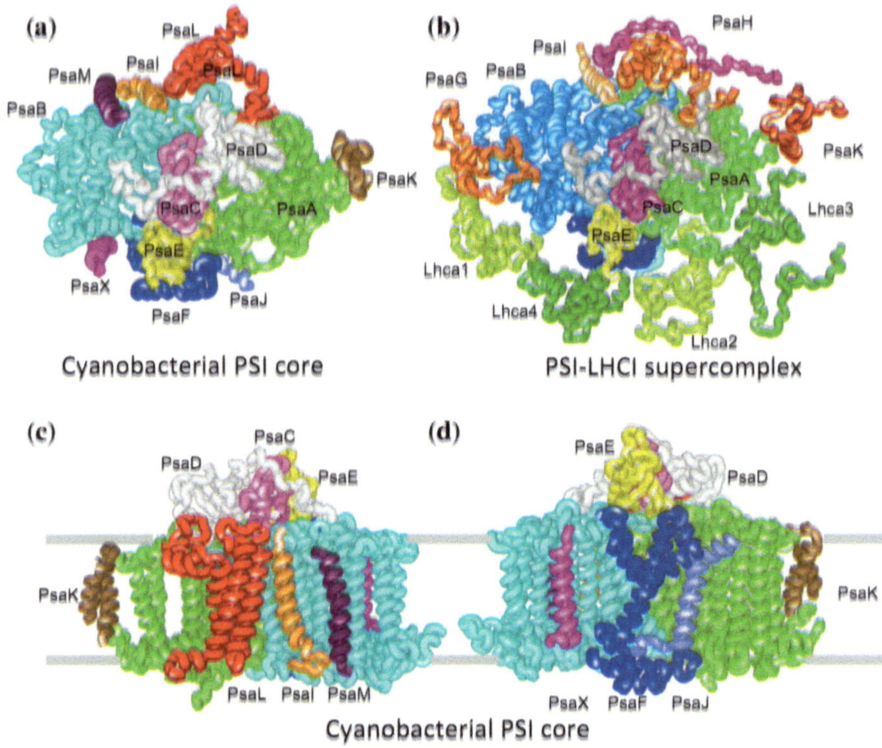

Fig. 3 Subunit organization of PSI core complex and PSI-LHCI supercomplex. **a** The subunit structure viewed from stromal side of cyanobacterial PSI core complex (PDB 1JB0). The structure of the monomeric PSI core is shown. **b** The subunit structure viewed from stromal side of plant PSI-LHCI supercomplex (PDB 2WSC). In addition to PSI core complex, four LHCI complexes are attached on the PsaF and PsaJ side of the PSI core complex. **c** The subunit structure viewed from PsaL/PsaI/PsaM side of the cyanobacterial PSI core complex. **d** The subunit structure viewed from PsaF/PsaJ/PsaX side of the cyanobacterial PSI core complex

PsaB, each with 11 transmembrane helices, form a heterodimer (Fig. 3a). On the surface of the stromal side, three subunits, PsaC, PsaD, and PsaE, are clustered to constitute the docking site for Fd (Fig. 3a). On a hydrophobic side of the RC, three intrinsic subunits, PsaI, PsaL, and PsaM, are intimately associated with one another (Fig. 3a, c). PsaL and Ca^{2+} are located at the center of PSI core trimer in cyanobacteria and are required for the stabilization of the trimer. In plants, PsaM is absent but PsaH is located in close proximity of PsaL (Fig. 3b). The presence of PsaH in plants and algae prevents the PSI core complex from the trimerization. Instead PsaL and PsaH are involved in reversible association of light-harvesting complexes II (LHCIIs) with PSI-LHCI supercomplex during the state transitions by which LHCIIs are shuttled between PSI and PSII complexes in order to balance excitation energy distribution. It is inferred that Chl*a* molecules bound to PsaL and PsaH may be involved in exciton transfer from LHCII to PSI RC [6]. On the other hydrophobic side of PsaI and PsaL, three intrinsic subunits, PsaF, PsaJ, and PsaX, are located (Fig. 3a, d). In plants, PsaX is absent but four LHCIs (Lhca1-4) are present on the side of PsaF and PsaJ. PsaK is attached to PsaA in cyanobacteria while PsaG is additionally associated with PsaB at a pseudo-symmetric position with respect to PsaK in plants and algae. These two subunits are homologous and are required for LHCI binding to PSI core complex in plants and algae [8]. In plants and algae, PsaF has two extended α-helices on the lumenal side of the PSI RC subunits, which are involved in Pc or Cyt *c* docking and are required for the efficient electron transfer from Pc or Cyt *c* to P700 [5].

4 Electron Transfer Reactions

The RC of PSI contains one electron donor and five electron acceptors involved in photochemical reactions (Fig. 4). The initial charge separation occurs between the primary electron donor, P700, and an intermediate electron acceptor, A_0, resulting in the radical pair, $P700^+/A_0^-$. The reduced A_0 subsequently donates the electron to another electron acceptor, A_1. A_1^- successively donates the electron to the three 4Fe-4S clusters, F_X, F_A, and F_B. The resulting charge separation $P700^+/F_B^-$ is significantly stable ($t_{1/2}$ of the back reaction from F_B^- to $P700^+$ is about 100 ms) so that $P700^+$ can oxidize Pc or Cyt *c* and F_B^- can reduce Fd. Finally the redox components of PSI return to the original redox state and carries out another round of photochemical reaction.

The crystal structure of PSI of the thermophilic cyanobacterium, *T. elongatus,* has also revealed the detailed organization of the redox components [4]. Figure 5 shows the configuration of the redox components in PSI core complex. Six Chl*a* and two naphthoquinone molecules are located pseudo-symmetrically on the heterodimer of PsaA and PsaB. It is of note that two branches of redox components from P700 to F_x (A and B branches) are present. The structure shows that P700 is Chl*a* heterodimer; one is Chl*a* epimer (Chl*a*′) coordinated by His-676 of PsaA and the other is Chl*a* coordinated by His-656 of PsaB. The Chl*a* molecules of the second pair, which

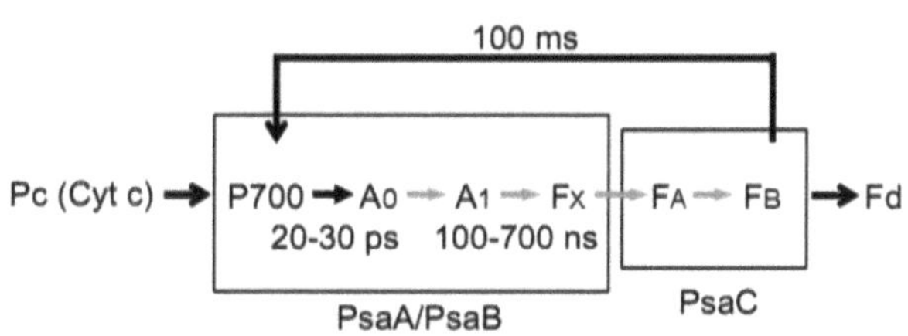

Fig. 4 Electron donor and acceptors of PSI. The primary electron donor, P700, and the three electron acceptors, A_0, A_1, and F_X, are localized on PsaA/PsaB heterodimer. The initial charge separation occurs at $t_{1/2}$ of 20–30 ps and the electron transfer from A_1 to F_X occurs at $t_{1/2}$ of 100 and 700 ns. The terminal electron acceptors, F_A and F_B, are located on PsaC. Pc (Cyt *c*) reduces $P700^+$, whereas Fd is reduced by F_B^-. The back reaction from F_B^- to $P700^+$ occurs at $t_{1/2}$ of about 100 ms

Fig. 5 The configuration of the PSI redox components. Chl*a* dimer, P700, and a pair of accessory chlorophylls, A_0, and A_1 form two branches of electron transfer chain (PDB 1JB0). The two branches are connected at F_X, and subsequently F_A and F_B successively accept electrons. Two His residues on PsaA and PsaB coordinate chlorophylls of P700. Two Trp residues on PsaA and PsaB are located between P700 and Pc (or Cyt c) binding site to facilitate electron transfer reaction from Pc (or Cyt *c*) to P700. Only chlorin ring of Chl*a* is shown

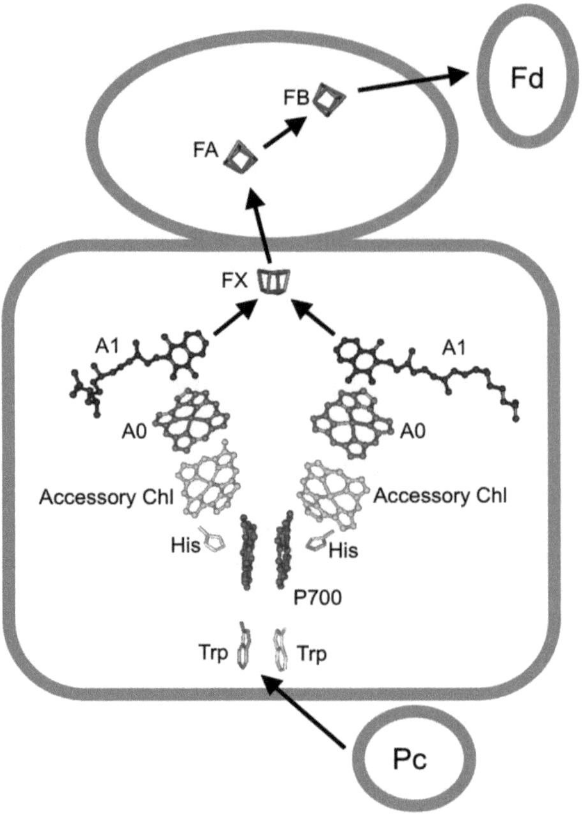

are designated as accessory chlorophylls, are located between P700 and A_0. They are coordinated by water molecules that are hydrogen-bonded to Asn-591 of PsaB and Asn-604 of PsaA, respectively. Biophysical measurements failed to detect a transient radical pair between P700 and the accessory chlorophyll probably because the lifetime of the reduced state of the accessory chlorophyll is too short to detect. The Chl*a* molecules of the third pair, which are coordinated by Met-688 of PsaA and Met-668 of PsaB, respectively, correspond to A_0. Naphthoquinone (phylloquinone) molecules of the fourth pair, which correspond to A_1, are hydrogen-bonded to NH of backbone of Leu-722 on PsaA and Leu-706 of PsaB, respectively. Additionally their quinone planes interact with indole rings of Trp-697 of PsaA and Trp-677 of PsaB, respectively. The chemical species of A_1 is naphthoquinone in the cyanobacterial and plant crystal structures. However three types of naphthoquinone have been reported to be present in PSI core complex depending on photosynthetic organisms (Fig. 6). A_1 is phylloquinone in higher plants, some algae, and some cyanobacteria.

Fig. 6 Chemical structures of the three naphthoquinones

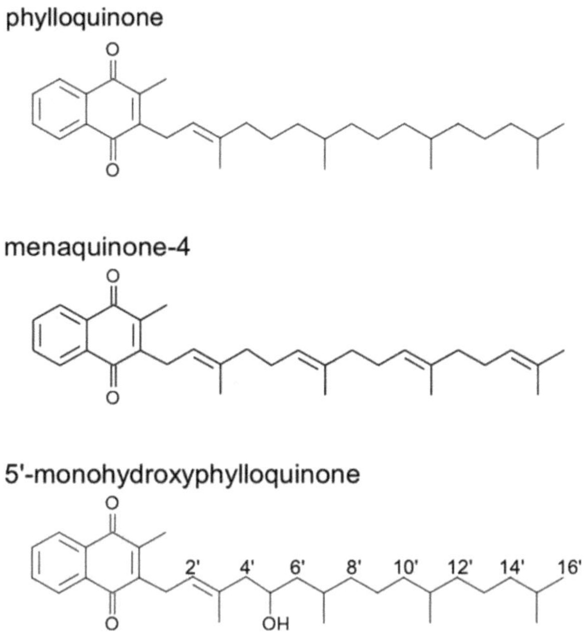

However, A_1 is 5′-monohydroxyphylloquinone or menaquinone-4 in some algae and some cyanobacteria [9]. It is not yet clear whether the difference of the naphthoquinone has any physiological significance.

Site-directed mutagenesis to replace His-676 of PsaA or His 656 of PsaA and biophysical characterization of the resulting mutants suggested that spin density on the $P700^+$ radical is unevenly distributed between the Chl*a* pair; more density is detected on Chl*a* coordinated by His-656 of PsaB [10, 11]. However, the two branches of the redox components are functional [12]. Kinetic measurements revealed that the reoxidation of A_1^- by F_X has two phases of $t_{1/2}$ = 13 ns and $t_{1/2}$ = 160 ns, suggesting that the two phylloquinone molecules are involved in the electron transfer from A_1 to F_X [13]. Site-directed mutagenesis to replace the Trp residue interacting with one of phylloquinone molecules revealed that the replacement of Trp residue of PsaB affected the fast phase while the replacement of Trp residue of PsaA affected the slow phase [14]. F_X is coordinated between PsaA and PsaB by four cysteine residues; Cys-575 and Cys-584 of PsaA and Cys-561 and Cys-570 of PsaB coordinate the 4Fe-4S cluster. Two 4Fe-4S clusters, F_A and F_B, are located on PsaC. Two Trp residues on PsaA and PsaB, respectively, are located between P700 and the docking site for Pc and Cyt *c*, and facilitate the efficient electron transfer from Pc or Cyt *c* to P700 [15].

5 Antenna Pigments in PSI Core and LHCI Complexes

The cyanobacterial PSI core complex associates 96 Chl*a* molecules; six Chl*a* molecules are required for the electron transfer reaction, whereas 90 Chl*a* molecules are involved in light harvesting as antenna chlorophylls. 79 Chl*a* molecules are bound to the PSI RC heterodimer; PsaA and PsaB associate 40 and 39 Chl*a* molecules, respectively, whereas the small subunits such as PsaJ, PsaK, PsaL, PsaM and PsaX, and a lipid, phosphatidylglycerol (PG), coordinate to Mg^{2+} of 11 Chl*a* molecules.

The arrangements of the 21 transmembrane helices of the PSI RC subunits (PsaA and PsaB) and the PSII core complex (PsbA, PsbB, PsbC, and PsbD) are well conserved. The domains consisting of six helices (a-f) of the N-termini of PsaA and PsaB (peripheral parts), which correspond to the six helices of CP47 and CP43, bind 27 Chl*a* molecules. By contrast, the domains consisting of five helices (g-k) of the C-termini of PsaA and PsaB (central parts), which correspond to the five helices of PsbA (D1) and PsbB (D2) of PSII RC, associate 13 and 12 Chl*a* molecules, respectively (Fig. 7). The antenna chlorophylls are not distributed uniformly so that the distances among adjacent Chl*a* molecules are close enough for efficient exciton transfer. Most Chl*a* molecules are coordinated to imidazole group of His residues, or oxygen atoms of the side chain of Asp, Gln, Glu, and Tyr residues, of peptide bonds, or of water molecules [4]. For example, the peripheral parts of PsaA and PsaB associate 18 Chl*a* molecules, which forms layer-like structures at the stromal and lumenal sides. However the antenna chlorophylls are separated from the six chlorophylls involved in the electron transfer except for the two Chl*a* molecules on PsaA and PsaB, which are located in the proximity of A_0 on PsaA and PsaB, respectively. It is proposed that excitons are transferred from the antenna pigments to RC mainly through these Chl*a* molecules.

One of the interesting features of the antenna chlorophylls of PSI complex is the presence of chlorophylls that absorb light at longer wavelengths than that of P700, which are designated as 'red chlorophylls' or 'red-shifted chlorophylls'. One trimer at the lumenal side of PsaB and three dimers of Chl*a* at the central parts of PsaA and PsaB are tentatively assigned as red chlorophylls [4]. The three Chl*a* molecules forming trimer are stacked like a staircase, whereas the chlorin planes of the three dimers are nearly parallel.

β-carotene molecules play roles in light harvesting and protecting RCs from photodamage by quenching excited Chl*a* triplet states generated by photochemical reactions. Most of β-carotene molecules are deeply inserted in the PSI complex although many Chl*a* molecules are present near the stromal or lumenal surface [1, 4, 7].

Recent progresses in the determination of the structure of PSI-LHCI have revealed detailed structure of four LHCIs (Lhca1-4) stably associated to PSI core complex in plant [5–7, 16]. PSI-LHCI associates 155 Chl*a*, 12 Chl*b*, 28 β-carotene, 5 lutein, and 4 violaxanthin molecules. The four LHCIs associate 45 Chl*a*, 12 Chl*b*, 4 β-carotene, 5 lutein, and 4 violaxanthin molecules [7]. Although the structures of the four LHCIs are well conserved, the distribution of Chl*b* is not uniform among

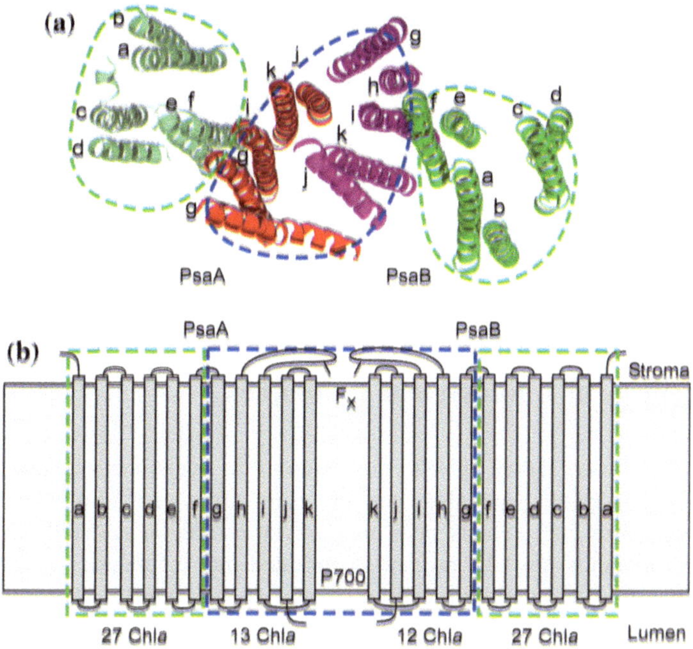

Fig. 7 The arrangement of 22 transmembrane helices of PsaA and PsaB. **a** Localization of 11 transmembrane helices (a-k) of PsaA and PsaB are shown. Six transmembrane helices of N-termini of PsaA and PsaB (a-f) are shown in green [4]. These domains form peripheral parts of PSI RC and function as light harvesting. Five transmembrane helices of C-termini of PsaA and PsaB (g-k) are shown in magenta and purple. These domains are interlocked to constitute central part involved in both light harvesting and photochemical reaction. **b** Cross-section model of 11 transmembrane helices of PsaA and PsaB. Two domains consisting of helices a-f of PsaA and PsaB and two domains consisting of helices g-k of PsaA and PsaB are shown. P700 is located between two k helices of PsaA and PsaB whereas F_X is located between two interhelical loops connecting h and i helices of PsaA and PsaB. The number of Chl*a* associated with the four domains is indicated

the four LHCIs as summarized in Table 2. Lhca2 and Lhca4, which are located in the center of the LHCI tetramer, associate more Chl*b* molecules than Lhca1 and Lhca3. Each LHCI associates one molecule of β-carotene, lutein, and violaxanthin in addition to Chl*a* and Chl*b*. However one additional lutein molecule is present

Table 2 Pigment compositions of PSI core and LHCI complexes[1, 2]

Pigments	PSI core		Lhca1-4[2]	Lhca1[2]	Lhca2[2]	Lhca3[2]	Lhca4[2]
	Cyanobacterium[1]	Plant[2]	Plant[2]				
Chl*a*	96	98	45 (52[3])	12	9	13	11
Chl*b*	0	0	12 (9[3])	2	5	1	4
β-carotene	22	22	4	1	1	1	1
Lutein	0	0	6	1.5[4]	1	1	1.5[4]
Violaxanthin	0	0	4	1	1	1	1

[1][4], [2][7], [3][6], [4]one lutein molecule is localized between Lhca1 and Lhca4

between Lhca1 and Lhca4. As shown in Fig. 3b, Lhca1, Lhca4, Lhca2, and Lhca3 form a tetramer on one side of PSI core complex (PsaF and PsaJ side), which is located between PsaG and PsaK. PsaG and PsaK, which are homologous and are located pseudo-symmetrically, are involved in stabilization of the LHCI tetramer.

Two main pathways for exciton transfer from LHCI to PSI core are proposed on the basis of the structure of PSI-LHCI supercomplex. PsaG and PsaK bind Chl*a* molecules and intimately associate with Lhca1 and Lhca3, respectively. It is inferred that excitons on the LHCI tetramer are first moved to Lhca1 or Lhca3, and are then transferred to PSI core antenna through Chl*a* molecules associated with PsaG and PsaK, respectively.

6 Biogenesis of Photosystem I Complex

The subunits and cofactors of PSI core complex in cyanobacteria and PSI-LHCI supercomplex in plants and algae are precisely organized as already described above. Thus these components need to be synthesized on demand and the synthesized components must be properly integrated into a functional complex. It is unlikely that this assembly process proceeds spontaneously. Accordingly it is expected that an assembly apparatus for PSI complex, which assists integration of the PSI components, should be present [17, 18]. However it remains elusive how and by which factors the constituent components are integrated into PSI complexes. However several factors that are necessary for proper assembly of PSI complex have been reported [17, 18]. Among them, the chloroplast-encoded Ycf3 and Ycf4 proteins are essential factors for PSI biogenesis in the green alga *Chlamydomonas*, because the knockout mutants of Ycf3 or Ycf4 accumulate no PSI complex [19]. Biochemical isolation and characterization of Ycf4 protein revealed that Ycf4 is part of a large complex, on which newly synthesized PSI subunits are assembled into a PSI core subcomplex. Subsequently the other small PSI subunits are successively integrated into the subcomplex. In the case of PSI-LHCI assembly, the integration of PsaG and PsaK are one of the latest steps in the PSI biogenesis [8]. However more detailed assembly process remain to be addressed.

7 Conclusion

The PSI complexes of cyanobacteria, algae, and plants have functionality for efficient light harvesting, charge separation, and electron transfer to oxidize Pc or Cyt *c* and reduce Fd. 11-14 subunits of PSI core complex and a number of cofactors involved in charge separation and electron transfer as well as in light harvesting are precisely organized in large chlorophyll-protein complexes. The determined structures of PSI core trimer and PSI-LHCI supercomplex have revealed the detailed structure responsible for the conversion of labile light energy into stable redox energy at quantum yield of nearly 1.0.

References

1. Jensen PE et al (2004) The PSI-O subunit of plant photosystem I is involved in balancing the excitation pressure between the two photosystems. J Biol Chem 279(23):24212–24217
2. Takahashi Y et al (2004) Comparison of the subunit compositions of the PSI-LHCI supercomplex and the LHCI in the green alga *Chlamydomonas reinhardtii*. Biochemistry 43 (24):7816–7823
3. Nelson N, Yocum CF (2006) Structure and function of photosystems I and II. Annu Rev Plant Biol 57:521–565
4. Jordan P et al (2001) Three-dimensional structure of cyanobacterial photosystem I at 2.5 Å resolution. Nature 411(6840):909–917
5. Ben-Shem A, Frolow F, Nelson N (2003) Crystal structure of plant photosystem I. Nature 426 (6967):630–635
6. Mazor Y, Borovikova A, Nelson N (2015) The structure of plant photosystem I super-complex at 2.8 A resolution, vol 4. Elife
7. Qin X et al (2015) Photosynthesis. structural basis for energy transfer pathways in the plant PSI-LHCI supercomplex. Science 348(6238):989–995
8. Ozawa S, Onishi T, Takahashi Y (2010) Identification and characterization of an assembly intermediate subcomplex of photosystem I in the green alga *Chlamydomonas reinhardtii*. J Biol Chem 285(26):20072–20079
9. Ozawa SI et al (2012) 5′-Monohydroxyphylloquinone is the dominant naphthoquinone of PSI in the green alga *Chlamydomonas reinhardtii*. Plant Cell Physiol 53(1):237–243
10. Redding K et al (1998) A systematic survey of conserved histidines in the core subunits of photosystem I by site-directed mutagenesis reveals the likely axial ligands of P700. EMBO J 17(1):50–60
11. Webber AN et al (1996) Site-directed mutations affecting the spectroscopic characteristics and midpoint potential of the primary donor in photosystem I. Biochemistry 35(39):12857–12863
12. Brettel K (1997) Electron transfer and arrangement of the redox cofactors in photosystem I. Biochimica Et Biophysica Acta-Bioenergetics 1318(3):322–373
13. Joliot P, Joliot A (1999) In vivo analysis of the electron transfer within photosystem I: are the two phylloquinones involved? Biochemistry 38(34):11130–11136
14. Guergova-Kuras M et al (2001) Evidence for two active branches for electron transfer in photosystem I. Proc Natl Acad Sci U S A 98(8):4437–4442
15. Sommer F et al (2004) The hydrophobic recognition site formed by residues PsaA-Trp(651) and PsaB-Trp(627) of photosystem I in Chlamydomonas reinhardtii confers distinct selectivity for binding of plastocyanin and cytochrome c(6). J Biol Chem 279(19):20009–20017
16. Amunts A, Drory O, Nelson N (2007) The structure of a plant photosystem I supercomplex at 3.4 A resolution. Nature 447(7140):58–63
17. Minagawa J, Takahashi Y (2004) Structure, function and assembly of Photosystem II and its light-harvesting proteins. Photosynth Res 82(3):241–263
18. Schottler MA, Albus CA, Bock R (2011) Photosystem I: Its biogenesis and function in higher plants. J Plant Physiol 168(12):1452–1461
19. Boudreau E et al (1997) The chloroplast ycf3 and ycf4 open reading frames of Chlamydomonas reinhardtii are required for the accumulation of the photosystem I complex. EMBO J 16(20):6095–6104

Part IX
Energy Conversion Using Photosynthesis Mechanism: Implementing Photosynthesis in Energy Storage Systems

PS-I and PS-II on Electrodes for Energy Generation and Photo-Sensor

Nao Terasaki

Abstract High performance of biocomponents such as photosynthetic protein, i.e. photosystem I (PSI) or II (PSII), can be attributed to well-designed spatial configurations (position, direction, etc.) and the environmental control of the functional molecules in the biocomponents. In order to introduce the utilization of the biocomponent in the context, in this chapter, we described the next four topics together with many reported examples; (1) PSI and PSII protein complex as the biocomponent, (2) oriented immobilization of PSI or PSII on electrode, (3) rapid and multilayered immobilization of PSI, (4) new approach; plugging a molecular wire into PSI as biocomponent for direct connection to artificial devices.

1 Introduction

Biocomponents from living body are very attractive nanomaterials because they have already achieved ultrahigh and ultimate performance as a result of many cycles of natural selection and mutation (Fig. 1). Especially in the case of photosynthesis, it is well-known that the quantum yield of the photo-electro conversion in the photosynthesis reaction center is almost unity [1, 2].

The high performance of biocomponents can be attributed to well-designed spatial configurations (position, direction, etc.) and the environmental control of the functional molecules in the biocomponents.

Many scientists have made efforts to mimic and model bio functions, and only some of them have developed materials with performance identical to or exceeding that of the original biocomponents. As to the photo-induced electron transfer, as shown in Fig. 2a, Moore and Gust et al. succeeded in achieving a long living charge separation using carotenoporphyrin–quinone triad system in which the distance

N. Terasaki (✉)
Advanced Manufacturing Research Institute, National Institute of Advanced Industrial Science and Technology (AIST), 807-1 Shuku-Machi, Tosu, Saga 841-0052, Japan
e-mail: nao-terasaki@aist.go.jp

M. Sugiyama et al. (eds.), *Solar to Chemical Energy Conversion*,
Lecture Notes in Energy 32, DOI 10.1007/978-3-319-25400-5_25

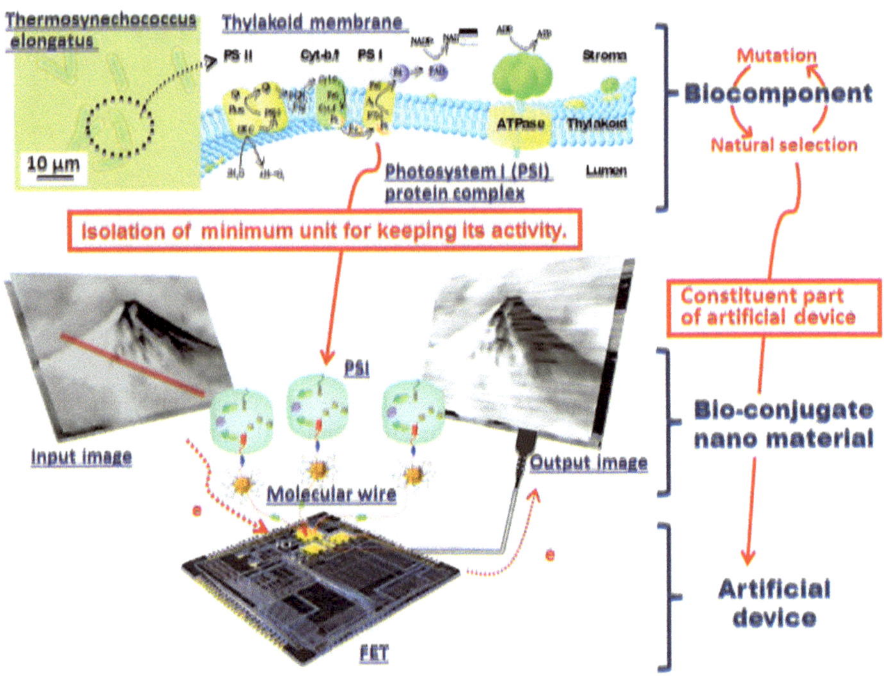

Fig. 1 Concept: biocomponent as constituent part of artificial device

between separated electron and hole was kept long to prevent backward electron transfer to reach charge recombination, the same strategy as photosynthetic electron relay system [3]. Furthermore Imahori and co-worker successfully developed high efficient photoelectric conversion device using gold electrodes modified with self-assembled monolayers of ferrocene-porphyrin-fullerene triads (Fig. 2b) [4, 5]. However, these successes have been limited to several examples because the fine tuning of multistep electron transfer reaction requires high skill and experience for organic synthesis and long time for development.

On the other hand, materials with ultimate and ultrahigh performance already exist in nature. From these viewpoints, we have proposed a new concept in which bio components are employed as vital constituents of the artificial devices; for this purpose, we designed a bio photo-sensor system consisting of a photosystem I (PSI) protein complex and electrode or FET device (Fig. 1). Here we discuss the next four topics; (1) photosystem I (PSI) and photosystem II (PSII) protein complex as the biocomponent, (2) oriented immobilization of the biocomponent, (3) rapid and multilayered immobilization, (4) new approach; plugging a molecular wire into biocomponent for direct connection to artificial devices.

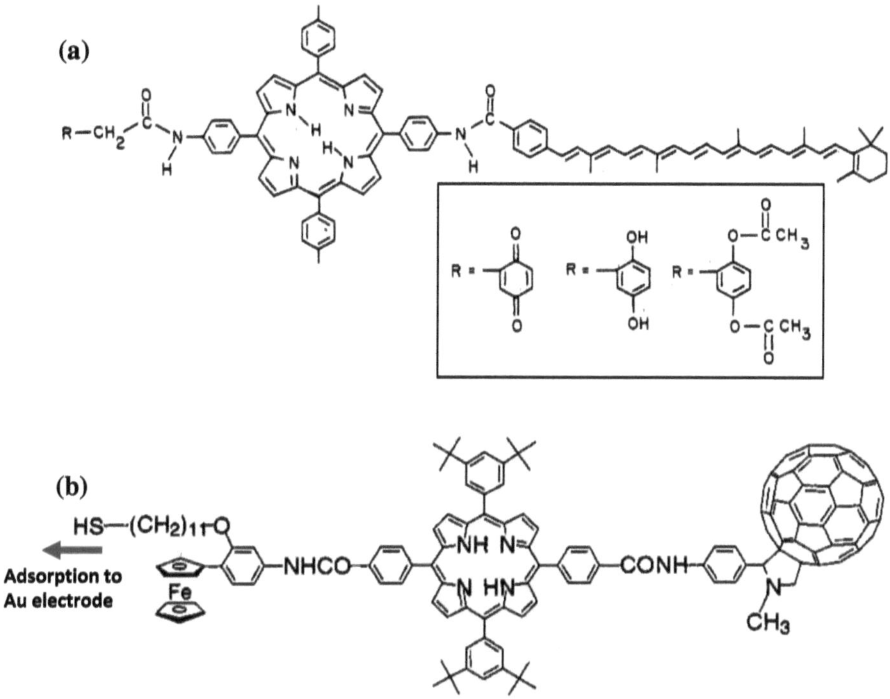

Fig. 2 Example of high performance artificial photosynthetic system using molecules reported by **a** Moore and Gust et al. [3] and **b** Imahori et al. [4]

2 Photosystem I and II as Biocomponent

As mentioned in the Introduction, the high performance of biocomponents such as PSI and PSII can be attributed to well-designed spatial configurations (position, direction, etc.) and the environmental control of the functional molecules in the biocomponents. Thus, first, the characteristics of PSI and PSII protein complex are introduced.

According to the crystal structure of PSI from the *thermophilic cyanobacterium Synechococcus elongatus* reported by Jordan et al. [6] and Fromme et al. [7, 8], we can obtain a picture at atomic detail of 12 protein subunits and 127 cofactors comprising 96 chlorophylls, 2 phylloquinones, 3 Fe_4S_4 clusters, 22 carotenoids, 4 lipids, a putative Ca^{2+} ion and 201 water molecules. The PSI protein complex is trimmer as shown in Fig. 3a, and the size is estimated as ca. 220 Å (22 nm) through the TEM image of PSI (Fig. 3b). In each monomer of PSI, there is an electron relay system as shown in Fig. 3d. The photo excitation of a chlorophyll hetero dimer (P700) induces its oxidation and a series of efficient electron transfer steps, as shown in Fig. 3c, d: P700 → (1–3 ps) → chlorophyll a(A_0) → (20–50 ns) → phylloquinone (VK_1 or A_1) → (20–200 ns) → iron-sulfur (Fe-S) clusters (F_X, F_A and F_B) [9, 10].

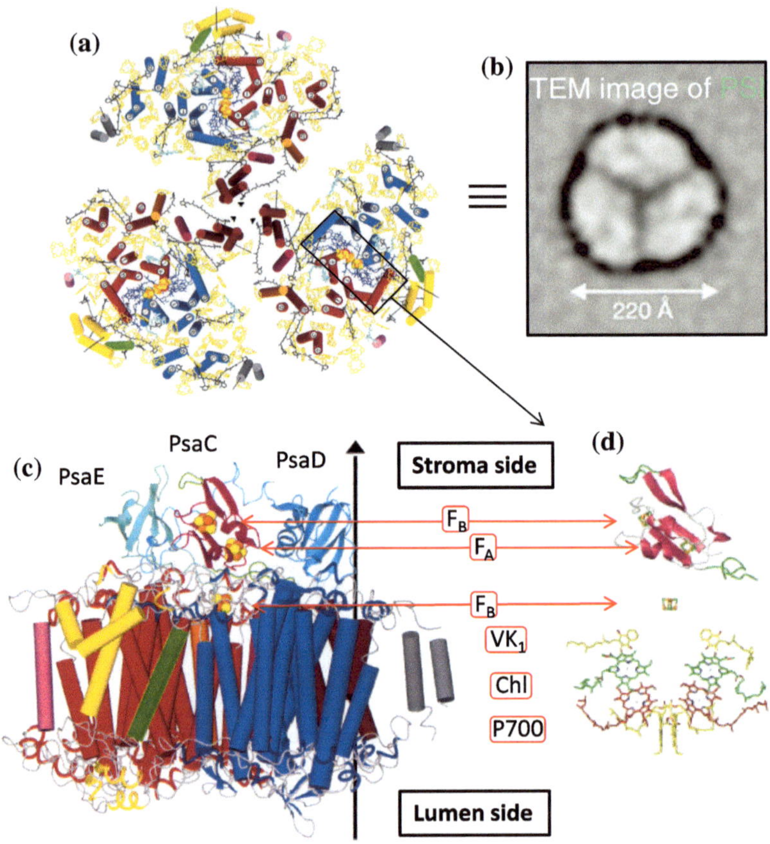

Fig. 3 Information of PSI reported by Jordan et al. [6] and Fromme et al. [7]. Structure of top view (**a**) and side view (**c**). TEM image (**b**) and electron relay system (**d**) of PSI trimmer protein complex

Meanwhile, in Fig. 4, the information of electron relay system (a) and structure (b) of cyanobacterial PSII are shown [11–13]. They show locations of and interactions between 20 protein subunits and 77 co factors per monomer. Assignment of 11 carotenes yields insights into electron and energy transfer and photo-protection mechanisms in the reaction centre and antenna subunits [11, 12].

The photo-activity including photo induced charge separation and subsequent reactions in PSII was reviewed by Rutherford and Boussac as below [13]. When a P680 (chlorophyll hetero dimer) is excited by light,

(1) charge separation (2–20 ps) occurs mainly between the chlorophyll (Chl_{D1}) in P680 and the pheophytin (Ph_{D1}). After that, the consequent electron transfer occurs;
(2) Ph_{D1} transfers an electron ($\sim$400 ps) to the quinone (Q_A).
(3) $P680^+$ is able to oxidize ($\sim$20 ns) the tyrosine-161 of D1.

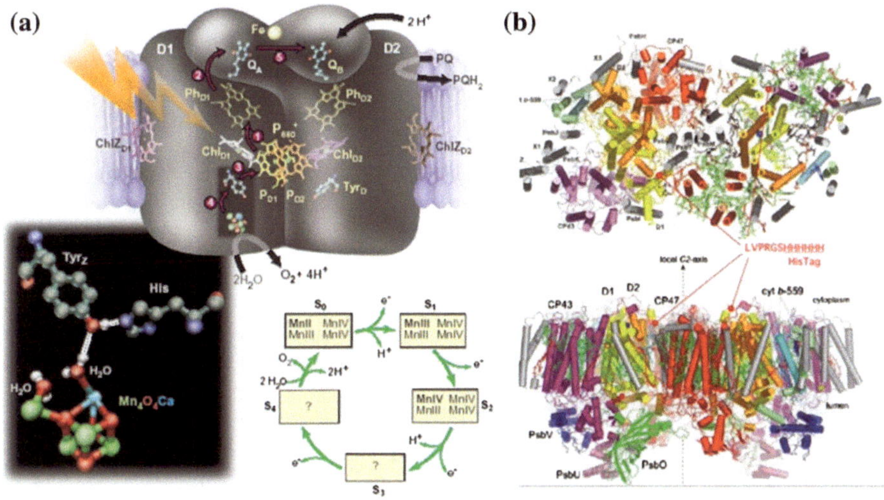

Fig. 4 Electron relay system [13] (**a**), and structure [12] (**b**) of cyanobacterial PSII

(4) TyrZ$^{\bullet}$ oxidizes ($\sim$30 μs) the Mn cluster (S_1–S_2).
(5) Q_A transfers an electron ($\sim$100 μs) to the second quinone (Q_B)

Through similar reactions but different kinetics, charges are accumulated on the Mn cluster and on Q_B. The accumulated four positive charge-equivalents on Mn cluster cause oxidization of $2H_2O$, releasing O_2. The photo-induced water splitting reaction is one of the most attractive points of PSII from the view point of photoelectric conversion with just water and fuel cell.

3 Oriented Immobilization of Biocomponent on Electrode

In order to employ bio-components in the artificial device, controlling the orientation and maintaining the activity of the bio-component must be needed for the efficient signal generation. Biocomponents such as PSI and PSII have a specific site on their surface such as hydrophilic and hydrophobic site [14–16], positively or negatively charged, amino acid which has functional group such as –SH, NH_2, –CO_2H for adsorption to gold electrode and/or binding reaction [14, 17–19]. Furthermore, some researchers prepared mutant biomaterials in order to specify the position for adsorption through attaching histidine tag (Histag) or removing steric hindrance [20, 21].

In order to highly control the orientation of biomaterial, the use of an appropriate interlayer (linker) has been efficient for attaching the specific site of biomaterial [14–21]. As the smart example as shown in Fig. 5a, Kalz et al. used a pyrene derivative with N-succinimidyl-[2, 3-3] propionate as the interlayer (linker)

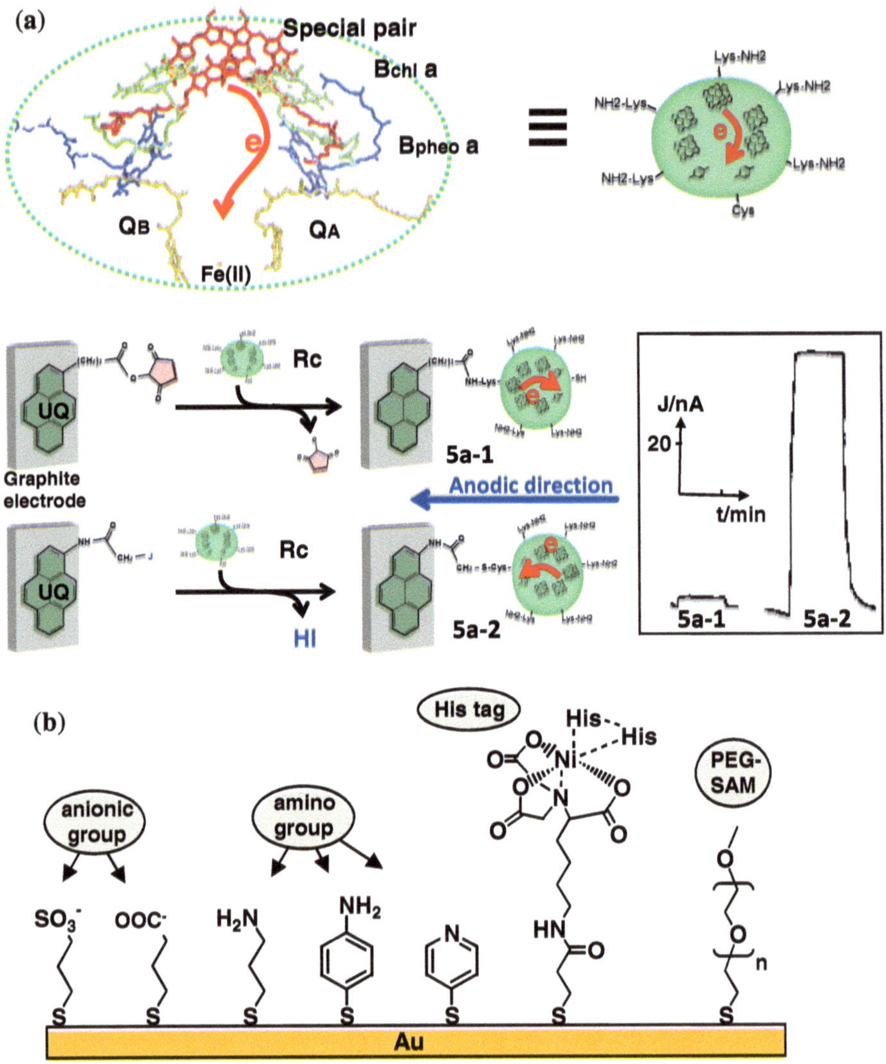

Fig. 5 Oriented immobilization of biocomponent on electrode. **a** Structure of photosynthetic protein (called RCs), Two types of RCs orientation on the graphite electrode via pyrene-containing linker molecule: bound through lysine and cystein residuals, and the corresponding photocurrent responses, reported by Kazt et al. [14]. **b** Chemical formulae of bifunctional linker molecules for immobilizing biocomponent on electrode

molecule, in which the pyrene group was chemically adsorbed on the graphite electrode [14]. Then, a photosynthetic protein complex (hereafter called reaction centers: RCs) was covalently bound to the linker by the coupling reaction (elimination of N-succinimidyl group). In this case, the orientation of RCs depends on the binding sites of RCs: lysine (Lys) or cysteine (Cys) residuals, denoted as orientations **5a-1** and **5a-2**, respectively. Photocurrent measurements were carried out in the presence of cytochrome c (10^{-2} M) as the electron donor for the RCs. A dramatic difference in the photocurrent intensity was observed for the different orientations of the RCs (Fig. 5a, right). The result clearly indicates that in the orientation **5a-2** the electron relay systems in the RCs are favorably directed to induce larger electron flow as if the quinone sites are located closer to the electrode. Accordingly, the self-assembling monolayer (SAM) as interlayer (linker) can serve as a promising method for preparing the electron flow pathway from the RCs.

Likewise, in Fig. 5b, some other SAMs of bifunctional thiols having sulfonate, carboxy, and amino group and others was shown, which have been used as the interlayers (linker) to bind the bio-components for further investigation of the photoelectric function of the RC [15, 17], PS I [16, 18–20], PSII [21] and so on.

At an early stage in our research, we investigated the possibility of PSI-based photosensor. Thus, we tried to immobilize the PSI on electrode through relatively convenient method, such as electrostatic interaction between a negatively charged self-assembling monolayer (SAM) and a positively charged ferredoxin-binding site which has $-NH_4^+$ group around the Fe-S clusters (F_X, F_A, and F_B) of PSI [18, 19]. Concretely, the PSI-modified gold electrode was prepared according to the scheme shown in Fig. 6a. The SAM of 3-mercapto-1-propane sulfonic acid sodium salt (MPS) was first assembled on the gold electrode surface. Then, the MPS-modified gold electrode was immersed in 2.8 mg/mL of PSI suspended in MES-NaOH buffer

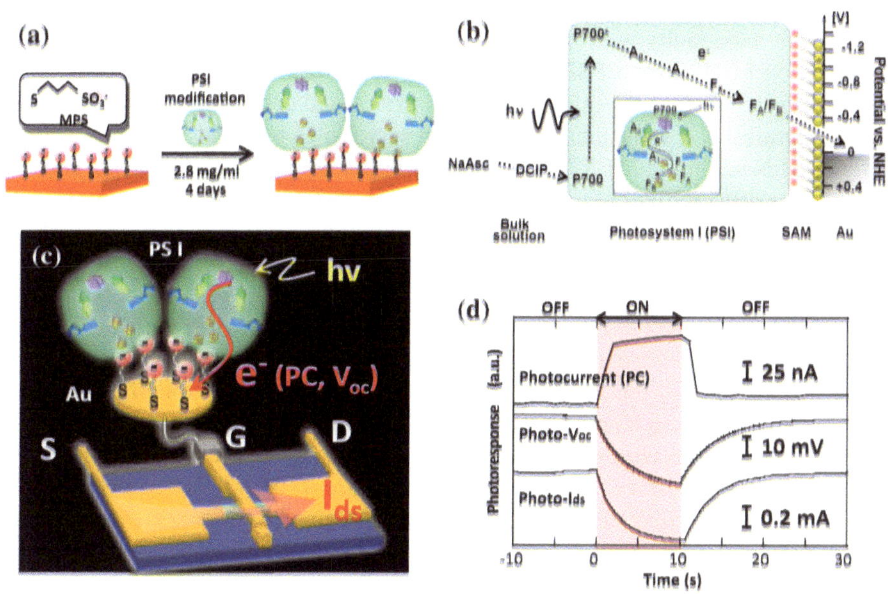

Fig. 6 PSI based biophotosensor. **a** Scheme for immobilization of PSI on electrode via negatively charged SAM. **b** Mechanism for photocurrent generation. **c** Schematic illustration **d** and photoresponse of PC, photo-Voc and photo-Ids of PSI based FET device

solution (pH6.4) for 4 days to obtain PSI-modified gold electrode. The experimental work flow for evaluating the series of photo-responses is described here. First, the photo response of the current (photocurrent, PC) through the mechanism as shown in Fig. 6b was carried out in the three-electrodes mode [counter electrode (CE): Pt wire; reference electrode (RE): Ag/AgCl (sat. KCl aq)]. The open-circuit voltage (photo-Voc; CE:Pt wire) of the PSI-modified gold electrode was evaluated by using same electro-chemical cell as the case of photocurrent measurement. Furthermore, after leading of the PSI-modified gold electrode to the gate of the FET as shown in Fig. 6c, the photo response of the source-drain current of the PSI-based FET (photo-Ids) was measured. As shown in Fig. 6d, the basic photoresponses originating from the excitation of PSI, including the photocurrent (106 nA) and the photoresponse of the open-circuit voltage (photo-Voc: 28.6 mV), were characterized. Then, the PSI-modified gold electrode was linked to the gate of the FET using a leadline, and the device was successfully driven by the photo electric signals from the PSI like a voltage follower circuit. Furthermore, we successfully demonstrated that the PSI-based FET acts as a photosensor in imaging devices.

The genetic introduction of a poly histidine tag (Histag) on the surface of a biocomponent has been investigated as a promising method to immobilize the biocomponent on the Ni-NTA-terminated SAM on electrodes with positional selectivity in a molecular order [20]. Based on this background, we prepared Histag-PSII (Fig. 2b) and immobilized the Histag-PSII on the Ni-NTA-terminated SAM through the scheme as shown in Fig. 7a, to obtain PSII/Au [21]. Before the experiment using PSII modified electrode, it was confirmed that the PSII used here can sufficiently cause a water-splitting reaction and the oxygen evolution; in other words, the bioactivity was maintained even after the purification and genetic introduction of Histag on PSII.

Photocurrent measurements of PSII/Au were carried out using the three-electrode electrochemical cell in the presence of 0.4 M sucrose, 15 mM $CaCl_2$, 15 mM $MgCl_2$, and 0.1 M NaCl in the MES-NaOH (pH6.4) buffer aqueous solution. This solution for the photocurrent measurements possessed almost the same composition as the solution in the oxygen-evolving experiment involving PSII; in particular, and there were no sacrificial electron donors. By subjecting the PSII/Au to light irradiation of 680 nm (3.3 mW), a stable anodic photocurrent was observed and the value was maintained constant for at least several tens of minutes (Fig. 7c). Figure 7d shows the applied potential dependence of the photocurrent in PSII/Au. The photocurrent values increased with the applied positive potential and the photocurrent direction was reversed at around 50 mV versus Ag/AgCl (sat.KCl), which is close to the redox potential of Q_B. The Histag, which provides the linkage to the gold electrode, was introduced on the surface of PSII at the point close to the Q_B molecule (Fig. 2b); therefore, electron injection from Q_B to electrode seemed to be easy. Based on these facts, the photocurrent must be generated through the following 4 steps (Fig. 7b):

(i) photoexcitation of PSII,
(ii) cascades of electron transfer reactions in PSII from P680 to Q_B,

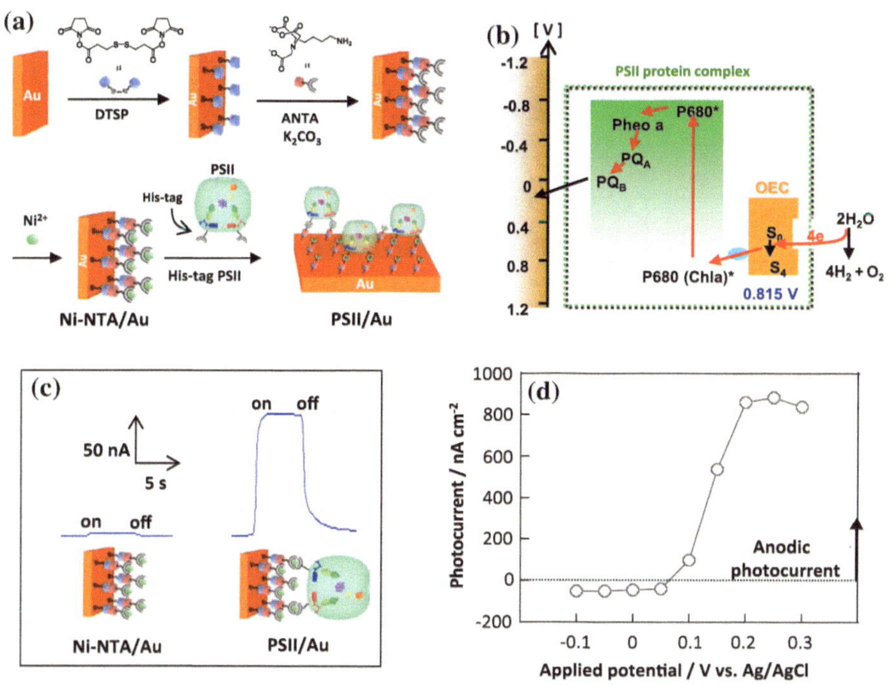

Fig. 7 PSII on electrode via Histag-Ni-NTA linker molecular. **a** Schematic illustration of procedure. **b** Mechanism, **c** photo-responses and **d** applied potential dependence on photocurrent of PSII modified electrode

(iii) electron injection from Q_B to gold electrode,
(iv) re-redox of P680^{+*} with the electron generated via the water-splitting reaction at the Mn cluster in PSII.

4 Rapid and Multilayered Immobilization on Electrode

The self-assembling method is very fascinating in the preparation of monolayer assemblies on the electrode. However, fabrication of SAMs on the planar electrode has been limited to the number of immobilized molecules. To overcome this limitation, the use of three-dimensional or multistructure electrodes is the most promising approach [22]. For this purpose, utilization of metal nanoparticle or nanostructure is convenient because the metal—S self-assembly can in principle be applicable, same as for the planner electrode. Thus, also for biocomponent modified electrode, Terasaki et al. utilized the nanostructured gold electrode prepared through electrolyte-induced precipitation of gold nanoparticle (so-called salting-out method) [19, 21]. They used ~15 nm gold nanoparticle that were comparable to the

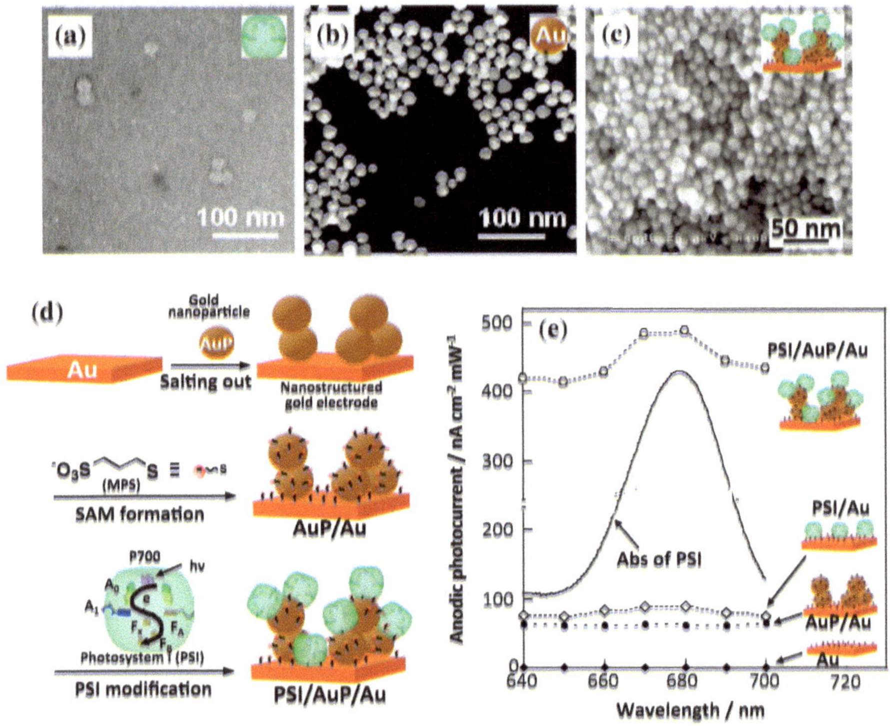

Fig. 8 Enhancement photocurrent using nanostructured gold electrode [19]. **a** TEM image of PSI. SEM images of **b** gold nanoparticle and **c** PSI/AuP/Au. **d** Schematic illustration of procedure. **e** Comparison of photocurrent between PSI modified planer and nanostructured gold electrode

size of PSI (Fig. 8a, b), and prepared the gold nanoparticle-PSI multistructure as shown in Fig. 8c through the procedure shown in Fig. 8d. As a result, they successfully obtained larger photocurrent responses with increasing the number of gold nanoparticle to precipitate for forming the nanostructured gold electrode (Fig. 8e).

Furthermore, Carmeli et al. fabricated an oriented multilayers of PSI proteins on solid surfaces by auto-metallization as shown in Fig. 9a [23]. The feature of this system is the next two points; (1) cysteine mutant PSI as binding site to gold electrode. The mutated amino acid is located near P700 and does not have stereo hindrance when placed on a gold surface, assuring the formation of sulfide bonds and close electron injection. (2) Direct platinum deposition at the endpoint of electron relay system of PSI by using own photoelectron transfer of PSI and reduction of Pt^{4+} ions. By utilizing the feature, they fabricated the oriented multilayers of PSI on gold electrode according to the cycles of the next two steps and successfully enhanced the photo-signals.

(i) Formation of oriented monolayer of PSI on gold electrode via the sulfide bond between the mutated cysteine and gold electrode.

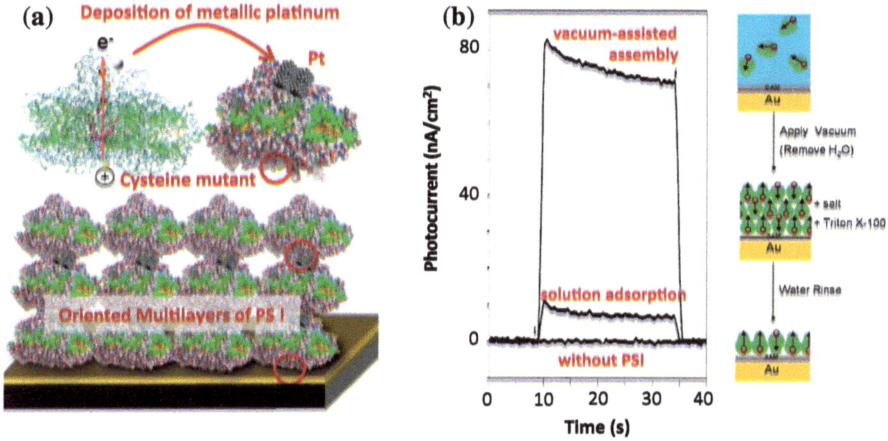

Fig. 9 Multidimensional PSI electrode. **a** Oriented multilayers of PSI on electrode by auto-metalization, reported by Cameli et al. [23]. **b** Photocell with vacuum deposited a dense PSI multilayer reported by Cliffel and Jennings et al. [24]

(ii) Photo-reduction of Pt^{4+} ions, which resulted in the deposition of Pt patches at the reducing end of each PSI molecule, is used to attach the next monolayer of PSI through the formation of sulfide bonds.

On the other hand, it requires a long time, i.e. a few days, to immobilize the biocomponent on electrode through interlayer (linker), and slow adsorption has also been an obstacle. To resolve this, Cliffel and Jennings et al. proposed a new method for preparing a dense PSI monolayer that reduces the time required for assembly by ~80-fold in comparison to that for adsorption from solution as shown in Fig. 9b [24]. This method consists of applying vacuum to an aqueous PSI solution during assembly process to concentrate PSI and precipitate it into a thick layer onto the self-assembled monolayers or directly onto the electrode surface. Rinsing with water yields a dense monolayer of PSI that draws ~100 nA/cm^2 of photocurrent from the gold electrode in the presence of appropriate mediators.

5 Plagging a Molecular Wire into PSI for Direct Connection to Arttificial Devices

In the previous paragraphs, PSI or PSII immobilization on electrode via interlayer (linker) molecules has been described. However, as mentioned earlier, the high performance of PSI is the result of its well-designed spatial configuration (position, direction, etc.); however, no investigations using the PSI photonic device in this context had been reported at that time. Thus, we have investigated and reported the first successful direct coupling between functional molecules of biocomponents

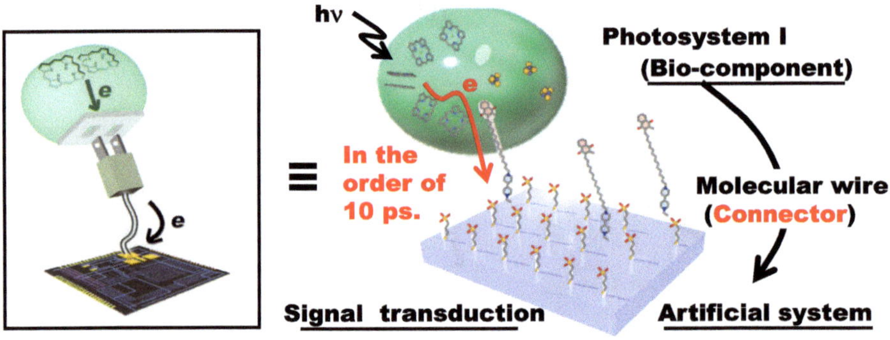

Fig. 10 strategy; direct coupling between functional molecules of PSI to artificial electronic devices via a molecular electric wire

(PSI) to artificial electronic devices (FET) via a molecular electric wire designed at the molecular order. This is similar to plugging an electric wire into power generator as shown in Fig. 10.

For the direct coupling, we focused on vitamin K_1 (VK_1, A_1 site of PSI, quinone derivative) by reason of the next three points;

(i) VK_1 can be extracted by treating organic solvent to obtain a quinone pocket.
(ii) Other quinone derivative can be reconstituted to the quinone pocket.
(iii) Electron transfer rate of each step of PSI, namely the first and second electron transfer reactions (from P700 to $A_1 = VK_1$) occur very rapidly (on a picosecond time scale), whereas the third step occurs at a much slower rate (in the order greater than approximately 10 ns).

Thus, the direct connection of an efficient molecular wire to the relay system at the A_1 site will guide the electrons in the direction of the wire. The main features of the design for molecular electric wire are as follows. First, the naphthoquinone unit functions as the binding site for the quinone pocket of PSI. Secondly, the lengths of the molecular wire and VK1 are kept identical to ensure that the end of the chain remains outside of the PSI system. Thirdly, the electron-accepting unit has the appropriate potential to relay electrons through the molecular wire. According to this strategy, we designed and synthesized two types of molecular wires, which are introduced below.

The first approach of the molecular electric wire was naphthoquinone-gold nanoparticle linked compound ($NQC_{15}SGNP$) as shown in Fig. 11 [25]. Certainly, although the gold nanoparticle does not show strong electron-accepting properties, $NQC_{15}SGNP$ is useful to visualize the reconstitution. First of all, we confirmed that the photo-activity of PSI disappeared once accompanied by extraction of VK_1 from PSI and then recovered after the reconstitution of $NQC_{15}SGNP$. Figure 11b shows TEM images of intact PSI and $NQC_{15}SGNP$ reconstituted PSI. A TEM image of the intact PSI exhibited gray circles with diameters of ca. 20 nm and ca. 10 nm. These sizes could be assigned as a trimmer (22 nm) and a monomer (10 nm) of PSI

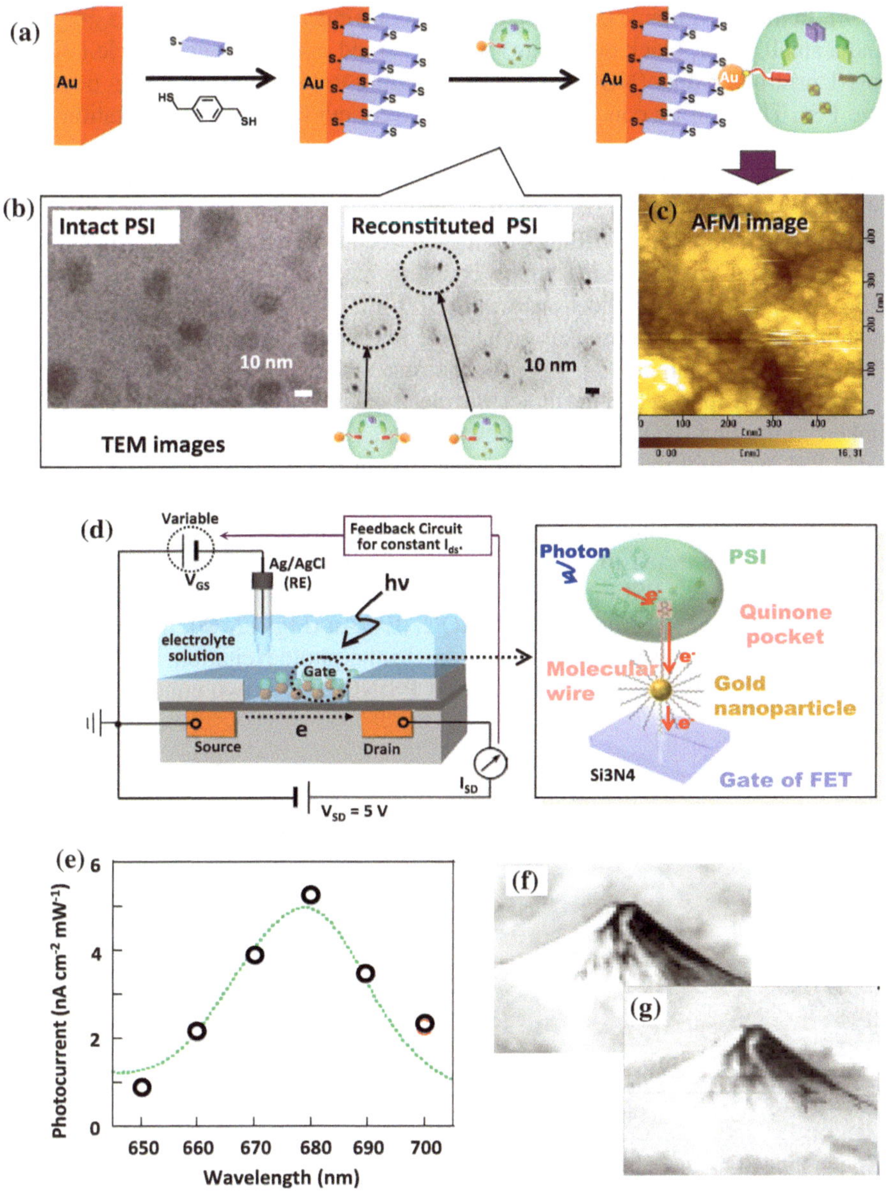

Fig. 11 Approach using $NQC_{15}GNP$ as molecularly electric wire, reported by Terasaki et al. [25]. **a** schematic illustration of reconstituted PSI modified gold electrode. **b** TEM images of intact and reconstituted PSI. **c** AFM image and **e** photocurrent action spectrum of the reconstituted PSI immobilized gold electrode. **d** Schematic structure of reconstituted PSI immobilized FET gate. **f** Input and **g** output image for demonstrating PSI based photosensor

[6–8]. In the TEM image of $NQC_{15}SGNP$ reconstituted PSI, both a large gray circle and a small, clear, black dot (2 nm) were observed, and every gray circle (PSI) has just 1 or 2 black dots ($NQC_{15}SGNP$) responding to the number of VK_1 pockets in one PSI. This can be concluded to be the direct evidence and visualization of the reconstitution. By means of connection between gold electrode and the gold nanoparticle of reconstituted PSI as shown in Fig. 11a, clear photocurrent was observed and confirmed that it was originated form photoexcitation of PSI (Fig. 11e). In addition, by immobilizing the reconstituted PSI on an Si_3N_4–Ta_2O_5 on the gate of an FET using silane coupling reaction and Au–S bond formation (Fig. 11d), the output of electrons from PSI to FET is demonstrated with the aim of obtaining a PSI-biophotosensor capable of interpreting the gradation in an image (input: Fig. 11f, output: Fig. 11g) for the first time.

The second approach of the molecular electric wire was Naphthoquinone-viologen linked compound ($NQC_{15}EV$) as shown in Fig. 12 [26]. In the $NQC_{15}EV$ reconstituted PSI, we confirmed that the photoelectron in PSI is transferred to the viologen along the electron relay system of PSI and the $NQC_{15}EV$ molecular wire within 10 ps by using laser flash photolysis. For the system, VK_1 extracted PSI (denoted as PSI (-VK_1)) was connected to $NQC_{15}EV$ molecule immobilized on gold electrode to obtain reconstructed PSI immobilized gold electrode (PSI@$NQC_{15}EV$/Au).

From the action spectrum of PSI@$NQC_{15}EV$/Au (Fig. 12, ●), it can be said that the photocurrent responses are mainly due to the photoexcitation. Furthermore, to clarify the effectiveness of electric wire of PSINQC_{15}EV, two control experiments

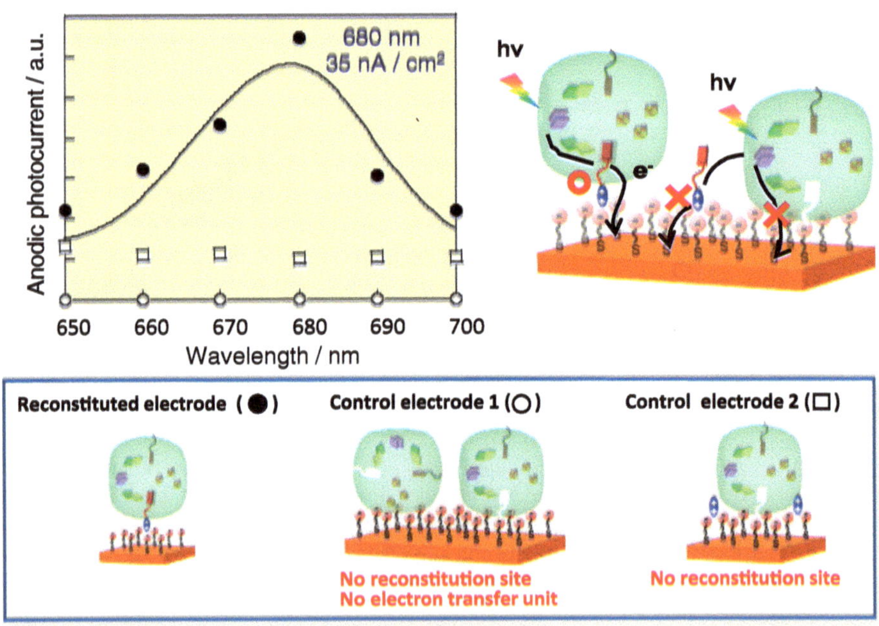

Fig. 12 Approach using NQC15EV as molecularly electric wire, reported by Terasaki et al. [26]

were carried out. For the control electrode 1, PSI(-VK_1) was directly immobilized an electrode through the same procedure, except that $NQC_{15}EV$ was not fixed, so that there were no binding sites for reconstitution and no electron acceptors on the electrode. Immobilization of PSI(-VK_1) was confirmed by surface plasmon resonance (SPR) measurements. Even though the conditions were the same as PSI@$NQC_{15}EV$/Au, photocurrent responses were not observed at all measured points (Fig. 12, ○). For control electrode 2, methyl viologen (MV) was used instead of $NQC_{15}EV$, so that there were no binding sites available for reconstitution with PSI(-VK_1), but there were electron acceptor units present around PSI on the gold electrode surface. In this case, the action spectrum shows no peaks around 680 nm (Fig. 12, □). This successfully demonstrated that the photocurrent responses are not a result of direct electron transfer from photoexcited P700 and/or other chlorophyll units to the viologen unit, but the result of electron transfer through the molecular electric wire $NQC_{15}EV$.

6 Conclusion

In this chapter, we introduced the following four topics together with many reported examples; (1) photosystem I (PSI) and photosystem II (PSII) protein complex as the biocomponent, (2) oriented immobilization of PSI or PSII on electrode, (3) rapid and multilayered immobilization of PSI, (4) new approach; plugging a molecular wire into PSI as biocomponent for direct connection to artificial devices.

How to achieve the ultimate performance of biocomponent remains a hot topic today, and many approaches have been reported on high impact factor journal; for example pseud dyad or triad biocomponent connected with graphen oxside (GO) [27], carbon nanotube (CNT), semiconductor such as p-doped Si [28] or GaAs [29] and so on. Furthermore, photocell [30] or photosensor [31] are considered promising practical applications. This trend should continue for the time being.

References

1. Emerson R, Chalmers RV, Cederstand C, Brody M (1956) Effect of temperature on the long-wave limit of photosynthesis. Science 123:673
2. Emerson R, Chalmers R, Cedrestrand ANDC (1957) Some factors influencing the long-wave light of photosynthesis. Natl Acad Sci USA 43:133–143
3. Moore TA, Gust D, Mathis P, Mialocq J-C, Chachaty C, Bensasson RV, Land EJ, Doizi D, Liddell PA, Lehmen WR, Nemeth GA, Moore AL (1984) Photodriven charge separation in a carotenoporphyrin–quinone triad. Nature **307**:630–632
4. Imahori H, Yamada H, Nishimura Y, Yamazaki I, Sakata Y (2000) Vectorial multistep electron transfer at the gold electrodes modified with self-assembled monolayers of ferrocene-porphyrin-fullerene triads. J Phys Chem B 104:2099–2108

5. Imahori H, Norieda H, Yamada H, Nishimura Y, Yamazaki I, Sakata Y, Fukuzumi S (2001) Light-harvesting and photocurrent generation by gold electrodes modified with mixed self-assembled monolayers of boron-dipyrrin and ferrocene-porphyrin-fullerene triad. J Am Chem Soc 123:100–110
6. Jordan P, Fromme P, Witt HT, Klukas O, Saenger W, Krauβ N (2001) Three-dimensional structure of cyanobacterial photosystem I at 2.5 Å resolution. Nature 411:909–917
7. Fromme P, Jordan P, Krauβ N (2001) Structure of photosystem I. Biochim Biophys Acta 1507:5–31
8. Bibby TS, Nield J, Barber J (2001) Iron deficiency induces the formation of an antenna ring around trimeric photosystem I in cyanobacteria. Nature 412:743–745
9. Brettel K (1997) Electron transfer and arrangement of the redox cofactors in photosystem I. Biochim Biophys Acta 1318:322–373
10. Poluektov OG, Paschenko SV, Utschig LM, Lakshmi KV, Thurnauer MC (2005) Bidirectional electron transfer in photosystem I: direct evidence from high-frequency time-resolved EPR spectroscopy. J Am Chem Soc 127:11910–11911
11. Loll B, Kern J, Saenger W, Zouni A, Biesiadka J (2005) Towards complete cofactor arrangement in the 3.0 Å resolution structure of photosystem II. Nature 438:1040–1044
12. Umena Y, Kawakami K, Shen J-R, Kamiya N (2011) Crystal structure of oxygen-evolving photosystem II at a resolution of 1.9 Å. Nature 473:55–61
13. Rutherford AW, Boussac A (2004) Water photolysis in biology. Science 303:1782–1784
14. Katz E (1994) Application of bifunctional reagents for on a carbon electrode surface: oriented of photosynthetic reaction centers. J Electroanal Chem 365:157–164
15. Lee I, Lee JW, Greenbaum E (1997) Biomolecular electronics: vectorial arrays of photosynthetic reaction centers, phys. Rev Lett 79:3294–3297
16. Ko BS, Babcock B, Jennings GK, Tilden SG, Peterson RR, Cliffel D (2004) Effect of surface composition on the adsorption of photosystem I onto alkanethiolate self-assembled monolayers on gold. Langmuir 20:4033–4038
17. Zha J, Zou Y, Liu B, Xu C, Kong J (2002) Differentiating the orientations of photosynthetic reaction centers on Au electrodes linked by different bifunctional reagents. Biosen Bioelectron 17:711–718
18. Terasaki N, Yamamoto N, Hattori M, Tanigaki N, Hiraga T, Ito K, Konno M, Iwai M, Inoue Y, Uno S, Nakazato K (2009) Photosensor based on an FET utilizing a biocomponent of photosystem I for use in imaging devices. Langmuir 25:11969–11974
19. Terasaki N, Yamamoto N, Hiraga T, Sato I, Inoue Y, Yamada S (2006) Fabrication of novel photosystem I—gold nanoparticle hybrids and their photocurrent enhancement. Thin Solid Films 499:153–156
20. Patrick RD, Kiley J, Segal M, Julie N, Yu AA, Wang L, Trammell SA, Reddick LE, Kumar R, Stellacci F, Lebedev N, Schnur J, Bruce BD, Zhang S, Baldo M (2004) Integration of photosynthetic protein molecular complexes in solid-state electronic devices. Nano Lett 4:1079–1083
21. Terasaki N, Iwai M, Yamamoto N, Hiraga T, Yamada S, Inoue Y (2008) Photocurrent generation properties of Histag-photosystem II immobilized on nanostructured gold electrode. Thin Solid Films 516:2553–2557
22. O'Regan B, Grätzel M (1991) A low-cost, high-efficiency solar cell based on dye-sensitized colloidal TiO_2 films. Nature 353:737–740
23. Frolov L, Wilner O, Carmeli C, Carmeli I (2008) Fabrication of oriented multilayers of photosystem I proteins on solid surfaces by auto-metallization. Adv Mater 20:263–266
24. Faulkner CJ, Lees S, Ciesielski PN, Cliffel DE, Jennings GK (2008) Rapid assembly of photosystem I monolayers on gold electrodes. Langmuir 24:8409–8412
25. Terasaki N, Yamamoto N, Tamada K, Hattori M, Hiraga T, Tohri A, Sato I, Iwai M, Iwai M, Taguchi S, Enami I, Inoue Y, Yamanoi Y, Yonezawa T, Mizuno K, Murata M, Nishihara H, Yoneyama S, Minakata M, Ohmori T, Sakai M, Fujii M (2007) Bio-photosensor: cyanobacterial photosystem I coupled with transistor via molecular wire. Biochim Biophys Acta 1767:653–659

26. Terasaki N, Yamamoto N, Hiraga T, Yamanoi Y, Yonezawa T, Nishihara H, Ohmori T, Sakai M, Fujii M, Tohri A, Iwai M, Inoue Y, Yoneyama S, Minakata M, Enami I (2009) Plugging a molecular wire into photosystem I: reconstitution of the photoelectric conversion system on a gold electrode. Angew Chem Int Ed 48:1585–1587
27. LeBlanc G, Winter KM, Crosby WB, Jennings GK, Cliffel DE (2014) Integration of photosystem I with graphene oxide for photocurrent enhancement. Adv Energy Mater 4:1301953
28. Sepunaru L, Tsimberov I, Forolov L, Carmeli C, Carmeli I, Rosenwaks Y (2009) Picosecond electron transfer from photosynthetic reaction center protein to GaAs. Nano Lett 9:2751–2755
29. LeBlanc G, Chen G, Gizzie EA, Jennings GK, Cliffel DE (2012) Enhanced photocurrents of photosystem I films on p-doped silicon. Adv Mater 24:5959–5962
30. Ciesielski PN, Hijazi FM, Scott AM, Faulkner CJ, Beard L, Emmett K, Rosenthal SJ, Cliffel D, Jennings GK (2010) Photosystem I—based biohybrid photoelectrochemical cells. Bioresource Tech 101:3047–3053
31. Matsumoto H, Tsukada J, Ozawa H, Uno S, Nakazato K, Terasaki N, Yamamoto N, Hiraga T, Iwai M, Konno M, Ito K, Inoue Y (2010) Integrated bio-imaging sensor array with complementary metal-oxide-semiconductor cascade source–drain follower. Jpn J App Phys 49:04DL01

Electronic Device Approach Using Photosynthesis Assembly of Photosynthetic Protein Complexes for the Development of Nanobiodevices

Masaharu Kondo, Takehisa Dewa and Mamoru Nango

Abstract Photosynthetic light-harvesting polypeptide/pigment complexes (LH) play an essential role in the primary process of an efficient solar energy-transduction in photosynthetic membrane. In our research, we aim to use the LH complex and control its direction and orientation on electrodes for the development of nanobiodevices from solar to fuel. Specifically, we focus on the construction of an array of the LH on electrodes using a modified photosynthetic protein complex prepared from modern biosynthetic manufacturing methods and in lipid membranes.

1 Introduction

There is a number of examples in nature where processes for energy conversion, storage, and transport are combined and optimized in a "smart system" that transforms energy from a molecular process to a cellular or higher organismal function. Using biological design principles, future biology-based photonics or synthetic organic materials could be a part of clean and inexpensive alternatives to current solar or fuel systems.

In the primary photosynthetic event, polypeptides/pigments complexes, that are LH and reaction centers (RC), cooperate to efficiently transducer of light energy.

M. Kondo
Department of Materials Science and Engineering, Graduate School of Engineering, Nagoya Institute of Technology, Gokiso-cho, Showa-ku, Nagoya 466-8555, Japan

T. Dewa · M. Nango
Department of Frontier Materials, Graduate School of Engineering, Nagoya Institute of Technology, Gokiso-cho, Showa-ku, Nagoya 466-8555, Japan

M. Nango (✉)
The OCU Advanced Research Institute for Natural Science and Technology (OCARINA), Osaka City University, 3-3-138, Sugimoto, Sumiyosi-ku, Osaka 558-8585, Japan
e-mail: nango@sci.osaka-cu.ac.jp

M. Sugiyama et al. (eds.), *Solar to Chemical Energy Conversion*, Lecture Notes in Energy 32, DOI 10.1007/978-3-319-25400-5_26

A photon is captured by a peripheral LH, funnels into a core LH complex, and subsequently transferred to an RC where charge separation takes place. Their light-harvesting complexes perform highly efficient photon capture followed by ultrafast excitation energy transfer between multiple chromophores. Understanding photosynthetic systems and mimicking their highly efficient mechanisms can be a very important approach to develop nanobiodevices for solar-energy conversion.

The past 10 years have seen tremendous progress in our understanding of the structure and function of the pigment-protein complexes involved in fundamental reactions in bacterial photosynthesis [1–9]. The structure of these reaction centers (RCs) consist of a nearly C_2 symmetrical arrangement of the redox centers. This system has been extensively studied with ultrafast laser spectroscopy. The structures of the LH2 complexes in photosynthetic bacteria have nonameric and octameric arrangements of repeating units consisting of two apoproteins, one or two carotenoids, and three bacteriochlorophylls (BChls) (Scheme 1). In contrast, recent evidence has shown that the crystal structure of the LH1-RC complex has the LH1 complex surrounding the contours of the RC; albeit, a high-resolution structure has not yet been determined for the LH1 complex (Scheme 1).

Some purple photosynthetic bacteria have peripheral LH2, whose beautiful cylindrical structures have been revealed by X-ray crystallography (Scheme 1) [5]. The subunit-type complexes bind two strongly coupled BChl *a* molecules absorbing at about 850 nm (B850), one BChl *a* molecule having an absorption band at 800 nm (B800), and one carotenoid molecule spanning the membrane in an all-*trans* conformation (Scheme 1). It is well known that light energy absorbed by

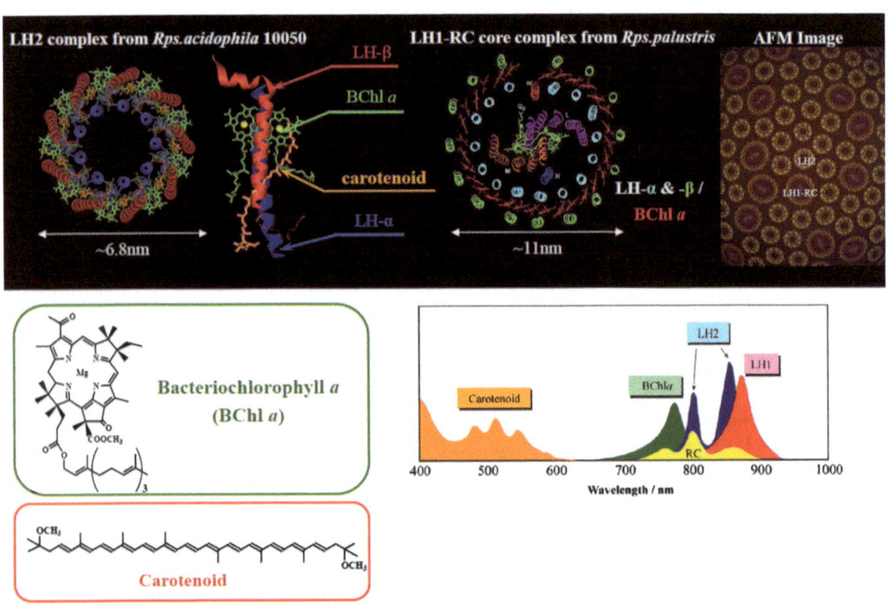

Scheme 1 X-ray analyses and AFM image of LH2 and LH1-RC, and the absorption spectra of bacterial chlorophyll *a* (BChl *a*) and carotenoids in the apoproteins (LH2, and LH1-RC)

B800 is transferred to B850 with a time constant of ~700 fs. Light energy absorbed by carotenoids is also efficiently transferred to B800 and B850. Consequently, the light energy is rapidly and efficiently collected by B850 in a sub-picosecond time domain. The excitation energy on B850 is then transferred to B850 of neighboring LH2 complexes, and finally funneled into a core complex composed of light-harvesting 1 antenna and reaction center complexes (LH1-RC) (Scheme 1), where charge separation follows [6]. LH2 and LH1-RC, therefore, form a highly efficient fused excitonic and photovoltaic device.

The light-harvesting mechanisms in these complexes have been studied both spectroscopically and theoretically [3, 4]. The knowledge has placed us in a unique position to design artificial photosynthetic antenna systems based on a biological blueprint. Our aim is to produce an antenna module, which acts as a sensitizer, and a light-induced redox component for nanobiodevices to convert solar energy to electricity. One of our system's unique features is that it functions over a large dynamic range of incident light intensities. It has a remarkable ability to efficiently capture photons even at very low light fluxes. However, the system can also withstand very high light fluxes by efficiently dissipating the excess photons, thereby protecting itself against the potentially harmful effects of over-excitation.

Recently, great advances have been made in the integration and miniaturization of organic devices that use these photosynthetic light-harvesting complexes in molecular electronics [10–36]. Atomic force microscopy (AFM) has been used to visualize membrane proteins, such as the LH2 complexes [7–10, 24, 34–36]. Further, currents produced by the RC films have been observed in multiple studies [14–20, 27–29, 34]. Recently scanning probe analyses, including scanning tunneling microscopy (STM) and conductive atomic force microscopy (C-AFM), have been very useful in the analysis of the electronic properties of individual molecules on conductive metal substrates. In particular, C-AFM allows contact resistance between the sample and tip to be controlled during current–voltage (I–V) measurements [28, 29, 34].

In this area of research, there are two fundamental challenges: (i) assembling protein pigment complexes with defined orientation onto the electrodes and (ii) establishing a good electrical contact between the pigments and the electrode. Genetically engineered mutations of RC and LH2 to produce His-tagged RCs or Cysteine-tagged LH2 s can be utilized to control their orientation on patterned surfaces [29, 33]. However, there have only been a few studies that investigated the relationship between the control of orientation and the electron transfer activity of these pigment complexes on electrodes. Establishing a fundamental understanding of photon-electron transfer reactions in a single protein pigment complex on a nanoscale apparatus will be key for developing nanobiodevices. We have found that the photocurrent response of LH1-RC complexes in self-assembled monolayers (SAMs) on a gold electrode depends on the methylene chain length of alkanethiol SAM and on the chemical nature of its termination group [27]. The LH1-RCs are adsorbed onto these SAMs via electrostatic interactions mediated by the native structure of the proteins and the alkanethiol's termination group. For this reason, we focus on the assembly of photosynthetic protein pigment complexes, LH1-RC, on electrodes modified with or without a lipid membrane [29, 34]. This method will be useful to study the energy transfer and electron transfer reactions between individual pigments from the supra-molecular complexes oriented on the electrode; this study will lay the foundation for the future development of nanobiodevices that convert solar energy to electricity.

2 Results and Discussion

2.1 Photocurrent and Electronic Activities of C- or N-His LH1-RC Complexes Assembled onto a Gold Electrode

In our study, C- or N- terminal His-tagged LH1-RCs from *Rhodobacter* (*Rb.*) *sphaeroides* were engineered (Fig. 1a, b) and attached to Ni-NTA-assembled substrates to assess their orientation and electric contact with the substrates (Fig. 1c) [29].

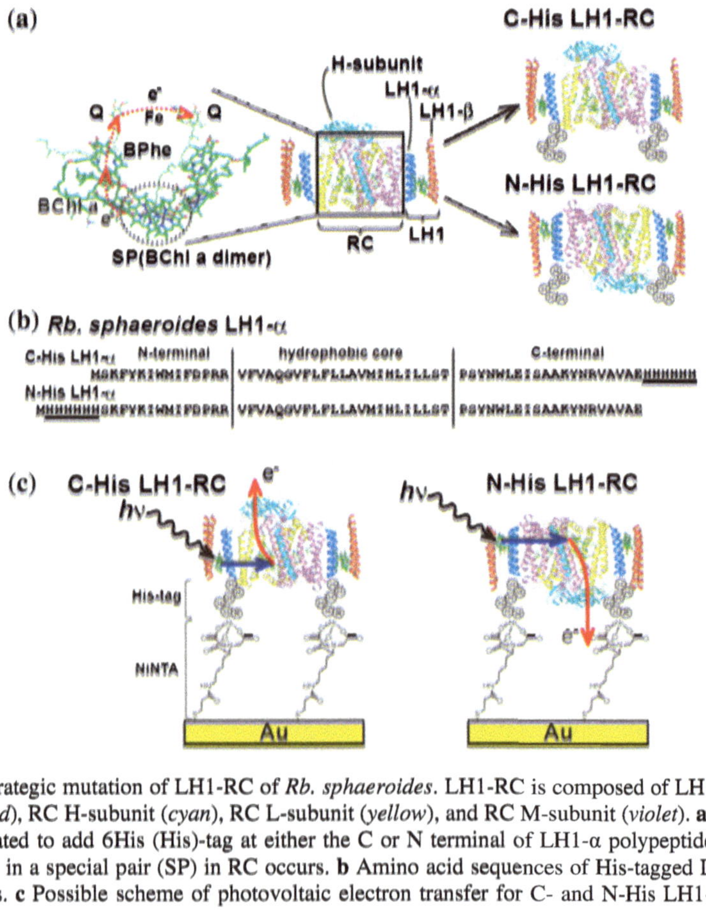

Fig. 1 Strategic mutation of LH1-RC of *Rb. sphaeroides*. LH1-RC is composed of LH1-α (*blue*), LH1-β (*red*), RC H-subunit (*cyan*), RC L-subunit (*yellow*), and RC M-subunit (*violet*). **a** LH1-RCs were mutated to add 6His (His)-tag at either the C or N terminal of LH1-α polypeptides. Charge separation in a special pair (SP) in RC occurs. **b** Amino acid sequences of His-tagged LH1-α, C-, and N-His. **c** Possible scheme of photovoltaic electron transfer for C- and N-His LH1-RC

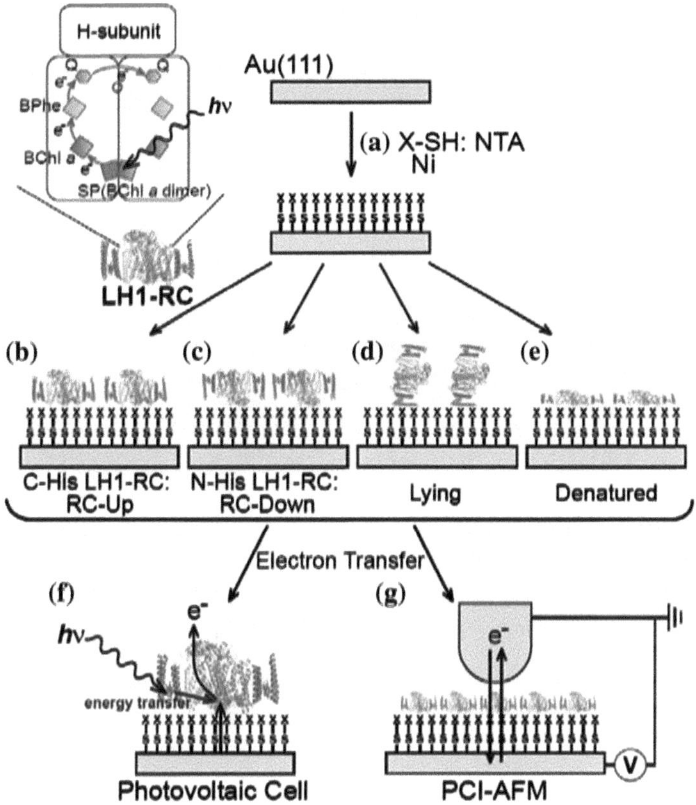

Fig. 2 Schematic strategy for the orientation of the His-tagged LH1-RC. **a** The gold substrates were modified by a SAM molecule (NTA) and Ni^{2+} ions complexes that were adsorbed to bind the His-tag side of the LH1-RC. **b** C-His LH1-RC is configured so that the special pair (SP) of RC is placed upwards on the substrate (**c**). N-His LH1-RC is configured so that the H subunit of RC is placed upwards on the substrate. **d** His-tagged LH1-RC is in a lying position **e** His-tagged LH1-RC is denatured. The obtained substrates will have **f** vectorial electron transfer of photocurrent in a photovoltaic cell and **g** point current imaging (PCI) AFM photocurrent

Point-contact current imaging (PCI) AFM was used to determine their topography and current/voltage characteristics (Fig. 2). This is the first approach to assemble the LH1-RC complex on electrodes with a defined orientation [27].

2.1.1 Photoinduced Electron Transfer from C- or N-His LH1-RC Complexes on SAM Modified Electrodes

Figure 3a shows the photocurrent response of the C- or N-His LH1-RC complexes on the gold electrodes modified with SAM (Ni-NTA and alkanethiol (1-decanethiol, $C_{10}SH$)) when these electrodes were illuminated with pulses of 880 nm light. The

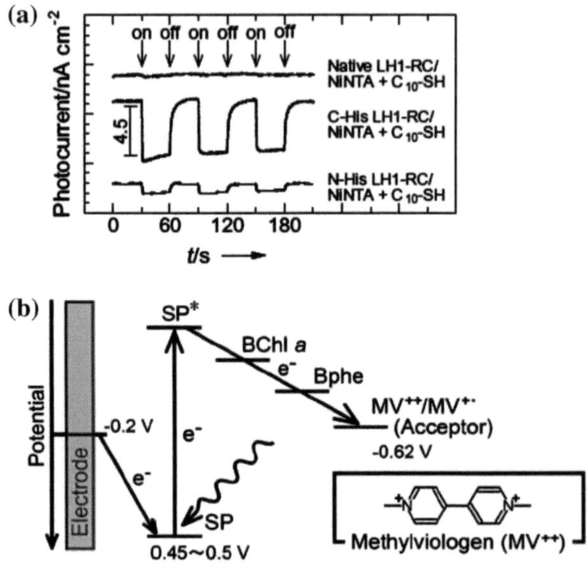

Fig. 3 **a** Photocurrent response of C- or N-His LH1-RC complex on a modified gold electrode when illuminated at 880 nm in phosphate buffer (pH 7) and methyl viologen (MV) at –0.2 V (vs. Ag/AgCl). **b** Electron state level of RC, methyl viologen, and equipment. The SP in RC is first excited by a photon. Charge separation occurs, and the electron is transferred from BChl *a* and Bphe to MV sequentially. The cationic SP captures an electron from the substrate to fill the ground state charge balance

dark current and noise significantly decreased when C_{10}-SH was present in the SAMs because C_{10}SH allows SAMs to be closely packed. This minimizes the direct electron transfer between the electrode and the solution. Interestingly, cathodic photocurrents were observed with both the C-His and N-His LH1-RCs. These results indicate that one-way electron transfer from pigments in the LH1-RC complex to methyl viologen (MV) occurred [27] (Fig. 3b). Furthermore, the photocurrent density, normalized by LH1-RC adsorbed, was greater for C-His LH1-RC than N-His LH1-RC. The magnitude of the photocurrent was, therefore, very sensitive to the orientation of LH1-RC complexes on the modified gold electrode.

The photocurrent response was wavelength-dependent and showed a maximum at the wavelength near the maximum absorption band of the complex in solution. For example, the action spectra of C-His and N-HisLH1-RC were similar to the absorption spectra in solution (data not shown).

Most likely the rate determining step in the electron transfer occurred between the electrode and the LH1-RC complex [27]. Since this electron transfer occurs by tunneling, the electron transfer rates should decrease exponentially as a function of the distance between the pigment in the RC and the surface of electrode [27]. Direct electron transfer from the electrode to methyl viologen can be ruled out simply because the tunneling distances that would be required to traverse the protein are too large (e.g., 7.3 nm across the LH1 or RC) [16]. Additionally, the applied voltage was too low to induce direct electron transfer to methyl viologen (−0.62 V vs. Ag/AgCl). The X-ray crystal structures of both the RCs and the LH1-RC complex have been evaluated, and this information can be used to propose a pathway for the multi-step electron transfer catalyzed by the RC from the electrode to methyl viologen (Fig. 3b).

2.1.2 Estimation of the Orientation of the C- or N-His LH1-RC Using PCI-AFM

Figure 4a shows an AFM topography image recorded under ambient (N_2) conditions of C-His LH1-RCs adsorbed onto the Ni-NTA substrate using an Au/Ir-coated Si probe. Current–voltage measurements with Au-coated probes were determined

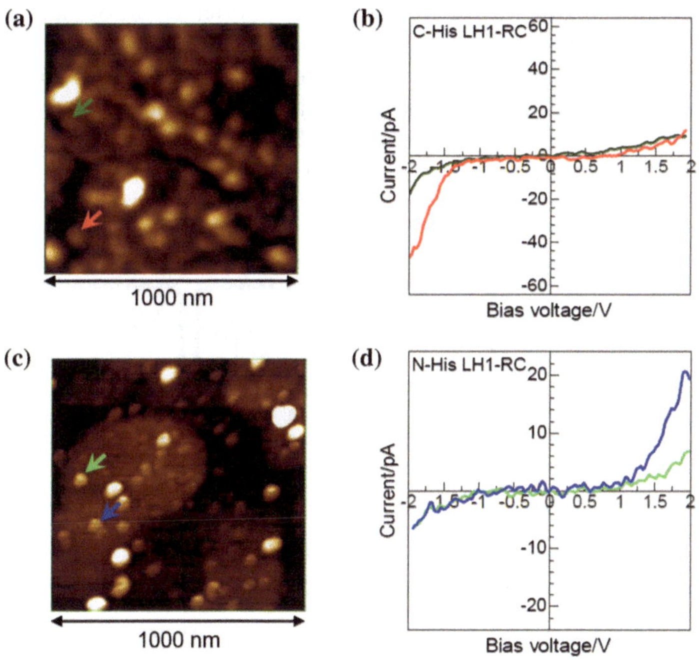

Fig. 4 **a** PCI-AFM topographic image of C-His LH1-RC. **b** *I–V* curves of C-His LH1-RC complex on Ni-NTA modified gold electrode. These *I–V* curves are measured on a single grain, which are marked by the same *colored arrows* in (**a**). *Red line* Clear rectification was obtained. *Green line* ohmic current response. (**c**) PCI-AFM topographic image of N-His LH1-RC. **d** *I–V* curves of N-His LH1-RC on Ni-NTA modified gold electrode. These *I–V* curves are also measured on a single grain, which are marked by the same *color arrows* in (**c**). The H chain of RC in LH1-RC is facing the substrate in the configuration in Fig. 2c. *Light green line* ohmic current response. *Blue line* Clear rectification was obtained. The *I–V* characteristics show that electrons propagate from the cantilever to the substrate, which ind icate that the C-His LH1-RC was oriented as expected, and electrons transferred based on th e energy states in Fig. 2g

on grains with diameters of 40 nm, i.e., single LH1-RCs (Fig. 4). The *I*–*V* curves shown were obtained from such single grains. Figure 4b shows the measured currents as a function of the applied voltage between the substrate and the tip. Interestingly, some of the adsorbed LH1-RCs displayed different *I*–*V* characteristics (e.g., red vs. green line) because the electron transfer pathway in the RC was directed toward the tip in the presence of a C-His tag terminal. Sixty five percent of the forty LH1-RCs adsorbed had the correct orientation, and the current showed a semiconducting *I*–*V* curve (the red line) that indicates rectification. Figure 4c, d show the AFM topography and *I*–*V* curves of N-His LH1-RCs adsorbe d onto a Ni-NTA substrate, respectively. The arrangement of N-His LH1-RCs observed was similar to that of the C-His LH1-RCs based on a comparison of the topography in Fig. 4a, c. The *I*–*V* curves in Fig. 4d show a reversed rectification direction relative to those of C-His LH1-RCs in Fig. 4b, indicating that approximately 62 % of the forty LH1-RCs were correctly oriented on the substrate, and 37 % were simply lying on the substrate as in Fig. 5. The data also show a semiconducting *I*–*V* curve (blue line). The relative orientations of the C- and N-His LH1-RCs based on the rectification curves are summarized in Table 1. These results show that His tags fused to the C or N termini of the LH1-RCs can control the orientation of the transmembrane protein complexes assembled on the substrate.

The *I*–*V* curves of LH1-RCs sandwiched between the two electrodes measured here clearly demonstrate that these proteins conduct electrons between the AFM tip and the electrode. Electrons may directly tunnel from one electrode (tip) to the

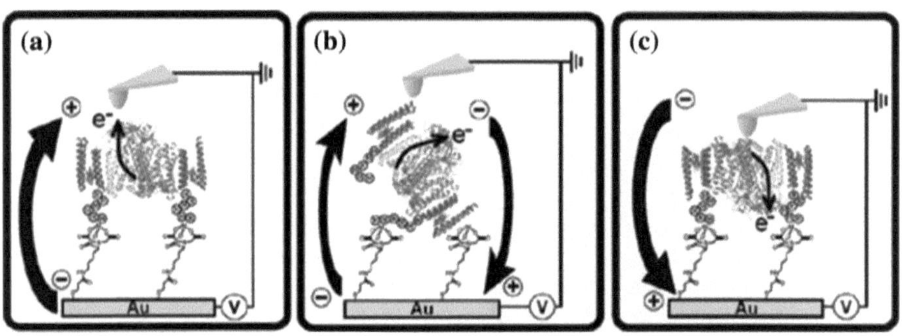

Fig. 5 Schematic images of electron transfer of C- or N-His LH1-RC s oriented on the Ni-NTA modified gold electrode. In nature, the electron transfer of RC is un idirectional due to the molecular wire, SP → BChl *a* → BPhe → Q, inside the RC molecule. **a** C-His LH1-RC. **b** Lying. **c** N-His LH1-RC

Table 1 Orientations of C- or N-His LH1-RC on a modified gold electrode. Forty LH1-RCs were measured

His-tagged position	Orientation of LH1-RC (%)		
	Down	Lying	Up
C-His LH-RC	65	35	0
N-His LH1-RC	0	37.5	62.5

other. For this case, this mode of charge propagation can be ruled out simply because the tunneling distances would be too large (e.g., 7.3 nm across the RC) [16]. In this study, the dependence of the direction of the rectification on the relative orientation of the LH1-RCs on the Au electrode SAM suggests that the electron from the gold coated probe transfers to the chemically modified Au substrate by the pathway shown in Figs. 1a and 5. Higher currents were only seen when positive voltages were applied to the Au substrate (or negative ones to the AFM tip) when N-His LH1-RCs were used.

In this study, electrons are injected into the special pair (SP) of the RC via the Au substrates or the AFM tip at negative or positive voltages, respectively, (Fig. 2g), and these electrons are then transferred along the pigments associated with the L-subunit of the RC from the SP to BChl a_L to bacteriopheophytin (Bphe o_L), and then finally to quinone, Q_L (Figs. 1a and 5). The reduced Q_L can then be directly oxidized by the substrate under the influence of the applied voltage. Modification of the electrode by Ni-NTA is useful for electron transfer reactions due to orientation of RC. The rate constants for each of these electron transfer steps in *Rb. sphaeroides* RCs have been determined. The rates of reverse electron transfer (Q_L to BChl a_L to $BChl_L$ to SP) are significantly smaller than the rates of forward transfer. This difference can then explain the observed asymmetry in the current output between the applied positive and negative voltages. The results described here complement those described for other devices using RCs with C60 [17] and cytochrome *c* [15].

2.1.3 Oriented Excited and Ground State Electron Transfer on the Electrode

The 3D structures of the LH1-RC and the electron transfer pathway, derived from spectroscopic experiments, are also shown in Fig. 1a, c. Direct excitation with light or indirect excitation by resonance energy transfer from an antenna complex first promotes an electron from the SP to an excited singlet state. Electrons are then transferred along the pigments associated with the L-chain of the RC, that is, from SP* to an accessory BChl *a* (0.47 nm transfer distance), to Bphe (0.38 nm transfer distance) and finally to Q (0.9 nm transfer distance). In this study, it is not clear where the linking with MV takes place. It could be via BChl *a* or Bphe since the redox potential of MV is much lower than that of Q in the RC. The oxidized SP^+ is then reduced by the electron transfer from the Au electrode. The distance from SP to the surface of RC is likely to 1.0 nm for C-His LH1-RC (Fig. 1c) [27]. Therefore, it is important for this model to immobilize the LH1-RC complex on various substrates with the defined orientation relative to the electrode, as shown in Fig. 1c.

2.2 *Electron Conduction and Photocurrent Generation of a LH-RC Complex on Electrodes Modified with a Lipid Membrane*

The electron conductivity and photocurrent of the LH1-RC complex embedded in a lipid membrane were measured using C-AFM and photoelectrochemical analyses [34]. AFM topography showed that LH1-RC molecules were well oriented with their H-subunits toward the membrane surface [6]. LH1-RC embedded in a membrane-generated photocurrent upon irradiation when assembled on an electrode. The observed action spectrum was consistent with the absorption spectrum of LH1-RC. The control of the orientation of LH1-RC by lipid membranes provided well-defined conductivity and photocurrent. This is the first approach to assemble the LH1-RC on electrode modified with lipid bilayers [34].

2.2.1 Electron Conduction of the LH1-RC Complex in Lipid Membrane Environments

Figure 6a shows an AFM image of reconstituted LH1-RC (lipid/LH1-RC = 35 mol/mol) in an aqueous condition. The height from the mica surface is ~7 nm, as shown in the height profile (Fig. 6ai) along the dashed line in Fig. 6. The height corresponds to that of LH1. The area of the plain bilayer (4 nm in height) is not shown. Because of the relatively high protein content, the LH1-RC complexes are densely packed with lipid molecules. The center-to-center distance between neighboring LH1-RC (whose long axis is 11 nm) was 12 nm ($n = 6$) [24], which indicates that lipid molecules exist between the complexes. Protrusions about 2 nm from the surface of the LH1 were clearly observed, and these are H-subunits of the RC (marked by the dotted yellow circle in Fig. 6a) [7, 8]. Many LH1-RC complexes without H-subunits were also observed (marked by the dotted white circle in Fig. 6a). We found that the number of H-subunits decreased after several imaging scans, which indicate that the H-subunits were pulled out by the AFM tip as described previously [9]. Considering this loss of H-subunits, LH1-RC complexes in the reconstituted membrane are predominantly oriented toward the H-subunit side (cytoplasmic side) as depicted in Fig. 6b. This is consistent with our past findings that showed a uniform orientation of LH1-RC complexes in a lipid membrane [30, 31]. Figure 6c shows an AFM image of reconstituted LH1-RC under ambient air conditions (similar to that of the C-AFM measurement). Patches of the LH1-RC-reconstituted membrane with a thickness of ~7 nm were observed (height profile ii). A high-magnification image revealed individual LH1-RC complexes in the membrane patch, in which LH1-RC complexes are densely packed (Fig. 6c, Inset). This height (7 nm) being lower than that of the intact LH1-RC complex (9 nm) is probably attributed to slight shrinkage of the hydrophilic domains of LH1-RC.

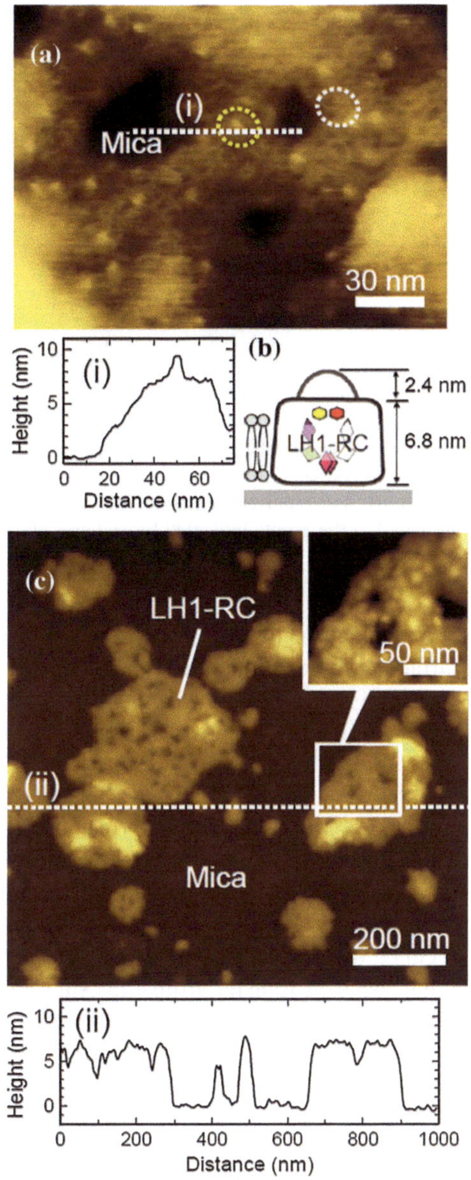

Fig. 6 **a** AFM images of LH1-RC-reconstituted membranes on mica under aqueous condition, **b** presumed orientation in the membrane environment, and **c** ambient air condition. Height profiles along the *white dotted lines* in (**a**) and (**c**) are shown in (*i*) and (*ii*), respectively. The *yellow* and *white dotted circles* in (**a**) show LH1-RC and LH1-RC without the H-subunit, respectively. H-subunits are dissociated during repeated scans. The *inset* image in (**c**) is a magnified image of the *white rectangle* area in (**c**)

A schematic illustration of the experimental configuration of C-AFM, which consists of an Au–mica substrate, LH1-RC embedded in the membrane, and a Pt-coated tip and cantilever, is shown in Fig. 7a. Prior to C-AFM measurement, an AFM topograph was acquired using the AC mode to find the appropriate target of an LH1-RC membrane patch on the Au–mica surface (Fig. 7b). The observed height of the membranes (7 nm) is consistent with that observed in Fig. 6c. This

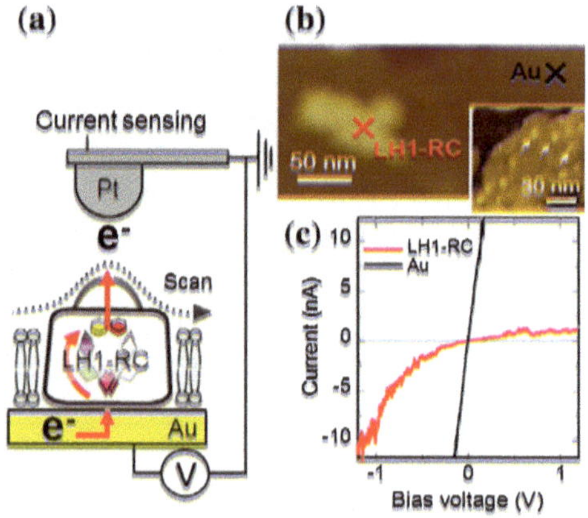

Fig. 7 **a** Schematic illustration of C-AFM measurement method, **b** topographic image of LH1-RC-reconstituted membrane on Au acquired under ambient air and aqueous (*inset*) conditions, and **c** current–voltage (*I*–*V*) curves of LH1-RC (*red*) and a bare Au surface (*black*) measured at the points indicated by *red* and *black crosses* in (**b**). A bias voltage is applied to the Au substrate and the current flowing through the LH1-RC is detected by the Pt-coated AFM tip

finding suggests that LH1-RC-reconstituted membranes are formed on the Au–mica surface in a manner similar to that for membranes formed on the mica surface. The inset of Fig. 7b shows an image of a LH1-RC membrane patch on the Au-mica where LH1-RC complexes orient the H-subunits upward (as indicated by white arrows).

When the tip electrode was placed on the Au surface, indicated by the black cross, the ohmic conductivity (black line in Fig. 7c) ranged from −1 to 1 V. No conductivity was observed when measurements were conducted in organic layers (composed of lipids or detergents remaining on the Au surface) (data not shown). On the LH1-RC membrane (marked by the red cross in Fig. 7b), an asymmetric *I*–*V* curve was observed (red line). The rectified conduction results from the preferential electron transfer from the Au substrate to the Pt tip electrode through the LH1-RC. A similar rectification property was observed on LH1-RC bearing the His-tag as mention above [29]. Electron transfer occurs along the alignment of pigments associated within the RC, that is, the transfer occurs from SP to accessory BChl *a* (0.47 nm transfer distance), to BPhe (0.38 nm transfer distance), and finally to quinone (Q_A) (0.9 nm transfer distance). In a membrane environment, it is reasonable to consider that electrons are transferred in the same way as that depicted in Fig. 7a. The clear rectification establishes the orientation of LH1-RC and vice versa. In contrast, for LH2, the *I*–*V* curve was symmetric at $V = 0$, which is consistent with the results of a study by Stamouli et al. [10]. They stated that carotenoids bound to LH2 are responsible for electron conduction. LH1 also possesses carotenoids, which may be one of conduction pathways; however, it can be negligible. Because a tip indentation on the H-subunit (2.4 nm of protrusion from LH1) is <0.5 nm, the tip electrode would be far enough from the position of carotenoids in LH1. The number of LH1-RC complexes in contact with the tip can be estimated to be not more than four on the basis of the tip radius (10 nm) and the distance between densely packed LH1-RC complexes ($\sim$11 nm).

We investigated the effect of the applied force on the rectification property of LH1-RC complex. The rectification property was evaluated as the fraction of the cathodic current in the total current. On the LH1-RC, a clear trend was found; a better rectification property of LH1-RC in the lipid membrane system is likely ascribed to the higher number of well-oriented LH1-RC molecules compared to those in the non-membrane system.

2.2.2 Photocurrent Generation of the LH1-RC Complex in Lipid Membrane Environments

To measure the photocurrent of LH1-RC in a lipid-bilayer environment, a LH1-RC-reconstituted planar membrane is formed on a positively charged ITO electrode modified with 3-aminopropyltriethoxysilane (APS). The LH1-RC-reconstituted membrane consisted of an anionic phospholipid (1,2-dioleoyl-*sn*-glycero-3-[phospho-*rac*-(1′-glycerol)]; DOPG) and was attached through electrostatic interaction [31, 32]. Figure 8a shows the absorption spectrum of the LH1-RC absorbed on the electrode (solid line), which is identical to that of a solution of an LH1-RC-reconstituted membrane (dotted line). The λ_{max} at 800 and 880 nm correspond to the accessory BChl *a* in RC and B880 of LH1 (Qy band of BChl *a*), respectively. Upon irradiation at 880 nm, the cathodic photocurrent was observed, and the response of the photocurrent to the presence and absence of irradiation using either an electron carrier ubiquinone-10 (UQ-10, red line) or ubiquinone-1 (UQ-1, black line) was analyzed (Fig. 8b). After illumination, it was confirmed that the absorption spectrum does not change; therefore, the LH1-RC complex is stably immobilized with the lipid membrane on the electrode. The action profile is observed in Fig. 8a is consistent with that of the absorption spectrum. This indicates that BChl*a* chromophores generate photocurrent from direct photoexcitation of SP and accessory BChl*a*, as well as the resonance energy transfer from LH1 to SP followed by electron transfer in RC. One of the rate-limiting steps is the electron transfer from the electrode to the SP in the RC [29]. Because the LH1-RC molecules orient the H-subunit in an upward configuration (as observed by AFM), the SP is located near the electrode ($\sim$1.2 nm) as depicted in Fig. 8c. Thus, facile electron transfer from the electrode to SP occurs. When the hydrophobic electron carrier UQ-10 was used, the resulting photocurrent was 2.3 times that measured for hydrophilic carrier UQ-1. This suggests that when compared with UQ-1, hydrophobic UQ-10 is more accessible to an RC embedded in a hydrophobic membrane for accepting electrons from the aligned pigments of the RC. Without the membrane, there was no photocurrent response in the presence of UQ-10, which implies that a fluidic membrane environment plays an important role in electron transport by the diffusive electron carrier UQ-10 as depicted in Fig. 8c. Overall, the system of light harvesting with the photovoltaic LH1-RC successfully functioned in a lipid membrane environment.

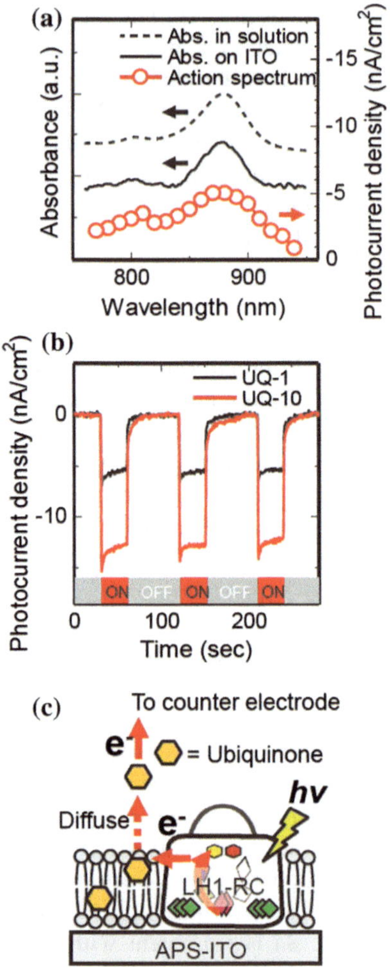

Fig. 8 Photocurrent generation by LH1-RC in a membrane environment. **a** NIR absorption and action spectrum of LH1-RC assembled on an APS-ITO electrode with a DOPG lipid membrane; the LH1-RC-reconstituted membrane was prepared with DOPG/LH1-RC = 500/1 (mol/mol). The electron mediator used was ubiquinone-10 (UQ-10). **b** The time course of photocurrent response in the presence and absence of irradiation at 880 nm. *Black* and *red lines* indicate responses with ubiquinone-1 (UQ-1) and UQ-10, respectively. **c** Schematic illustration of photocurrent generation by LH1-RC on an APS-ITO electrode

Thus, we have demonstrated electron conductivity and photocurrent generation of LH1-RC in a lipid-bilayer [34]. With respect to electron conductivity, we successfully detected the well-defined rectification of LH1-RC and noted that an appropriately controlled applied force is a critical parameter for the measurement of conductivity at the molecular level. In a membrane environment, hydrophobic UQ-10 plays an important role as an electron carrier, which is readily accept electron in the hydrophobic environment of a lipid bilayer. There are reports on the immobilization and photocurrent generation of LH1-RC and RC using various immobilization techniques [20]. Our reconstitution strategy is advantageous to bottom-up fabrication of supra-molecular structures composed of LH2 and LH1-RC [36]. A combination of C-AFM and photochemical techniques is expected to be effective for addressing the structure-function relationship of the supra-molecular assembly of LH2 and LH1-RC, which cooperatively function as light-harvesting and electron-transfer devices. The establishment of this experimental setup will provide useful information about this relationship as well as the excitonic and photovoltaic functions.

3 Summary

The His tag was successfully fused to the C- or N- terminus of the light-harvesting (LH1)-α chain of the photosynthetic antenna core complex, LH1-RC, from *Rhodobacter sphaeroides*. The His-tagged LH1-RCs were adsorbed onto an Au electrode modified with Ni-NTA. A relatively strong photovoltaic response of the engineered LH1-RC complexes on SAMs was observed upon illumination and depended on their orientation. Using PCI-AFM, we measured the electrical conduction properties of RCs sandwiched between an Au-coated probe and an Au substrate. Higher currents observed at the applied voltage suggested that approximately 65 % of the LH1-RCs have their His-tagged side facing the electrode. These results provide key information about how specific surface modification of the electrodes controls the performance of LH1-RC complexes in systems suitable for the development of solar energy converters and other types of energy harvesting biomaterials.

Further, the electron conductivity and photocurrent generation of the LH1-RC complex embedded in a lipid membrane was measured using C-AFM and photoelectrochemical analyses. AFM topographs showed that LH1-RC molecules were well oriented with their H-subunits toward the membrane surface. Rectified conductivity was observed in LH1-RC under precise control of the applied force on the probe electrode. LH1-RC embedded in a membrane generated photocurrent upon irradiation when assembled on an electrode. The observed action spectrum was consistent with the absorption spectrum of LH1-RC. The control of the orientation of LH1-RC by lipid membranes provided well-defined conductivity and photocurrent.

This method of using self-assembly of photosynthetic protein complexes to study electron transfer reactions of LH on electrodes is promising for the development of nanobiodevices from solar to fuel. Based on biological design principles, future biology-based or synthetic organic photonics could potentially provide clean and inexpensive energy alternatives [37].

Acknowledgements The authors thank to Dr. Ayumi Sumino and Dr. Kouji Iida for performing experiments and helpful discussions. M.N. and T.D. thank AOARD for funding.

References

1. Blankenship RE (2002) Molecular mechanisms of photosynthesis. Blackwell Science, Oxford
2. Deisenhofer J, Epp O, Miki K, Huber R, Michel H (1985) Structure of the protein subunits in the photosynthetic reaction centre of Rhodopseudomonas viridis at 3 Å resolution. Nature 318:618–624

3. Vos MH, Rappaport F (1993) Visualization of coherent nuclear motion in a membrane protein by femtosecond spectroscopy. Nature 363:320–325
4. Fleming GR, Martin JL, Breton J (1988) Rates of primary electron transfer in photosynthetic reaction centres and their mechanistic implications. Nature 333:190–192
5. McDermott G, Prince SM, Freer AA, Hawthornthwaite-Lawless AM, Papiz MZ, Cogdell RJ, Isaacs NW (1995) Crystal structure of an integral membrane light-harvesting complex from photosynthetic bacteria. Nature 377:517–521
6. Roszak AW, Howard TD, Southall J, Gardiner AT; Law CJ, Isaacs NW, Cogdell RJ (2003) Crystal structure of the RC-LH1 core complex from *Rhodopseudomonas palustris*. Science 302:1969–1972
7. Scheuring S, Gonçalves RP, Prima V, Sturgis JN (2006) The photosynthetic apparatus of Rhodopseudomonas palustris: structures and organization. J Mol Biol 358:83–96
8. Scheuring S, Sturgis JN (2005) Chromatic adaptation of photosynthetic membranes. Science 309:484–487
9. Bahatyrova S, Frese RN, Siebert CA, Olsen JD, van der Werf KO, van Grondelle RA, Niederman RA, Bullough PA, Otto C, Hunter CN (2004) The native architecture of a photosynthetic membrane. Nature 430:1058–1062
10. Stamouli A, Frenken JWM, Oosterkamp TH, Cogdell RJ, Aartsma TJ (2004) The electron conduction of photosynthetic protein complexes embedded in a membrane. FEBS Lett 560:109–114
11. Bumm LA (2008) Measuring molecular junctions: what is the standard? ACS Nano 2:403–407
12. Joachim C, Ratner MA (2005) Molecular electronics special feature—perspective, molecular electronics special feature: molecular electronics: some views on transport junctions and beyond. Proc Natl Acad Sci USA 102:8801–8808
13. Elbing M, Ochs R, Koentopp M, Fischer M, von Hänisch C, Weigend F, Evers F, Weber HB, Mayor M (2005) Physical sciences—chemistry—molecular electronics special feature, molecular electronics special feature: a single-molecule diode. Proc Natl Acad Sci USA 102:8815–8820
14. Das R, Kiley PJ, Segal M, Norville J, Yu AA, Wang L, Trammell SA, Reddick LE, Kumar R, Stellacci F, Lebedev N, Schnur J, Bruce BD, Zhang S, Baldo M (2004) Integration of photosynthetic protein molecular complexes in solid-state electronic devices. Nano Lett 4:1079–1083
15. Lebedev N, Trammell SA, Spano A, Lukashev E, Griva I, Schnur J (2006) Conductive wiring of immobilized photosynthetic reaction center to electrode by cytochrome c. J Am Chem Soc 128:12044–12045
16. Trammel SA, Griva I, Spano A, Tsoi S, Tender LM, Schnur S, Lebedev M (2004) Effects of distance and driving force on photoinduced electron transfer between photosynthetic reaction centers and gold electrodes. J Phys Chem C 111:17122–17130
17. Lebedev N, Trammel SA, Tsoi S, Spano A, Kim JH, Xu J, Twigg ME, Schnur JM (2008) Increasing efficiency of photoelectronic conversion by encapsulation of photosynthetic reaction center proteins in arrayed carbon nanotube electrode. Langmuir 24:8871–8876
18. Lee I, Lee JW, Greenbaum E (2008) Biomolecular electronics: vectorial arrays of photosynthetic reaction centers. Phys Rev Lett 79:3294–3297
19. Ron I, Pecht I, Sheves M, Cahen D (2010) Proteins as solid-state electronic conductors. Acc Chem Res 43:945–953
20. den Hollander M-J, Magis JG, Fuchsenberger P, Aartsma TJ, Jones MR, Frese RN (2011) Enhanced photocurrent generation by photosynthetic bacterial reaction centers through molecular relays, light-harvesting complexes, and direct protein-gold interactions. Langmuir 27:10282–10294
21. Ogawa M, Kanda R, Dewa T, Iida K, Nango M (2002) Molecular assembly of light-harvesting antenna complex on ITO electrode. Chem Lett 31:466–467
22. Nagata M, Nango M, Kashiwada A, Yamada S, Ito S, Sawa N, Ogawa M, Iida K, Kurono Y, Ohtsuka T (2003) Construction of photosynthetic antenna complex using light-harvesting

polypeptide-α from photosynthetic bacteria, *R. rubrum* with Zinc Substituted Bacteriochlorophyll *a*. Chem Lett 32:216–217
23. Ogawa M, Shinohara K, Nakamura Y, Suemori Y, Nagata M, Iida K, Gardiner AT, Cogdell RJ, Nango M (2004) Self-assembled monolayer of light-harvesting 1 and reaction center (LH1-RC) complexes isolated from rhodospirillum rubrum on an amino-terminated ITO electrode. Chem Lett 33:772–773
24. Iida K, Inagaki J, Shinohara K, Suemori Y, Ogawa M, Dewa T, Nango M (2005) Near-IR absorption and fluorescence spectra and AFM observation of the light-harvesting 1 complex on a mica substrate refolded from the subunit light-harvesting 1 complexes of photosynthetic bacteria *Rhodospirillum rubrum*. Langmuir 21:3069–3075
25. Dewa T, Yamada T, Ogawa M, Sugimoto M, Mizuno T, Yoshida K, Nakao Y, Kondo M, Iida K, Yamashita K, Tanaka T, Nango M (2005) Design and expression of cysteine-bearing hydrophobic polypeptides and their self-assembling properties with bacteriochlorophyll *a* derivatives as a mimic of bacterial photosynthetic antenna complexes. effect of steric confinement and orientation of the polypeptides on the pigment/polypeptide assembly process. Biochemistry 44:5129–5139
26. Dewa T, Sugiura R, Suemori Y, Sugimoto M, Takeuchi T, Hiro A, Iida K, Gardiner AT, Cogdell RJ, Nango M (2006) Lateral organization of a membrane protein in a supported binary lipid domain: direct observation of the organization of bacterial light-harvesting complex 2 by total internal reflection fluorescence microscopy. Langmuir 22:5412–5418
27. Kondo M, Nakamura Y, Fujii K, Nagata M, Suemori Y, Dewa T, Iida K, Gardiner AT, Cogdell RJ, Nango M (2007) Self-assembled monolayer of light-harvesting core complexes from photosynthetic bacteria on a gold electrode modified with alkanethiols. Biomacromolecules 8:2457–2463
28. Mikayama T, Iida K, Suemori Y, Dewa T, Miyashita T, Nango M, Gardiner AT (2008) The electronic behavior of a photosynthetic reaction center monitored by conductive atomic force microscopy. J Nanosci Nanotechnol 8:1–11
29. Kondo M, Iida K, Dewa T, Tanaka H, Ogawa T, Nagashima S, Nagashima KVP, Shimada K, Hashimoto H, Gardiner AT, Cogdell RJ, Nango M (2012) Photocurrent and electronic activities of oriented-His-tagged photosynthetic light-harvesting/reaction center core complexes assembled onto a gold electrode. Biomacromolecules 13:432–438
30. Sumino A, Dewa T, Takeuchi T, Sugiura R, Sasaki N, Misawa N, Tero R, Urisu T, Gardiner AT, Cogdell RJ, Hashimoto H, Nango M (2011) Construction and structural analysis of tethered lipid bilayer containing photosynthetic antenna proteins for functional analysis. Biomacromolecules 12:2850–2858
31. Sumino A, Dewa T, Kondo M, Morii T, Hashimoto H, Gardiner AT, Cogdell RJ, Nango M (2011) Selective assembly of photosynthetic antenna proteins into a domain-structured lipid bilayer for the construction of artificial photosynthetic antenna systems: structural analysis of the assembly using surface plasmon resonance and atomic force microscopy. Langmuir 27:1092–1099
32. Dewa T, Sugiura R, Suemori Y, Sugimoto M, Takeuchi T, Hiro A, Iida K, Gardiner AT, Cogdell RJ, Nango M (2006) Lateral organization of a membrane protein in a supported binary lipid domain: direct observation of the organization of bacterial light-harvesting complex 2 by total internal reflection fluorescence microscopy. Langmuir 22:5412–5418
33. Yajima S, Furukawa RA, Nagata M, Sakai S, Kond M, Iida K, Dewa T, Nango M (2012) Two-dimensional patterning of bacterial light-harvesting 2 complexes on lipid-modified gold surface. Appl Phys Lett 100:233701
34. Sumino A, Dewa T, Sasaki N, Kondo M, Nango M (2013) Electron conduction and photocurrent generation of light-harvesting/reaction center core complex in lipid membrane environments. J Phys Chem Lett 4:1087–1092

35. Sumino A, Dewa T, Noji T, Nakano Y, Watanabe N, Hildner R, Bösch N, Köhler J, Nango M (2013) Phoospholipids modulate self-assembled nanostructure and energy transfer of the light-harvesting complex 2 in lipid bilayers. J Phys Chem B 117:10395–10404
36. Dewa T, Sumino A, Watanabe N, Noji T, Nango M (2013) Energy transfer and clustering of photosynthetic light-harvesting complexes in reconstituted lipid membranes. Chem Phys 419:200–204
37. Yoneda Y, Noji T, Katayama T, Mizutani N, Komori D, Nango M, Miyasaka H, Itoh S, Nagasawa Y, Dewa T (2015) Extension of light-harvesting ability of photosynthetic light-harvesting complex 2 (LH2) through ultrafast energy transfer from covalently attached artificial chromophores. J Am Chem Soc 137:13121–13129

Solar Energy Storage Using Algae

Midori Kurahashi

Abstract Energy consumption in contemporary society continues to increase at a rapid pace, and global warming due to the resulting use of fossil fuels is growing increasingly severe. We need to search desperately for a practical source of renewable energy to replace fossil fuels. Most of the renewable energy sources available for use on earth are limited by factors such as energy from the sun or the inside of the earth, or gravitational attraction between the earth and the moon. However, tracing the origins of fossil fuels shows that they contain concentrated energy originally from the sun and inside the earth. Accordingly, production of fuel oil from microalgae is an attempt to artificially reproduce an instant version of this process. While microalgae fuel oil does emit carbon dioxide when used, it can be considered carbon neutral because the microalgae absorb carbon dioxide as they grow. Through photosynthesis of microalgae, solar energy is converted into and stored as chemical substances. While fats and oils extracted from them can be described as an energy source, they also can be used as a source of energy for human beings, fish, and shellfish—that is, as food. It is projected that in the future the world will face an increasingly severe food crisis due to causes including rapid population growth and climate change caused by global warming. The photosynthetic organisms of microalgae, not very well known until now, have the potential to make great contributions to solving both energy and food problems simultaneously.

1 Introduction

While it is said that the birth of the human race (*Homo sapiens*) took place 200000 years ago, the 18th-century industrial revolution that brought about the use of petroleum, coal, and electricity led to an explosion in the world population over

M. Kurahashi (✉)
Biotechnology Research Centre, The University of Tokyo, 1-1-1 Yayoi, Bunkyo-ku, Tokyo 113-0032, Japan
e-mail: utkura@mail.ecc.u-tokyo.ac.jp

M. Sugiyama et al. (eds.), *Solar to Chemical Energy Conversion*,
Lecture Notes in Energy 32, DOI 10.1007/978-3-319-25400-5_27

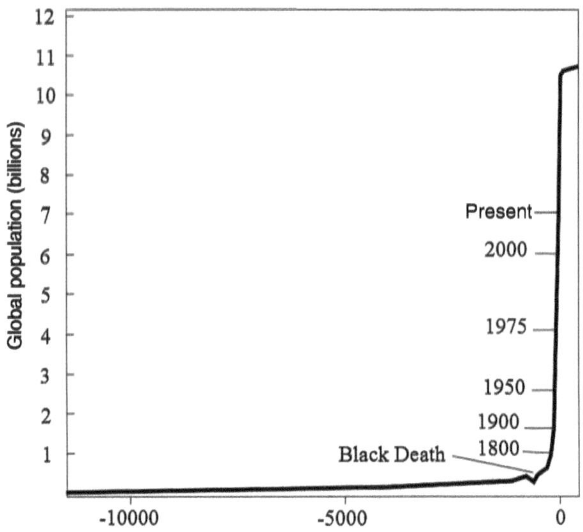

Fig. 1 Ten thousand years of human population growth redrawn from population reference bureau, 1984 [1]

just the past 300 years, as seen in Fig. 1. The issue of population, which continues to grow along a steep S-shaped curve, is closely related to issues related to food supply, drinking water, energy, climate, and the environment, all of which are projected to become major issues on a global scale in the future. The root cause of these issues, which could shake contemporary society at its foundations, must be said to be found in the cerebrum of *Homo sapiens,* which has undergone massive, if incomplete, development. We describe its development as incomplete because it is questionable whether it will be able to resolve these major issues related to the survival of contemporary society. Humanity continues to use nuclear power without understanding how ultimately to dispose of the waste it generates. It hopes that in genetic engineering it will not turn out to have opened a Pandora's box. It also has released massive volumes of chemicals and carbon dioxide into the environment. Unceasing growth in the concentration of carbon dioxide in the atmosphere is bringing about extreme weather. This increase in extreme weather is clear from a look at the way numbers of people dying from weather-related causes are increasing rapidly worldwide. But can we humans put the carbon dioxide that we have released into the atmosphere back into its Pandora's box?

2 Photosynthesis

Viewed in chemical terms, photosynthesis refers to the process by which plants such as those on land and phytoplankton produce glucose, a monosaccharide, from water and carbon dioxide. Glucose is a carbohydrate. As their name implies, carbohydrates are produced by combining the carbon from carbon dioxide with water.

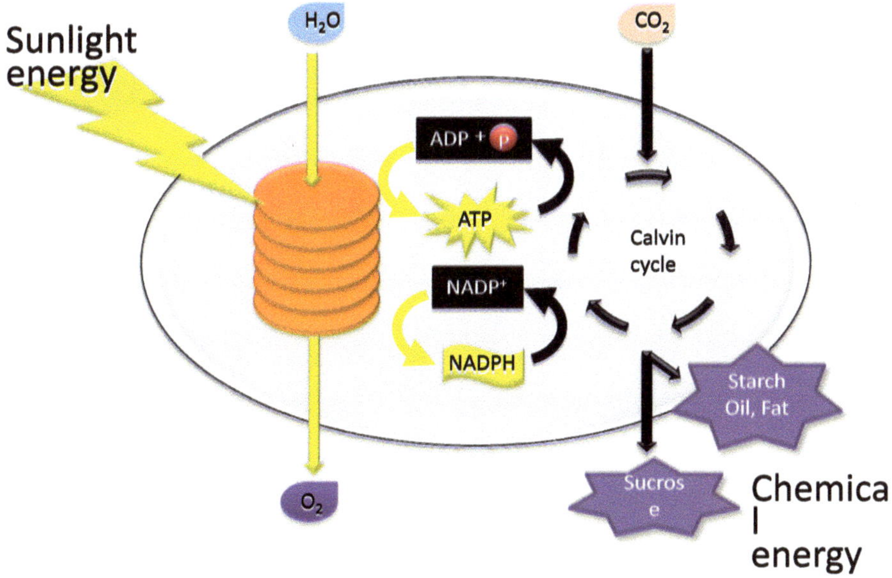

Fig. 2 Solar to chemical energy conversion in biochemical factory

At the same time, when viewed in terms of energy photosynthesis can be described as a biochemical reaction in which leaves convert energy from sunlight to chemical energy. These two points of view can be integrated to understand photosynthesis as the process by which light energy absorbed by leaves is converted to the kinetic energy of electrons within water molecules and transmitted between compounds through transfer of electrons (Fig. 2). In other words, the chemical reaction of oxidation reduction takes place, as a result storing the energy from sunlight in the chemical compounds of carbohydrates through the mechanism of photosynthesis. Photosynthesis reduces 1 mol of CO_2 and fixates 114 kcal of solar energy.

$$CO_2 + H_2O + \text{energy}(114\ \text{kcal}) \Leftrightarrow 1/6(C_6H_{12}O_6) + O_2$$

3 Pigment and Wavelength

Photosynthesis starts with the absorption of visible light with wavelengths of 400–700 nm by photosynthetic pigments. Excitation energy moves between antenna pigments and is collected by reaction center pigments having low levels of excitation energy. Pigments that demonstrate this antenna function are referred to as light-harvesting pigments, main examples of which are chlorophyll, carotenoid, and phycobilin. Microalgae evolved through a process by which pigmented prokaryotes caused intracellular symbiosis (primary endosymbiosis) in eukaryotes and then

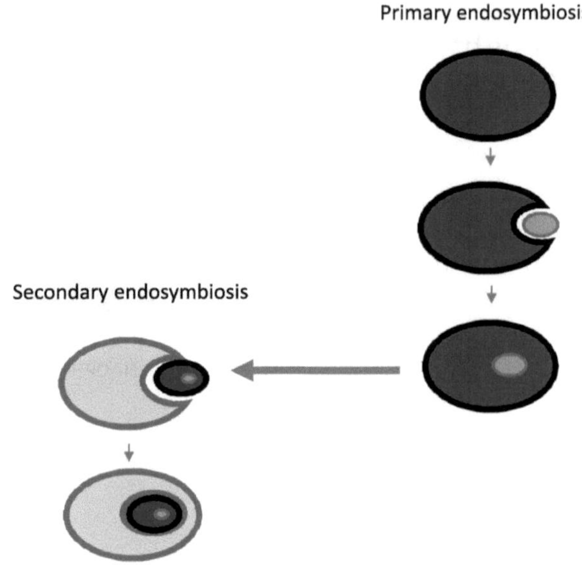

Fig. 3 The endosymbiosis theory

those eukaryotes further caused intracellular symbiosis (secondary endosymbiosis) in other eukaryotes (Fig. 3). Each time, a plastid was obtained in a parallel manner (Fig. 4). For this reason, since pigments vary by taxonomic group, the efficiency of

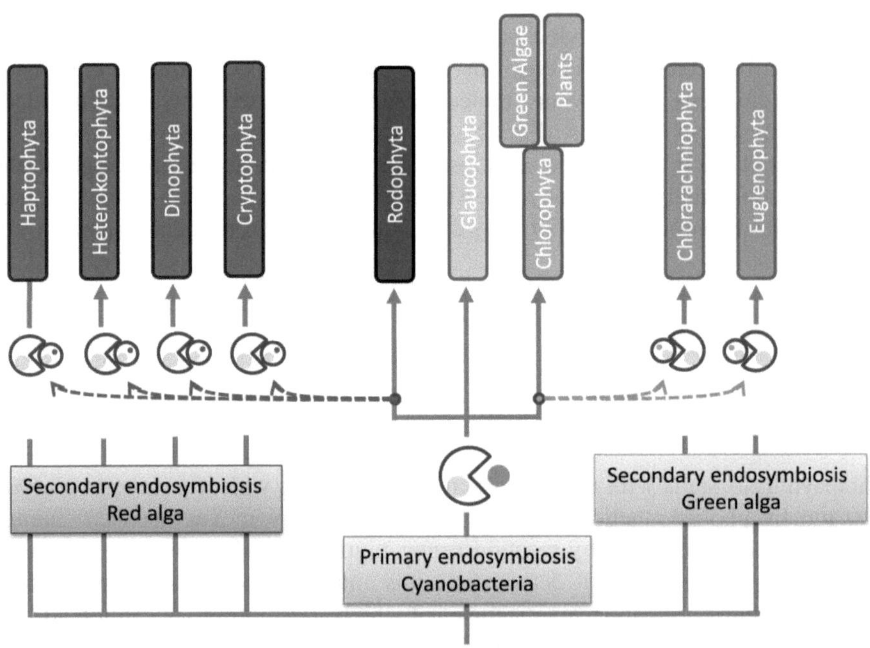

Fig. 4 View showing a frame format of multiple endosymbiosis events

use of light by wavelength varies by type of microalgae. Sensitivity to strength of light also varies by species of microalgae, leading to the possibility that when cultivating microalgae out of doors problems could arise with damage caused by strong light in the cases of many species.

4 About Microalgae

What kinds of creatures are referred to by the term microalgae in the first place? In general the terms phytoplankton or, even more familiar, "green water" make it easier to understand.

About 4.6 billion years ago the earth was formed in the gaseous primordial solar nebula, and about 3.8 billion years ago the first primitive life appeared. It is thought that bacilli capable of photosynthesis appeared 2.7 million years ago. As these bacilli became symbiotic with eukaryotes (primary endosymbiosis), they gradually came to incorporate within their cells the photosynthesis organelle called a chloroplast. Furthermore, in many cases a phenomenon occurred in which the photosynthetic eukaryotes that contained the photosynthesis organelles inside their cells became symbiotic with other eukaryotes (secondary endosymbiosis). Roughly, the creatures capable of photosynthesis that appeared as a result of these evolutionary events included land plants and large marine plants along with one other category: microalgae. For this reason, microalgae include numerous strains that differ considerably in evolutionary terms, and it is not the name of a single taxonomic group.

It should be noted here that these creatures have a surprising quality by which they transformed the earth's environment. About 2.2 billion years ago, there was a massive increase in microalgae cyanobacteria capable of the photosynthesis that generates oxygen. This dramatically changed the earth's environment by releasing large quantities of oxygen into the atmosphere. Of course, this event was calamitous to the creatures that had adapted to the earth's environment as it had been until then, and these were differentiated clearly into those that went extinct and those that adapted through evolution. On today's earth, which is dominated by the ecosystem of photosynthesis, nearly all living creatures depend on the land plants, large marine plants, and microalgae that produce oxygen through photosynthesis. It goes without saying that we human beings are no exception. While microalgae are a group of creatures one doesn't usually hear about, without a doubt they are an important group essential to life on earth.

Today, the creature *Homo sapiens* also truly is in the process of greatly transforming the earth's environment. However, since this change is so drastic the possibility cannot be denied that it could face extinction as a species before it can adapt to the changes it has brought about itself. To prevent this from happening, carbon-dioxide concentrations need to be returned to their previous levels quickly. To accomplish this difficult task the idea naturally occurs of choosing microalgae, which have proven their ability to dramatically transform the earth's environment in the past, as our counterparts on this job.

5 Microalgae Productivity

In general, the starting point of an ecosystem is absorption of solar energy, by the higher plants on land and by microalgae in the sea. Ordinarily, microalgae is visible only in forms such as red tide or blue-green algae, and it is difficult to ascertain its productivity using the senses. Here we will consider its productivity in comparison to land plants. While the biomass of land plants worldwide expressed in terms of carbon is 500–600 billion tons-C, the biomass of microalgae is only 1 billion tons-C. It is true that when viewed from an airplane the land appears to be covered by greenery while green pools can almost never be spotted in the sea. However, comparison of the biomass of animals at one step higher than photosynthesis creatures on the food chain unexpectedly shows that biomass figures largely are equal between land and sea, at 500–600 million tons-C on the former and 400–500 million tons-C in the latter. The key to understanding this mystery is the rate of turnover. That is, plants on land range in lifespan from annual herbs to trees that can live for 1000 of years, storing organic matter produced over these periods. On the other hand, a generation of microalgae lasts only about 1 week. Organisms soon are consumed or break down, so that little remains as biomass. Surprisingly, the annual production of organic matter by creatures using photosynthesis is estimated at 50–60 billion tons-C on land and 40–50 billion tons-C in the sea, and these estimates are not inconsistent with the figures on animal biomass seen above [2]. Every day, tiny microscopic-sized creatures in the sea use photosynthesis to carry out production at a scale greater than we would imagine.

6 The Choice of Bio-fuels

The main products of creatures that employ photosynthesis are carbohydrates (cellulose and polysaccharides such as starch), proteins, and lipids. Of these, carbohydrates and lipids can be used as bio-fuels.

Loosely speaking, lipid can be extracted and used as-is as a fuel, while carbohydrates (whether polysaccharides such as starch or cellulose) can be used as fuels only after first extracting them and then converting them to fuels such as ethanol or butanol through a process of fermentation by microorganisms. Ultimately, this difference in production processes can be considered to result in the differences in energy profit ratio (EPR) between alcohol fuels and diesel oil fuels (Fig. 5). However, undeniably even alcohol fuels can have value for use, with no need for concern regarding their EPRs, under favorable conditions. Examples include cases in which large volumes of carbohydrates already have been concentrated in forms such as those of byproducts or wastes.

Creatures that employ photosynthesis use it to produce monosaccharides such as glucoses and then form polysaccharides by combining glucoses. One of the main purposes of forming polysaccharides is to store them as a source of energy, while

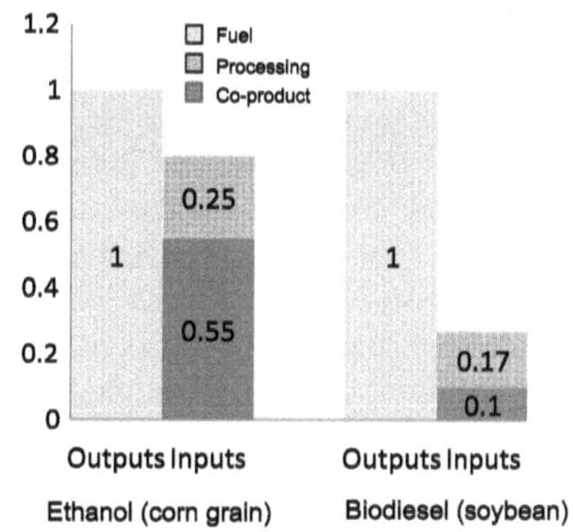

Fig. 5 Input and output energies for corn-ethanol and biodiesel. *Source* Ref. [3]

another is to form the structure of the plant itself. In the former case, α-glucoses are connected together to form starches, while in the latter case β-glucoses are used as materials to form cellulose. Land plants strengthen their structures by converting most carbohydrates to cellulose, enabling them to resist the force of gravity by broadening their leaves to bathe in sunlight. In contrast, since microalgae need to absorb sunlight while floating on the water, in general they have low percentages of cellulose and high percentages of lipids. Comparison of lipid production per unit of surface area shows that microalgae produces anywhere from several times to several tens of times the lipid that land plants produce [4]. Also, since land plants concentrate lipids in parts such as their seeds and fruit, it takes some effort to harvest the lipid, while industrial harvesting is simple in the case of microalgae thanks to their slurry forms. Furthermore, since microalgae can be produced using deserts and sea surfaces, they do not require the use of precious food-producing land.

For the above reasons, it can be considered useful to target lipids from microalgae in order to produce bio-fuels. Hereinafter, this chapter will look mainly at the portions of the bio-fuel field involved in production of fuel oil using microalgae.

7 Positioning of Fuel Production Through Microalgae

Solar power generation and production of fuel using microalgae are similar in that they both use a conversion mechanism to turn energy from sunlight either into electric energy directly or into chemical energy. However, if these two were applied to the same surface area then solar power generation would generate about 10 times the energy available from microalgae [5]. The reason fuel production using microalgae is being advanced even despite this disparity is outlined below.

While most renewable energy sources (examples of such sources include solar power, wind power, hydroelectric power, geothermal power, and power generation by temperature difference) are intended to produce electric power, a secondary energy, a precondition for the use of electric power is that the necessary infrastructure be in place. It also is difficult to use electricity in moving objects. In addition, fuel from microalgae, which makes it possible to obtain heat directly through burning the fuel, is highly valuable as a means of obtaining thermal energy, for which there is high demand as a means of use of energy in industry. Furthermore, since it is simple to store, such fuel makes it easy to achieve stable supplies and enables the immobilization of carbon dioxide during storage. One also could add that oil from microalgae offers a high level of value for non-energy uses as well, since it can be used as a raw material for chemical products (Fig. 6).

Fuel from microalgae also can be considered from the following approach. Today, when humanity is confronted with the need to reduce its carbon dioxide emissions, it is clear that a paradigm shift in energy needs to occur. However, the fossil fuels that humanity already has discovered constitute an easy-to-use source of energy with much higher energy densities than renewable energy. The hurdles faced can be said to be very high in trying to get humanity, once it has experienced such a taste of honey, to shift rapidly from high-density fossil fuels to low-density renewables. Depending on the method used, production of fuel from microalgae can be expected to have the effect of lowering these hurdles to some extent. It can do so by adopting the concept of solely using renewable energy to cover the energy needed to produce fuel from microalgae (Fig. 7). Fuel from microalgae can be seen as a transfer point for overcoming these high hurdles, since it is what could be called a fast fossil fuel.

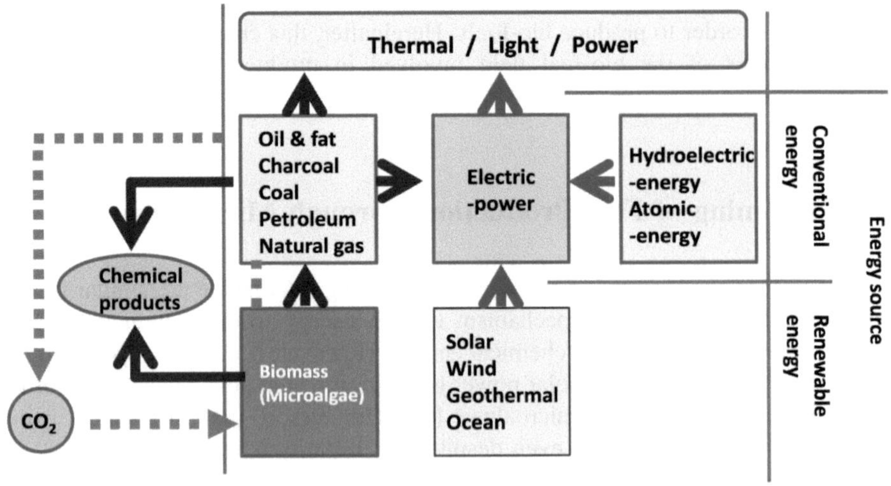

Fig. 6 Positioning of fuel production through microalgae

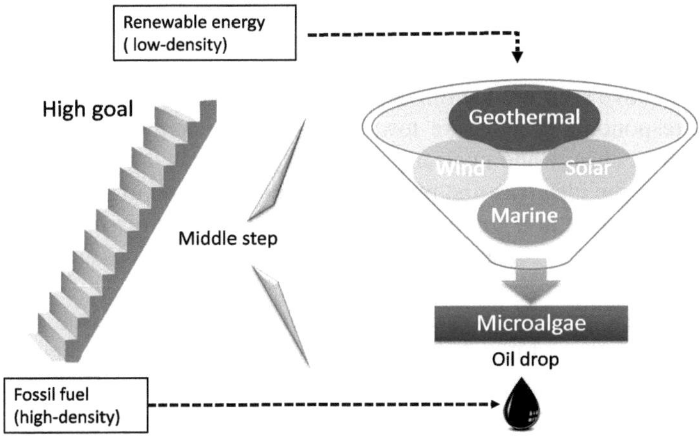

Fig. 7 The example to illustrate the concept of the usage of renewable energy

8 History of Fuel Production Through Microalgae

The history of industrial use of microalgae and development of related applications is very short. Its applications were first discussed seriously during World War II, when Germany looked at microalgae diatoms as a substitute for plant oil sources. At that time a very unique method was attempted in which diatoms were cultivated on a solid surface containing water, actually harvesting oil at a rate of 1 g/m^2/day in outdoor experiments. It has been estimated that deploying this method on agricultural land could result in production volumes at least 10 times those of soybeans or peanuts.

Analysis of the lipid content of a bloom of *Botryococcus braunii* that had appeared in Oak Mere, England in the 1960s initially showed unexpectedly that its lipid content, thought to consist of fatty acids and their derivatives as well as lipids, contained hydrocarbon content as high as several tenths. Furthermore, in 1976 a similar bloom appeared in a reservoir in northern Australia, where analysis by Melbourne aerospace and materials laboratories showed that dried bacterial cells consisted as much as 30 % of hydrocarbons. Australian researchers then began R&D efforts targeting this microalgae, suggesting that it could serve as a renewable energy source. Since this coincided with the oil crisis of the 1970s, the then-European Community (EC) saw future potential for these studies and research into producing oil from algae began in France and Belgium.

Later, in addition to *Botryococcus braunii* microalgae was discovered that produced hydrocarbons rather than lipids, but at present these are few in number [6]. While their industrial use would require cultivation in massive volumes, as yet no hydrocarbon-producing microalgae has been discovered that would combine together conditions favorable for such production at one time.

9 Choosing Types of Microalgae

Recent years have seen production of fuel from microalgae get a closer look as one means of responding to pressure toward renewable energy in response to issues such as climate change and nuclear power safety. While the goal differs somewhat, as noted above the idea was first proposed around 1970 and experiments have been conducted repeatedly in a number of countries. The largest reason for the inability, despite such efforts, to reach the stage of full-fledged practical application is production cost.

We are developing a conceptual design for resolving this issue of production cost and moving toward its practical application. In doing so, we first developed a list of the necessary conditions, in terms of types of microalgae, as needed to realize production of fuel from microalgae. While in general discussion on this issue has focused on comparison of lipid production capacity per unit of surface area exposed to light, we believe that there is no need to stick exclusively to such a line of discussion. Rather, such considerations should be given a lower priority because they are not fatal to the possibilities of such technology.

The first necessary condition is that the microalgae be of a type that can be targeted at lipids. The reason for this condition was described in the preceding paragraph. The second necessary condition is that the microalgae be a marine type. Since microalgae float on the water, cultivating them in large volumes requires a massive amount of water. Water such as that in rivers, lakes, and marshes accounts for only 0.01 % of the water on earth. What's more, it is said that desertification is advancing on a global level, so that it would not be practical to use freshwater microalgae to produce fuel. The third necessary condition is that it be possible to implement anticontamination measures that would not be too costly. The fourth is that the microalgae be of a species that demonstrates toughness against differences in temperature and against strong light.

10 Yields

Expectations regarding production of materials using microalgae have reached an unprecedented level for various reasons including their ability to use sunlight to produce chemical compounds, their ability to reduce carbon dioxide at the same time as production, and the fact that they can be produced using non-arable land. However, it is a fact that there are too many points that remain unclear to discuss the practicality of such production. Accordingly, we attempted to list figures in as practical terms as possible with regard to biomass volumes and lipid production volumes. The table lists (Table 1) the results of studying biomass and lipid production efficiency when actually cultivating microalgae [7].

There are various methods of cultivation, including use of flasks, tubes, panels, and outdoor raceway ponds. In addition, media used and cultivation conditions

Table 1 Lipid content and productivities of different microalgae species [7]

	Lipid content (% dry weight biomass)	Lipid productivity (mg/L/day)	Volumetric productivity of biomass (g/L/day)	A real productivity of biomass (g/m^2/day)
Ankistrodesmus sp.	24.0–31.0			11.5–17.4
Botryococcus braunii	25.0–75.0		0.02	3.00
Chaetoceros muelleri	33.6	21.8	0.07	
Chaetoceros calcitrans	14.4–16.4/39.8	17.6	0.04	
Chlorella emersonii	25.0–63.0	10.3–50.0	0.036–0.041	0.91–0.97
Chlorella protothecoides	14.6–57.8	1214	2.00–7.70	
Chlorella sorokiniana	19.0–22.0	44.7	0.23–1.47	
Chlorella vulgaris	5.0–58.0	11.2–40.0	0.02–0.20	0.57–0.95
Chlorella sp.	10.0–48.0	42.1	0.02–2.5	1.61–16.47/25
Chlorella pyrenoidosa	2.0		2.90–3.64	72.5/130
Chlorella	18.0–57.0	18.7		3.50–13.90
Chlorococcum sp.	19.3	53.7	0.28	
Crypthecodinium cohnii	20.0–51.1		10	
Dunaliella salina	6.0–25.0	116.0	0.22–0.34	1.6–3.5/20–38
Dunaliella primolecta	23.1		0.09	14
Dunaliella tertiolecta	16.7–71.0		0.12	
Dunaliella sp.	17.5–67.0	33.5		
Ellipsoidion sp.	27.4	47.3	0.17	
Euglena gracilis	14.0–20.0		7.70	
Haematococcus pluvialis	25.0		0.05–0.06	10.2–36.4
Isochrysis galbana	7.0–40.0		0.32–1.60	
Isochrysis sp.	7.1–33	37.8	0.08–0.17	
Monodus subterraneus	16.0	30.4	0.19	
Monallanthus salina	20.0–22.0		0.08	12
Nannochloris sp.	20.0–56.0	60.9–76.5	0.17–0.51	
Nannochloropsis oculata	22.7–29.7	84.0–142.0	0.37–0.48	
Nannochloropsis sp.	12.0–53.0	37.6–90.0	0.17–1.43	1.9–5.3

(continued)

Table 1 (continued)

	Lipid content (% dry weight biomass)	Lipid productivity (mg/L/day)	Volumetric productivity of biomass (g/L/day)	A real productivity of biomass ($g/m^2/day$)
Neochloris oleoabundans	29.0–65.0	90.0–134.0		
Nitzschia sp.	16.0–47.0			8.8–21.6
Oocystis pusilla	10.5			40.6–45.8
Pavlova salina	30.9	49.4	0.16	
Pavlova lutheri	35.5	40.2	0.14	
Phaeodactylum tricornutum	18.0–57.0	44.8	0.003–1.9	2.4–21
Porphyridium cruentum	9.0–18.8/60.7	34.8	0.36–1.50	25
Scenedesmus obliquus	11.0–55.0		0.004–0.74	
Scenedesmus quadricauda	1.9–18.4	35.1	0.19	
Scenedesmus sp.	19.6–21.1	40.8–53.9	0.03–0.26	2.43–13.52
Skeletonema sp.	13.3–31.8	27.3	0.09	
Skeletonema costatum	13.5–51.3	17.4	0.08	
Spirulina platensis	4.0–16.6		0.06–4.3	1.5–14.5/24–51
Spirulina maxima	4.0–9.0		0.21–0.25	25
Thalassiosira pseudonana	20.6	17.4	0.08	
Tetraselmis suecica	8.5–23.0	27.0–36.4	0.12–0.32	19
Tetraselmis sp.	12.6–14.7	43.4	0.30	

varied by experiment. Since for these reasons the figures shown cannot be used unconditionally, in order to ascertain in rough terms the lipid production capacity of microalgae we took averages after removing the extremely large outliers from this table. Results showed lipid content of approximately 27 %, lipid production efficiency of approximately 42 mg/L/day, biomass production efficiency by volume of approximately 0.4 g/L/day, and biomass production efficiency by surface area of approximately 16.5 $g/m^2/day$. The figures shown in this table were derived from the results of experiments, and one should expect even smaller figures in the case of practical commercial production. Moody argues that about one-third of the figures in most contemporary literature are realistic and proposes figures of 24 and 27 $m^3/ha/year$ (corresponding biomass yields of 13–15 $g/m^2/day$) for lipid production efficiency [8].

For now, an accepted target for purposes of practical commercial production of fuel oil from algae is biomass production efficiency of at least 20 $g/m^2/day$. However, in fact most algae species that show biomass production efficiency of at least 20 $g/m^2/day$ tend in fact to have low lipid content. In addition, while in general

Table 2 Comparison of oil feedstocks [7]

	Seed oil content (% oil by wt in biomass)	Oil yield (L oil/ha/year)
Corn/Maize (*Zea mays* L.)	44	172
Hemp (*Cannabis sativa* L.)	33	363
Soybean (*Glycine max* L.)	18	636
Jatropha (*Jatropha curcas* L.)	28	741
Camelina (*Camelina sativa* L.)	42	915
Canola/Rapeseed (*Brassica napus* L.)	41	974
Sunflower (*Helianthus annuus* L.)	40	1070
Castor (*Ricinus communis*)	48	1307
Palm oil (*Elaeis guineensis*)	36	5366
Microalgae (low oil content)	30	58,700
Microalgae (medium oil content)	50	97,800
Microalgae (high oil content)	70	1,36,900

lipid content has been shown to increase when cultivating algae in media from which nitrogen has been removed, biomass production efficiency also decreases. For these reasons, there is a need to focus on lipid production efficiency when choosing species of algae from the point of view of production efficiency.

Now let's compare the above figures on production efficiency shown by microalgae with oils from land plants, generally known as vegetable oils (Table 2). Microalgae's oil production efficiency per unit of surface area is likely to be three to five times that of palm oil, which has the highest production efficiency of any vegetable oil. While comparison of this figure with the figures often reported in the literature and elsewhere may lead to a sudden dashing of hopes, if we think of microalgae fuel oil production as a form of agriculture then these definitely are not bad figures. To the contrary, the feasibility of such production becomes clearer thanks to the realistic feel of these figures.

11 Carbon Dioxide

In 2014, the United Nations Intergovernmental Panel on Climate Change (IPCC) issued its first Synthesis Report in 7 years. The report warned that a failure to take stronger action to reduce greenhouse-gas emissions would lead to an even more severe situation due to warming, with a very high risk that the situation would become irreversible. Furthermore, in October 2014 the White House in the U.S. issued a report that warned that a delay of 10 years in taking bold action to counter global warming would increase the costs of such measures by 40 % and might men it was too late to take action.

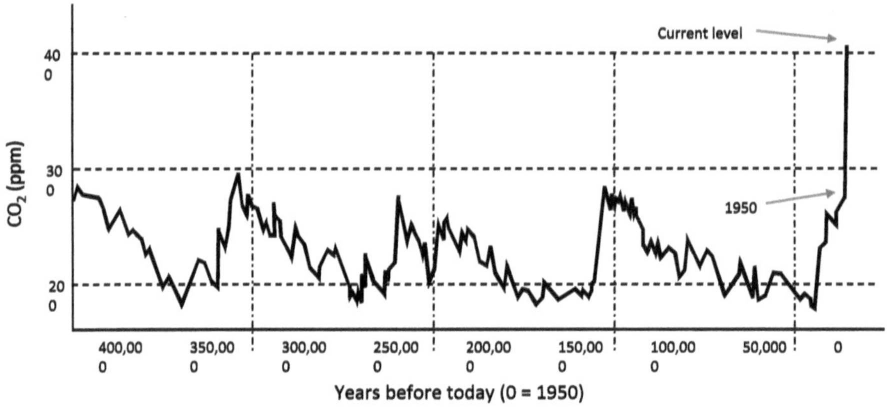

Fig. 8 CO_2 levels over the last 400000 years. *Source* NDAA (National Oceanic and Atmospheric Administration)

The Illustration (Fig. 8) shows changes in carbon-dioxide concentrations over the past 400000 years from analysis of Antarctic ice core samples. They show that carbon-dioxide concentrations have repeatedly risen and fallen within the range 180–300 ppm at a cycle of roughly 100000 years. However, in April 2014 the concentration of carbon dioxide in the atmosphere exceeded 400 ppm at all observation sites in the northern hemisphere. While around the end of the last ice age it took at least 5000 years for the concentration to rise by 100 ppm, it has risen by much more than 100 ppm over the most recent 100 years. We should be concerned not only by the fact that carbon-dioxide concentrations continue to set new records for the past 800000-year period but also by the too-rapid rate of increase and the fact that the earth is deviating considerably from the past pattern of rising and falling concentrations. The IPCC makes projections for the future based on six possible scenarios. According to these, if things are left largely as they are now, then carbon-dioxide concentrations would reach 1000 ppm around 2100. It is projected that such rapid changes in carbon-dioxide concentrations would bring about dramatic changes in climate and the environment, and in fact we have the opportunity to see some of these projected changes before our very eyes. It is unclear to what extent our human race can adapt to such drastic environmental changes.

While most sources of renewable energy such as hydroelectric power, solar power, wind power, and geothermal power can contribute to achieving carbon neutrality, they cannot actively reduce the volumes of carbon dioxide already released into the atmosphere today. At the same time, already carbon capture and storage (CCS) technology is seeing practical use in capturing high-density carbon dioxide (7–50 %) from thermoelectric power plants and steel mills and storing it underground or under the sea. However, since CCS technology consumes large volumes of energy during carbon capture and has very low levels of efficiency for capturing low-density carbon dioxide such as that in the atmosphere, and it involves other elements of uncertainty such as the possibility of leakage of CO_2, at present it is limited to temporary use.

As seen above, if we assume production efficiency by surface area of approximately 15 g/m^2/day then artificial mass cultivation of microalgae on non-arable land would have the capacity to fixate approximately 25 g/m^2/day of carbon dioxide. Incidentally, a forest in a temperate zone is said to fixate approximately 5 g/m^2/day.

While a massive increase in microalgae cyanobacteria 2.2 billion years ago dramatically changed the composition of the atmosphere, there is another type of microalgae that also left a major mark on the history of the earth. This is the coccolithophore of the Cretaceous period. A coccolithophore is a phytoplankton belonging to the phylum Haptophyta with a disc-shaped crystal of calcium carbonate on its cell surface. First appearing during the Jurassic period, it was the most thriving microalgae during the Cretaceous period (Fig. 9). At the same time it uses photosynthesis to fixate carbon dioxide as organic carbon, it also fixates inorganic carbon as the calcium carbonate crystal. Numerous species such as *Emiliania huxleyi* are known to cause autonomous blooms on the earth today, and they are thought to play a role in the earth's carbon cycle. While the organic carbon they synthesize breaks down and enters the carbon cycle, their inorganic carbon sometimes accumulates as

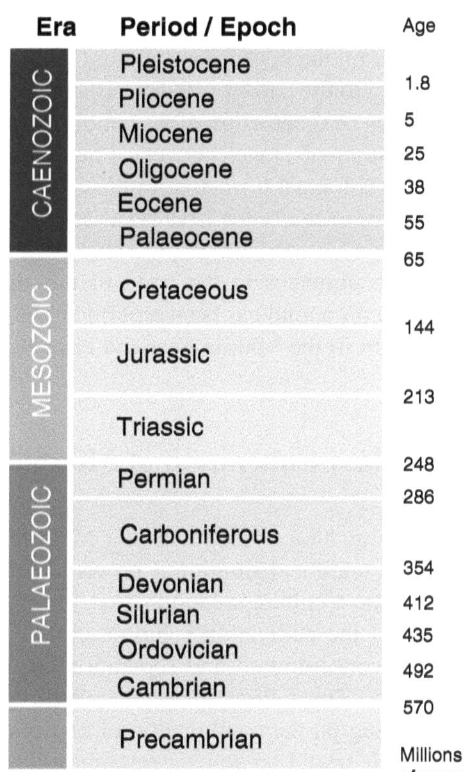

Fig. 9 The standard geologic column

Fig. 10 A view of chalk at the white cliffs of dover

sediment on the ocean floor without dissolving. In fact, the name Cretaceous period (referring to the period 140–65 million years ago) comes from the layer of limestone caused by sedimentation of coccolithophore. The white limestone cliffs on both sides of the Strait of Dover between France and England, called chalk (Fig. 10), are a typical geological formation of the Cretaceous period. It is known that some species belonging to the order Isochrysidales in this group of coccolithophores synthesize long-chain unsaturated alkyl ketone with a carbon number of 37–39, and that 20–30 % accumulates in the body of the algae. Another interesting fact is that this organic compound has been reported to be the source of the underground deposits of petroleum in the Middle East and elsewhere [9].

12 Mass Cultivation of Microalgae

Mass cultivation of microalgae is conducted mainly through the following process. First, an undiluted solution is prepared for cultivation, and fertilizer is injected into it to create a culture solution. Next, cultivation begins with the mixture in a culture vessel of the cultivation solution and the source microalgae. After the cultivation period, concentration and separation are conducted in preparation for subsequent processes. This cultivation process involves a number of choices to make (Fig. 11). Depending on the combination of choices, cost and production efficiency can vary widely. It would be no exaggeration to describe the mass cultivation of microalgae

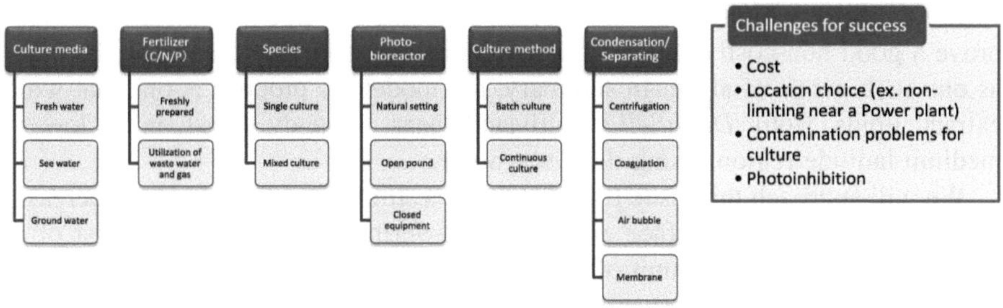

Fig. 11 Main options for microalgae mass cultivation process

as an issue of system structuring. The first hurdle that must be cleared when structuring the system is, of course, that of cost. It must be possible to achieve a price that will motivate development in comparison to current oil prices. Also, since the most important reason for beginning production of fuel oil form microalgae is to decrease carbon dioxide in the atmosphere, a quantitative perspective must not be overlooked. In addition, contamination that would impede cultivation must be avoided. Another issue on which most microalgae species require care is that of damage from strong light. Of course, it goes without saying that in any case the best system would be one that achieves high performance at low cost.

The question of whether it will be possible to shift to a low-carbon society for the earth's future is the same as that of whether we can put low-density energy to full use. Fundamentally, microalgae are creatures of low energy and low material densities, so that their photosynthesis activity is impeded in a high-energy material environment. For this reason, the correct choice for the human race is to overcome this difficult challenge together with microalgae, living creatures that have put low-density energy to its fullest use.

13 A Conceptual Design to Benefit Both the Economy and Carbon Balance, and Basic Strategic Model

As mentioned above, renewable energy offers surprisingly low levels of energy density. Production of lipids from creatures that employ photosynthesis is no exception to this tendency. Accordingly, in order to use biomass to supply a considerable amount of energy a broad surface exposed to sunlight is needed for cultivation. Deserts and sea surfaces can satisfy this requirement. In consideration of initial investment costs, the choice of the desert would have an overwhelming advantage. Next, we would propose the use of deep sea water to supply the large volumes of water and fertilizer needed. Such water can be considered a form of seawater that includes natural fertilizers. As will be described in greater detail below, deep sea water also can be used in power generation by temperature difference and in desalination processes that also use temperature difference [10]. As such, it has a very high utility value. Also, choosing extreme halophilic algae might prove a good non-costly anticontamination measure. We are looking at *Dunaliella* as one such genus of algae. In summary, the model we propose is one that would extract lipids from *Dunaliella* cultivated near a sandy coast in a low- to medium-latitude region, using deep sea water.

We will approach this issue from two angles simultaneously: those of increasing production per unit of surface area and of lowering production cost. First of all, increasing production per unit of surface area would be pointless if achieving this goal were to require the swelling of energy and cost inputs. Accordingly, we propose the following three improvements. The first of these involves harvesting lipids twice through exposure to sunlight. Although *Dunaliella* has no cell wall, it can grow in high-salinity water. It is known to regulate osmotic pressure through producing glycerin and regulating the density of glycerin within the cell. Accordingly, after first extracting the lipids and glycerin simultaneously from *Dunaliella* lipids can be harvested a second time by then fermenting the glycerin in yeast or some other culture. The second improvement involves increasing production volumes through use of deep sea water instead of surface sea water as the culture fluid. Characteristics of deep sea water include its low temperature, high purity, and high content of nutritive salts and minerals [11]. For example, compared to the surrounding surface sea water the deep sea water near Japan's Izu Oshima Island contains about 30 times the nitrate nitrogen content and about 60 times the content of phosphorus in the form of phosphates. Our experiment showed that use of deep sea water more than doubled the production of *Dunaliella.* The third improvement involves further increasing production by returning the dregs from fermentation to the culture fluid, as fertilizer.

The other angle is that of cutting costs. Here, we have devised two types of savings. We are considering the costs of fuel production using microalgae split broadly into the culture phase and downstream phases. Of these, the cultivation phase would not be controlled artificially, in order to lower costs related to cultivation. Only natural light would be used as a source of light energy, and mixing would be left to natural wind. Since neither temperature nor light would be controlled, the microalgae chosen needs to be a tough variety such as *Dunaliella.* Next, how should we reduce major downstream costs? Actually, most downstream costs are energy costs. Accordingly, we propose the concept of using renewable energy from local sources for all such energy. It is expected that there is considerable potential for use of renewable energy sources available in the area near a sandy coast in a low- to medium-latitude region, such as solar power, solar thermoelectric power generation, and wind power. Of these, my first choice of deep sea water can be used not only for power generation by temperature difference but for desalination processes that use temperature difference as well, making it possible to obtain large volumes of freshwater, a precious resource in a desert. Furthermore, the high purity characteristics is valuable aid in contamination control measures. If it were to become possible to produce energy and water in a desert considered barren until now, then it would be more than just a dream to imagine further stimulation of the agriculture and fisheries industries as well.

14 Uses of Chemical Compounds from Algae

Since the group of microalgae is not a monophyletic group there is no scientific meaning to direct comparison with the higher plants on land. But still, the Illustration (Fig. 4) can provide a feel for the diversity of microalgae. Photosynthetic creatures mainly use carbon dioxide as a source of carbon and water as an electron donor to synthesize a variety of compounds, starting with glucose. While some of their key products are polysaccharides, lipids, and proteins, their rates of production and types of products vary widely, reflecting their diversity. It would be no exaggeration to describe the subjects for future research in this area as a vast frontier.

15 Energy

Since most of the lipids accumulated by microalgae are in the form of triglycerides (TAGs), in discussing use of microalgae for energy above we kept TAGs in mind. While various methods, including transesterification, pyrohydrolysis, catalytic cracking, and hydrocracking, have been proposed for use of TAG as a fuel oil (at present, transesterification (Fig. 12) is the main method in actual use), each of these requires chemical treatment. On the other hand, algae such as *Botryococcus* are known to produce hydrocarbons that can be used as-is as fuel [12]. In addition, some algae produce large quantities of polysaccharides such as starch. For example, *Chlorella vulgaris* has been reported to accumulate starch equivalent to 37 % of its dry weight. This makes it possible to produce ethanol through a process of fermentation with yeast, just as is done using corn [13]. As various methods have been tested in recent years in efforts toward practical use of renewable energy, hydrogen has attracted attention as a secondary energy source. The microalgae cyanobacteria *Anabaena variabilis* has been reported to show hydrogen production efficiency of

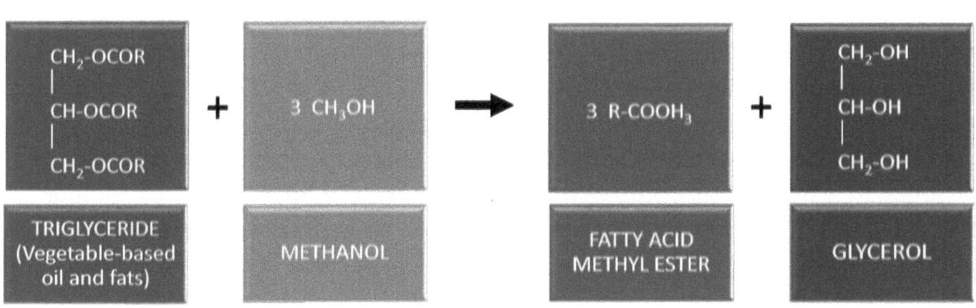

Fig. 12 Esterification. Fatty acid methyl esters and glycerin are produced through the methyl esterification of methanol and triglyceride, such as palm oil

up to 167.6 μmol/mg Chl-a/h at the same time it fixates nitrogen [14]. This nitrogen fixation function has the ability to capture free nitrogen in the air and convert it to nitrogen compounds (such as ammonia, nitrates, and nitrogen dioxide. Since this microalgae uses nitrogen in the air as a source of nutrition, it does not require fertilizer as a source of nitrogen during cultivation. Also, Hemschemeier has reported that achieving a hydrogen production efficiency of 500 μmol/mg Chl-a/h for the green algae *Chalmydomonas reinharadtii* under anaerobic conditions [15].

16 Food

While herbivores such as cattle use microorganisms living symbiotically inside their digestive tract to break down cellulose, human beings are unable to digest cellulose. For this reason, nutritionally speaking it would not be strange to think of algae, which is rich in lipids and proteins and low in cellulose, as a prospective source of human food. In fact, large seaweed has an established status as an ordinary food in East Asia and Southeast Asia. In 2008 seaweed cultivation worldwide produced an annual harvest of 15.8 million tons. In addition, in recent years seaweed has come to be consumed in Western countries as well, out of increasing health consciousness. However, unlike seaweed microalgae is something the existence of which, as a microorganism, most people are unaware and most of humanity has had no past experience consuming it as food.

However, around the world there are, in fact, peoples who have consumed these special microorganisms as food. The African nation of Chad is home to a large salt lake called Lake Chad that has produced large volumes of the microalgae *Spirulina* since ancient times. The Kanembu people living on the shores of the lake harvest blooms of this microalgae, which is visible to the naked eye, dry it in the sunlight on the sand, and use it as a dried powdered food called Dihé. In addition, from the 14th through the 16th centuries there was a shallow salt lake with a narrow strait in its center on the Mexican Plateau, spanning 300 km east to west and 100 km north to south. North of the strait it was called Lake Texcoco, while south of the strait it was called Lake Xochimilco. The Aztecs, who flourished in the region at that time, ate the *Spirulina* that bloomed on the salt lake as a source of dietary protein, which they called techui. This is described in records kept by the Spanish military when Mexico was a Spanish colony. There is a tradition of consuming microalgae as a food in Japan as well. The microalgae called suizenjinori belonging to the genus *Aphanothece* is used in traditional Japanese cuisine. Other examples include the microalgae ashitsuki belonging to the genus *Nostoc* and the microalgae ishikurage, another species in the same genus. In China, a species belonging to the genus *Nostoc* long has been eaten as a good-luck charm on auspicious occasions.

17 Feed

Already today more than 40 species of microalgae are cultivated for use in aquaculture, as a nutritional supplement for zooplankton in cultivation of young shrimp, young fish, or bivalves. Determining whether a microalgae is suitable for aquaculture requires studying it to see how large its cells are, how easy it is to digest, and whether it is capable of mass cultivation. In doing so, assessment of its nutritional value is important, and Brown et al. have conducted nutritional analysis of 16 species of microalgae (Table 3).

Microalgae are living creatures that convert solar energy into biomass, supporting oceanic life at the bottom of the food chain. There is a high likelihood that demand for microalgae will grow in the future for use as feed for aquaculture and

Table 3 Concentrations of chlorophyl α, protein, carbohydrate and lipid in 16 species of micro-algae commonly used in aquaculture

	Dry weight (pg/cell)	Chlorophyll α	Protein	Carbohydrate	Lipid
		Percentage of dry weight			
Bacillariophyceae					
Chaetoceros calcitrans	11.3	3.01	34	6.0	16
Chaetoceros gracilis	74.8	1.04	12	4.7	7.2
Nitzschia closterium	–	–	26	9.8	13
Phaeodactylum tricornutum	76.7	0.53	30	8.4	14
Skeletonema costatum	52.2	1.21	25	4.6	10
Thalassiosira pseudonana	28.4	0.95	34	8.8	19
Chlorophyceae					
Dunaliella tertiolecta	99.9	1.73	20	12.2	15
Nannochloris atomus	21.4	0.37	30	23.0	21
Cryptophyceae					
Chroomonas salina	122.5	0.80	29	9.1	12
Eustigmatophyceae					
Nannochloropsis oculata	6.1	0.89	35	7.8	18
Parsinophyceae					
Tetraselmis chui	269.0	1.42	31	12.1	17
Tetraselmis suecica	168.2	0.97	31	12.0	10
Prymnesiophyceae					
Isochrysis galbana	30.5	0.98	29	12.9	23
Isochrysis aff. *galbana* (T-iso)	29.7	0.98	23	6.0	20
Pavlova lutheri	102.3	0.84	29	9.0	12
Pavlova salina	93.1	0.98	26	7.4	12

Modified from Refs. [16, 17]

fish farming. Fisheries industries worldwide are seeing catches at the sea surface decrease due to the effects of factors such as climate change and environmental degradation in the ocean. At the same time, demand for seafood is increasing rapidly due to factors such as global population increases, advances in transportation networks and freezing technology, and rising health awareness. As a result, the fisheries industries are shifting from ones based on catching wild fish to ones aiming to raise or breed seafood. It is likely that there will be moves to decrease the volume of fish caught from the sea in the future in order to protect the environment and resources, and in fact natural fisheries resources cannot be relied on even today. For this reason, attention needs to be paid to the small fish that will need to be caught from the sea repeatedly in the process of producing a single bluefin tuna through the aquaculture and fish farming of the future. What we as residents of the earth probably should aim for is aquaculture and fish farming that start with the cultivation of microalgae. Just as the food chain on land has plants at the base of its pyramid shape, so it is the producers of the sea, or the microalgae, who support the base of the food chain in the ocean. In light of this fact, cultivation of microalgae will be essential to the achievement of sustainable aquaculture.

18 Other

While the carbohydrates produced by photosynthetic creatures consist mainly of cellulose in the case of the higher plants on land, in the case of seaweed they consist mainly of polysaccharides. These are grouped into the categories of structural polysaccharides (such as cellulose and mannan), intercellular adhesion polysaccharides (such as alginic acid and carrageenan), and storage polysaccharides (such as amylose and starch). Since each of these polysaccharides has the nature of a viscous fluid that solidifies under certain conditions such as those of temperature, pH, and concentration, they are useful for a very wide range of applications. Of these, alginic acid, agar, and carrageenan are in commercial use around the world as industrial raw materials made from polysaccharides contained in seaweed (Fig. 13). Alginic acid is

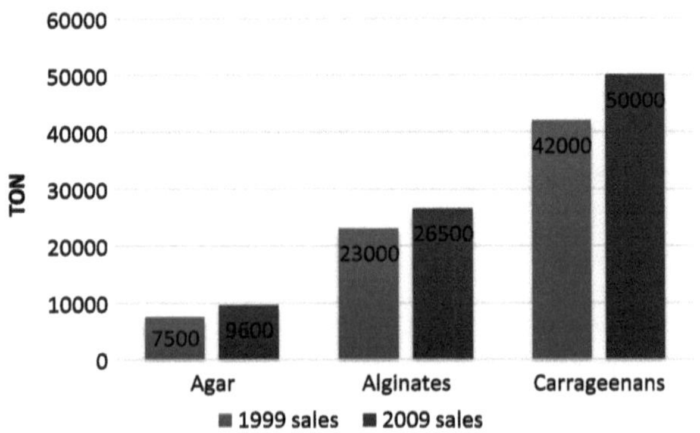

Fig. 13 Seaweed hydrocolloid sales volume [23]

an adhesion polysaccharide extracted mainly from brown algae and an acid of which uronic acid is a constituent element. Today it is produced in volumes of roughly 30,000–60,000 tons/year, with a wide range of main uses including in textile dyes, food and drink, pharmaceuticals, and pet food. Agar is a straight-chain polysaccharide hydrothermally extracted from red algae, with a basic structure of galactose. It long has been used as a traditional food, and it is used mainly in food products that use its galling function. It has been confirmed to have a laxative effect as a physiological function, and in Japan it is recognized as a medicinal ingredient for medicines for chronic constipation. Carrageenan is a species of straight-chain polysaccharide extracted from red algae such as *Eucheuma muricatum*, an anionic high polymer. It is used in products such as foods and cosmetics thanks to its stable thickening properties. Numerous interesting reports have been made concerning the anticoagulant [18, 19], anticancer [20, 21], and other properties of fucoidan, part of the cell wall structure of brown algae. It is the subject of active research and development for use in pharmaceuticals and health foods since it is easily extractable from sources such as the traditional Japanese foods of mozuku seaweed and kombu sea tangle. *Aphanothece sacrum* is a species of cyanobacteria endemic to Japan, where it is used in food. A. sacrum produces the anionic polysaccharide sacran, which contains sulfate and carboxylate groups. Recent studies have shown that at least 11 monosaccharides make up the polysaccharide sacran, which has an extremely diverse and complex structure, and that it contains more than 10 million molecules, making it a massive natural high polymer unlike any other. It also is being shown to have very high moisture retention capabilities [22]. It is expected that applications will be developed in the future to use these features.

References

1. Kent M, Haub C (1984) 1984 World Population Data Sheet. Population Reference Bureau Inc, Washington
2. Naganuma T (2006) Deep ocean water (upwelling) cultivates an ocean [in Japanese]. Shueisha, Tokyo, p 25
3. Hill J, Nelson E, Tilman D, Polasky S, Tiffany D (2006) Environmental, economic, and energetic costs and benefits of biodiesel and ethanol biofuels. Proc Natl Acad Sci 103 (30):11206–11210
4. Chisti Y (2007) Biodiesel from microalgae. Biotechnol Adv 25(3):294–306
5. Tsuchiya Y, Imamura Y (2013) An analysis of energy and cost efficiencies of Jatropha co-firing power generation [in Japanese]. CRIEPI Environ Sci Res Lab Rep, No. V12009
6. Satoh A, Kato M, Yamato K, Ishibashi M, Sekiguchi H, Kurano N, Miyachi S (2010) Characterization of the lipid accumulation in a new microalgal species, *Pseudochoricystis ellipsoidea* (*Trebouxiophyceae*). J Jpn Inst Energy 89(9):909–913
7. Mata TM, Martins AA, Caetano NS (2010) Microalgae for biodiesel production and other applications: a review. Renew Sustain Energy Rev 14(1):217–232
8. Moody JW, McGinty CM, Quinn JC (2014) Global evaluation of biofuel potential from microalgae. In: Proceedings of the National Academy of Sciences, pp 8691–8696

9. Jordan RW, Kleijne A (1994) A classification system for living coccolithophores. In: Winter A, Siesser WG (eds) Coccolithophores. Cambridge University Press, Cambridge, pp 83–105
10. Garcia-Rodriguez L (2002) Seawater desalination driven by renewable energies: a review. Desalination 143(2):103–113
11. Othmer DF, Roels OA (1973) Power, fresh water, and food from cold, deep sea water. Science 182(4108):121–125
12. Metzger P, Largeau C, Casadevall E (1991) Lipids and macromolecular lipids of the hydrocarbon-rich microalga *Botryococcus braunii*. Chemical structure and biosynthesis. Geochemical and biotechnological importance. Fortschritte der Chemie organischer Naturstoffe/Progress in the Chemistry of Organic Natural Products, Springer Vienna, pp 1–70
13. Metzger P, Largeau C (2005) *Botryococcus braunii*: a rich source for hydrocarbons and related ether lipids. Appl Microbiol Biotechnol 66(5):486–496
14. Dutta D, De D, Chaudhuri S, Bhattacharya SK (2005) Hydrogen production by Cyanobacteria. Microb Cell Fact 4(1):36
15. Hemschemeier A, Melis A, Happe T (2009) Analytical approaches to photobiological hydrogen production in unicellular green algae. Photosynth Res 102(2–3):23–540
16. Brown MR (1991) The amino-acid and sugar composition of 16 species of microalgae used in mariculture. J Exp Mar Biol Ecol 145(1):79–99
17. Lavens P, Sorgeloos P (1996) Manual on the production and use of live food for aquaculture. FAO Fisheries Technical Paper, 361
18. Bernardi G, Springer GF (1962) Properties of highly purified fucan. J Biol Chem 237 (1):75–80
19. Usui T, Asari K, Mizuno T (1980) Isolation of highly purified "Fucoidan" from Eisenia bicyclis and its anticoagulant and antitumor activities. Agric Biol Chem 44(8):1965–1966
20. Noda H, Amano H, Arashima K, Nisizawa K (1990) Antitumor activity of marine algae. Hydrobiologia 204(1):577–584
21. Zhuang C, Itoh H, Mizuno T, Ito H (1995) Antitumor active fucoidan from the brown seaweed, umitoranoo (*Sargassum thunbergii*). Biosci Biotechnol Biochem 59(4):563–567
22. Okajima MK, Bamba T, Kaneso Y, Hirata K, Fukusaki E, Kajiyama SI, Kaneko T (2008) Supergiant ampholytic sugar chains with imbalanced charge ratio form saline ultra-absorbent hydrogels. Macromolecules 41(12):4061–4064
23. Bixler HJ, Porse H (2011) A decade of change in the seaweed hydrocolloids industry. J Appl Phycol 23(3):321–335

Future Perspective

This book focuses on technologies for converting solar energy into chemical energy, in particular, hydrogen as a main form of chemical energy. To date, various methods have been proposed in a wide area of research, ranging from photoelectrochemical cells, photocatalysts, solar cells combined with electrolysis, solar thermal chemical reactions, photosynthesis-mimicking approaches to algae breeding, as discussed in detail in this book. The advancement of studies in these fields promises numerous possibilities of utilizing solar energy, but at the same time also reveals challenges and difficulties that need to be overcome.

The importance of developing solar energy conversion technologies is increasingly being recognized. Energy is the most basic element for human and industrial activities. Today, society uses enormous amounts of energy, leading to growing demands for abundant inexpensive energy supplies. For centuries, civilization has relied on the use of natural energy resources buried in the earth such as natural gas, oil, coal and uranium. However, the continuous use of these natural energy resources in massive amounts is causing severe damage to the global environment. First of all, natural energy resources buried in the earth are limited and will thus become completely depleted one day. Accordingly, the development of technologies for sustainable energy supply such as solar energy conversion is indispensable in the long run. It is no exaggeration to say that our future may be bleak if we do not succeed in the large-scale and practical application of solar energy conversion.

So how can we achieve our goal? The total amount of solar energy reaching the earth surface is sufficiently large but there are some obstacles in its application. First, the density of solar energy per unit area and unit time is not necessarily large, which requires the use of solar energy conversion devices over a large area. Secondly, the solar energy density incessantly changes with time. In particular, solar light is only available during the day. Thirdly, the solar energy density changes according to location. In general, well populated areas, where energy usage is high, have many cloudy and rainy days, indicating that they are not suitable places for solar energy conversion. On the other hand, deserts have abundant sunlight and most are far away from areas of dense population. It should be mentioned also that solar light has a wide spectral distribution, which makes high conversion efficiency difficult. This situation suggests that the development of a

M. Sugiyama et al. (eds.), *Solar to Chemical Energy Conversion*,
Lecture Notes in Energy 32, DOI 10.1007/978-3-319-25400-5

single efficient technique is not sufficient, and there is a need to develop a diversity of techniques from various perspectives and review their appropriate combinations. The diverse discussions presented in this book will contribute to this end.

In relation to this situation, we also have to note that electric energy as well as light energy cannot be stored effectively. Therefore, it is necessary to include techniques for energy storage and transport such as electric to chemical energy conversion and vice versa in solar energy conversion technologies. In particular, finding efficient electrodes for electrochemical oxygen evolution and oxygen reduction is of key importance because both reactions have high over-voltages at present and lead to large energy losses though they are fundamental reactions in mutual conversion between electrical and chemical energy. This book also attaches high importance to this topic.

Another serious problem arising from the above situation is that it imposes severe requirements on solar energy conversion technologies. Namely, high efficiency, low cost and long-term durability are fundamental conditions for the successful application of solar energy conversion technologies. These conditions are, however, quite difficult to meet to a sufficient extent, because there is a trade-off between high efficiency and low cost or between long-term durability and low cost. In fact, this is why large-scale practical applications of solar energy conversion technologies are delayed. Moreover, in actual researches, we often face various other dilemmas. To overcome such difficulties, evidently further studies with deep insight into the physics and chemistry of materials are needed. Needless to say, it is of crucial importance to find new stable efficient materials with the further development of properties of available promising materials and their effective and skillful utilization for solar energy conversion.

Finally, it is worth noting that we may have a means of working out a new strategy for achieving a breakthrough in the above difficulties. The photosynthesis system in natural plants is able to repair damaged parts and restore the original state. It also has the ability to organize itself so that it realizes the highest possible efficiency. Highly functional arrangements of electron-transferring molecules in the lipid bilayer membrane prove that natural plants really have such ability. In principle, it is possible to realize similar functions and abilities in artificial devices. There is no reason why they cannot be achieved. Thus, it is a challenging goal to clarify the principles of the self-restoring and self-improving functions and abilities of the natural photosynthesis system and apply them to artificial solar energy conversion devices. Chemical devices such as (photo)electrochemical electrodes and photocatalysts will be suitable sites for incorporating such functions and abilities. Successful incorporation will not only lead to solutions to the above trade-off but also to the development of new technologies which can be applied to other fields extensively. In this respect, studies on solar energy conversion technologies may play a leading role in the development of novel cutting edge technologies. Indeed, challenges and obstacles can serve as opportunities for making such new discoveries and achievements.

Index

M. Sugiyama et al. (eds.), *Solar to Chemical Energy Conversion*,
Lecture Notes in Energy 32, DOI 10.1007/978-3-319-25400-5

U

V

W

Y

Z

MIX
Papier aus verantwortungsvollen Quellen
Paper from responsible sources
FSC® C105338

If you have any concerns about our products,
you can contact us on
ProductSafety@springernature.com

In case Publisher is established outside the EU,
the EU authorized representative is:
Springer Nature Customer Service Center GmbH
Europaplatz 3, 69115 Heidelberg, Germany

Printed by Libri Plureos GmbH
in Hamburg, Germany